AF358635

Anticancer Activities of Dietary Phytochemicals

Anticancer Activities of Dietary Phytochemicals

Editors

Ching-Hsein Chen
Yi-Wen Liu

Basel • Beijing • Wuhan • Barcelona • Belgrade • Novi Sad • Cluj • Manchester

Editors
Ching-Hsein Chen
Microbiology, Immunology
and Biopharmaceuticals
National Chiayi University
Chiayi City
Taiwan

Yi-Wen Liu
Microbiology, Immunology
and Biopharmaceuticals
National Chiayi University
Chiayi City
Taiwan

Editorial Office
MDPI AG
Grosspeteranlage 5
4052 Basel, Switzerland

This is a reprint of articles from the Special Issue published online in the open access journal *Nutrients* (ISSN 2072-6643) (available at: https://www.mdpi.com/journal/nutrients/special_issues/64AVO88A04).

For citation purposes, cite each article independently as indicated on the article page online and as indicated below:

Lastname, A.A.; Lastname, B.B. Article Title. *Journal Name* **Year**, *Volume Number*, Page Range.

ISBN 978-3-7258-2513-4 (Hbk)
ISBN 978-3-7258-2514-1 (PDF)
doi.org/10.3390/books978-3-7258-2514-1

Contents

About the Editors

Ching-Hsein Chen

Ching-Hsein Chen's research field of interest is studying the anti-cancer, antioxidant, and anti-inflammatory effects of various natural product extracts. In terms of anti-cancer treatment, the apoptosis induced by various natural product extracts and the mechanism of oxidative stress in various human cancer cells were studied. In terms of antioxidants, Professor Chen has established a damage model of oxidative stress caused by various oxidizing substances, such as H2O2, cumene peroxide or menadione, on human cells, and then screened the antioxidant capacity of various extracted ingredients, using ingredients with an antioxidant capacity for an antioxidant discussion on the mechanisms of action. In terms of anti-inflammation, Professor Chen has established various experimental models for inducing inflammatory reactions, screened the ability of various extracted ingredients to inhibit cyclooxygenase 2 and prostaglandin E2, and has committed to exploring the mechanism of anti-inflammation. In the future, the research results can be provided for the advancement of basic medicine or clinical medicine, thereby improving the health of the world.

Yi-Wen Liu

Yi-Wen Liu, The research field of Professor Liu involves investigating the anti-cancer effect of pure compounds and natural product extracts and examining immune-related anti-cancer effects. In addition, he has also studied tumor biomarkers. Professor Liu's has done etensive work in the field of bladder cancer, and he has helped to establish a special animal model for the study of this condition. Bladder cancer is highly recurrent following specific transurethral resection and intravesical chemotherapy, which has prompted continuing efforts to develop novel therapeutic agents and early-stage diagnostic tools. Professor Liu's laboratory features the first local research team to establish a mouse orthotropic bladder tumor model, and some international publications on therapeutic chemical studies have already utilized this model. His work has also focused on the gene expression change in bladder tumorigenesis in order to examine tumor biomarkers. Additionally, Professor Liu has contributed to international publications in this field. In his work on immune-related anti-cancer studies, Professor Liu has examined tumor microenvironment macrophages and the effect of pure compounds or natural product extracts on the microenvironment. His team is dedicated to uncovering new methods of diagnosis, chemotherapy, immunotherapy, and chemoprevention of bladder cancer.

Preface

In recent years, knowledge regarding the anti-cancer effect by dietary phytochemicals has rapidly expanded. The dietary phytochemicals mainly inhibit the growth of cancer cells through many biological effects: (1) promoting cancer cell apoptosis and controlling their proliferation; (2) inhibiting cancer angiogenesis; (3) inhibiting angiogenesis and blocking the blood supply of nutrients that help cancer cells to grow, stop growing and avoid metastasis; (4) turning cancer cells from malignant to benign and preventing them from dividing and growing; (5) with its antioxidant effect, free radicals that are always present in the human body will not cause damage to the genes of normal cells, thereby reducing the formation of cancer cells; (6) inhibiting the signaling transduction can delay the cancerization process and inhibit the division and growth of cancer cells, causing cancer cells to transform from malignant to benign and no longer divide and grow.

To reflect the newest findings of dietary phytochemicals on anti-cancer effects, this Special Issue collects 15 research articles, including 10 research articles and 5 review articles. This Special Issue focuses on the anticancer activities of dietary phytochemicals.

The research articles in this Special Issue have two parts. The first part is about the anti-cancer activity and anti-cancer mechanism of dietary phytochemicals. The main contents include the following: (1) *Hibiscus Anthocyanins* extracts induce intrinsic and extrinsic apoptosis by activating AMP-activated protein kinase and inhibiting Akt in human colorectal cancer cells. (2) A combination of selected Qatari medicinal plant crude extracts inhibits proliferation and metastasis and induces apoptosis in breast cancer cells. (3) The ability of flavonoids, particularly luteolin, to modulate the intestinal microbiota in an animal model for colorectal cancer to contribute to an improvement in the health of individuals. The anti-colorectal cancer potential of luteolin, manifested through a modulation of the intestinal microbiota and a reduction in the number of tumors. (4) Fractionated leaf extracts of *Ocimum gratissimum* inhibit the proliferation and induce apoptosis of lung adenocarcinoma A549 cells. (5) The aqueous extract of the fungus *Fomitopsis betulina* induced significant cytotoxic effects at lower concentrations in lung cancer, melanoma, and colon cancer cells, showing promise as a potential anticancer agent. (6) Customized nutritional plans to potentially reduce gastric cancer risk could include increasing coffee and polyphenol-rich food consumption, especially for individuals with high polygenic risk scores. Theaflavate, rugosin E, vitisifuran B, and plantacyanin can be recommended regardless of SEMA3C_rs1527482 variant status.

The second part is about the anti-cancer effect and mechanism of dietary phytochemicals enhancement and/or the synergism clinical drugs. The main contents include the following: (1) Amurensin G enhances cell death by sensitizing cholangiocarcinoma to gemcitabine. Treatment of amurensin G reduces colony formation and cell migration by inhibiting cancer stem-like properties. Inhibition of Sirt1 reduces the expression of Nrf2, leading to apoptosis through increased ROS formation. (2) Synergistic anti-cancer effects of lovastatin and *Antrodia camphorata* extract in combination against PC3 androgen-refractory prostate cancer cells, accompanied by AXL and stemness molecules inhibition. (3) A combined resveratrol and parecoxib treatment to enhance the anti-cancer activity in DLD-1 colorectal cancer cells via the inhibition of TXNDC5 and Akt signaling and through the enhancement of JNK/p38 MAPK pathways. (4) Methanolic extract of *Cimicifuga foetida* induces G1 cell cycle arrest and apoptosis and inhibits the metastasis of glioma cells. Combining a ethanolic extract of *Cimicifuga foetida* with temozolomide further enhanced its therapeutic potential.

The main contents of the review articles in this Special Issue include the following: (1) Pharmacological features and the therapeutic implications of plumbagin in cancer and metabolic disorders. (2) The anticancer effects and therapeutic potential of kaempferol in triple-negative breast cancer. (3) The therapeutic effects of natural products, including polysaccharides, flavonoids, terpenoids, alkaloids, polyphenol compounds, quinone, and volatile oils, on liver cancer and their potential mechanisms. (4) Organosulfur compounds in colorectal cancer prevention and progression through modulating effects on the intestinal microbiota or their positive effects on intestinal mucosal health. (5) The histone acyl code in precision oncology: mechanistic insights from dietary and metabolic factors.

This Special Issue is offered to expand and renew our knowledge on the anti-cancer activities of dietary phytochemicals. It is focused on the molecular, biochemical, and signal transduction mechanisms of action in regard to dietary phytochemicals and anti-cancer treatment.

The Special Issue collects topics regarding the effects of dietary phytochemicals on cancer chemoprevention, cancer chemotherapy, and cancer adjuvant treatment.

The Special Issue provides a foundation of scientific knowledge for the interpretation and evaluation of future advances in dietary phytochemicals in regard to anti-cancer and human health.

This Special Issue is relevant to any students or practitioners interested in how dietary phytochemicals work against cancers, including in the fields of nutrition, biology and life sciences, oncology, medicine and pharmacology, and public health and healthcare.

Ching-Hsein Chen and Yi-Wen Liu

Editors

Review

Histone Acyl Code in Precision Oncology: Mechanistic Insights from Dietary and Metabolic Factors

Sultan Neja [1], Wan Mohaiza Dashwood [1], Roderick H. Dashwood [1,2,*] and Praveen Rajendran [1,2,3,*]

[1] Center for Epigenetics & Disease Prevention, Texas A&M Health, Houston, TX 77030, USA; sultanabda@tamu.edu (S.N.); wdashwood@tamu.edu (W.M.D.)

[2] Department of Translational Medical Sciences, Texas A&M College of Medicine, Houston, TX 77030, USA

[3] Antibody & Biopharmaceuticals Core, Texas A&M Health, Houston, TX 77030, USA

[*] Correspondence: rdashwood@tamu.edu (R.H.D.); prajendran@tamu.edu (P.R.); Tel.: +1-713-677-7806 (R.H.D.); +1-713-677-7803 (P.R.)

Abstract: Cancer etiology involves complex interactions between genetic and non-genetic factors, with epigenetic mechanisms serving as key regulators at multiple stages of pathogenesis. Poor dietary habits contribute to cancer predisposition by impacting DNA methylation patterns, non-coding RNA expression, and histone epigenetic landscapes. Histone post-translational modifications (PTMs), including acyl marks, act as a molecular code and play a crucial role in translating changes in cellular metabolism into enduring patterns of gene expression. As cancer cells undergo metabolic reprogramming to support rapid growth and proliferation, nuanced roles have emerged for dietary- and metabolism-derived histone acylation changes in cancer progression. Specific types and mechanisms of histone acylation, beyond the standard acetylation marks, shed light on how dietary metabolites reshape the gut microbiome, influencing the dynamics of histone acyl repertoires. Given the reversible nature of histone PTMs, the corresponding acyl readers, writers, and erasers are discussed in this review in the context of cancer prevention and treatment. The evolving 'acyl code' provides for improved biomarker assessment and clinical validation in cancer diagnosis and prognosis.

Keywords: acyl code; cancer; diet; epigenetics; histones; metabolism

Citation: Neja, S.; Dashwood, W.M.; Dashwood, R.H.; Rajendran, P. Histone Acyl Code in Precision Oncology: Mechanistic Insights from Dietary and Metabolic Factors. *Nutrients* **2024**, *16*, 396. https://doi.org/10.3390/nu16030396

Academic Editors: Yi-Wen Liu and Ching-Hsein Chen

Received: 30 December 2023
Revised: 26 January 2024
Accepted: 27 January 2024
Published: 30 January 2024

1. Introduction

Cancer is a complex disease characterized by the uncontrolled growth and dissemination of aberrant cells, and stands among the foremost contributors to global mortality, resulting in approximately 10 million deaths in 2020 [1]. The development and progression of cancer involves genetic and epigenetic alterations. Epigenetics, the study of heritable changes in gene expression via alterations in DNA methylation, histone modifications, and non-coding RNAs, exerts a crucial role in regulating gene expression and cellular differentiation during development [2]. Aberrant epigenetic changes have been linked to the pathogenesis of several diseases, including cancer [3].

Histone PTMs are reversible covalent alterations that affect the structure and accessibility of chromatin, and thereby regulate gene expression. Among these PTMs, histone acetylation and methylation have been extensively reviewed [4], and will not be discussed in detail herein. Recent research has uncovered several new histone modifications, collectively referred to as acyl marks or the 'acyl code', which include crotonylation, propionylation, butyrylation, malonylation, succinylation, glutarylation, hydroxybutyrylation, benzoylation, and lactylation (Table 1). Histone acyl marks possess distinct functional characteristics based on their chemical structure, polarity, and reactivity. While the epigenetic 'writers' may be shared in some cases, histone acyl marks exhibit preferences for certain 'readers' and associated chromatin remodelers [5]. Additionally, the cellular metabolic state can produce distinct acyl-CoA substrates, such as succinyl-CoA or butyryl-CoA. These substrates bind to specific histone regions and regulate gene expression patterns, illustrating the significance of different acyl marks in the interplay between metabolism and epigenetics [6–8]. Understanding the roles and mechanisms by which dietary metabolites and metabolism-derived intermediates affect the acyl code and their pathogenic consequences is of paramount importance.

Table 1. Overview of histone acyl modifications in cancer.

Type of Acylation	Chemical Nature	Dietary/Metabolic Source	Writers	Readers	Erasers	References
Acetylation (Ac)	Hydrophobic	CHO & SCFA from gut microbes, glycolysis, TCA	p300/CBP, HAT, GNATs	BRD3, BRD4, PBRM1	All HDAC family	[8–10]
Propionylation (Pr)	Hydrophobic	SCFA from dietary fiber & gut microbes, TCA	p300/CBP, GNATs, MYSTs	YEATS, DPF	SIRT1,2,3	[11–15]
Butyrylation (Bu)	Hydrophobic	SCFA from dietary fiber & gut microbes, TCA	p300/CBP, GNATs, HBO1	YEATS, DPF	SIRT1,2,3	[9,15–17]
Crotonylation (Cr)	Hydrophobic	SCFA from dietary fiber & gut microbes, TCA	p300/CBP	YEATS, DPF	SIRT1,2,3, HDAC3	[9,18–21]
Benzoylation (Bz)	Hydrophobic	N/A	HBO1	YEATS, DPF		[5,9,22]
β-Hydroxybutyrylation (Bhb)	Polar	Ketogenic diet, starvation,	p300/CBP	YEATS, DPF	SIRT3, HDAC1,2,3	[23–25]
2-Hydroxyisobutyrylation (Bhib)	Polar	SCFA, Amino acid metabolism	P300, MYSTs	YEATS, DPF	N/A	[16,26]
Lactylation (La)	Acidic	Glycolysis, lactate from exercise, LGSH	p300	N/A	HDAC1,3	[16]
Malonylation (Mal)	Acidic	Citrate metabolism, FAO	N/A	N/A	SIRT2,5	[27,28]
Succinylation (Succ)	Acidic	TCA	p300/CBP, GNATs, CPT1A, GCN5	YEATS	SIRT5, 7	[17,29,30]
Glutarylation (Glu)	Acidic	TCA, amino acid metabolism	p300, GCN5	N/A	N/A	[9,26]
O-GlcNacylation (GlcNac)	Polar	Pentose–phosphate pathway	N/A	N/A	N/A	[31]
Palmitoylation (Pal)	Hydrophilic	Edible oils, HFD	LPCAT1	N/A	APT, PPT SIRT6	[32,33]
Myristoylation (Myr)	Hydrophilic	Edible oils, HFD	N/A	N/A	SIRT2, 6	[33,34]

APT: acyl protein thioesterases, PPT: palmitoyl–protein thioesterase, CHO: carbohydrate, TCA: tricarboxylic acid, SCFA: short-chain fatty acid, GNATs: Gcn5-related *N*-acetyltransferases, FAO: fatty acid oxidation, HFD: high-fat diet, LGSH: lactoylglutathione, LPCAT1: lysophosphatidylcholine acyltransferase I, CPT1A: carnitine acyltransferase I, MYST: lysine acetyltransferases (Moz, Ybf2/Sas3, Sas2, and Tip60), HBO1: histone acetyltransferase binding to ORC1, N/A: data not available.

Aberrant histone modifications have been implicated in cancer development and progression. Such modifications occur via various mechanisms, including mutation or deregulation of the enzymes involved in epigenetic control, altered expression levels of the regulatory factors, or via oncogenic substrates produced by cancer metabolism. For example, histone deacetylases (HDACs) are overexpressed in many types of cancer, leading to hypoacetylation of histone and non-histone proteins and repression of tumor suppressor genes [35,36]. Deregulation of chromatin remodelers can also lead to aberrant gene expression and genomic instability, which are among the hallmarks of cancer [37]. In addition to poor dietary habits and associated oncometabolites that increase the risk of cancer [38–43], cancer cells undergo metabolic reprograming, triggering epigenetic imbalances to fuel tumor growth and proliferation [44–46].

It is well established that the digestion and fermentation of dietary fibers by gut microbes, as well as energy metabolism within cells, produce diverse metabolites that can enter cells and generate various acyl-CoAs [47,48]. These acyl-CoAs not only serve as substrates for ATP production but also contribute to PTMs of histone and non-histone proteins [49,50]. These PTMs play vital roles in governing the transcriptome, proteome, and metabolome, thereby exerting significant control over multiple cellular processes [49]. Furthermore, the intricate array of PTMs play a key role in fine-tuning chromatin structure and function. The notion of a 'histone code' has gained substantial credibility, linked to PTMs on histone proteins that predominantly regulate DNA transcription [51]. However, the concept of a histone acyl code is relatively new, and it demonstrates that in cancer cells, alterations in metabolic pathways lead to changes in the levels and ratios of acetyl-CoA to acyl-CoA, which, in turn, affect histone acetylation and/or acylation patterns and gene expression. Although some oncometabolites drive tumorigenesis through non-histone protein acylation [16,52–54], histone acylation is the predominant epigenetic mechanism. Histone acylation often serves as a marker of chromatin activity, facilitating increased transcriptional output and influencing cellular metabolism through alterations in chromatin structure and function [11,55–63]. Additionally, genome-wide analysis has revealed that acyl modifications are associated with gene activation [64]. Notably, the histone acyl code derived from diet and metabolism constitutes an evolving field in epigenetics, offering new insights into the interplay between cellular metabolism and gene regulation [8]. For example, the histone acyl code can influence chromatin remodeling complexes by modifying the structure and function of histones and their interaction with other chromatin-associated proteins.

No independent acyl writer or eraser has been reported to date. Thus, histone acylation marks are reversibly regulated by the opposing actions of well-documented histone acyltransferases (HATs) and HDACs [65,66]. Just as with acetyl marks, HATs add acyl groups to lysine residues whereas HDACs remove them. The balance between HATs and HDACs is crucial for the proper functioning of cells and tissues, and deregulation of this balance has been linked to various diseases, including cancer [67,68]. Therefore, studying how the intricate repertoire of the acyl code produced from dietary and cellular metabolism is linked to changes in epigenetic regulation could provide new insights into the mechanisms underlying cancer. As epigenetic deregulation is often associated with cancer progression and metastasis [69], understanding the mechanisms and functions of these modifications could lead to the design of new diagnostic and therapeutic strategies. In this regard, histone acylation could provide novel biomarkers for cancer diagnosis, prognosis, and therapy response prediction. Changes in histone acyl code alter chromatin structure and function, leading to changes in gene expression patterns [70]. Understanding this interplay provides insights into metabolic adaptations of cancer cells and their contributions to cancer progression. Targeting acyl readers and enzymes responsible for generating and removing histone acyl marks could provide novel cancer therapies. In this review, we outline current progress in the understanding of histone acyl marks in the realm of cancer biology, as well as the potential therapeutic prospects.

2. Histone Acylation

As mentioned above, the histone acyl code encompasses a range of reversible modifications, such as propionylation, butyrylation, crotonylation, succinylation, malonylation, glutarylation, β-hydroxybutyrylation, and benzoylation, among others (Table 1: Overview of histone acyl modifications in cancer). These acylations are dynamic and can be regulated by metabolic changes, providing a diverse repertoire of acyl moieties at any given time. Notably, as previously mentioned, acyl marks are predominantly gene activation marks. They exhibit non-redundancy with histone acetylation [71], and differ in polarity and reactivity, allowing them to differentially regulate gene expression and chromatin structure [50,70]. For instance, specific histone butyrylation, propionylation and β-hydroxybutyrylation marks were associated with the activation of genes involved in lipid metabolism and the response to starvation [11,59], while lysine benzoylation designates promoters of glycerophospholipid metabolism-related genes [62]. In this way, each form of acylation may preferentially recognize specific genomic loci and regulate the expression of distinct sets of genes. While histone acetylation is predominantly recognized by bromodomain readers, acyl marks are often recognized by readers containing DPF and YEATS domains [5]. Histone acylation serves as a distinguishing feature for transcriptional activation during various physiological processes, such as signal-induced gene activation, spermatogenesis, tissue injury, and metabolic stress [8]. Understanding diet- and metabolism-associated acyl code regulation is an evolving new field of precision nutrition [8,72].

The body's physiological and metabolic state influences precursor molecule availability for the histone acyl code. Factors such as disrupted metabolism of glucose and fatty acids, and the associated regulatory enzymes, play a role in determining the levels of acyl-CoA metabolites [73]. Additionally, rapidly growing cancer cells undergo metabolic alterations that impact acyl-CoA metabolite levels [74]. The availability of acyl-CoA synthases is a crucial factor influencing the diversity and dynamics of histone acylation [75]. These enzymes convert precursor molecules into specific acyl-CoA metabolites, and their levels and activity influence the availability of different acyl-CoA metabolites and subsequently lead to the formation of different types of acylated histones. Additionally, the enzymatic action of HATs and the proclivity of each acyl-CoA metabolite for non-enzymatic acylation are key factors that can impact the diversity and dynamics of histone acylation [76].

Histone acetylation is one of the most well-studied PTMs in chromatin. It entails the transfer of an acetyl group from acetyl-CoA to the ε-amino group of lysine residues in histone tails. This modification is catalyzed by HATs, including p300 and lysine acetyltransferase 2A (KAT2A), which exhibit relatively low binding affinity for catalyzing histone crotonylation and succinylation as compared to the canonical acetyl-CoA substrate [58,77]. On the other hand, histone crotonylation is a modification that involves the transfer of a crotonyl group to lysine residues, and is facilitated by p300/CBP-associated factor (PCAF) [58]. Like histone acetylation, histone crotonylation has also been found to be particularly enriched in active gene promoters and enhancers [78]. Nevertheless histone crotonylation and succinylation contribute to chromatin relaxation, rendering DNA more accessible to transcription factors and other regulatory proteins [79,80]. Conversely, histone deacetylation, carried out by HDACs, leads to chromatin compaction and gene repression.

Histone propionylation and butyrylation involve the transfer of propionyl and butyryl groups, respectively, to lysine residues in histone tails. These modifications are mainly catalyzed by sirtuins (SIRTs) and exert similar effects on chromatin structure and gene expression as acetylation [81]. Succinylation, malonylation, and glutarylation involve the transfer of succinyl, malonyl, and glutaryl groups, respectively, to lysine residues on histone tails. As summarized in Table 1: Overview of histone acyl modifications in cancer, these modifications are also mediated by HATs, such as KAT2A (hGCN5), which possesses corresponding acyl-transferase activity. Given the reactivity of acyl-CoA metabolites like succinyl-CoA and malonyl-CoA towards lysine residues, they can also undergo non-enzymatic histone acylation [47]. On the other hand, β-hydroxybutyrylation involves the transfer of a β-hydroxybutyryl group to lysine residues and is carried out by the enzyme p300/CBP, which exhibits β-hydroxybutyrate dehydrogenase activity [82]. This modification is notably enriched in the liver and is believed to play a role in metabolic regulation [82].

2.1. Mechanisms of Histone Acylation

Histone acylation involves the interplay between enzymatic activity, metabolic signaling, chromatin structure, and effects on gene expression. For instance, histone acetylation is regulated by the availability of acetyl-CoA, a metabolite produced during cellular metabolism [83–86]. Acetyl-CoA levels are influenced by metabolic state, nutrient availability, and stress responses [87]. Relative to acetylation, other histone acyl marks are 1–5% as abundant and correlate with cellular levels of the respective acyl-CoA donors [7].

Histone acylation associated HATs and HDACs are subject to regulation by metabolic signaling pathways and other factors [88]. Concomitant to the well-studied HATs and classical family of zinc-dependent HDACs, other enzymes can also regulate histone acylation. For example, SIRT5 relies on NAD^+ and can remove malonyl and succinyl groups from histone lysine residues through demalonylation and desuccinylation, respectively [89,90]. These modifications are thought to play important roles in regulating metabolism and mitochondrial function. Acetylation, propionylation, and butyrylation can relax chromatin structure, making DNA more accessible for transcription factors and other regulatory proteins [91,92]. Conversely, deacetylation, depropionylation, and debutyrylation can lead to chromatin compaction and gene repression [64,93].

Lysine acylation occurs through enzymatic as well as non-enzymatic actions of acyl CoA-thioesters, acyl phosphates, and α-dicarbonyls [94]. Unlike enzymatic acylation, non-enzymatic acyl lysine modifications accumulate in various proteins, particularly in the aging process [95]. The mechanism of other histone acylations, such as crotonylation, succinylation, malonylation, and glutarylation, are less well understood but are thought to play roles in the regulation of gene expression and chromatin structure [60,61], although their precise roles in gene regulation are still being studied.

2.2. Histone Acylation Writers and Erasers

KATs are a diverse group of HAT enzymes that transfer acetyl groups from acetyl-CoA to the ε-amino group of lysine residues present in proteins, including histone tails. Several KAT families, including GNAT (Gcn5-related *N*-acetyltransferase), MYST (Moz, Ybf2/Sas3, Sas2, and Tip60), p300/CBP, and TAF (TATA-binding protein-associated factor) [96] are multi-functional enzymes. They are responsible for transferring various acyl groups, such as acetyl, crotonyl, butyryl, propionyl, malonyl, succinyl, and others, to the lysine residues of histone proteins, with decreased acyl-transferase activity for bulkier acyl-CoAs [97].

Several metabolic enzymes that catalyze the interconversion of acyl substrates contribute to the diversity of the histone acyl code (Figure 1). For instance, malonyl-CoA is produced in the cytosol during the synthesis of fatty acids from citrate. This process involves the enzymatic action of ATP citrate lyase (ACL), which converts citrate into acetyl-CoA, followed by the conversion of acetyl-CoA to malonyl-CoA by acetyl-CoA carboxylase (ACC) [98]. Another enzyme, propionyl-CoA carboxylase (PCC), generates D-methylmalonyl-CoA from propionyl-CoA, which is then transformed into L-methylmalonyl-CoA by methylmalonyl-CoA epimerase (MCEE) [99]. Lastly, methylmalonyl-CoA mutase facilitates the formation of succinyl-CoA [99,100]. Gut microbiome-produced 3-hydroxybutyryl-CoA dehydrogenase (crotonase) converts betahydroxybutyryl-coa into crotonyl-CoA [101], highlighting the role of the microbiota in regulating the diversity of acyl-CoA substrates.

The HDACs that remove acetyl/acyl groups from histone and non-histone proteins include four classes. Class I consists of Rpd3-like proteins and includes HDAC1, HDAC2, HDAC3, and HDAC8. Class II, Hda1-like proteins, includes HDAC4, HDAC5, HDAC6, HDAC7, HDAC9, and HDAC10. Class III, NAD^+-dependent Sir2-like proteins, comprises SIRT1, SIRT2, SIRT3, SIRT4, SIRT5, SIRT6, and SIRT7. Lastly, there is a single enzyme in the Class IV protein, which is HDAC11 [102]. These HDACs exhibit differences in subcellular localization, function, and regulatory mechanisms, allowing them to participate in diverse cellular processes and contribute to gene expression regulation, chromatin remodeling, and other important cellular functions. The specificity or preference of an HDAC for acetyl versus other histone acyl marks has not been adequately addressed in the current literature.

2.3. Histone Acylation Readers

Histone modifications are read and interpreted by a diverse array of proteins called 'readers', which recognize and bind to specific PTMs. Based on the structural domains, the main families of readers are bromodomain-containing proteins (BRDs), chromodomain-containing proteins (CRDs), tudor domain-containing proteins, PHD finger-containing proteins, and YEATS domain-containing proteins [103,104]. These epigenetic readers play diverse essential cellular functions. They can directly modify histone marks or serve as effector proteins, influencing the functional consequences of histone modifications by translating the histone code into actionable changes. Readers can also recognize and bind to specific epigenetic marks, thereby enabling the recruitment of molecular machinery to modify chromatin structure [105,106].

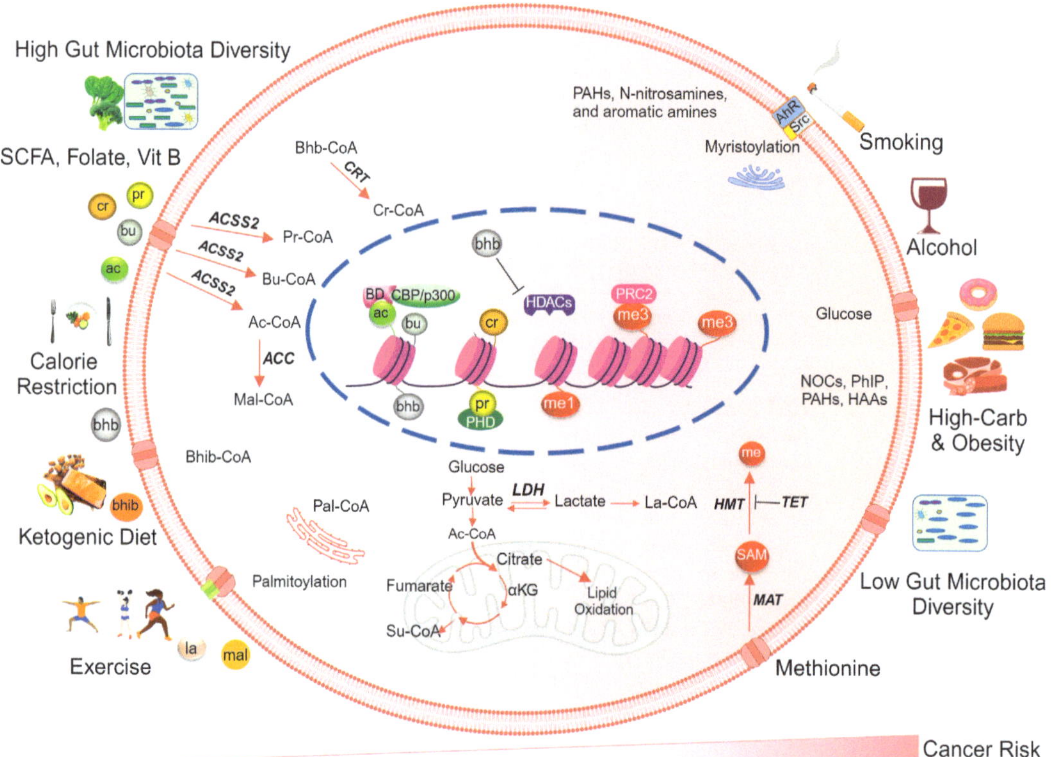

Figure 1. Dietary metabolites regulate cancer risk by modulating the histone acylation landscape. Dietary and metabolic alterations represent non-genetic/environmental risk factors influencing histone acylation during cancer development and progression. The right side of the diagram illustrates risk factors that increase cancer development through epigenetic mechanisms, such as an increase in histone methylation and a low acetylation/acylation ratio. On the left side, the diagram demonstrates how a healthy lifestyle and dietary metabolites that increase the diversity of the gut microbiota alter histone acetylation/acylation patterns. Increased levels of histone acyl-CoA precursors from both dietary and cellular metabolism contribute to chromatin decondensation through HDAC inhibition and the removal of repressive histone marks, thereby reducing the risk of cancer. Key abbreviations: MAT: methionine adenosyltransferase, HMT: histone methyltransferase, TET: ten-eleven translocation, ACC: acetyl-CoA carboxylase, ACSS2: acetyl-CoA synthetase 2, BD: bromodomain reader, PHD: plant homeodomain reader, HDACs: histone deacetylases, SAM: S-adenosyl methionine, PRC2: polycomb repressive complex 2, αKG: alpha ketoglutarate, PAHs: polycyclic aromatic hydrocarbons, HAAs: heterocyclic aromatic amines, NOCs: *N*-nitroso compounds, PhIP: 2-amino-1-methyl-6-phenylimidazo[4,5-b]pyridine, me3: H3K27me3, me1: H3K4me1, AhR: aryl hydrocarbon receptor, CRT: crotonase, me: methyl. Created with BioRender.com, accessed on 26 January 2024.

Recent studies revealed preferential reader selection between histone acetyl and acyl marks [103]. BRDs and CRD-containing proteins typically recognize acetylated and methylated lysine residues on histones, respectively [107]. However, DPF and YEATS domain-containing readers preferentially recognize longer acyl forms of lysine residues, such as crotonylation, butyrylation, and propionylation [103]. The binding specificity of these readers relies on their affinity in binding to aromatic acyl groups [18,30].

As noted above, currently there are no specific writers or erasers exclusively dedicated to 'non-acetyl' acylation. Therefore, recruitment of histone acylation readers becomes a crucial determinant of chromatin accessibility in modulating the expression of specific genes. Interestingly, under various metabolic, developmental, or disease-related conditions, YEATS and DPF domain-containing readers interacted with larger acyl groups, remaining bound to chromatin, while BRD-containing proteins were excluded [103]. One of the DPF-domain-containing proteins, DPF2, serves as an accessory component of the BAF-family chromatin remodeler has been reported to exert a repressive role in myeloid differentiation [108].

3. Dietary Metabolites Regulating Histone Acylation

Evidence has accrued for diet-associated bioactive compounds and intermediary metabolites affecting histone acylation marks, including the following:

- Butyrate, propionate, and acetate are short-chain fatty acids (SCFAs) produced by gut microbiome-mediated fermentation of dietary fiber. These metabolites inhibit HDAC activity and increase histone acetylation status [109];
- Polyphenols found in various fruits, vegetables, and beverages, including resveratrol and curcumin, act on SIRTs and other HDACs to alter histone acetylation status and gene expression [110–113];
- Omega-3 fatty acids abundant in fatty fish, flaxseeds, and walnuts, have been implicated in regulating histone acetylation [114]. They influence the activity of HATs and HDACs, promoting a favorable balance between histone acetylation and deacetylation [115,116];
- Vitamin B3 (niacin) is involved in energy metabolism as a precursor for the coenzyme nicotinamide adenine dinucleotide (NAD^+), which is required by SIRTs for deacetylase activity. By affecting NAD^+ availability, niacin can indirectly modulate histone acylation [117];
- Glucose utilization, microbiota-derived SCFAs, or dietary fat metabolism can impact acetyl/acyl-CoA ratios, thereby affecting overall histone acetylation patterns [118–121]. Since most histone acylation competes for the same HATs, the acetyl/acyl-CoA ratios in different cellular pools dictate which acylation pattern occurs on histones [118,121];
- Dietary antioxidants such as vitamins C and E, and certain polyphenols, modulate cellular redox status and signaling pathways involved in histone acetylation [122]. Additionally, nutrient-sensing pathways, such as the mammalian target of rapamycin (mTOR) pathway, can integrate dietary and metabolic signals to influence histone acylation [123]. Among the nutrient-sensing signaling pathways that govern histone PTMs, the sucrose non-fermenting/AMP-activated protein kinase (AMPK/Snf1) and carbohydrate response element binding protein (ChREBP) pathways play pivotal roles. For instance, AMPK/Snf1 acts as a histone kinase [124], not only phosphorylating but also regulating the activity of several HATs and HDACs through enzyme phosphorylation [125]. Moreover, this pathway influences histone acetylation and deacetylation by controlling levels of acetyl CoA and NAD^+ levels [125].

3.1. The Role of Dietary and Metabolism-Derived Histone Acylation in Cancer Development and Progression

Many cancers are influenced by non-genetic/environmental risk factors. For instance, tobacco products, tanning beds, UV exposure, alcohol consumption, toxin exposure, and poor dietary habits have been reported to increase the risks of lung cancer, skin cancer, liver cancer, and colorectal cancer, respectively [126,127]. Non-genetic risk factors often cause epigenetic derangements that underlie cancer development. Once oncogenesis is established, cancer cells are also known for their metabolic reprogramming and adaptability, which enable survival and proliferation within the tumor microenvironment [128,129], for which altered metabolism and subsequent epigenetic deregulation play roles [130,131]. In this section of the review, we explore how the metabolic rewiring of cancer cells contributes to the development and progression of tumors by altering histone acylation.

The mechanisms by which poor dietary habits influence histone acylation patterns and cancer risk are complex (Figure 1). Briefly, lifestyle and dietary habit driven alterations in nutrient availability and oncogenic-driven metabolic reprogramming within cancer cells influence the levels of crucial metabolites that govern signaling pathways and epigenetic processes [132]. Consequently, an altered metabolic state and changes in acyl modifications lead to the recruitment of epigenetic writers, chromatin remodelers, erasers, and readers, resulting in a distinct histone acylation landscape that connects cellular metabolism to the epigenome [133].

Deficiencies in essential nutrients like folate [134,135], vitamin B12 [136], and iron [137], which function as critical cellular substrates and cofactors, can impair the function of enzymes involved in histone acylation, leading to abnormal histone modifications and potentially promoting oncogenesis. Altered metabolite levels and imbalanced nutrient utilization can also affect the activity of HATs and HDACs, which regulate histone acylation. Cancer cells, in contrast to normal cells, exhibit elevated methionine cycle activity and rely on external or dietary methionine for sustained growth [138]. The significance of methionine metabolism in cancer biology is linked to its role in GSH biosynthesis, Polyamine Synthesis, and as a donor of methyl groups for DNA and histone modification [138]. Additionally, there is an increasing risk of cancer susceptibility due to exposure to carcinogens such as NOCs, PhIP, PAHs, and HAAs, which can result from the consumption of thermally processed meat [139], as well as other carcinogens associated with smoking [140]. Poor dietary habits, including regular consumption of highly processed foods and insufficient fruits and vegetables, can promote chronic inflammation and oxidative stress in the body [141], which can induce changes in histone acylation marks leading to deregulated cellular processes promoting oncogenesis [142].

Metabolic reprogramming is a critical factor driving cancer progression, supporting energy generation, the biosynthesis of anabolic molecules [143], and maintaining the optimal cellular redox states within cancer cells [144]. Solid tumors often exhibit the Warburg effect and hypoxia, contributing to cancer cell reprogramming [145]. Unlike normal cells, cancer cells rely on aerobic glycolysis, the pentose phosphate pathway, the hexosamine pathway, and the serine biosynthesis pathway. This increased glycolytic activity can overwhelm mitochondria, leading to the production of reactive oxygen species [45]. Lactate dehydrogenase plays a pivotal role by converting pyruvate to lactate, preventing mitochondrial import, and maintaining NAD^+ homeostasis [146]. Lactate, with roles in reversing the Warburg effect and modifying histones [63,147], holds significant importance. Cancer cells also rapidly consume glutamine, utilizing it as a nitrogen donor and carbon source for anabolic pathways [148]. Oncogenic signals further drive metabolic reprogramming by enhancing glucose and glutamine transporters [149,150] and modulating metabolic enzyme activity [151]. These altered metabolic pathways are vital conduits, supplying the necessary metabolic intermediates and cofactors for epigenetic modifiers. Consequently, cancer metabolism, marked by significant changes in cellular metabolite levels compared to normal conditions, intricately intertwines with cancer epigenetics [45,152]. For example, the acetylation of histone H3 lysine 9 (H3K9ac) and histone H3 lysine 27 (H3K27ac) have been shown to be regulated by the activity of the acetyl-CoA synthetase enzyme (ACSS2), which catalyzes the conversion of acetate to acetyl-CoA [153]. ACSS2 is upregulated in various cancers, and increased levels of H3K9ac and H3K27ac have been observed in cancer cells [154,155]. These modifications are associated with increased expression of oncogenes. For example in colorectal cancer, an enzyme called ACL, responsible for converting citrate to acetyl-CoA, is suppressed, leading to a decreased nuclear acetyl-CoA reservoir [83]. During glucose deprivation, cancer cells also utilize glutamine as a substrate for the production of acetyl-CoA in the tricarboxylic acid (TCA) cycle, which, in turn, increases histone acetylation to support the proliferation and growth of tumor cells [156–158]. Similarly, propionylation and butyrylation of histones have also been implicated in cancer development.

The enzymes responsible for these modifications are regulated by metabolic pathways involved in the breakdown of fatty acids [47]. Dysregulation of these pathways leads to altered histone propionylation and butyrylation patterns, which can affect gene expression and contribute to tumorigenesis.

The cross-talk between diet, microbes and host cells also influences cancer outcomes [159,160]. For example, green leafy vegetables such as spinach can alter gut microbes [161], potentially resulting in the generation of microbial enzymes and metabolites serving as molecular messengers. Microbial metabolites can also alter the tumor microenvironment, comprising of a variety of cell types and inflammatory mediators, thereby influencing epigenetic events that play a role in cancer progression and the effectiveness of immunotherapy [160,162]. One of the most common fermentation products of gut microbiome is SCFAs and its metabolites that inhibits HDACs, exerting an epigenetically mediated anti-cancer function [163–166].

Histone acylation can also change chromatin structure by modulating the interaction between histones and other chromatin-associated proteins. For example, the acetylation of H3K56 has been shown to enhance the binding of the chromatin assembly factor CAF-1, which is involved in nucleosome formation [167]. Other studies have demonstrated that the H4K5 acylation/acetylation ratio fine-tunes BRD4–chromatin interactions highlighting the balance between histone acetylation and acylation [77]. This balance, regulated by metabolic processes, may serve as a widespread mechanism that governs the functional genomic distribution of bromodomain factors [168]. Thus, the recruitment of chromatin-remodeling complexes, associated readers, and eraser complexes that, in turn alter chromatin structure and gene expression, relies on metabolic dynamics within the tumor microenvironment [103,169].

Protein acylation also plays a significant role in shaping the immunosuppressive tumor microenvironment by influencing immune cell exhaustion, activation, and infiltration [170]. Through its regulation of immune cell activation, infiltration, and antigen presentation, protein acylation can influence the formation of an immunosuppressive tumor microenvironment [170]. Cancer cells also utilize histone lactylation as a mediator of immunosuppression [63,171]. Lactate increases histone lactylation and leads to heightened expression of Arg1 and other genes that mediate the transition toward the immunosuppressive M2 macrophage phenotype, thereby restraining immune cell activity in the tumor microenvironment [63,171,172]. Epigenetic rewiring that is intimately connected to cancer metabolism could be one of the mechanisms of cancer immune escape.

3.2. Metabolism-Derived Histone Acyl Codes as Cancer Biomarkers

Global histone hypoacetylation is a biomarker of cancer etiology [173,174], and the deregulation of metabolic pathways can lead to alterations in histone acylation patterns that contribute to cancer development and progression (Table 2). Such changes can function as biomarkers for diagnosis and prognosis, as exhibited for breast, prostate, and colorectal cancer [173,175]. For instance, histone H3K9ac and histone H3K27ac, which are considered active histone mark for normal cells, are aberrantly elevated in prostate cancer [175]. ACSS2 is often upregulated in various cancers [154,155] along with altered expression of HATs, HDACs and associated epigenetic reader proteins. For instance, the binding of acetylation reader ENL to H3K9ac and H3K27ac has been observed in acute myeloid leukemia and is associated with increased expression of oncogenes that can be used as biomarkers for diagnosis or prognosis [176]. In another study, increased acetylation of H2BK120, H3.3K18, and H4K77 in liver cancer tissues were biomarkers of unfavorable prognosis. In an independent clinical cohort of hepatocellular carcinoma (HCC) patients, these markers correlated with decreased survival rates and increased recurrence rates [177].

Table 2. Aberrant histone acylations linked to cancer.

Histone Acylation Type	Cancer Type	Association with Cancer	References
Global H3K18ac, H3K9ac, H3K12ac	Prostate	Elevated levels correlate with prostate cancer risk	[175]
Global losses of H3K16ac	Leukemia, lymphoma, breast, colorectal, lung, prostate, cervical	A hallmark of human tumor cells	[173]
H3K23pr	Medulloblastoma, leukemia, glioma, colorectal	Low H3K23pr contributes to cancer development	[178]
Global histone Kcr	Esophageal, colon, pancreatic, lung	Low Kcr is associated with cancer	[179]
	HCC	Kcr levels correlate with HCC progression	[180]
	Prostate	Kcr levels correlate with prostate cancer malignancy	[181]
H3K9bhb	HCC	High H3K9bhb correlates with HCC progression	[182]
Global Khib	Pancreatic	Khib is a tumor promoter in pancreatic cancer	[183]
H3K18la	Melanoma	High H3K18la enhances melanoma	[184]
H3K9la and H3K56la	HCC	High H3K9la and H3K56la increase the proliferation and migration of liver cancer stem cells	[185]
H3K79succ, H3K122succ	Glioblastoma	High H3K79succ promotes the proliferation and development of glioma cells	[77]
Global histone Kbz	HCC	Kbz is involved in HCC progression	[22]

H3K18ac: histone H3 lysine 18 acetylation, H3K9ac: histone H3 lysine 9 acetylation, H3K12ac: histone H3 lysine 12 acetylation, H3K16ac: histone H3 lysine 16 acetylation, H3K23pr: histone H3 lysine 23 propionylation, histone Kcr: global histone lysine crotonylation, HCC: hepatocellular carcinoma, H3K9bhb: histone H3 lysine 9 β-hydroxybutyrylation, Khib: global histone lysine 2-hydroxyisobutyrylation, H3K18la: histone H3 lysine 18 lactylation, H3K9la: histone H3 lysine 9 lactylation, H3K56la: histone H3 lysine 56 lactylation, H3K79succ: histone H3 lysine 79 succinylation, H3K122succ: histone H3 lysine 122 succinylation, Kbz: global histone lysine benzoylation.

Changes in other histone acylation marks, such as propionylation and butyrylation, were identified in cancer cells and could be promising biomarkers. For example, the propionylation of histone H3K23 in U937 leukemia cells surpass those in non-leukemia cells by at least six-fold. Furthermore, a significant drop in propionylation levels occurred during monocyte differentiation of U937 cells, suggesting that the initial hyperpropionylation in U937 cells might serve as a specific marker of leukemia development [186].

The levels of certain metabolites, such as 2-hydroxyglutarate (2HG), were associated with histone acylation patterns and may serve as potential biomarkers for cancer diagnosis and prognosis [187]. Increased levels of 2HG have been observed in several types of cancer, and this metabolite has been shown to inhibit the activity of histone demethylases, leading to altered histone methylation and gene expression [187–189].

Gene mutation in adenomatous polyposis coli (APC) may dictate the onset of colorectal cancer (CRC), but recent epidemiological studies have shown that the majority of young adults diagnosed with CRC do not possess hereditary syndromes or germline mutations typically associated with CRC [190,191]. Remarkably, the conventional clinical criteria used to identify individuals at higher risk of CRC often prove inadequate in these cases [192,193], indicating the need for epigenetic-based biomarkers for screening specific cancers.

4. Targeting Histone Acylation for Cancer Prevention and Therapy

Cancer prevention involves proactive measures to reduce cancer risk through lifestyle choices, avoidance of carcinogens, and possibly utilizing medications or vaccines. Environmental and lifestyle factors, encompassing radiation, toxins, pollutants, infectious agents, and diet, influence epigenetic events [194,195]. Disruption of these events leads to abnormal gene expression, notably contributing to severe diseases like cancer. Fortunately unlike genetic mutations, epigenetic changes are potentially reversible, offering a crucial avenue for cancer prevention and therapy [196]. The anticancer role of dietary bioactive compounds and phytochemicals mediated by histone PTMs has been previously reviewed [197–201]. Several clinical trials involving natural products and diet interventions for cancer therapy have also been extensively reviewed [202–206]. This section explores how dietary and metabolism-derived histone acylation events can be used as an attractive target for cancer prevention and therapy.

4.1. Targeting Histone Acylation for Cancer Prevention

It is well documented that dietary and lifestyle factors can affect the metabolism-derived histone acyl code and modify cancer risk. For example, certain dietary compounds such as butanoates [161], which are produced by gut microbiota from dietary fiber, can promote histone acetylation and reduce cancer risk [164]. Similarly, exercise and physical activity can affect the metabolism of fatty acids and improve histone acylation patterns [156,207,208]. Several bioactive compounds from the diet have been reported to play a role in preventing cancer through epigenetic mechanisms [209,210] (Figure 1). Hence, reversing the impact of aberrant histone acylation is one approach to preventing the early development and progression of cancer.

Essential nutrients like folate, vitamin B-12, selenium, and zinc, alongside dietary compounds such as sulforaphane, tea polyphenols, curcumin, and allyl sulfur compounds, are part of an expanding arsenal that influences epigenetic processes [210,211] by targeting enzymes involved in histone acylation, such as HATs and HDACs [164,212]. Emerging evidence also suggests that metabolism-derived histone acylations may be involved in regulating gene expression in response to nutrient availability, oxidative stress, and other environmental cues [8,11,213]. For example, the levels of histone acetylation, butyrylation, and succinylation have been shown to change in response to caloric restriction, fasting, or high-fat diets [59,213]. Additionally, targeting metabolic pathways that produce acyl-CoA metabolites, such as fatty acid metabolism, can also be addressed with dietary or pharmacological agents to modify histone acylation patterns and reduce cancer risk. Specifically, fatty acid synthesis inhibitors such as soraphen A, cerulenin, orlistat, TOFA, GSK165, and UB006 have demonstrated antitumor efficacy in cancers such as neuroblastoma [214], prostate cancer [215], and colorectal cancer [216–218]. Although targeting histone acylation has been suggested to potentially prevent or slow down cancer development and progression [8,170], additional investigation is warranted to comprehensively elucidate the mechanisms that underlie the connection between altered histone acylation and cancer. This research is crucial for developing effective interventions that can be used in clinical settings.

4.2. Targeting Histone Acylation for Cancer Therapy

Histone acylation has been implicated in the pathogenesis of various diseases, including cancer (Table 2). These marks will be discussed in brief.

4.2.1. Targeting Acylation Writer and Eraser

Numerous HDAC inhibitors have undergone development and evaluation in both preclinical and clinical investigations for treating cancer, inflammatory diseases, and metabolic disorders [219,220]. These inhibitors have demonstrated the ability to trigger apoptosis and inhibit tumor growth across diverse cancer types such as lymphoma, leukemia, breast cancer, prostate cancer, and lung cancer [221]. The mode of action of HDAC inhibitors entails suppressing HDAC activity, elevating histone acylation levels and yielding anticancer effects, as outlined in several reviews [221–223]. Gut microbiota butyrate production also triggers HDAC inhibition, leading to elevated expression of IFN-γ and granzyme B in cytotoxic T cells (CTLs) [165]. New mechanistic insight was reported for HDAC inhibition by linoleate and butyrate metabolites acting via the IFN-γ pathway to mediate reactivation of immune related genes for antitumor response in a preclinical model of CRC [164]. The working hypothesis was that epigenetic suppression of MHC cell surface presentation could be rectified by correctly positioning neoepitopes to engage the host immune system at the adenoma stage. Exploring the types of histone acylation change required to restore functional MHC complexes on the surface of cancer cells warrants further investigation.

Dietary HDAC inhibitors have also demonstrated anticancer effects linked to histone acetylation status while minimizing the likelihood of adverse effects [224], including sulforaphane [225], polyphenols [226], and spinach metabolites [164]. Research has pinpointed other histone acylation marks as potential therapeutic targets. For instance, the levels of histone succinylation were elevated in certain types of cancer [14], and inhibition of the desuccinylase SIRT5 reduced tumor growth in preclinical models [227,228]. Similarly, histone butyrylation levels linked to insulin resistance and diabetes were addressed by inhibiting the butyryltransferase CBP, improving glucose homeostasis in mouse models. In vivo, gut microbiota-derived butyrate inhibited class I HDACs, thereby affecting histone decrotonylation in mice colons [21]. In vitro studies also indicated that HDAC3 possessed decrotonylase activity [229], suggesting a potential target for HDAC3-specific inhibitors.

Targeting HDACs can also work for nonhistone protein acylation, like palmitoylation. Specifically, the palmitoylation of interferon gamma receptor 1 (IFNGR1) alters its protein–protein interactions. Instead of associating with optineurin, palmitoylated IFNGR1 binds to the adaptor protein complex 3 subunit delta-1 (AP3D1), leading to lysosomal degradation of IFNGR1. This degradation hinders the IFNγ and MHC-I pathways, contributing to immune evasion [230]. Conversely, depalmitoylation of IFNGR1 promotes its stability and the functionality of downstream MHC-I signaling, crucial for effective antigen presentation to T cells [170,230]. Additionally, HDAC2 inhibits PD-L1 acetylation, enhancing nuclear localization and immune checkpoint activation [231]. Meanwhile, P300 promotes MEF2D acetylation, boosting PD-L1 transcription [232]. PD-L1 palmitoylation, facilitated by ZDHHC3 and ZDHHC9, prevents lysosomal degradation, contributing to T cell exhaustion [233,234]. Furthermore, PCAF and GCN5-mediated acetylation enhances Rae-1 stability, activating NK/T cell killing ability. Conversely, P300-driven TRIB3 acetylation hinders T cell infiltration by dampening CXCL10 transcription. SIRT1-mediated deacetylation of p53 promotes TAM infiltration through CXCL12 secretion [235]. Leveraging the immune system through epigenetic drug intervention holds promise for both cancer prevention and therapy.

In addition to targeting epigenetic erasers, other inhibitors of histone acyl-modifying enzymes are also being developed as potential cancer therapies. For example, inhibitors of the histone acetyltransferase (HAT) writer CBP/p300 suppress tumor growth in preclinical models of breast and lung cancer [236]. Consequently, there is a growing interest in CBP/p300 inhibitors and protein degraders as promising therapeutic agents for cancer treatment, with the potential for translation into clinical settings [237]. In the case of breast and prostate cancer, CBP/p300 regulate nuclear hormone receptor signaling [238]. Targeting CBP/p300 may be tissue specific and context dependent, adding to the paradoxical roles in tumor suppression and oncogene actions [238,239].

4.2.2. Targeting Acylation Readers

Recent studies reported improved antitumor outcomes through epigenetic combination therapy via HDAC plus acetyl reader inhibition [240–242]. Small-molecule BET inhibitors, such as JQ1, have entered clinical trials [243,244]. Tea and soy polyphenols have been shown to inhibit the non-BET family member BRD9, triggering DNA damage and apoptosis in colon cancer cells [245]. Human MOZ (KAT6A) and DPF2 (BAF45d) use their double PHD finger domains to bind various histone lysine acylations, favoring Kcr, followed by Kpr and Ku [18,30,103,246]. The existence of distinct acylation readers with a preference for specific histone modifications [18,103,246] underscores the significance of these approaches in future investigations of histone acylation reader-targeted therapeutics.

4.3. Current Approaches and Future Directions in Targeting Histone Acylation for Cancer Interception

Despite the significant challenges, identifying and addressing the critical hurdles outlined above holds the promise for effective new acyl code-based anticancer therapies [247]. Developing site-specific and more precise molecular tools for targeted acylation or deacylation to control the expression of anticancer therapy genes remain an aspirational pursuit [248,249].

Presently, HDAC inhibitors stand out as the most extensively studied compounds targeting histone acyl-modifications. Several HDAC inhibitors, including vorinostat, romidepsin, and belinostat, have been approved by the FDA for the treatment of various types of cancer [250]. Other approaches include the development of HAT inhibitors and other histone acyl-modifying enzymes such as butyryltransferases, crotonyltransferases, and propionyltransferases [247]. Inhibitors of the HAT p300/CBP showed promise in breast and lung cancer models [236]. Additionally, inhibitors for butyryltransferase GCN5 and the crotonyltransferase PCAF demonstrated antitumor effects in preclinical models [237,251]. Although this review focuses only on histone acylation, there are also non-histone protein acylation and acyltransferases linked to cancer. For instance, a homologous recombination (HR) protein MRE11 is lactylated by CBP in response to DNA damage. High lactate levels in cancer cells lead to MRE11 lactylation and chemoresistance, providing insights into the role of cellular metabolism in DSB repair and chemotherapeutic response [252]. A succinyl transferase OXT1-mediated succinylation of beta-lactamase-like protein (LACTB) inhibits its proteolytic activity, leading to HCC progression [253] indicating potential use of OXCT1 inhibitors for such cancers [254].

The diversity and complexity of histone acyl modifications and their biological functions are areas under intense exploration. Innovative drug delivery methods, such as nanotechnology-based approaches, show promise in improving the bioavailability and efficacy of histone acyl-modifying enzyme inhibitors [255]. Addressing possible toxicity concerns and resistance mechanisms might necessitate combination strategies with immune-based therapies [256–258]. Finally, understanding the pharmacokinetic properties of histone acyl-modifying enzyme inhibitors, such as bioavailability and metabolism, is needed for optimizing efficacy and safety in vivo.

4.4. Challenges in Targeting Histone Acylation

Translating histone acylation-based treatments faces several challenges, most notably the requirement for extensive proteomic-based screening of histone and non-histone protein acylation in bodily fluids like peripheral blood, fecal samples, or saliva. Such screening methodologies aim to identify early diagnostic markers or therapeutic targets for various diseases, including cancer [259]. Moreover, the heterogeneous nature of cancer cells demands the identification of specific histone acyl codes tailored to various cancer types and individual patient profiles. Ensuring specificity in targeting these modifications is complicated by the overlapping substrate specificity and activity of modifying enzymes, and a consideration of non-enzymatic lysine acylation on non-histone proteins [67,260]. Additionally, delivery of therapeutic agents to cancer cells within the tumor microenvironment is complex, while minimizing off-target effects [260]. Furthermore, the rapid adaptability of cancer cells to changes in their metabolic milieu can result in resistance to therapies targeting histone acylation [261]. Concerns regarding toxicity, including hematological and cardiac adverse effects, along with the influence of pharmacokinetic properties on efficacy and safety in vivo, further complicates the development and clinical use of histone acyl-modifying enzyme inhibitors for cancer therapy.

5. Conclusions

Diet- and metabolism-derived histone acylation marks have been implicated in cancer epigenetics, but their relative contributions to overall disease pathogenesis remain underexplored. Complex and dynamic change in histone modifications, catalyzed by specific enzymes, influence gene expression, chromatin structure, and cellular behavior. Understanding the significance and role of dietary and metabolism-derived histone acyl code in cancer epigenetics has the potential to unveil new cancer biomarkers and therapeutic targets. While targeting histone acylation holds promise for cancer therapy, challenges such as the lack of inhibitors for the specific enzymes need to be addressed. Future research endeavors should focus on unraveling the mechanisms of diet and metabolism-derived histone acylation changes, aiming to develop more effective cancer therapies and precision immunoprevention.

Author Contributions: Conceptualization, S.N., P.R. and R.H.D.; writing—original draft preparation, S.N. and P.R.; writing—review and editing, S.N., P.R., W.M.D. and R.H.D.; supervision, P.R. and R.H.D. All authors have read and agreed to the published version of the manuscript.

Funding: S.N. was supported by a Cancer Therapeutics Training Program (CTTP) of the Gulf Coast Consortium (GCC) (CPRIT-RP210043). This work was supported by grants CA122959 and CA257559 from the National Cancer Institute, by the John S. Dunn Foundation, and by a Chancellor's Research Initiative. P.R. was also supported by an Alkek Fellowship Award from the Texas A&M Institute for Biosciences and Technology (IBT).

Acknowledgments: We thank members of the Dashwood and Rajendran labs for valuable discussion of topics related to the review article, including Sabeeta Kapoor, Nive Mohan, Jorge Tovar Perez, and Ahmed Muhsin.

Conflicts of Interest: The authors declare no conflicts of interest.

References

1. Ferlay, J.; Ervik, M.; Lam, F.; Colombet, M.; Mery, L.; Piñeros, M.; Znaor, A.; Soerjomataram, I.; Bray, F. *Global Cancer Observatory: Cancer Today*; International Agency for Research on Cancer: Lyon, France, 2020.
2. Ilango, S.; Paital, B.; Jayachandran, P.; Padma, P.R.; Nirmaladevi, R. Epigenetic alterations in cancer. *Front. Biosci.-Landmark* **2020**, *25*, 1058–1109.
3. Baylin, S.B.; Jones, P.A. Epigenetic determinants of cancer. *Cold Spring Harb. Perspect. Biol.* **2016**, *8*, a019505. [CrossRef] [PubMed]
4. Fu, L.N.; Tan, J.; Chen, Y.X.; Fang, J.Y. Genetic variants in the histone methylation and acetylation pathway and their risks in eight types of cancers. *J. Dig. Dis.* **2018**, *19*, 102–111. [CrossRef] [PubMed]
5. Ren, X.; Zhou, Y.; Xue, Z.; Hao, N.; Li, Y.; Guo, X.; Wang, D.; Shi, X.; Li, H. Histone benzoylation serves as an epigenetic mark for DPF and YEATS family proteins. *Nucleic Acids Res.* **2021**, *49*, 114–126. [CrossRef] [PubMed]
6. Etchegaray, J.P.; Mostoslavsky, R. Interplay between Metabolism and Epigenetics: A Nuclear Adaptation to Environmental Changes. *Mol. Cell.* **2016**, *62*, 695–711. [CrossRef] [PubMed]
7. Simithy, J.; Sidoli, S.; Yuan, Z.-F.; Coradin, M.; Bhanu, N.V.; Marchione, D.M.; Klein, B.J.; Bazilevsky, G.A.; McCullough, C.E.; Magin, R.S. Characterization of histone acylations links chromatin modifications with metabolism. *Nat. Commun.* **2017**, *8*, 1141. [CrossRef] [PubMed]
8. Sabari, B.R.; Zhang, D.; Allis, C.D.; Zhao, Y. Metabolic regulation of gene expression through histone acylations. *Nat. Rev. Mol. Cell Biol.* **2017**, *18*, 90–101. [CrossRef]
9. Xu, Y.; Shi, Z.; Bao, L. An expanding repertoire of protein acylations. *Mol. Cell. Proteom.* **2022**, *21*, 100193. [CrossRef]
10. Liu, X.; Cooper, D.E.; Cluntun, A.A.; Warmoes, M.O.; Zhao, S.; Reid, M.A.; Liu, J.; Lund, P.J.; Lopes, M.; Garcia, B.A.; et al. Acetate Production from Glucose and Coupling to Mitochondrial Metabolism in Mammals. *Cell* **2018**, *175*, 502–513.e513. [CrossRef]
11. Kebede, A.F.; Nieborak, A.; Shahidian, L.Z.; Le Gras, S.; Richter, F.; Gomez, D.A.; Baltissen, M.P.; Meszaros, G.; Magliarelli, H.d.F.; Taudt, A. Histone propionylation is a mark of active chromatin. *Nat. Struct. Mol. Biol.* **2017**, *24*, 1048–1056. [CrossRef]
12. Han, Z.; Wu, H.; Kim, S.; Yang, X.; Li, Q.; Huang, H.; Cai, H.; Bartlett, M.G.; Dong, A.; Zeng, H. Revealing the protein propionylation activity of the histone acetyltransferase MOF (males absent on the first). *J. Biol. Chem.* **2018**, *293*, 3410–3420. [CrossRef]
13. Vollmuth, F.; Geyer, M. Interaction of propionylated and butyrylated histone H3 lysine marks with Brd4 bromodomains. *Angew. Chem.* **2010**, *122*, 6920–6924. [CrossRef]
14. Zhao, S.; Zhang, X.; Li, H. Beyond histone acetylation—Writing and erasing histone acylations. *Curr. Opin. Struct. Biol.* **2018**, *53*, 169–177. [CrossRef] [PubMed]

15. Son, M.Y.; Cho, H.S. Anticancer Effects of Gut Microbiota-Derived Short-Chain Fatty Acids in Cancers. *J. Microbiol. Biotechnol.* **2023**, *33*, 849–856. [CrossRef] [PubMed]

16. Fu, Y.; Yu, J.; Li, F.; Ge, S. Oncometabolites drive tumorigenesis by enhancing protein acylation: From chromosomal remodelling to nonhistone modification. *J. Exp. Clin. Cancer Res.* **2022**, *41*, 144. [CrossRef]

17. Wang, Y.; Jin, J.; Chung, M.W.H.; Feng, L.; Sun, H.; Hao, Q. Identification of the YEATS domain of GAS41 as a pH-dependent reader of histone succinylation. *Proc. Natl. Acad. Sci. USA* **2018**, *115*, 2365–2370. [CrossRef] [PubMed]

18. Andrews, F.H.; Shinsky, S.A.; Shanle, E.K.; Bridgers, J.B.; Gest, A.; Tsun, I.K.; Krajewski, K.; Shi, X.; Strahl, B.D.; Kutateladze, T.G. The Taf14 YEATS domain is a reader of histone crotonylation. *Nat. Chem. Biol.* **2016**, *12*, 396–398. [CrossRef]

19. Du, J.; Zhou, Y.; Su, X.; Yu, J.J.; Khan, S.; Jiang, H.; Kim, J.; Woo, J.; Kim, J.H.; Choi, B.H. Sirt5 is a NAD-dependent protein lysine demalonylase and desuccinylase. *Science* **2011**, *334*, 806–809. [CrossRef]

20. Zhao, D.; Guan, H.; Zhao, S.; Mi, W.; Wen, H.; Li, Y.; Zhao, Y.; Allis, C.D.; Shi, X.; Li, H. YEATS2 is a selective histone crotonylation reader. *Cell Res.* **2016**, *26*, 629–632. [CrossRef]

21. Fellows, R.; Denizot, J.; Stellato, C.; Cuomo, A.; Jain, P.; Stoyanova, E.; Balázsi, S.; Hajnády, Z.; Liebert, A.; Kazakevych, J.; et al. Microbiota derived short chain fatty acids promote histone crotonylation in the colon through histone deacetylases. *Nat. Commun.* **2018**, *9*, 105. [CrossRef]

22. Tan, D.; Wei, W.; Han, Z.; Ren, X.; Yan, C.; Qi, S.; Song, X.; Zheng, Y.G.; Wong, J.; Huang, H. HBO1 catalyzes lysine benzoylation in mammalian cells. *iScience* **2022**, *25*, 105443. [CrossRef] [PubMed]

23. Zhang, X.; Cao, R.; Niu, J.; Yang, S.; Ma, H.; Zhao, S.; Li, H. Molecular basis for hierarchical histone de-β-hydroxybutyrylation by SIRT3. *Cell Discov.* **2019**, *5*, 35. [CrossRef]

24. Gaffney, D.O.; Jennings, E.Q.; Anderson, C.C.; Marentette, J.O.; Shi, T.; Oxvig, A.-M.S.; Streeter, M.D.; Johannsen, M.; Spiegel, D.A.; Chapman, E. Non-enzymatic lysine lactoylation of glycolytic enzymes. *Cell Chem. Biol.* **2020**, *27*, 206–213.e206. [CrossRef]

25. Zhou, T.; Cheng, X.; He, Y.; Xie, Y.; Xu, F.; Xu, Y.; Huang, W. Function and mechanism of histone β-hydroxybutyrylation in health and disease. *Front. Immunol.* **2022**, *13*, 981285. [CrossRef]

26. Huang, H.; Luo, Z.; Qi, S.; Huang, J.; Xu, P.; Wang, X.; Gao, L.; Li, F.; Wang, J.; Zhao, W. Landscape of the regulatory elements for lysine 2-hydroxyisobutyrylation pathway. *Cell Res.* **2018**, *28*, 111–125. [CrossRef]

27. Zhang, R.; Bons, J.; Scheidemantle, G.; Liu, X.; Bielska, O.; Carrico, C.; Rose, J.; Heckenbach, I.; Scheibye-Knudsen, M.; Schilling, B.; et al. Histone malonylation is regulated by SIRT5 and KAT2A. *iScience* **2023**, *26*, 106193. [CrossRef]

28. Xie, Z.; Dai, J.; Dai, L.; Tan, M.; Cheng, Z.; Wu, Y.; Boeke, J.D.; Zhao, Y. Lysine succinylation and lysine malonylation in histones. *Mol. Cell. Proteom.* **2012**, *11*, 100–107. [CrossRef]

29. Park, J.; Chen, Y.; Tishkoff, D.X.; Peng, C.; Tan, M.; Dai, L.; Xie, Z.; Zhang, Y.; Zwaans, B.M.; Skinner, M.E. SIRT5-mediated lysine desuccinylation impacts diverse metabolic pathways. *Mol. Cell* **2013**, *50*, 919–930. [CrossRef]

30. Li, Y.; Sabari, B.R.; Panchenko, T.; Wen, H.; Zhao, D.; Guan, H.; Wan, L.; Huang, H.; Tang, Z.; Zhao, Y. Molecular coupling of histone crotonylation and active transcription by AF9 YEATS domain. *Mol. Cell* **2016**, *62*, 181–193. [CrossRef]

31. Sakabe, K.; Wang, Z.; Hart, G.W. β-N-acetylglucosamine (O-GlcNAc) is part of the histone code. *Proc. Natl. Acad. Sci. USA* **2010**, *107*, 19915–19920. [CrossRef]

32. Zou, C.; Ellis, B.M.; Smith, R.M.; Chen, B.B.; Zhao, Y.; Mallampalli, R.K. Acyl-CoA: Lysophosphatidylcholine acyltransferase I (Lpcat1) catalyzes histone protein O-palmitoylation to regulate mRNA synthesis. *J. Biol. Chem.* **2011**, *286*, 28019–28025. [CrossRef]

33. Jiang, H.; Khan, S.; Wang, Y.; Charron, G.; He, B.; Sebastian, C.; Du, J.; Kim, R.; Ge, E.; Mostoslavsky, R.; et al. SIRT6 regulates TNF-α secretion through hydrolysis of long-chain fatty acyl lysine. *Nature* **2013**, *496*, 110–113. [CrossRef]

34. Liu, Z.; Yang, T.; Li, X.; Peng, T.; Hang, H.C.; Li, X.D. Integrative chemical biology approaches to examine 'erasers' for protein lysine fatty-acylation. *Angew. Chem. (Int. Ed. Engl.)* **2015**, *54*, 1149. [CrossRef]

35. Kim, E.; H Bisson, W.; V Lohr, C.; E Williams, D.; Ho, E.; H Dashwood, R.; Rajendran, P. Histone and non-histone targets of dietary deacetylase inhibitors. *Curr. Top. Med. Chem.* **2016**, *16*, 714–731. [CrossRef]

36. Kaur, J.; Daoud, A.; Eblen, S.T. Targeting chromatin remodeling for cancer therapy. *Curr. Mol. Pharmacol.* **2019**, *12*, 215. [CrossRef]

37. Farooqi, A.A.; Fayyaz, S.; Poltronieri, P.; Calin, G.; Mallardo, M. Epigenetic deregulation in cancer: Enzyme players and non-coding RNAs. *Semin. Cancer Biol.* **2022**, *83*, 197–207. [CrossRef]

38. Blot, W.J.; Tarone, R.E. Doll and Peto's quantitative estimates of cancer risks: Holding generally true for 35 years. *JNCI J. Natl. Cancer Inst.* **2015**, *107*, djv044. [CrossRef]

39. Doll, R.; Peto, R. The causes of cancer: Quantitative estimates of avoidable risks of cancer in the United States today. *JNCI J. Natl. Cancer Inst.* **1981**, *66*, 1192–1308. [CrossRef]

40. Guo, D.; Horio, D.T.; Grove, J.S.; Dashwood, R.H. Inhibition by chlorophyllin of 2-amino-3-methylimidazo-[4,5-f] quinoline-induced tumorigenesis in the male F344 rat. *Cancer Lett.* **1995**, *95*, 161–165. [CrossRef]

41. Wang, R.; Dashwood, W.M.; Nian, H.; Löhr, C.V.; Fischer, K.A.; Tsuchiya, N.; Nakagama, H.; Ashktorab, H.; Dashwood, R.H. NADPH oxidase overexpression in human colon cancers and rat colon tumors induced by 2-amino-1-methyl-6-phenylimidazo [4, 5-b] pyridine (PhIP). *Int. J. Cancer* **2011**, *128*, 2581–2590. [CrossRef] [PubMed]

42. Kang, Y.; Nian, H.; Rajendran, P.; Kim, E.; Dashwood, W.; Pinto, J.; Boardman, L.; Thibodeau, S.; Limburg, P.; Löhr, C. HDAC8 and STAT3 repress BMF gene activity in colon cancer cells. *Cell Death Dis.* **2014**, *5*, e1476. [CrossRef]

43. Blum, C.A.; Xu, M.; Orner, G.A.; Fong, A.T.; Bailey, G.S.; Stoner, G.D.; Horio, D.T.; Dashwood, R.H. β-Catenin mutation in rat colon tumors initiated by 1, 2-dimethylhydrazine and 2-amino-3-methylimidazo [4, 5-f] quinoline, and the effect of post-initiation treatment with chlorophyllin and indole-3-carbinol. *Carcinogenesis* **2001**, *22*, 315–320. [CrossRef] [PubMed]

44. Ferrer, C.M.; Lynch, T.P.; Sodi, V.L.; Falcone, J.N.; Schwab, L.P.; Peacock, D.L.; Vocadlo, D.J.; Seagroves, T.N.; Reginato, M.J. O-GlcNAcylation regulates cancer metabolism and survival stress signaling via regulation of the HIF-1 pathway. *Mol. Cell* **2014**, *54*, 820–831. [CrossRef] [PubMed]

45. Kim, J.-A.; Yeom, Y.I. Metabolic signaling to epigenetic alterations in cancer. *Biomol. Ther.* **2018**, *26*, 69. [CrossRef] [PubMed]

46. Kaelin, W.G.; McKnight, S.L. Influence of metabolism on epigenetics and disease. *Cell* **2013**, *153*, 56–69. [CrossRef]

47. Trefely, S.; Lovell, C.D.; Snyder, N.W.; Wellen, K.E. Compartmentalised acyl-CoA metabolism and roles in chromatin regulation. *Mol. Metab.* **2020**, *38*, 100941. [CrossRef]

48. Lund, P.J.; Gates, L.A.; Leboeuf, M.; Smith, S.A.; Chau, L.; Lopes, M.; Friedman, E.S.; Saiman, Y.; Kim, M.S.; Shoffler, C.A.; et al. Stable isotope tracing in vivo reveals a metabolic bridge linking the microbiota to host histone acetylation. *Cell Rep.* **2022**, *41*, 111809. [CrossRef]

49. Sun, X.; Zhang, Y.; Chen, X.-F.; Tang, X. Acylations in cardiovascular biology and diseases, what's beyond acetylation. *Ebiomedicine* **2023**, *87*, 104418. [CrossRef]

50. Trub, A.G.; Hirschey, M.D. Reactive acyl-CoA species modify proteins and induce carbon stress. *Trends Biochem. Sci.* **2018**, *43*, 369–379. [CrossRef]

51. Strahl, B.D.; Allis, C.D. The language of covalent histone modifications. *Nature* **2000**, *403*, 41–45. [CrossRef]

52. Singh, B.N.; Zhang, G.; Hwa, Y.L.; Li, J.; Dowdy, S.C.; Jiang, S.-W. Nonhistone protein acetylation as cancer therapy targets. *Expert Rev. Anticancer. Ther.* **2010**, *10*, 935–954. [CrossRef]

53. Li, T.; Zhang, C.; Hassan, S.; Liu, X.; Song, F.; Chen, K.; Zhang, W.; Yang, J. Histone deacetylase 6 in cancer. *J. Hematol. Oncol.* **2018**, *11*, 111. [CrossRef]

54. Deakin, N.O.; Turner, C.E. Paxillin inhibits HDAC6 to regulate microtubule acetylation, Golgi structure, and polarized migration. *J. Cell Biol.* **2014**, *206*, 395–413. [CrossRef]

55. Klein, B.J.; Jang, S.M.; Lachance, C.; Mi, W.; Lyu, J.; Sakuraba, S.; Krajewski, K.; Wang, W.W.; Sidoli, S.; Liu, J. Histone H3K23-specific acetylation by MORF is coupled to H3K14 acylation. *Nat. Commun.* **2019**, *10*, 4724. [CrossRef]

56. Dai, L.; Peng, C.; Montellier, E.; Lu, Z.; Chen, Y.; Ishii, H.; Debernardi, A.; Buchou, T.; Rousseaux, S.; Jin, F.; et al. Lysine 2-hydroxyisobutyrylation is a widely distributed active histone mark. *Nat. Chem. Biol.* **2014**, *10*, 365–370. [CrossRef]

57. Goudarzi, A.; Zhang, D.; Huang, H.; Barral, S.; Kwon, O.K.; Qi, S.; Tang, Z.; Buchou, T.; Vitte, A.-L.; He, T.; et al. Dynamic Competing Histone H4 K5K8 Acetylation and Butyrylation Are Hallmarks of Highly Active Gene Promoters. *Mol. Cell* **2016**, *62*, 169–180. [CrossRef]

58. Sabari, B.R.; Tang, Z.; Huang, H.; Yong-Gonzalez, V.; Molina, H.; Kong, H.E.; Dai, L.; Shimada, M.; Cross, J.R.; Zhao, Y.; et al. Intracellular Crotonyl-CoA Stimulates Transcription through p300-Catalyzed Histone Crotonylation. *Mol. Cell* **2015**, *58*, 203–215. [CrossRef]

59. Xie, Z.; Zhang, D.; Chung, D.; Tang, Z.; Huang, H.; Dai, L.; Qi, S.; Li, J.; Colak, G.; Chen, Y.; et al. Metabolic Regulation of Gene Expression by Histone Lysine β-Hydroxybutyrylation. *Mol. Cell* **2016**, *62*, 194–206. [CrossRef]

60. Smestad, J.; Erber, L.; Chen, Y.; Maher, L.J. Chromatin succinylation correlates with active gene expression and is perturbed by defective TCA cycle metabolism. *Iscience* **2018**, *2*, 63–75. [CrossRef]

61. Bao, X.; Liu, Z.; Zhang, W.; Gladysz, K.; Fung, Y.M.E.; Tian, G.; Xiong, Y.; Wong, J.W.H.; Yuen, K.W.Y.; Li, X.D. Glutarylation of Histone H4 Lysine 91 Regulates Chromatin Dynamics. *Mol. Cell* **2019**, *76*, 660–675.e669. [CrossRef]

62. Huang, H.; Zhang, D.; Wang, Y.; Perez-Neut, M.; Han, Z.; Zheng, Y.G.; Hao, Q.; Zhao, Y. Lysine benzoylation is a histone mark regulated by SIRT2. *Nat. Commun.* **2018**, *9*, 3374. [CrossRef]

63. Zhang, D.; Tang, Z.; Huang, H.; Zhou, G.; Cui, C.; Weng, Y.; Liu, W.; Kim, S.; Lee, S.; Perez-Neut, M.; et al. Metabolic regulation of gene expression by histone lactylation. *Nature* **2019**, *574*, 575–580. [CrossRef]

64. Jo, C.; Park, S.; Oh, S.; Choi, J.; Kim, E.-K.; Youn, H.-D.; Cho, E.-J. Histone acylation marks respond to metabolic perturbations and enable cellular adaptation. *Exp. Mol. Med.* **2020**, *52*, 2005–2019. [CrossRef]

65. Zhou, Y.; Peng, J.; Jiang, S. Role of histone acetyltransferases and histone deacetylases in adipocyte differentiation and adipogenesis. *Eur. J. Cell Biol.* **2014**, *93*, 170–177. [CrossRef]

66. Peserico, A.; Simone, C. Physical and functional HAT/HDAC interplay regulates protein acetylation balance. *J. Biomed. Biotechnol.* **2011**, *2011*, 371832. [CrossRef]

67. Wapenaar, H.; Dekker, F.J. Histone acetyltransferases: Challenges in targeting bi-substrate enzymes. *Clin. Epigenetics* **2016**, *8*, 59. [CrossRef] [PubMed]

68. Audia, J.E.; Campbell, R.M. Histone modifications and cancer. *Cold Spring Harb. Perspect. Biol.* **2016**, *8*, a019521. [CrossRef] [PubMed]

69. Chatterjee, A.; Rodger, E.J.; Eccles, M.R. Epigenetic drivers of tumourigenesis and cancer metastasis. *Semin. Cancer Biol.* **2018**, *51*, 149–159. [CrossRef] [PubMed]

70. Dutta, A.; Abmayr, S.M.; Workman, J.L. Diverse Activities of Histone Acylations Connect Metabolism to Chromatin Function. *Mol. Cell* **2016**, *63*, 547–552. [CrossRef] [PubMed]

71. Cavalieri, V. The expanding constellation of histone post-translational modifications in the epigenetic landscape. *Genes* **2021**, *12*, 1596. [CrossRef] [PubMed]
72. Fu, Q.; Cat, A.; Zheng, Y.G. New Histone Lysine Acylation Biomarkers and Their Roles in Epigenetic Regulation. *Curr. Protoc.* **2023**, *3*, e746. [CrossRef]
73. Grevengoed, T.J.; Klett, E.L.; Coleman, R.A. Acyl-CoA metabolism and partitioning. *Annu. Rev. Nutr.* **2014**, *34*, 1–30. [CrossRef]
74. Morrison, A.J. Cancer cell metabolism connects epigenetic modifications to transcriptional regulation. *FEBS J.* **2022**, *289*, 1302–1314. [CrossRef]
75. Pietrocola, F.; Galluzzi, L.; Bravo-San Pedro, J.M.; Madeo, F.; Kroemer, G. Acetyl coenzyme A: A central metabolite and second messenger. *Cell Metab.* **2015**, *21*, 805–821. [CrossRef]
76. Dai, Z.; Ramesh, V.; Locasale, J.W. The evolving metabolic landscape of chromatin biology and epigenetics. *Nat. Rev. Genet.* **2020**, *21*, 737–753. [CrossRef]
77. Wang, Y.; Guo, Y.R.; Liu, K.; Yin, Z.; Liu, R.; Xia, Y.; Tan, L.; Yang, P.; Lee, J.-H.; Li, X.-j. KAT2A coupled with the α-KGDH complex acts as a histone H3 succinyltransferase. *Nature* **2017**, *552*, 273–277. [CrossRef]
78. Tan, M.; Luo, H.; Lee, S.; Jin, F.; Yang, J.S.; Montellier, E.; Buchou, T.; Cheng, Z.; Rousseaux, S.; Rajagopal, N. Identification of 67 histone marks and histone lysine crotonylation as a new type of histone modification. *Cell* **2011**, *146*, 1016–1028. [CrossRef] [PubMed]
79. Ntorla, A.; Burgoyne, J.R. The regulation and function of histone crotonylation. *Front. Cell Dev. Biol.* **2021**, *9*, 624914. [CrossRef] [PubMed]
80. Zorro Shahidian, L.; Haas, M.; Le Gras, S.; Nitsch, S.; Mourão, A.; Geerlof, A.; Margueron, R.; Michaelis, J.; Daujat, S.; Schneider, R. Succinylation of H3K122 destabilizes nucleosomes and enhances transcription. *EMBO Rep.* **2021**, *22*, e51009. [CrossRef] [PubMed]
81. Chen, Y.; Sprung, R.; Tang, Y.; Ball, H.; Sangras, B.; Kim, S.C.; Falck, J.R.; Peng, J.; Gu, W.; Zhao, Y. Lysine propionylation and butyrylation are novel post-translational modifications in histones. *Mol. Cell. Proteom.* **2007**, *6*, 812–819. [CrossRef] [PubMed]
82. Huang, H.; Zhang, D.; Weng, Y.; Delaney, K.; Tang, Z.; Yan, C.; Qi, S.; Peng, C.; Cole, P.A.; Roeder, R.G. The regulatory enzymes and protein substrates for the lysine β-hydroxybutyrylation pathway. *Sci. Adv.* **2021**, *7*, eabe2771. [CrossRef] [PubMed]
83. Wellen, K.E.; Hatzivassiliou, G.; Sachdeva, U.M.; Bui, T.V.; Cross, J.R.; Thompson, C.B. ATP-citrate lyase links cellular metabolism to histone acetylation. *Science* **2009**, *324*, 1076–1080. [CrossRef] [PubMed]
84. Takahashi, H.; McCaffery, J.M.; Irizarry, R.A.; Boeke, J.D. Nucleocytosolic acetyl-coenzyme a synthetase is required for histone acetylation and global transcription. *Mol. Cell* **2006**, *23*, 207–217. [CrossRef] [PubMed]
85. Lee, J.V.; Carrer, A.; Shah, S.; Snyder, N.W.; Wei, S.; Venneti, S.; Worth, A.J.; Yuan, Z.-F.; Lim, H.-W.; Liu, S. Akt-dependent metabolic reprogramming regulates tumor cell histone acetylation. *Cell Metab.* **2014**, *20*, 306–319. [CrossRef] [PubMed]
86. Cai, L.; Sutter, B.M.; Li, B.; Tu, B.P. Acetyl-CoA induces cell growth and proliferation by promoting the acetylation of histones at growth genes. *Mol. Cell* **2011**, *42*, 426–437. [CrossRef] [PubMed]
87. Moffett, J.R.; Puthillathu, N.; Vengilote, R.; Jaworski, D.M.; Namboodiri, A.M. Acetate revisited: A key biomolecule at the nexus of metabolism, epigenetics and oncogenesis—Part 1: Acetyl-CoA, acetogenesis and acyl-CoA short-chain synthetases. *Front. Physiol.* **2020**, *11*, 580167. [CrossRef]
88. Legube, G.; Trouche, D. Regulating histone acetyltransferases and deacetylases. *EMBO Rep* **2003**, *4*, 944–947. [CrossRef]
89. Roessler, C.; Tüting, C.; Meleshin, M.; Steegborn, C.; Schutkowski, M. A novel continuous assay for the deacylase sirtuin 5 and other deacylases. *J. Med. Chem.* **2015**, *58*, 7217–7223. [CrossRef]
90. Yang, L.; Ma, X.; He, Y.; Yuan, C.; Chen, Q.; Li, G.; Chen, X. Sirtuin 5: A review of structure, known inhibitors and clues for developing new inhibitors. *Sci. China Life Sci.* **2017**, *60*, 249–256. [CrossRef]
91. Andrews, A.J.; Chen, X.; Zevin, A.; Stargell, L.A.; Luger, K. The histone chaperone Nap1 promotes nucleosome assembly by eliminating nonnucleosomal histone DNA interactions. *Mol. Cell* **2010**, *37*, 834–842. [CrossRef]
92. Manohar, M.; Mooney, A.M.; North, J.A.; Nakkula, R.J.; Picking, J.W.; Edon, A.; Fishel, R.; Poirier, M.G.; Ottesen, J.J. Acetylation of histone H3 at the nucleosome dyad alters DNA-histone binding. *J. Biol. Chem.* **2009**, *284*, 23312–23321. [CrossRef] [PubMed]
93. Khochbin, S.; Verdel, A.; Lemercier, C.; Seigneurin-Berny, D. Functional significance of histone deacetylase diversity. *Curr. Opin. Genet. Dev.* **2001**, *11*, 162–166. [CrossRef]
94. Baldensperger, T.; Glomb, M.A. Pathways of non-enzymatic lysine acylation. *Front. Cell Dev. Biol.* **2021**, *9*, 664553. [CrossRef]
95. Baldensperger, T.; Eggen, M.; Kappen, J.; Winterhalter, P.R.; Pfirrmann, T.; Glomb, M.A. Comprehensive analysis of posttranslational protein modifications in aging of subcellular compartments. *Sci. Rep.* **2020**, *10*, 7596. [CrossRef] [PubMed]
96. Fukuda, H.; Sano, N.; Muto, S.; Horikoshi, M. Simple histone acetylation plays a complex role in the regulation of gene expression. *Brief. Funct. Genom.* **2006**, *5*, 190–208. [CrossRef]
97. Kaczmarska, Z.; Ortega, E.; Goudarzi, A.; Huang, H.; Kim, S.; Márquez, J.A.; Zhao, Y.; Khochbin, S.; Panne, D. Structure of p300 in complex with acyl-CoA variants. *Nat. Chem. Biol.* **2017**, *13*, 21–29. [CrossRef] [PubMed]
98. Guay, C.; Madiraju, S.R.M.; Aumais, A.; Joly, É.; Prentki, M. A Role for ATP-Citrate Lyase, Malic Enzyme, and Pyruvate/Citrate Cycling in Glucose-induced Insulin Secretion*. *J. Biol. Chem.* **2007**, *282*, 35657–35665. [CrossRef]
99. Koren, D.; Palladino, A. Chapter 3—Hypoglycemia. In *Genetic Diagnosis of Endocrine Disorders*, 2nd ed.; Weiss, R.E., Refetoff, S., Eds.; Academic Press: San Diego, CA, USA, 2016; pp. 31–75. [CrossRef]
100. Blasl, A.-T.; Schulze, S.; Qin, C.; Graf, L.G.; Vogt, R.; Lammers, M. Post-translational lysine ac (et) ylation in health, ageing and disease. *Biol. Chem.* **2022**, *403*, 151–194. [CrossRef]

101. Miller, T.L.; Jenesel, S.E. Enzymology of butyrate formation by Butyrivibrio fibrisolvens. *J. Bacteriol.* **1979**, *138*, 99–104. [CrossRef]
102. Seto, E.; Yoshida, M. Erasers of histone acetylation: The histone deacetylase enzymes. *Cold Spring Harb. Perspect. Biol.* **2014**, *6*, a018713. [CrossRef]
103. Khan, A.; Bridgers, J.B.; Strahl, B.D. Expanding the Reader Landscape of Histone Acylation. *Structure* **2017**, *25*, 571–573. [CrossRef]
104. Kim, J.; Daniel, J.; Espejo, A.; Lake, A.; Krishna, M.; Xia, L.; Zhang, Y.; Bedford, M.T. Tudor, MBT and chromo domains gauge the degree of lysine methylation. *EMBO Rep.* **2006**, *7*, 397–403. [CrossRef]
105. Scheid, R.; Chen, J.; Zhong, X. Biological role and mechanism of chromatin readers in plants. *Curr. Opin. Plant Biol.* **2021**, *61*, 102008. [CrossRef] [PubMed]
106. Bennett, R.L.; Licht, J.D. Targeting epigenetics in cancer. *Annu. Rev. Pharmacol. Toxicol.* **2018**, *58*, 187–207. [CrossRef]
107. Josling, G.A.; Selvarajah, S.A.; Petter, M.; Duffy, M.F. The role of bromodomain proteins in regulating gene expression. *Genes* **2012**, *3*, 320–343. [CrossRef] [PubMed]
108. Huber, F.M.; Greenblatt, S.M.; Davenport, A.M.; Martinez, C.; Xu, Y.; Vu, L.P.; Nimer, S.D.; Hoelz, A. Histone-binding of DPF2 mediates its repressive role in myeloid differentiation. *Proc. Natl. Acad. Sci. USA* **2017**, *114*, 6016–6021. [CrossRef] [PubMed]
109. Donohoe, D.R.; Collins, L.B.; Wali, A.; Bigler, R.; Sun, W.; Bultman, S.J. The Warburg effect dictates the mechanism of butyrate-mediated histone acetylation and cell proliferation. *Mol. Cell* **2012**, *48*, 612–626. [CrossRef] [PubMed]
110. Rajendran, P.; Abdelsalam, S.A.; Renu, K.; Veeraraghavan, V.; Ben Ammar, R.; Ahmed, E.A. Polyphenols as potent epigenetics agents for cancer. *Int. J. Mol. Sci.* **2022**, *23*, 11712. [CrossRef] [PubMed]
111. Bonkowski, M.S.; Sinclair, D.A. Slowing ageing by design: The rise of NAD$^+$ and sirtuin-activating compounds. *Nat. Rev. Mol. Cell Biol.* **2016**, *17*, 679–690. [CrossRef] [PubMed]
112. Milne, J.C.; Denu, J.M. The Sirtuin family: Therapeutic targets to treat diseases of aging. *Curr. Opin. Chem. Biol.* **2008**, *12*, 11–17. [CrossRef]
113. Lee, Y.-H.; Song, N.-Y.; Suh, J.; Kim, D.-H.; Kim, W.; Ann, J.; Lee, J.; Baek, J.-H.; Na, H.-K.; Surh, Y.-J. Curcumin suppresses oncogenicity of human colon cancer cells by covalently modifying the cysteine 67 residue of SIRT1. *Cancer Lett.* **2018**, *431*, 219–229. [CrossRef]
114. Isaac, A.R.; da Silva, E.A.N.; de Matos, R.J.B.; Augusto, R.L.; Moreno, G.M.M.; Mendonça, I.P.; de Souza, R.F.; Cabral-Filho, P.E.; Rodrigues, C.G.; Gonçalves-Pimentel, C.; et al. Low omega-6/omega-3 ratio in a maternal protein-deficient diet promotes histone-3 changes in progeny neural cells and favors leukemia inhibitory factor gene transcription. *J. Nutr. Biochem.* **2018**, *55*, 229–242. [CrossRef]
115. Sadli, N.; Ackland, M.L.; De Mel, D.; Sinclair, A.J.; Suphioglu, C. Effects of zinc and DHA on the epigenetic regulation of human neuronal cells. *Cell. Physiol. Biochem.* **2012**, *29*, 87–98. [CrossRef]
116. Bakhshi, T.J.; Way, T.; Muncy, B.; Georgel, P.T. Effects of the omega-3 fatty acid DHA on histone and p53 acetylation in diffuse large B cell lymphoma. *Biochem. Cell Biol.* **2023**, *101*, 172–191. [CrossRef]
117. Merzvinskyte, R.; Treigyte, G.; Savickiene, J.; Magnusson, K.E.; Navakauskiene, R. Effects of histone deacetylase inhibitors, sodium phenyl butyrate and vitamin B3, in combination with retinoic acid on granulocytic differentiation of human promyelocytic leukemia HL-60 cells. *Ann. N. Y. Acad. Sci.* **2006**, *1091*, 356–367. [CrossRef]
118. Fellows, R.C. Microbiota Regulate Short Chain Fatty Acids and Influence Histone Acylations in Intestinal Epithelial Cells. Ph.D. Thesis, University of Cambridge, Cambridge, UK, 2020.
119. Bradshaw, P.C. Acetyl-CoA Metabolism and Histone Acetylation in the Regulation of Aging and Lifespan. *Antioxidants* **2021**, *10*, 572. [CrossRef] [PubMed]
120. Sidoli, S.; Trefely, S.; Garcia, B.A.; Carrer, A. Integrated Analysis of Acetyl-CoA and Histone Modification via Mass Spectrometry to Investigate Metabolically Driven Acetylation. *Methods Mol. Biol.* **2019**, *1928*, 125–147. [CrossRef] [PubMed]
121. Carrer, A.; Parris, J.L.; Trefely, S.; Henry, R.A.; Montgomery, D.C.; Torres, A.; Viola, J.M.; Kuo, Y.-M.; Blair, I.A.; Meier, J.L. Impact of a high-fat diet on tissue Acyl-CoA and histone acetylation levels. *J. Biol. Chem.* **2017**, *292*, 3312–3322. [CrossRef] [PubMed]
122. Khajebishak, Y.; Alivand, M.; Faghfouri, A.H.; Moludi, J.; Payahoo, L. The effects of vitamins and dietary pattern on epigenetic modification of non-communicable diseases. *Int. J. Vitam. Nutr. Res.* **2021**, *93*, 362–377. [CrossRef] [PubMed]
123. Rohde, J.R.; Cardenas, M.E. The Tor Pathway Regulates Gene Expression by Linking Nutrient Sensing to Histone Acetylation. *Mol. Cell. Biol.* **2003**, *23*, 629–635. [CrossRef] [PubMed]
124. Bungard, D.; Fuerth, B.J.; Zeng, P.Y.; Faubert, B.; Maas, N.L.; Viollet, B.; Carling, D.; Thompson, C.B.; Jones, R.G.; Berger, S.L. Signaling kinase AMPK activates stress-promoted transcription via histone H2B phosphorylation. *Science* **2010**, *329*, 1201–1205. [CrossRef]
125. Salminen, A.; Kauppinen, A.; Kaarniranta, K. AMPK/Snf1 signaling regulates histone acetylation: Impact on gene expression and epigenetic functions. *Cell. Signal.* **2016**, *28*, 887–895. [CrossRef]
126. Anand, P.; Kunnumakkara, A.B.; Sundaram, C.; Harikumar, K.B.; Tharakan, S.T.; Lai, O.S.; Sung, B.; Aggarwal, B.B. Cancer is a preventable disease that requires major lifestyle changes. *Pharm. Res.* **2008**, *25*, 2097–2116. [CrossRef] [PubMed]
127. Irigaray, P.; Newby, J.A.; Clapp, R.; Hardell, L.; Howard, V.; Montagnier, L.; Epstein, S.; Belpomme, D. Lifestyle-related factors and environmental agents causing cancer: An overview. *Biomed. Pharmacother.* **2007**, *61*, 640–658. [CrossRef] [PubMed]
128. Dey, P.; Kimmelman, A.C.; DePinho, R.A. Metabolic Codependencies in the Tumor Microenvironment. *Cancer Discov.* **2021**, *11*, 1067–1081. [CrossRef]
129. Campbell, S.L.; Wellen, K.E. Metabolic signaling to the nucleus in cancer. *Mol. Cell* **2018**, *71*, 398–408. [CrossRef] [PubMed]

130. Hui, L.; Chen, Y. Tumor microenvironment: Sanctuary of the devil. *Cancer Lett.* **2015**, *368*, 7–13. [CrossRef]
131. Thakur, C.; Chen, F. Connections between metabolism and epigenetics in cancers. *Semin. Cancer Biol.* **2019**, *57*, 52–58. [CrossRef]
132. Min, H.Y.; Lee, H.Y. Oncogene-Driven Metabolic Alterations in Cancer. *Biomol. Ther.* **2018**, *26*, 45–56. [CrossRef]
133. Ge, T.; Gu, X.; Jia, R.; Ge, S.; Chai, P.; Zhuang, A.; Fan, X. Crosstalk between metabolic reprogramming and epigenetics in cancer: Updates on mechanisms and therapeutic opportunities. *Cancer Commun.* **2022**, *42*, 1049–1082. [CrossRef]
134. Kim, Y.I. Folate and colorectal cancer: An evidence-based critical review. *Mol. Nutr. Food Res.* **2007**, *51*, 267–292.
135. Ulrich, C.M.; Potter, J.D. Folate and cancer—Timing is everything. *JAMA* **2007**, *297*, 2408–2409. [CrossRef] [PubMed]
136. Loedin, A.K.; Speijer, D. Is There a Carcinogenic Risk Attached to Vitamin B(12) Deficient Diets and What Should We Do About It? Reviewing the Facts. *Mol. Nutr. Food Res.* **2021**, *65*, e2000945. [CrossRef] [PubMed]
137. Aksan, A.; Farrag, K.; Aksan, S.; Schroeder, O.; Stein, J. Flipside of the coin: Iron deficiency and colorectal cancer. *Front. Immunol.* **2021**, *12*, 635899. [CrossRef] [PubMed]
138. Wanders, D.; Hobson, K.; Ji, X. Methionine Restriction and Cancer Biology. *Nutrients* **2020**, *12*, 684. [CrossRef] [PubMed]
139. Bulanda, S.; Janoszka, B. Consumption of Thermally Processed Meat Containing Carcinogenic Compounds (Polycyclic Aromatic Hydrocarbons and Heterocyclic Aromatic Amines) versus a Risk of Some Cancers in Humans and the Possibility of Reducing Their Formation by Natural Food Additives-A Literature Review. *Int. J. Environ. Res. Public Health* **2022**, *19*, 4781. [CrossRef]
140. Li, Y.; Hecht, S.S. Carcinogenic components of tobacco and tobacco smoke: A 2022 update. *Food Chem. Toxicol.* **2022**, *165*, 113179. [CrossRef]
141. Reuter, S.; Gupta, S.C.; Chaturvedi, M.M.; Aggarwal, B.B. Oxidative stress, inflammation, and cancer: How are they linked? *Free. Radic. Biol. Med.* **2010**, *49*, 1603–1616. [CrossRef]
142. García-Giménez, J.-L.; Garcés, C.; Romá-Mateo, C.; Pallardó, F.V. Oxidative stress-mediated alterations in histone post-translational modifications. *Free. Radic. Biol. Med.* **2021**, *170*, 6–18. [CrossRef]
143. McDonald, O.G.; Li, X.; Saunders, T.; Tryggvadottir, R.; Mentch, S.J.; Warmoes, M.O.; Word, A.E.; Carrer, A.; Salz, T.H.; Natsume, S. Epigenomic reprogramming during pancreatic cancer progression links anabolic glucose metabolism to distant metastasis. *Nat. Genet.* **2017**, *49*, 367–376. [CrossRef]
144. Vander Heiden, M.G.; DeBerardinis, R.J. Understanding the intersections between metabolism and cancer biology. *Cell* **2017**, *168*, 657–669. [CrossRef]
145. Warburg, O.; Wind, F.; Negelein, E. The metabolism of tumors in the body. *J. Gen. Physiol.* **1927**, *8*, 519. [CrossRef] [PubMed]
146. Fan, J.; Hitosugi, T.; Chung, T.-W.; Xie, J.; Ge, Q.; Gu, T.-L.; Polakiewicz, R.D.; Chen, G.Z.; Boggon, T.J.; Lonial, S. Tyrosine phosphorylation of lactate dehydrogenase A is important for NADH/NAD$^+$ redox homeostasis in cancer cells. *Mol. Cell. Biol.* **2011**, *31*, 4938–4950. [CrossRef]
147. Chen, A.-N.; Luo, Y.; Yang, Y.-H.; Fu, J.-T.; Geng, X.-M.; Shi, J.-P.; Yang, J. Lactylation, a novel metabolic reprogramming code: Current status and prospects. *Front. Immunol.* **2021**, *12*, 688910. [CrossRef] [PubMed]
148. Pavlova, N.N.; Thompson, C.B. The emerging hallmarks of cancer metabolism. *Cell Metab.* **2016**, *23*, 27–47. [CrossRef] [PubMed]
149. Osthus, R.C.; Shim, H.; Kim, S.; Li, Q.; Reddy, R.; Mukherjee, M.; Xu, Y.; Wonsey, D.; Lee, L.A.; Dang, C.V. Deregulation of glucose transporter 1 and glycolytic gene expression by c-Myc. *J. Biol. Chem.* **2000**, *275*, 21797–21800. [CrossRef] [PubMed]
150. Wise, D.R.; DeBerardinis, R.J.; Mancuso, A.; Sayed, N.; Zhang, X.-Y.; Pfeiffer, H.K.; Nissim, I.; Daikhin, E.; Yudkoff, M.; McMahon, S.B. Myc regulates a transcriptional program that stimulates mitochondrial glutaminolysis and leads to glutamine addiction. *Proc. Natl. Acad. Sci. USA* **2008**, *105*, 18782–18787. [CrossRef] [PubMed]
151. Hitosugi, T.; Kang, S.; Vander Heiden, M.G.; Chung, T.-W.; Elf, S.; Lythgoe, K.; Dong, S.; Lonial, S.; Wang, X.; Chen, G.Z. Tyrosine phosphorylation inhibits PKM2 to promote the Warburg effect and tumor growth. *Sci. Signal.* **2009**, *2*, ra73. [CrossRef]
152. Miranda-Gonçalves, V.; Lameirinhas, A.; Henrique, R.; Jerónimo, C. Metabolism and Epigenetic Interplay in Cancer: Regulation and Putative Therapeutic Targets. *Front. Genet.* **2018**, *9*, 427. [CrossRef]
153. Ling, R.; Chen, G.; Tang, X.; Liu, N.; Zhou, Y.; Chen, D. Acetyl-CoA synthetase 2(ACSS2): A review with a focus on metabolism and tumor development. *Discov. Oncol.* **2022**, *13*, 58. [CrossRef]
154. Schug, Z.T.; Peck, B.; Jones, D.T.; Zhang, Q.; Grosskurth, S.; Alam, I.S.; Goodwin, L.M.; Smethurst, E.; Mason, S.; Blyth, K. Acetyl-CoA synthetase 2 promotes acetate utilization and maintains cancer cell growth under metabolic stress. *Cancer Cell* **2015**, *27*, 57–71. [CrossRef] [PubMed]
155. Gao, X.; Lin, S.H.; Ren, F.; Li, J.T.; Chen, J.J.; Yao, C.B.; Yang, H.B.; Jiang, S.X.; Yan, G.Q.; Wang, D.; et al. Acetate functions as an epigenetic metabolite to promote lipid synthesis under hypoxia. *Nat. Commun.* **2016**, *7*, 11960. [CrossRef]
156. McDonnell, E.; Crown, S.B.; Fox, D.B.; Kitir, B.; Ilkayeva, O.R.; Olsen, C.A.; Grimsrud, P.A.; Hirschey, M.D. Lipids reprogram metabolism to become a major carbon source for histone acetylation. *Cell Rep.* **2016**, *17*, 1463–1472. [CrossRef] [PubMed]
157. Rabhi, N.; Hannou, S.A.; Froguel, P.; Annicotte, J.-S. Cofactors as metabolic sensors driving cell adaptation in physiology and disease. *Front. Endocrinol.* **2017**, *8*, 304. [CrossRef] [PubMed]
158. Le, A.; Lane, A.N.; Hamaker, M.; Bose, S.; Gouw, A.; Barbi, J.; Tsukamoto, T.; Rojas, C.J.; Slusher, B.S.; Zhang, H. Glucose-independent glutamine metabolism via TCA cycling for proliferation and survival in B cells. *Cell Metab.* **2012**, *15*, 110–121. [CrossRef] [PubMed]
159. Levy, M.; Thaiss, C.A.; Elinav, E. Metabolites: Messengers between the microbiota and the immune system. *Genes Dev.* **2016**, *30*, 1589–1597. [CrossRef] [PubMed]

160. Rossi, T.; Vergara, D.; Fanini, F.; Maffia, M.; Bravaccini, S.; Pirini, F. Microbiota-derived metabolites in tumor progression and metastasis. *Int. J. Mol. Sci.* **2020**, *21*, 5786. [CrossRef]
161. Chen, Y.-S.; Li, J.; Menon, R.; Jayaraman, A.; Lee, K.; Huang, Y.; Dashwood, W.M.; Zhang, K.; Sun, D.; Dashwood, R.H. Dietary spinach reshapes the gut microbiome in an Apc-mutant genetic background: Mechanistic insights from integrated multi-omics. *Gut Microbes* **2021**, *13*, 1972756. [CrossRef]
162. Hanus, M.; Parada-Venegas, D.; Landskron, G.; Wielandt, A.M.; Hurtado, C.; Alvarez, K.; Hermoso, M.A.; López-Köstner, F.; De la Fuente, M. Immune system, microbiota, and microbial metabolites: The unresolved triad in colorectal cancer microenvironment. *Front. Immunol.* **2021**, *12*, 612826. [CrossRef]
163. Kespohl, M.; Vachharajani, N.; Luu, M.; Harb, H.; Pautz, S.; Wolff, S.; Sillner, N.; Walker, A.; Schmitt-Kopplin, P.; Boettger, T. The microbial metabolite butyrate induces expression of Th1-associated factors in CD4+ T cells. *Front. Immunol.* **2017**, *8*, 1036. [CrossRef] [PubMed]
164. Chen, Y.-S.; Li, J.; Neja, S.; Kapoor, S.; Tovar Perez, J.E.; Tripathi, C.; Menon, R.; Jayaraman, A.; Lee, K.; Dashwood, W.M. Metabolomics of Acute vs. Chronic Spinach Intake in an Apc–Mutant Genetic Background: Linoleate and Butanoate Metabolites Targeting HDAC Activity and IFN–γ Signaling. *Cells* **2022**, *11*, 573. [CrossRef] [PubMed]
165. Luu, M.; Weigand, K.; Wedi, F.; Breidenbend, C.; Leister, H.; Pautz, S.; Adhikary, T.; Visekruna, A. Regulation of the effector function of CD8+ T cells by gut microbiota-derived metabolite butyrate. *Sci. Rep.* **2018**, *8*, 14430. [CrossRef] [PubMed]
166. Kiweler, N.; Wünsch, D.; Wirth, M.; Mahendrarajah, N.; Schneider, G.; Stauber, R.H.; Brenner, W.; Butter, F.; Krämer, O.H. Histone deacetylase inhibitors dysregulate DNA repair proteins and antagonize metastasis-associated processes. *J. Cancer Res. Clin. Oncol.* **2020**, *146*, 343–356. [CrossRef] [PubMed]
167. Li, Q.; Zhou, H.; Wurtele, H.; Davies, B.; Horazdovsky, B.; Verreault, A.; Zhang, Z. Acetylation of histone H3 lysine 56 regulates replication-coupled nucleosome assembly. *Cell* **2008**, *134*, 244–255. [CrossRef] [PubMed]
168. Gao, M.; Wang, J.; Rousseaux, S.; Tan, M.; Pan, L.; Peng, L.; Wang, S.; Xu, W.; Ren, J.; Liu, Y. Metabolically controlled histone H4K5 acylation/acetylation ratio drives BRD4 genomic distribution. *Cell Rep.* **2021**, *36*, 109460. [CrossRef] [PubMed]
169. Nitsch, S.; Zorro Shahidian, L.; Schneider, R. Histone acylations and chromatin dynamics: Concepts, challenges, and links to metabolism. *EMBO Rep.* **2021**, *22*, e52774. [CrossRef] [PubMed]
170. Shang, S.; Liu, J.; Hua, F. Protein acylation: Mechanisms, biological functions and therapeutic targets. *Signal Transduct. Target. Ther.* **2022**, *7*, 396. [CrossRef] [PubMed]
171. Liberti, M.V.; Locasale, J.W. The Warburg effect: How does it benefit cancer cells? *Trends Biochem. Sci.* **2016**, *41*, 211–218. [CrossRef]
172. Chen, L.; Huang, L.; Gu, Y.; Cang, W.; Sun, P.; Xiang, Y. Lactate-lactylation hands between metabolic reprogramming and immunosuppression. *Int. J. Mol. Sci.* **2022**, *23*, 11943. [CrossRef]
173. Fraga, M.F.; Ballestar, E.; Villar-Garea, A.; Boix-Chornet, M.; Espada, J.; Schotta, G.; Bonaldi, T.; Haydon, C.; Ropero, S.; Petrie, K. Loss of acetylation at Lys16 and trimethylation at Lys20 of histone H4 is a common hallmark of human cancer. *Nat. Genet.* **2005**, *37*, 391–400. [CrossRef]
174. Kamińska, K.; Nalejska, E.; Kubiak, M.; Wojtysiak, J.; Żołna, Ł.; Kowalewski, J.; Lewandowska, M.A. Prognostic and predictive epigenetic biomarkers in oncology. *Mol. Diagn. Ther.* **2019**, *23*, 83–95. [CrossRef]
175. Seligson, D.B.; Horvath, S.; Shi, T.; Yu, H.; Tze, S.; Grunstein, M.; Kurdistani, S.K. Global histone modification patterns predict risk of prostate cancer recurrence. *Nature* **2005**, *435*, 1262–1266. [CrossRef]
176. Wan, L.; Wen, H.; Li, Y.; Lyu, J.; Xi, Y.; Hoshii, T.; Joseph, J.K.; Wang, X.; Loh, Y.-H.E.; Erb, M.A.; et al. ENL links histone acetylation to oncogenic gene expression in acute myeloid leukaemia. *Nature* **2017**, *543*, 265–269. [CrossRef]
177. Chai, X.; Guo, J.; Dong, R.; Yang, X.; Deng, C.; Wei, C.; Xu, J.; Han, W.; Lu, J.; Gao, C. Quantitative acetylome analysis reveals histone modifications that may predict prognosis in hepatitis B-related hepatocellular carcinoma. *Clin. Transl. Med.* **2021**, *11*, e313. [CrossRef] [PubMed]
178. Yan, K.; Rousseau, J.; Machol, K.; Cross, L.A.; Agre, K.E.; Gibson, C.F.; Goverde, A.; Engleman, K.L.; Verdin, H.; De Baere, E. Deficient histone H3 propionylation by BRPF1-KAT6 complexes in neurodevelopmental disorders and cancer. *Sci. Adv.* **2020**, *6*, eaax0021. [CrossRef] [PubMed]
179. Li, N.; Hamor, C.; An, Y.; Zhu, L.; Gong, Y.; Toh, Y.; Guo, Y.R. Biological functions and therapeutic potential of acylation by histone acetyltransferases. *Acta Mater. Medica* **2023**, *2*, 228–254. [CrossRef]
180. Wan, J.; Liu, H.; Ming, L. Lysine crotonylation is involved in hepatocellular carcinoma progression. *Biomed. Pharmacother.* **2019**, *111*, 976–982. [CrossRef]
181. Xu, X.; Zhu, X.; Liu, F.; Lu, W.; Wang, Y.; Yu, J. The effects of histone crotonylation and bromodomain protein 4 on prostate cancer cell lines. *Transl. Androl. Urol.* **2021**, *10*, 900–914. [CrossRef] [PubMed]
182. Zhang, H.; Chang, Z.; Qin, L.-n.; Liang, B.; Han, J.-x.; Qiao, K.-l.; Yang, C.; Liu, Y.-r.; Zhou, H.-g.; Sun, T. MTA2 triggered R-loop trans-regulates BDH1-mediated β-hydroxybutyrylation and potentiates propagation of hepatocellular carcinoma stem cells. *Signal Transduct. Target. Ther.* **2021**, *6*, 135. [CrossRef] [PubMed]
183. Lu, Y.; Li, X.; Zhao, K.; Qiu, P.; Deng, Z.; Yao, W.; Wang, J. Global landscape of 2-hydroxyisobutyrylation in human pancreatic cancer. *Front. Oncol.* **2022**, *12*, 1001807. [CrossRef] [PubMed]
184. Liu, K.; Li, F.; Sun, Q.; Lin, N.; Han, H.; You, K.; Tian, F.; Mao, Z.; Li, T.; Tong, T. p53 β-hydroxybutyrylation attenuates p53 activity. *Cell Death Dis.* **2019**, *10*, 243. [CrossRef] [PubMed]

185. Pan, L.; Feng, F.; Wu, J.; Fan, S.; Han, J.; Wang, S.; Yang, L.; Liu, W.; Wang, C.; Xu, K. Demethylzeylasteral targets lactate by inhibiting histone lactylation to suppress the tumorigenicity of liver cancer stem cells. *Pharmacol. Res.* **2022**, *181*, 106270. [CrossRef] [PubMed]
186. Liu, B.; Lin, Y.; Darwanto, A.; Song, X.; Xu, G.; Zhang, K. Identification and characterization of propionylation at histone H3 lysine 23 in mammalian cells. *J. Biol. Chem.* **2009**, *284*, 32288–32295. [CrossRef] [PubMed]
187. Ježek, P. 2-Hydroxyglutarate in Cancer Cells. *Antioxid Redox Signal* **2020**, *33*, 903–926. [CrossRef] [PubMed]
188. Dang, L.; White, D.W.; Gross, S.; Bennett, B.D.; Bittinger, M.A.; Driggers, E.M.; Fantin, V.R.; Jang, H.G.; Jin, S.; Keenan, M.C.; et al. Cancer-associated IDH1 mutations produce 2-hydroxyglutarate. *Nature* **2009**, *462*, 739–744. [CrossRef] [PubMed]
189. Janke, R.; Iavarone, A.T.; Rine, J. Oncometabolite D-2-Hydroxyglutarate enhances gene silencing through inhibition of specific H3K36 histone demethylases. *eLife* **2017**, *6*, e22451. [CrossRef] [PubMed]
190. Stoffel, E.M.; Koeppe, E.; Everett, J.; Ulintz, P.; Kiel, M.; Osborne, J.; Williams, L.; Hanson, K.; Gruber, S.B.; Rozek, L.S. Germline genetic features of young individuals with colorectal cancer. *Gastroenterology* **2018**, *154*, 897–905.e891. [CrossRef]
191. Stoffel, E.M.; Murphy, C.C. Epidemiology and Mechanisms of the Increasing Incidence of Colon and Rectal Cancers in Young Adults. *Gastroenterology* **2020**, *158*, 341–353. [CrossRef]
192. Vilar, E.; Stoffel, E.M. Universal genetic testing for younger patients with colorectal cancer. *JAMA Oncol.* **2017**, *3*, 448–449. [CrossRef]
193. Stoffel, E.M. Colorectal Cancer in Young Individuals: Opportunities for Prevention. *J. Clin. Oncol.* **2015**, *33*, 3525–3527. [CrossRef] [PubMed]
194. Kanherkar, R.R.; Bhatia-Dey, N.; Csoka, A.B. Epigenetics across the human lifespan. *Front. Cell Dev. Biol.* **2014**, *2*, 49. [CrossRef] [PubMed]
195. Busch, C.; Burkard, M.; Leischner, C.; Lauer, U.M.; Frank, J.; Venturelli, S. Epigenetic activities of flavonoids in the prevention and treatment of cancer. *Clin. Epigenetics* **2015**, *7*, 64. [CrossRef] [PubMed]
196. Pehlivan, F.E. Diet-Epigenome interactions: Epi-drugs modulating the epigenetic machinery during cancer prevention. In *Epigenetics to Optogenetics—A New Paradigm in the Study of Biology*; IntechOpen: London, UK, 2021.
197. Szarc vel Szic, K.; Ndlovu, M.N.; Haegeman, G.; Vanden Berghe, W. Nature or nurture: Let food be your epigenetic medicine in chronic inflammatory disorders. *Biochem. Pharmacol.* **2010**, *80*, 1816–1832. [CrossRef] [PubMed]
198. Vahid, F.; Zand, H.; Nosrat–Mirshekarlou, E.; Najafi, R.; Hekmatdoost, A. The role dietary of bioactive compounds on the regulation of histone acetylases and deacetylases: A review. *Gene* **2015**, *562*, 8–15. [CrossRef] [PubMed]
199. Evans, L.W.; Ferguson, B.S. Food Bioactive HDAC Inhibitors in the Epigenetic Regulation of Heart Failure. *Nutrients* **2018**, *10*, 1120. [CrossRef] [PubMed]
200. Shankar, E.; Kanwal, R.; Candamo, M.; Gupta, S. Dietary phytochemicals as epigenetic modifiers in cancer: Promise and challenges. *Semin. Cancer Biol.* **2016**, *40–41*, 82–99. [CrossRef] [PubMed]
201. Özyalçin, B.; Sanlier, N. The effect of diet components on cancer with epigenetic mechanisms. *Trends Food Sci. Technol.* **2020**, *102*, 138–145. [CrossRef]
202. Lévesque, S.; Pol, J.G.; Ferrere, G.; Galluzzi, L.; Zitvogel, L.; Kroemer, G. Trial watch: Dietary interventions for cancer therapy. *Oncoimmunology* **2019**, *8*, 1591878. [CrossRef]
203. Mercier, B.D.; Tizpa, E.; Philip, E.J.; Feng, Q.; Huang, Z.; Thomas, R.M.; Pal, S.K.; Dorff, T.B.; Li, Y.R. Dietary Interventions in Cancer Treatment and Response: A Comprehensive Review. *Cancers* **2022**, *14*, 5149. [CrossRef]
204. Mottamal, M.; Zheng, S.; Huang, T.L.; Wang, G. Histone deacetylase inhibitors in clinical studies as templates for new anticancer agents. *Molecules* **2015**, *20*, 3898–3941. [CrossRef]
205. Pal, D.; Raj, K.; Nandi, S.S.; Sinha, S.; Mishra, A.; Mondal, A.; Lagoa, R.; Burcher, J.T.; Bishayee, A. Potential of Synthetic and Natural Compounds as Novel Histone Deacetylase Inhibitors for the Treatment of Hematological Malignancies. *Cancers* **2023**, *15*, 2808. [CrossRef]
206. Sauter, E.R.; Mohammed, A. Natural Products for Cancer Prevention and Interception: Preclinical and Clinical Studies and Funding Opportunities. *Pharmaceuticals* **2024**, *17*, 136. [CrossRef]
207. Taylor, E.M.; Jones, A.D.; Henagan, T.M. A review of mitochondrial-derived fatty acids in epigenetic regulation of obesity and type 2 diabetes. *J. Nutr. Health Food Sci.* **2014**, *2*, 1.
208. McGee, S.L.; Fairlie, E.; Garnham, A.P.; Hargreaves, M. Exercise-induced histone modifications in human skeletal muscle. *J. Physiol.* **2009**, *587*, 5951–5958. [CrossRef] [PubMed]
209. Nasir, A.; Bullo, M.M.H.; Ahmed, Z.; Imtiaz, A.; Yaqoob, E.; Jadoon, M.; Ahmed, H.; Afreen, A.; Yaqoob, S. Nutrigenomics: Epigenetics and cancer prevention: A comprehensive review. *Crit. Rev. Food Sci. Nutr.* **2020**, *60*, 1375–1387. [CrossRef] [PubMed]
210. Hardy, T.M.; Tollefsbol, T.O. Epigenetic diet: Impact on the epigenome and cancer. *Epigenomics* **2011**, *3*, 503–518. [CrossRef] [PubMed]
211. Ho, E.; Beaver, L.M.; Williams, D.E.; Dashwood, R.H. Dietary factors and epigenetic regulation for prostate cancer prevention. *Adv. Nutr.* **2011**, *2*, 497–510. [CrossRef]
212. Eckschlager, T.; Plch, J.; Stiborova, M.; Hrabeta, J. Histone Deacetylase Inhibitors as Anticancer Drugs. *Int. J. Mol. Sci.* **2017**, *18*, 1414. [CrossRef] [PubMed]
213. Nieborak, A.; Schneider, R. Metabolic intermediates–cellular messengers talking to chromatin modifiers. *Mol. Metab.* **2018**, *14*, 39–52. [CrossRef] [PubMed]

214. Ruiz-Pérez, M.V.; Sainero-Alcolado, L.; Oliynyk, G.; Matuschek, I.; Balboni, N.; Ubhayasekera, S.; Snaebjornsson, M.T.; Makowski, K.; Aaltonen, K.; Bexell, D.; et al. Inhibition of fatty acid synthesis induces differentiation and reduces tumor burden in childhood neuroblastoma. *iScience* **2021**, *24*, 102128. [CrossRef]
215. Sena, L.A.; Denmeade, S.R. Fatty Acid Synthesis in Prostate Cancer: Vulnerability or Epiphenomenon? *Cancer Res.* **2021**, *81*, 4385–4393. [CrossRef] [PubMed]
216. Zhou, Y.; Bollu, L.R.; Tozzi, F.; Ye, X.; Bhattacharya, R.; Gao, G.; Dupre, E.; Xia, L.; Lu, J.; Fan, F. ATP citrate lyase mediates resistance of colorectal cancer cells to SN38. *Mol. Cancer Ther.* **2013**, *12*, 2782–2791. [CrossRef] [PubMed]
217. Huang, P.L.; Zhu, S.N.; Lu, S.L.; Dai, Z.S.; Jin, Y.L. Inhibitor of fatty acid synthase induced apoptosis in human colonic cancer cells. *World J. Gastroenterol.* **2000**, *6*, 295. [PubMed]
218. Shiragami, R.; Murata, S.; Kosugi, C.; Tezuka, T.; Yamazaki, M.; Hirano, A.; Yoshimura, Y.; Suzuki, M.; Shuto, K.; Koda, K. Enhanced antitumor activity of cerulenin combined with oxaliplatin in human colon cancer cells. *Int. J. Oncol.* **2013**, *43*, 431–438. [CrossRef] [PubMed]
219. Hull, E.E.; Montgomery, M.R.; Leyva, K.J. HDAC inhibitors as epigenetic regulators of the immune system: Impacts on cancer therapy and inflammatory diseases. *BioMed Res. Int.* **2016**, *2016*, 8797206. [CrossRef]
220. Su, M.; Gong, X.; Liu, F. An update on the emerging approaches for histone deacetylase (HDAC) inhibitor drug discovery and future perspectives. *Expert Opin. Drug Discov.* **2021**, *16*, 745–761. [CrossRef] [PubMed]
221. Li, Y.; Seto, E. HDACs and HDAC Inhibitors in Cancer Development and Therapy. *Cold Spring Harb. Perspect. Med.* **2016**, *6*, a026831. [CrossRef] [PubMed]
222. Li, G.; Tian, Y.; Zhu, W.-G. The roles of histone deacetylases and their inhibitors in cancer therapy. *Front. Cell Dev. Biol.* **2020**, *8*, 576946. [CrossRef]
223. Bose, P.; Dai, Y.; Grant, S. Histone deacetylase inhibitor (HDACI) mechanisms of action: Emerging insights. *Pharmacol. Ther.* **2014**, *143*, 323–336. [CrossRef]
224. Bassett, S.A.; Barnett, M.P. The role of dietary histone deacetylases (HDACs) inhibitors in health and disease. *Nutrients* **2014**, *6*, 4273–4301. [CrossRef]
225. Dashwood, R.H.; Ho, E. Dietary histone deacetylase inhibitors: From cells to mice to man. *Semin. Cancer Biol.* **2007**, *17*, 363–369. [CrossRef]
226. Rajendran, P.; Ho, E.; Williams, D.E.; Dashwood, R.H. Dietary phytochemicals, HDAC inhibition, and DNA damage/repair defects in cancer cells. *Clin. Epigenetics* **2011**, *3*, 4. [CrossRef]
227. Fiorentino, F.; Castiello, C.; Mai, A.; Rotili, D. Therapeutic Potential and Activity Modulation of the Protein Lysine Deacylase Sirtuin 5. *J. Med. Chem.* **2022**, *65*, 9580–9606. [CrossRef] [PubMed]
228. Xiangyun, Y.; Xiaomin, N.; Yunhua, X.; Ziming, L.; Yongfeng, Y.; Zhiwei, C.; Shun, L. Desuccinylation of pyruvate kinase M2 by SIRT5 contributes to antioxidant response and tumor growth. *Oncotarget* **2017**, *8*, 6984. [CrossRef] [PubMed]
229. Madsen, A.S.; Olsen, C.A. Profiling of substrates for zinc-dependent lysine deacylase enzymes: HDAC3 exhibits decrotonylase activity in vitro. *Angew. Chem. (Int. Ed. Engl.)* **2012**, *51*, 9083–9087. [CrossRef]
230. Du, W.; Hua, F.; Li, X.; Zhang, J.; Li, S.; Wang, W.; Zhou, J.; Wang, W.; Liao, P.; Yan, Y. Loss of optineurin drives cancer immune evasion via palmitoylation-dependent IFNGR1 lysosomal sorting and degradation. *Cancer Discov.* **2021**, *11*, 1826–1843. [CrossRef] [PubMed]
231. Gao, Y.; Nihira, N.T.; Bu, X.; Chu, C.; Zhang, J.; Kolodziejczyk, A.; Fan, Y.; Chan, N.T.; Ma, L.; Liu, J. Acetylation-dependent regulation of PD-L1 nuclear translocation dictates the efficacy of anti-PD-1 immunotherapy. *Nat. Cell Biol.* **2020**, *22*, 1064–1075. [CrossRef]
232. Ma, L.; Liu, J.; Liu, L.; Duan, G.; Wang, Q.; Xu, Y.; Xia, F.; Shan, J.; Shen, J.; Yang, Z. Overexpression of the Transcription Factor MEF2D in Hepatocellular Carcinoma Sustains Malignant Character by Suppressing G2–M Transition GenesPromotion of Cancer Cell Growth by MEF2D in HCC. *Cancer Res.* **2014**, *74*, 1452–1462. [CrossRef]
233. Yao, H.; Lan, J.; Li, C.; Shi, H.; Brosseau, J.-P.; Wang, H.; Lu, H.; Fang, C.; Zhang, Y.; Liang, L. Inhibiting PD-L1 palmitoylation enhances T-cell immune responses against tumours. *Nat. Biomed. Eng.* **2019**, *3*, 306–317. [CrossRef]
234. Yang, Y.; Hsu, J.-M.; Sun, L.; Chan, L.-C.; Li, C.-W.; Hsu, J.L.; Wei, Y.; Xia, W.; Hou, J.; Qiu, Y. Palmitoylation stabilizes PD-L1 to promote breast tumor growth. *Cell Res.* **2019**, *29*, 83–86. [CrossRef]
235. Fang, H.; Huang, Y.; Luo, Y.; Tang, J.; Yu, M.; Zhang, Y.; Zhong, M. SIRT1 induces the accumulation of TAMs at colorectal cancer tumor sites via the CXCR4/CXCL12 axis. *Cell. Immunol.* **2022**, *371*, 104458. [CrossRef]
236. Lasko, L.M.; Jakob, C.G.; Edalji, R.P.; Qiu, W.; Montgomery, D.; Digiammarino, E.L.; Hansen, T.M.; Risi, R.M.; Frey, R.; Manaves, V.; et al. Discovery of a selective catalytic p300/CBP inhibitor that targets lineage-specific tumours. *Nature* **2017**, *550*, 128–132. [CrossRef]
237. Chen, Q.; Yang, B.; Liu, X.; Zhang, X.D.; Zhang, L.; Liu, T. Histone acetyltransferases CBP/p300 in tumorigenesis and CBP/p300 inhibitors as promising novel anticancer agents. *Theranostics* **2022**, *12*, 4935. [CrossRef] [PubMed]
238. Waddell, A.R.; Huang, H.; Liao, D. CBP/p300: Critical Co-Activators for Nuclear Steroid Hormone Receptors and Emerging Therapeutic Targets in Prostate and Breast Cancers. *Cancers* **2021**, *13*, 2872. [CrossRef] [PubMed]
239. Goodman, R.H.; Smolik, S. CBP/p300 in cell growth, transformation, and development. *Genes Dev.* **2000**, *14*, 1553–1577. [CrossRef] [PubMed]

240. Kapoor, S.; Gustafson, T.; Zhang, M.; Chen, Y.-S.; Li, J.; Nguyen, N.; Perez, J.E.T.; Dashwood, W.M.; Rajendran, P.; Dashwood, R.H. Deacetylase plus bromodomain inhibition downregulates ERCC2 and suppresses the growth of metastatic colon cancer cells. *Cancers* **2021**, *13*, 1438. [CrossRef] [PubMed]

241. Mazur, P.K.; Herner, A.; Mello, S.S.; Wirth, M.; Hausmann, S.; Sánchez-Rivera, F.J.; Lofgren, S.M.; Kuschma, T.; Hahn, S.A.; Vangala, D. Combined inhibition of BET family proteins and histone deacetylases as a potential epigenetics-based therapy for pancreatic ductal adenocarcinoma. *Nat. Med.* **2015**, *21*, 1163–1171. [CrossRef]

242. Rajendran, P.; Johnson, G.; Li, L.; Chen, Y.-S.; Dashwood, M.; Nguyen, N.; Ulusan, A.; Ertem, F.; Zhang, M.; Li, J. Acetylation of CCAR2 establishes a BET/BRD9 acetyl switch in response to combined deacetylase and bromodomain inhibition. *Cancer Res.* **2019**, *79*, 918–927. [CrossRef] [PubMed]

243. Zhu, H.; Wei, T.; Cai, Y.; Jin, J. Small molecules targeting the specific domains of histone-mark readers in cancer therapy. *Molecules* **2020**, *25*, 578. [CrossRef] [PubMed]

244. Shorstova, T.; Foulkes, W.D.; Witcher, M. Achieving clinical success with BET inhibitors as anti-cancer agents. *Br. J. Cancer* **2021**, *124*, 1478–1490. [CrossRef]

245. Kapoor, S.; Damiani, E.; Wang, S.; Dharmanand, R.; Tripathi, C.; Tovar Perez, J.E.; Dashwood, W.M.; Rajendran, P.; Dashwood, R.H. BRD9 Inhibition by Natural Polyphenols Targets DNA Damage/Repair and Apoptosis in Human Colon Cancer Cells. *Nutrients* **2022**, *14*, 4317. [CrossRef] [PubMed]

246. Xiong, X.; Panchenko, T.; Yang, S.; Zhao, S.; Yan, P.; Zhang, W.; Xie, W.; Li, Y.; Zhao, Y.; Allis, C.D. Selective recognition of histone crotonylation by double PHD fingers of MOZ and DPF2. *Nat. Chem. Biol.* **2016**, *12*, 1111–1118. [CrossRef]

247. Cheng, Y.; He, C.; Wang, M.; Ma, X.; Mo, F.; Yang, S.; Han, J.; Wei, X. Targeting epigenetic regulators for cancer therapy: Mechanisms and advances in clinical trials. *Signal Transduct. Target. Ther.* **2019**, *4*, 62. [CrossRef]

248. Neja, S.A. Site-Specific DNA Demethylation as a Potential Target for Cancer Epigenetic Therapy. *Epigenetics Insights* **2020**, *13*, 2516865720964808. [CrossRef]

249. Kwon, D.Y.; Zhao, Y.T.; Lamonica, J.M.; Zhou, Z. Locus-specific histone deacetylation using a synthetic CRISPR-Cas9-based HDAC. *Nat. Commun.* **2017**, *8*, 15315. [CrossRef] [PubMed]

250. Yoon, S.; Eom, G.H. HDAC and HDAC Inhibitor: From Cancer to Cardiovascular Diseases. *Chonnam. Med. J.* **2016**, *52*, 1–11. [CrossRef]

251. Stimson, L.; Rowlands, M.G.; Newbatt, Y.M.; Smith, N.F.; Raynaud, F.I.; Rogers, P.; Bavetsias, V.; Gorsuch, S.; Jarman, M.; Bannister, A.; et al. Isothiazolones as inhibitors of PCAF and p300 histone acetyltransferase activity. *Mol. Cancer Ther.* **2005**, *4*, 1521–1532. [CrossRef] [PubMed]

252. Chen, Y.; Wu, J.; Zhai, L.; Zhang, T.; Yin, H.; Gao, H.; Zhao, F.; Wang, Z.; Yang, X.; Jin, M.; et al. Metabolic regulation of homologous recombination repair by MRE11 lactylation. *Cell* **2024**, *187*, 294–311.e221. [CrossRef]

253. Ma, W.; Sun, Y.; Yan, R.; Zhang, P.; Shen, S.; Lu, H.; Zhou, Z.; Jiang, Z.; Ye, L.; Mao, Q.; et al. OXCT1 functions as a succinyltransferase, contributing to hepatocellular carcinoma via succinylating LACTB. *Mol. Cell*, 2024; *in press*. [CrossRef]

254. Ozsvari, B.; Sotgia, F.; Simmons, K.; Trowbridge, R.; Foster, R.; Lisanti, M.P. Mitoketoscins: Novel mitochondrial inhibitors for targeting ketone metabolism in cancer stem cells (CSCs). *Oncotarget* **2017**, *8*, 78340–78350. [CrossRef]

255. Tu, B.; Zhang, M.; Liu, T.; Huang, Y. Nanotechnology-based histone deacetylase inhibitors for cancer therapy. *Front. Cell Dev. Biol.* **2020**, *8*, 400. [CrossRef] [PubMed]

256. Liu, Z.; Ren, Y.; Weng, S.; Xu, H.; Li, L.; Han, X. A New Trend in Cancer Treatment: The Combination of Epigenetics and Immunotherapy. *Front. Immunol.* **2022**, *13*, 809761. [CrossRef] [PubMed]

257. Gomez, S.; Tabernacki, T.; Kobyra, J.; Roberts, P.; Chiappinelli, K.B. Combining epigenetic and immune therapy to overcome cancer resistance. *Semin. Cancer Biol.* **2020**, *65*, 99–113. [CrossRef] [PubMed]

258. Villanueva, L.; Álvarez-Errico, D.; Esteller, M. The contribution of epigenetics to cancer immunotherapy. *Trends Immunol.* **2020**, *41*, 676–691. [CrossRef]

259. Huang, H.; Lin, S.; Garcia, B.A.; Zhao, Y. Quantitative proteomic analysis of histone modifications. *Chem. Rev.* **2015**, *115*, 2376–2418. [CrossRef]

260. Zhao, Z.; Shilatifard, A. Epigenetic modifications of histones in cancer. *Genome Biol.* **2019**, *20*, 245. [CrossRef]

261. Dragic, H.; Chaveroux, C.; Cosset, E.; Manie, S.N. Modelling cancer metabolism in vitro: Current improvements and future challenges. *FEBS J.* **2022**. *early view*. [CrossRef]

Review

Organosulfur Compounds in Colorectal Cancer Prevention and Progression

Patrick L. McAlpine [1,2,3], Javier Fernández [1,2,3,*], Claudio J. Villar [1,2,3] and Felipe Lombó [1,2,3,*]

[1] Research Group BIONUC (Biotechnology of Nutraceuticals and Bioactive Compounds), Departamento de Biología Funcional, Área de Microbiología, Universidad de Oviedo, 33006 Oviedo, Spain; mcalpineatsantaclara@gmail.com (P.L.M.); cjvg@uniovi.es (C.J.V.)
[2] IUOPA (Instituto Universitario de Oncología del Principado de Asturias), 33006 Oviedo, Spain
[3] ISPA (Instituto de Investigación Sanitaria del Principado de Asturias), 33011 Oviedo, Spain
* Correspondence: fernandezfjavier@uniovi.es (J.F.); lombofelipe@uniovi.es (F.L.)

Abstract: This work represents an overview of the current investigations involving organosulfur compounds and colorectal cancer. The molecules discussed in this review have been investigated regarding their impact on colorectal cancer directly, at the in vitro, in vivo, and clinical stages. Organosulfur compounds may have indirect effects on colorectal cancer, such as due to their modulating effects on the intestinal microbiota or their positive effects on intestinal mucosal health. Here, we focus on their direct effects via the repression of multidrug resistance proteins, triggering of apoptosis (via the inhibition of histone deacetylases, increases in reactive oxygen species, p53 activation, β-catenin inhibition, damage in the mitochondrial membrane, etc.), activation of TGF-β, binding to tubulin, inhibition of angiogenesis and metastasis mechanisms, and inhibition of cancer stem cells, among others. In general, the interesting positive effects of these nutraceuticals in in vitro tests must be further analyzed with more in vivo models before conducting clinical trials.

Keywords: allicin; sulforaphane; glucosinolate; indol-3-carbinol; isothiocyanate

Citation: McAlpine, P.L.; Fernández, J.; Villar, C.J.; Lombó, F. Organosulfur Compounds in Colorectal Cancer Prevention and Progression. *Nutrients* 2024, 16, 802. https://doi.org/ 10.3390/nu16060802

Academic Editors: Yi-Wen Liu and Ching-Hsein Chen

Received: 2 February 2024
Revised: 7 March 2024
Accepted: 9 March 2024
Published: 11 March 2024

1. Introduction

Colorectal cancer (CRC) is one of the most common cancers in Europe after only lung and prostate cancers in incidence in men (incidence of 35–42 cases per 100,000 people) and after breast cancer in women (24–32 cases per 100,000). At the global level, CRC generates 1.85 million new cases and 881,000 deaths annually [1]. The main risk factors associated with this type of cancer are the consumption of alcohol, tobacco, processed meat, saturated fat, and red meat. The increased risk from processed meat consumption is due to its content of nitrosamine-generating preservatives, such as nitrates and nitrites. From red meat consumption, it is due to its high content of heme iron, a cause of oxidative stress in the digestive tract. On the contrary, protective factors include high consumption of whole grains, vegetables and fruits with prebiotic fibers and other nutraceuticals, calcium-rich foods (such as milk), and foods rich in vitamin D [2–4]. Some other risk factors include an enrichment in pro-inflammatory gut microbiota taxons, such as *Fusobacterium*, *Porphyromonas*, *Atopobium*, and *Bilophila*, as well as the presence of inflammatory conditions, such as Crohn's disease [5–7]. These pro-inflammatory conditions induce high levels of oxidative stress in the colon mucosa, as well as an impairment in immune response, aberrant cell signaling and upregulation of proliferative pathways, angiogenesis, and migration. A common characteristic under these circumstances is the overexpression of NF-κB, STAT, and HIF1α transcription factors [8,9].

Most CRC cases (about 70%) are due to sporadic mutations in colonocyte genes, associated with chromosomal instability due to *APC* mutations (in addition to mutations in *KRAS*, *TP53*, and other genes), and experience slow progression rates (over 10 years to generate large polyps) [10,11]. The majority of remaining cases of this digestive neoplasia are due to *BRAF* mutation and MAPK activation, giving rise to promoter hypermethylation and gene silencing (such as in *MLH1* or in *p16*). This causes serrated adenomas with crypt branching, flat structure, irregular limits, and a mucus cap that reduces the probability of bleeding. These cases progress rapidly after acquiring microsatellite instability features [12,13]. Surveillance programs (occult fecal blood tests and colonoscopy) reduce the number of advanced tumors, which otherwise may need surgery plus chemotherapy to improve patients' survival [14].

Some nutraceuticals, such as prebiotic fibers or polyphenols, have demonstrated in vitro and in vivo CRC prevention, by inducing the colonic production of antitumor and anti-inflammatory compounds by the intestinal microbiota, such as short-chain fatty acids (propionate, butyrate) or hydroxycinnamic acids, which block histone deacetylases and promote tumor colonocyte apoptosis [7,15–21]. These bioactive molecules modulate signaling pathways and the expression of genes involved in apoptosis, cell cycle regulation, and differentiation [15,22]. Another family of plant nutraceuticals are organosulfur compounds, such as diallyl trisulfide (DATS), methylsulfonylmethane, and isothiocyanates. They have been tested as antitumor or chemosensitizers in the co-therapy of CRC, due to their inhibition of matrix metalloproteases, inhibition of carcinogen activation enzymes, and induction of apoptosis [23–25]. In the human diet, these organosulfur compounds are present in vegetables such as garlic, onion, broccoli, cabbage, and others from those plant families (*Amaryllidaceae*, *Brassicaceae*) [26,27].

2. Organosulfur Compounds Derived from Cruciferous Vegetables

The organosulfur compounds found in cruciferous vegetables (*Brassicaceae* family) can be subdivided into two primary groups, isothiocyanates and indoles. Isothiocyanates can be further divided by those derived from L-methionine and those derived from L-phenylalanine [28–30]. The L-methionine-derived isothiocyanates include allyl isothiocyanate (AITC), sulforaphane, sulforaphene, and iberin. The first step in the biosynthesis of these molecules is the variable side-chain elongation of L-methionine [31]. One cycle of elongation gives rise to homomethionine, which then undergoes a series of enzyme-catalyzed reactions to form a desulfoglucosinolate. At this point, the pathway branches into two distinct parts. In one direction, the sulfur atom bound to carbon C4 in desulfoglucosinolate has a carbonyl group added to it and the nitrogen atom has a sulfuric acid added to it in place of the hydroxyl group, forming glucoiberverin. From here, another series of enzyme-catalyzed reactions lead to sinigrin formation, which is the glucosinolate precursor to allyl isothiocyanate (AITC). Upon cell damage, sinigrin comes into contact with the enzyme myrosinase, which hydrolyzes it to form allyl isothiocyanate (AITC) [32]. Glucosinolates are stored in plant cell vacuoles while myrosinase is stored in separate myrosinase cells. It is only upon physical damage to plant tissues that myrosinase is released and comes into contact with glucosinolates to form isothiocyanates [33]. In the second branch of the pathway, the desulfoglucosinolate is converted into glucoiberverin when the sulfur that is bound to carbon C4 has two carbonyl groups added to it and the nitrogen atom has its hydroxyl group replaced with sulfuric acid. Glucoiberverin is then converted into iberin by myrosinase [34].

Alternatively, L-methionine can undergo two cycles of side-chain elongation to form dihomomethionine. From there, it is converted to desulfoglucosinolate over the course of multiple reactions. The desulfoglucosinolate then has the hydroxy group on its nitrogen atom replaced with sulfuric acid to form glucoerucin. The sulfur atom on carbon C5 then has a carbonyl group attached to it to form glucoraphanin. Upon cell damage, myrosinase converts glucoraphanin into either sulforaphane or sulforaphene depending on the presence of a double bond between carbons C3 and C4 [33] (Figure 1).

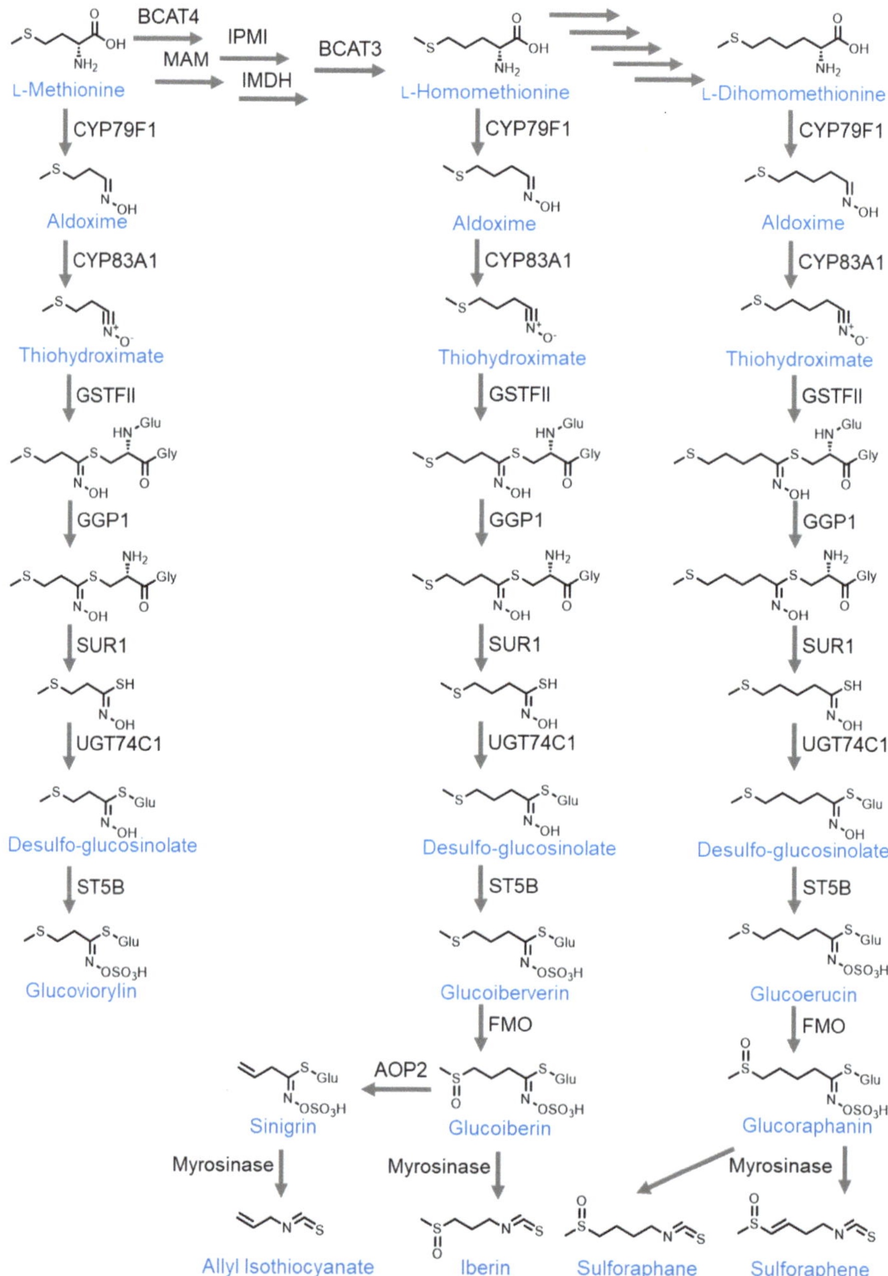

Figure 1. Biosynthesis of glucosinolates from methionine. BCAT4: branched-chain amino transferase 4, MAM: Methylthioalkylmalate synthase, IPMI: Isopropylmalate isomerase, IMDH: Inosine-5′-monophosphate dehydrogenase, BCAT3: branched-chain amino transferase 3, CYP: Cytochrome, GSTFII: Glutathione-S-transferase II, GGP1: Glucosinolate-γ-glutamyl peptidase 1, SUR1: Alkyl-thiohydroxymate C-S lyase, UGT: UDP-glucosyl transferase, ST5B: Aliphatic desulfoglucosinolate sulfotransferase B, FMO: Flavin-containing monooxygenase, AOP2: 2-Oxoglutarate-dependent dioxygenase.

The second group of isothiocyanates are those that are derived from L-phenylalanine. Unlike L-methionine, L-phenylalanine does not require side-chain elongation and can directly undergo a series of enzymatic reactions to form benzyl desulfoglucosinolate. This then has the hydroxy group on its nitrogen atom replaced with sulfuric acid to form glucosinolate glucotropaeolin. Upon tissue damage, glucotropaeolin is converted into benzyl isothiocyante (BITC) by myrosinase [35]. Alternatively, L-phenylalanine can undergo one cycle of side-chain elongation to form homophenylalanine. This then undergoes a series of enzymatic reactions to form glucosinolate gluconasturtiin. Upon tissue damage, myrosinase then converts gluconasturtiin into phenethyl isothiocyanate (PEITC) [36] (Figure 2).

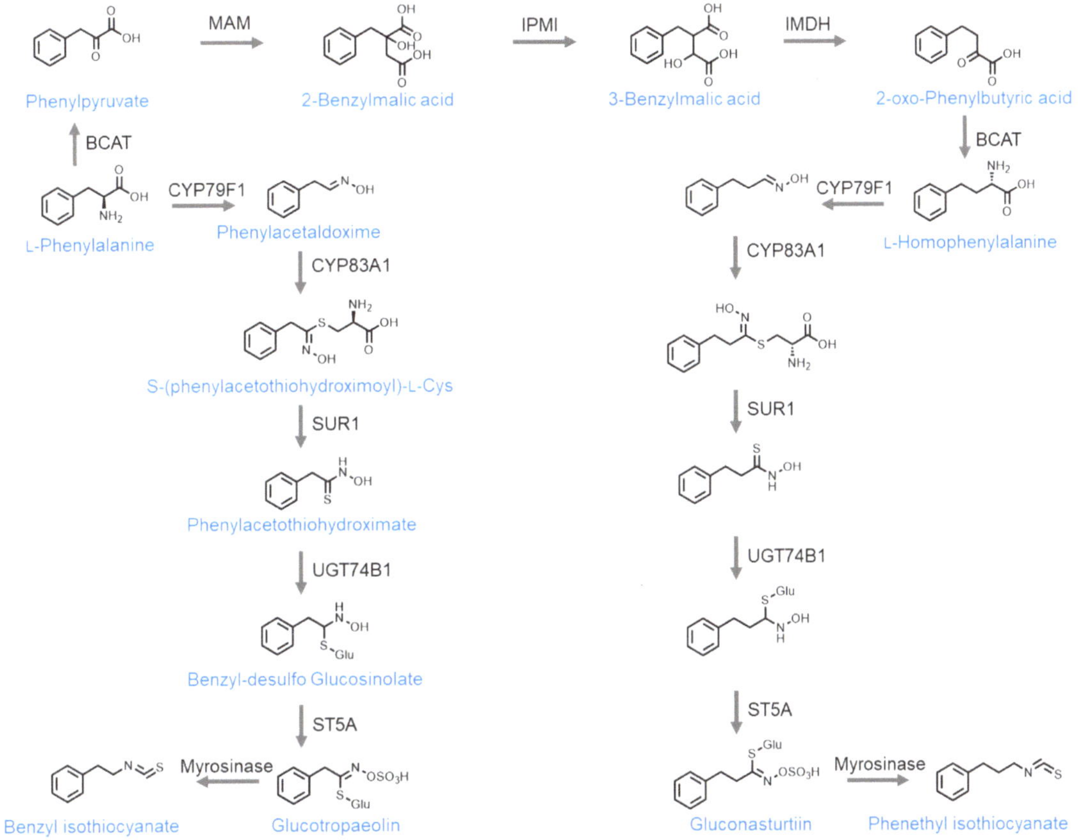

Figure 2. Biosynthesis of glucosinolates from phenylalanine. BCAT: branched-chain amino transferase 4, MAM: Methylthioalkylmalate synthase, IPMI: Isopropylmalate isomerase, IMDH: Inosine-5′-monophosphate dehydrogenase, CYP: Cytochrome, SUR1: Alkyl-thiohydroxymate C-S lyase, UGT: UDP-glucosyl transferase, ST5B: Aliphatic desulfoglucosinolate sulfotransferase B.

The third major group of organosulfur compounds found in cruciferous vegetables includes indoles. These are derived from the amino acid L-tryptophan and require no side-chain elongation. Over a series of enzymatic reactions, L-tryptophan is converted into indolylmethyl desulfoglucosinolate. This then has the hydroxy group on the nitrogen atom replaced with sulfuric acid to form glucobrassicin, the glucosinolate precursor to indoles. From here, upon tissue damage, myrosinase converts glucobrassicin to 3-indolylmethylisothiocyanate, which is unstable and hydrolyzes to form indole-3-carbinol [37]. Indole-3-carbinol is itself a potent organosulfur compound, which will be discussed further in this review, but it can also undergo acid condensation in the stomach to form 3,3-diindolylmethane [38] (Figure 3).

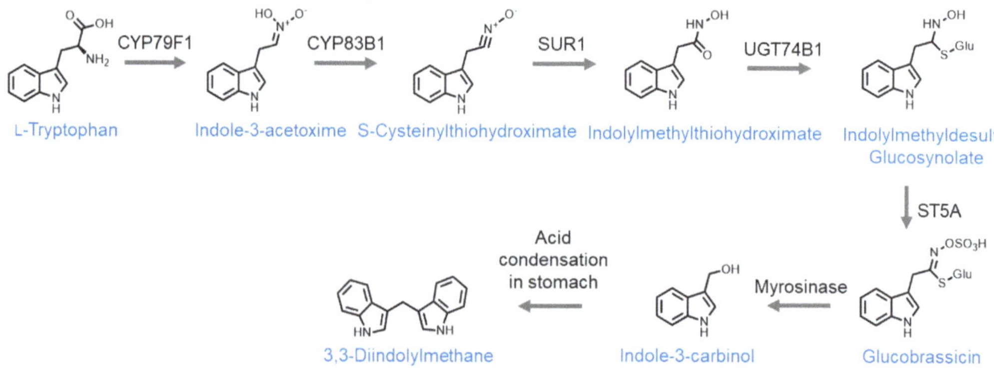

Figure 3. Biosynthesis of indoles from tryptophan. CYP: Cytochrome, SUR1: Alkyl-thiohydroxymate C-S lyase, UGT: UDP-glucosyl transferase, ST5B: Aliphatic desulfoglucosinolate sulfotransferase B.

2.1. Cruciferous Vegetable Consumption

In the last decade, numerous epidemiological studies have indicated the protective effect that cruciferous vegetable consumption has on colorectal cancer development, and this effect has been further confirmed by numerous meta-analyses [39–42]. The protective effect is still observed when total fiber intake is accounted for, which indicates that this protection comes from molecules found intrinsically in cruciferous vegetables. It is generally regarded that the molecules that induce the antitumoral effect are glucosinolates and their derivatives, indoles, and isothiocyanates. Even so, some epidemiological studies and meta-analyses have not recapitulated these positive results [43,44], indicating that the protective effects of the organosulfur compounds found in cruciferous vegetables may not be readily achieved at dietary concentrations. For this reason, the use of pure compounds and extracts containing glucosinolates at higher, pharmacologically relevant concentrations has been a major area of study.

In addition to the epidemiological studies listed previously, other groups have attempted to confirm the protective effects of cruciferous vegetable consumption in vivo. Studies in rodent models of colorectal cancer have demonstrated the protective effects of including cruciferous vegetables in the diet by detecting a reduced formation of aberrant crypt foci in cruciferous vegetable-eating animals [45,46]. Furthermore, these studies have demonstrated that the protective effects are not recapitulated when other classes of vegetables are included in the diet in place of cruciferous vegetables. With this said, other studies in pigs have demonstrated that broccoli consumption induced detectable levels of nuclear damage in colonic mucosal cells, but that this effect was dependent upon how the broccoli was prepared [47]. Thus, how the vegetables are manipulated may affect their molecular composition, which in turn may modify their impact upon the colorectal epithelium.

In contrast to these animal studies, the consumption of large amounts of cruciferous vegetables has not been found to be protective against colorectal cancer development in human trials. One study included patients who were diagnosed with colonic adenomas and consumed 50 g/day of raw broccoli sprouts every other day for 6 months. They found no statistically significant difference in the numbers of aberrant crypt foci before and after treatment [48]. It should be noted that this investigation took place after CRC diagnosis and thus the protective effects of broccoli consumption prior to CRC development were not investigated.

2.2. Cruciferous Vegetable Extracts

Multiple in vitro studies have demonstrated the efficacy of cruciferous vegetable extracts in reducing colorectal cancer cell viability. One study demonstrated that broccoli extract significantly inhibited cell growth in both Caco-2 and HT-29 colorectal cancer cell lines, even more so than pure sulforaphane. The IC 50 of broccoli extract was reached at 1.6 and 3.2 μM of sulforaphane content in each respective cell line, while the IC 50 of pure sulforaphane was reached at 37.5 and 50.9 μM, respectively. This indicates the potential for synergistic effects from multiple antitumoral compounds within the same extract [49]. A second study demonstrated that both broccoli and watercress extracts significantly reduced HT-29 cell viability in the monolayer (IC 50 at 14.8 μM sulforaphane content and 33.9 μM phenethyl isothiocyanate content in broccoli and watercress, respectively) as well as 3D spheroid models (IC 50 at 44.4 μM and 135.6 μM, respectively). This decrease in viability was induced by cell cycle arrest at the G2/M phase. Furthermore, treatment with both extracts decreased the expression of *PROM1* and *LGR5*, two markers of stemness [50]. Thus, cruciferous vegetable extracts may not only reduce cancer cell viability but also reduce the development of cancer stem cells.

In vivo studies in rodents have confirmed the antitumorigenic effects that have been observed in vitro. Kassie et al. showed that the addition of brussels sprouts juices to the drinking water (5% of total volume) of F344 rats in a 2-amino-3-methylimidazo[4,5-*f*] quinoline-induced model of colorectal cancer, starting 5 days before and during cancer induction, significantly reduced the number of colonic aberrant crypt foci 16 weeks after induction. A similar effect was not observed with red cabbage juices at the same concentration [51]. In a similar study by the same group, the addition of brussels sprouts juices to the drinking water of rats in the same model and same concentration, for 25 days, but beginning after tumor initiation, caused a downward trend in the number of colonic aberrant crypt foci, but not a statistically significant decrease at 16 weeks post-induction [52].

A human trial of dietary supplementation with broccoli extract containing 200 μmol of sulforaphane demonstrated a rapid increase in sulforaphane concentration in the participant's blood plasma (from 0 to 9 μM after 3 h). Supplementation also decreased HDAC3 protein levels and increased p16 protein levels in peripheral blood mononuclear cells (PBMCs) [53]. HDACs (histone deacetylase enzymes) are known to be overexpressed in several cancers and their downregulation is associated with cell cycle arrest and apoptosis [54]. Furthermore, p16 is a well-known tumor suppressor protein and its overexpression is associated with cell cycle arrest. Thus, although broccoli extract has not been demonstrated to directly reduce colorectal cancer development in human trials, it does demonstrate great potential in cancer chemoprevention.

2.3. Glucosinolates

2.3.1. Sulforaphane

Of the isothiocyanates derived from cruciferous vegetables, the most studied one in colorectal cancer prevention and treatment is sulforaphane. Numerous in vitro studies in CRC cell lines (Caco-2, HCT116, HT-29, SW480, RKO, DLD-1, LoVo, SW48) as well as patient-derived primary cell cultures have demonstrated its antiproliferative properties via cell cycle arrest, apoptosis, increasing levels of reactive oxygen species, DNA damage, and altered histone acetylation patterns. These effects were achieved at concentrations ranging from 7 to 60 μM depending on the cell type and desired level of toxicity (Table 1) [49,55–65]. Furthermore, treating PBMCs with as little as 5 μM sulforaphane increased their secretion of cytokines and increased the apoptosis of colorectal cancer cells in co-culture [66,67]. Sulforaphane treatment has even been shown to prevent angiogenesis and migration at a concentration of 12.5 μM by inhibiting HIF-1a and VEGF [68]. Finally, sulforaphane has been demonstrated to be cytoprotective of healthy colon cells [47] and to increase the efficacy of other chemotherapeutic agents such as PR-104A (2.5 μM sulforaphane and PR-104A at all concentrations), 5 fluorouracil (7.1 μM of 5 fluorouracil with 5.8 μM sulforaphane derivative), and salinomycin (5 μM salinomycin with 10 μM sulforaphane)

(Table 1) [69–71]. Sulforaphane at a concentration of 10 µM has also been shown to increase the effectiveness of folinic acid (FOLFOX) against cancer cell lines, but it also increased the expression of multidrug resistance protein 2 (MRP2), meaning that it may reduce the activity of some other chemotherapeutic agents due to their expulsion from the cell [72].

Table 1. Summary of bioactive-tested concentrations for the different organosulfur compounds in in vitro and in vivo assays.

Compound	Summary	Positive Dosing	Negative Dosing	Citations
Sulforaphane	Induced cell cycle arrest and apoptosis, and reduced angiogenesis and migration in vitro. Reduced xenograft size and tumor initiation in vivo.	7–60 µM in cell culture, 0.08 µmoles via daily intraperitoneal injection, 300 ppm in diet (1.6 mg/day).		[47–49,55–73]
Sulforaphene	Induced cell cycle arrest at the G2/M phase, upregulation of the JNK pathway, inhibition of microtubule polymerization, and increase in intracellular reactive oxygen species in vitro. Reduced xenograft size in vivo.	5 µM in cell culture, 5 mg/kg of body weight daily intraperitoneal injection.		[74,75]
Allyl isothiocyanate (AITC)	Induced cell cycle arrest and apoptosis, and reduced migration in vitro.	10–20 µM in cell culture.		[76,77]
Phenethyl isothiocyanate (PEITC)	Reduced viability, stemness, and spheroid formation in vitro. Reduced xenograft size, and tumor initiation in vivo.	12–88 µM in vitro, 60 mg/kg of body weight daily intraperitoneal injection, and 20 mg/kg of body weight daily oral administration.		[78–80]
Benzyl isothiocyanate (BITC)	Induced cell cycle arrest and apoptosis in vitro. Increased tumor initiation in vivo.	10–20 µM in cell culture.	0.5 g/kg of diet	[81–83]
Iberin	Reduced proliferation, methyl guanine methyl transferase methylation, and increased miRNA expression in vitro.	8–10 µM in cell culture.		[84,85]
Indole-3-carbinol	Reduced viability and proliferation, plus p53 upregulation in vitro. Both reduced and increased tumor initiation in vivo.	500 µM to 1 mM in cell culture, 1 g/kg of diet.	1 g/kg of diet.	[65,86–91]
3,3′-Diindolylmethane	Reduced viability, cell cycle arrest, and COX1/2 and ERK1/2 protein inhibition in vitro. Reduced xenograft size in vivo.	25–56 µM in cell culture, 40 mg/kg of body weight via intraperitoneal injection.	40 mg/kg of body weight via oral administration.	[92–97]

Table 1. *Cont.*

Compound	Summary	Positive Dosing	Negative Dosing	Citations
Allicin	Reduced viability, proliferation, and migration in vitro. Reduced xenograft size and tumor initiation in vivo.	25–50 µM for viability and 4 µg/mL for migration in cell culture. Intraperitoneal injection of 5 mg/kg of body weight and 48 mg/kg of diet.		[98–101]
Alliin/S-Allyl-L-cysteine sulfoxide (SACS)	Reduced viability and EGFR (epithelial growth factor receptor) expression in vitro. Reduced tumor initiation in vivo.	100 µg/mL in cell culture, 125 mg/kg of diet, 200 mg/kg of body weight administered orally.		[102–105]
Diallyl sulfide (DAS)	Induced apoptosis and inhibited migration/metastasis in vitro. Non-significant reductions in xenograft size and tumor initiation in vivo.	50 µM in cell culture.	Intraperitoneal injection of 6 mg/kg of body weight and 300 ppm in diet.	[106–109]
Diallyl disulfide (DADS)	Reduced viability, migration, and invasion, and increased apoptosis in vitro. Reduced metastasis, xenograft size (best in TRAIL overexpressing tumors), and tumor initiation in vivo.	5–50 µM for viability and 25 µM for migration in cell culture. Intraperitoneal injection of 100 mg/kg of body weight and 42 ppm in diet.		[107–116]
Diallyl trisulfide (DATS)	Induced apoptosis, reduced stem cell viability, and reduced migration/invasion in vitro. Reduced xenograft size and tumor initiation in vivo.	30–40 µM for apoptosis, 40 µM for stem cells, 10 µM for migration. Intraperitoneal injection of 6 mg/kg of body weight and oral administration of 50 mg/kg of body weight.		[23,107,108,117–120]
Diallyl tetrasulfide	Induced cell cycle arrest, apoptosis, and reduced spheroid formation in vitro. Modified molecule reduced xenograft size in vivo.	10–50 µM in cell culture. An amount of 50 µM dibenzyl tetrasulfide in zebrafish.		[121]
Methylsulfonylmethane (MSM) ((dimethyl sulfone (DMSO2))	Reduced viability, stemness, and spheroid formation in vitro.	100–250 mM in cell culture.		[24]

Table 1. *Cont.*

Compound	Summary	Positive Dosing	Negative Dosing	Citations
(Z)-ajoene	Reduced viability and Wnt/β-catenin pathway inhibition in vitro.	30 μM for 72 h of treatment.		[122]
S-allylmercaptocysteine (SAMC)	Reduced proliferation, induced apoptosis, and reduced migration in vitro. Reduced xenograft size in combination treatment in vivo.	200–450 μM for apoptosis, 400 μM for migration, and 300 mg/kg body weight/day administered orally in combination with rapamycin treatment.		[123–127]

Adding to the in vitro investigations, numerous in vivo studies in murine models have further confirmed sulforaphane's efficacy in colorectal cancer treatment and prevention. Two recent studies tested sulforaphane's efficacy against primary CRC cell cultures and Caco-2 cells, and then confirmed their positive in vitro results by testing sulforaphane treatment against xenografts derived from the same primary cells in SCID/nude mice [64,70]. Both studies found that the treatment significantly reduced the size of the xenografts compared with controls, with one study observing a 70% reduction in xenograft size with daily intraperitoneal injections of 0.08 μmoles of sulforaphane (Table 1) [64]. With intraperitoneal administration, these compounds bypass initial structural changes that may occur with oral administration since they are readily absorbed at mesentery capillaries and directed toward the liver via the poral vein. From there, they reenter the digestive tract at the duodenum via bile secretions. Thus, the compounds are protected from modification due to stomach acids and small intestine enzymes, some of which are of microbial origin. For this reason, intraperitoneal injection leads to greater concentrations of the compounds in tissues and plasma prior to possible modifications when compared with oral administration. Furthermore, numerous studies have observed that dietary sulforaphane supplementation at 300 to 400 ppm/day before colorectal cancer initiation significantly reduced the number and size of macroscopic tumors as well as the number of aberrant crypt foci (Table 1) [48,53,73]. These results have been confirmed in APC^{min} mice, azoxymethane-induced CRC mice, and DMH-induced CRC mice and have not displayed any major negative impacts on rodent health.

Shockingly, despite all the promise sulforaphane has shown in CRC treatment and prevention in vitro and in vivo, no clinical trials on the topic have been performed or are in progress as of the writing of this paper, per ClinicalTrials.gov. That said, as of writing, 92 clinical trials have been completed or are in progress testing sulforaphane treatment against a wide variety of other diseases, such as chronic kidney disease, autism, chronic obstructive pulmonary disease, and Parkinson's disease, indicating its promise for human intake.

2.3.2. Sulforaphene

In contrast to sulforaphane, sulforaphene as a treatment for CRC has been studied far less. One group treated multiple CRC cell lines (HCT116, HT-29, KM12, SNU-1040, and DLD-1) and found that 5 μM sulforaphene was effective in reducing proliferation by arresting the cell cycle at the G2/M phase, upregulating the JNK pathway, inhibiting microtubule polymerization, and increasing intracellular reactive oxygen species. This antiproliferative effect was further confirmed in vivo by observing that daily intraperitoneal injections of sulforaphene at 5 mg/kg of body weight (Table 1) significantly reduced the growth of HCT116 cell-derived xenografts in a nude mouse model [74]. RNA seq analysis of SW480 cell lines treated with sulforaphene or sulforaphane showed remarkable similarity in the gene expression alterations induced by both molecules. In both conditions, genes associated with the p53 signaling pathway, endoplasmic reticulum protein processing, MAPK signaling pathways, and FoxO signaling pathways were overexpressed compared to untreated control cells. Sulforaphene uniquely induced overexpression of genes associated with ubiquitin-mediated proteolysis and estrogen signaling pathways [75]. Considering that the in vitro results were confirmed using multiple cell lines, as well as the similarity in gene expression alterations between sulforaphene and sulforaphane, it seems likely that sulforaphene could have potential uses in CRC treatment.

2.3.3. Allyl Isothiocyanate (AITC)

Allyl isothiocyanate (AITC) has received relatively little attention as a potential treatment for CRC, with only a few in vitro studies published in the last decade and zero in vivo studies. Chiang et al. found that HT-29 cells treated with 20 μM (AITC) exhibited cell cycle arrest at the G2/M phase and increased apoptosis (Table 1). This apoptosis was induced by reactive oxygen species (ROS) related to mitochondrial and endoplasmic reticulum stress. On top of finding increased ROS in the cells, they also found that the growth arrest and DNA damage-inducible protein 153 (DDIT3, also known as CHOP and GADD153) levels were increased, which is a clear marker of endoplasmic reticulum stress. Furthermore, they observed a loss of mitochondrial membrane potential and an increase in cytosolic Ca^{2+}, indicating mitochondrial stress [76].

A second study investigating the invasive and migrative capabilities of HT-29 cell lines found that 10 μM AITC treatment inhibited cell migration and invasion via transwell and wound healing assays. Investigation of the protein levels indicated that this inhibition was induced by the downregulation of matrix metalloproteinases 2 and 9, as well as the downregulation of MAP kinases [77]. Thus, AITC shows great potential as a chemotherapeutic agent against CRC, by inhibiting tumor growth and metastasis. Unfortunately, so few studies have been conducted that further investigation is severely needed. The current study only tested the effects of AITC against the HT-29 cell line, so future studies must confirm AITC's effectiveness against a wider variety of CRC cell lines.

2.3.4. Phenethyl Isothiocyanate (PEITC)

In vitro studies testing phenethyl isothiocyanate (PEITC) against CRC cell lines have demonstrated that PEITC reduces CRC cell viability in the HT-29, DLD-1, and SW480 cell models, while also being effective against CRC stem cells at concentrations from 12 to 88 μM (Table 1) [78,79]. This treatment reduced the number and size of CRC spheroids and appears to be related to Wnt/β-catenin pathway inhibition. Furthermore, Western blot analysis revealed the significant downregulation of proteins associated with stemness, including NANOG, Oct4, and Sox2 [80].

In vivo studies have further confirmed these results. Sprague Dawley rats that had CRC induced by 1,2-dimethylhydrazine (DMH) and received daily intraperitoneal injections of PEITC at 60 mg/kg of body weight developed significantly fewer aberrant crypt foci than untreated controls. Colonic tissue HDAC1, IL-6, and TNF-a levels were also significantly reduced, indicating a reduction in inflammation, and possibly altered histone methylation patterns. In contrast, 20 mg/kg of body weight of 5-fluorouracil (5-FU) administered in the same way was markedly more effective in reducing the numbers of aberrant crypt foci than PEITC and both showed similar levels of hepatotoxicity and nephrotoxicity (Table 1). Interestingly, a combination treatment of low doses of PEITC and the anthraquinone laccaic acid (30 mg/kg and 100 mg/kg, respectively) was even more effective than the 5-FU treatment in reducing the number of aberrant crypt foci while also inducing significantly less hepatotoxicity and nephrotoxicity. Thus, PEITC may have limited use as a sole chemotherapeutic agent due to its inferior performance to 5-FU (the current standard of care), but its use in combination treatments should be further explored [79].

Furthermore, oral PEITC administration at 20 mg/kg of body weight, five times per week for two weeks before HCT116 cell-derived xenograft transplantation in Balb/c nude male mice significantly reduced the xenograft's growth compared to untreated controls. PEITC treatment beginning after xenograft transplantation also reduced xenograft growth but not as significantly as pretreatment did. Crucially, this was achieved without any negative impact on mouse weight or any induction of hepatotoxicity [80]. Thus, on top of its potential as part of combination treatments, PEITC could have potential use in CRC chemoprevention.

2.3.5. Benzyl Isothiocyanate (BITC)

Benzyl isothiocyanate (BITC) is a compound that has received some attention as a potential chemotherapeutic agent in vitro but has been sparsely investigated in vivo. One study demonstrated that treating a CRC cell line (HCT116) with 20 μM BITC reduced cell viability, induced apoptosis, and induced cell cycle arrest at the sub-G1 phase (Table 1). Interestingly, these effects coincided with the activation of the PI3K/Akt/FoxO pathway. This pathway drives cell growth and proliferation and is associated with oncogenesis, differentiation, and drug resistance. Further experiments combining BITC treatment with the use of PI3K inhibitors demonstrated significantly greater cytotoxicity than BITC alone. For this reason, the authors hypothesize that the observed activation of the PI3K/Akt/FoxO pathway serves as a resistance mechanism for the cell against BITC treatment [81]. Thus, BITC may be of limited use as a chemotherapeutic agent on its own, but it also has the potential to reduce the efficacy of other drugs in combination treatments due to its activation of the PI3K/Akt/FoxO pathway.

Abe et al. found that 10 μM BITC inhibits proliferation by activating p65. When activated, p65 upregulates nuclear factor-κB (NF-κB) transcription and inhibits β-catenin activity. This causes the downregulation of cyclin D1 expression and arrests the cell cycle. Interestingly, BITC's antiproliferative effects are only observed in HT-29 cells with mutated p53 in this study [82]. BITC's antiproliferative effects were significantly reduced in HCT116 cells with wild-type p53, contrasting the results of the previously described study.

In sharp contrast to the positive results of the in vitro studies, the in vivo results are much less promising. In a large study of CRC chemoprevention in azoxymethane-induced F344 rats, Wargovich et al. found that BITC dietary supplementation as low as 0.5 g/kg of diet increased the formation of aberrant crypt foci when compared to animals fed a standard diet (Table 1) [83]. Seeing as the positive effects of BITC treatment vary based on the cell line assayed and are p53 mutation-dependent, BITC could have a possible function as a chemotherapeutic agent against CRCs with specific genotypes. The in vivo study demonstrates that BITC is likely not effective as a chemopreventive agent where its function would be to prevent tumorigenesis in healthy cells, but this does not take away from its potential in treating specific CRC genotypes.

2.3.6. Iberin

Iberin is an organosulfur compound that has been studied very little in the context of CRC treatment and prevention. One study found that treating Caco-2 CRC cells with 8 μM iberin resulted in a significant decrease in cell proliferation. DNA methylation analysis found that iberin treatment did not affect p16, ESR1, and LINE-1 methylation patterns but did significantly decrease the percent methylation of methyl guanine methyl transferases (MGMTs). This coincided with an increase in DNMT1, 3, and 3B mRNA levels. The alterations in the methylation patterns were further confirmed in a second CRC cell line (HCT116), although changes in proliferation were not investigated in this line. It is known that DNMT mRNA levels vary depending on the cell cycle stage, which indicates a possible connection between MGMT expression levels and iberin's antiproliferative effect, but this cannot be confirmed since cell cycle analysis was not conducted in this study [84]. In a separate study, Slaby et al. analyzed the miRNA profiles of non-transformed colonic epithelial cell lines treated with 10 μM iberin (Table 1). They found that total miRNA levels were higher in treated vs. untreated cells, which indicates a potential antitumoral effect since global downregulation of miRNA levels has been observed in CRC tumors. Furthermore, treatment induced an increase in miR-23b levels. miR-23b is a known tumor suppressor that regulates the epithelial–mesenchymal transition and TGF-β signaling [85]. In both studies, iberin and sulforaphane were compared, with very few differences between their effects. As such, iberin may have some potential as a chemotherapeutic compound against CRC, but a great deal of research remains to be conducted, especially in in vivo models, which are entirely lacking. Furthermore, the impact of iberin on CRC cell viability should be investigated in a wider variety of cell lines. It could be argued that iberin's effects on healthy cells' miRNA levels make it a candidate for chemoprevention.

2.4. Indoles

2.4.1. Indole-3-Carbinol

Indole-3-carbinol (I3C) is an organosulfur compound found in cruciferous vegetables and its antitumorigenic properties against CRC have been well characterized both in vivo and in vitro. Treating a wide variety of CRC cell lines (DLD1, HCT116, HT-29, LS513, RKO, LoVo, and SW480) has confirmed its ability to reduce cancer cell viability and proliferation by inducing apoptosis, although with large IC 50s, at 500 μM at 48 h and 1 mM at 24 h [86,87]. Of more promise is the finding that the I3C derivative 3-(2-bromoethyl)-indole had an IC 50 of only 50 μM at 24 h (Table 1) [88]. Mechanistically, one study found that I3C treatment induced the upregulation of p53, which induced apoptosis and inhibited cell migration [65]. Furthermore, a separate study found that I3C is an aryl hydrocarbon receptor (AHR) agonist. Activation of AHR appears to be critical to I3C's apoptosis-inducing activity since AHR downregulation protected the cells and prevented apoptosis [87].

In contrast to the positive in vitro results, in vivo studies in rodents have yielded mixed results. In one study, rats that were fed hemin (a potential carcinogen in red meat) and had CRC induced by DMH experienced less tumor incidence and growth when I3C was supplemented into their diets at a concentration of 1 g/kg, leading to an average consumption of 23.43 mg/rat/day [89]. At the same time, supplementing the diets with I3C, probiotics, and prebiotics increased the size and incidence of the tumors relative to I3C supplementation alone, indicating potential metabolization by the gut microbiota. In direct contrast to these results, a subsequent study by the same group found that rats with CRC induced by DMH experienced increased tumor incidence and size when I3C was supplemented into their diets at a concentration of 1 g/kg (Table 1). The addition of prebiotics and probiotics to the diet along with I3C canceled out the effects of I3C and gave similar results to the un-supplemented controls [90]. Furthermore, a third study by a different group found that C57BL/6J mice that were infected with *Citrobacter rodentium* and treated with dietary I3C at 1 mmol/kg (0.15 g/kg) of diet experienced reduced colonic *C. rodentium* colonization, reduced crypt hyperplasia, and reduced inflammatory biomarkers IL-17A, IL-6, and IL-1β. They also found that treating Caco-2 cells in culture with 25 μM I3C reduced *C. rodentium* growth and prevented it from binding to the cells [91].

What these three studies have in common is that they attempted to modulate the intestinal microbiomes of the rodent models that they used. It is well known that intestinal dysbiosis is associated with the development of colitis-associated CRC. For this reason, investigation into the effects of any supplement on CRC development should be conducted in the context of the intestinal microbiome. Studies have shown a large amount of variation in the intestinal microbiota between rodents from different vendors and even rodents housed in different cages for prolonged periods of time [128]. This could be one of the reasons for the conflicting in vivo results. Due to the positive in vitro results in numerous cell lines, as well as some positive results in vivo, I3C should be considered a promising chemotherapeutic agent in CRC treatment or prevention. With this said, a more in-depth characterization of the microbiome is necessary to understand its confounding effects.

2.4.2. 3,3′-Diindolylmethane

3,3′-Diindolylmethane (DIM) is an organosulfur compound generated from the dimerization of I3C under the gastric acidic pH that has received quite a bit of attention as a potential treatment for CRC. In vitro studies have been extremely promising and have confirmed DIM's anticancer properties in a wide variety of cell lines. One study found that treatment of multiple CRC cell lines (HCT116, HT-29, HCT15, and DLD1) with 40 μM DIM reduced their viability and induced cell cycle arrest at the G1 phase (Table 1). This was found to be induced by COX1/2 and ERK1/2 protein inhibition [92]. Furthermore, other studies have confirmed the findings that a wide variety of CRC cell lines (HCT116, SW480, HT-29, LoVo, Caco-2, and Colo-320) showed increased apoptosis, cell cycle arrest, and reduced viability after DIM treatment at 25–56 μM concentrations. (Table 1). These results are attributed to endoplasmic reticulum stress and cyclin D1 downregulation [93], upregulation of the N-Myc pathway [94], ATF-3 upregulation [95], and inhibition of calcium channels [96].

In contrast to the positive in vitro results, in vivo studies have yielded mixed findings. Two studies have demonstrated DIM's efficacy in reducing the size of cell- and patient-derived xenografts in Balb/C mice. One study found that DIM treatment by intraperitoneal injection of 40 mg/kg of body weight increased the sensitivity of DLD-1 and HCT116 cell-derived xenografts to 5-Fluorouracil treatment, and this was found to be associated with the inhibition of pyrimidine metabolism [97]. A second study found that daily oral DIM treatment at 40 mg/kg body weight significantly reduced the growth of patient-derived xenografts and did so without the induction of any detectable side effects. In contrast to these positive results, this same study also found that daily oral DIM treatment at the same concentration did not reduce the metastatic ability of HT-29 cells injected into the tail vein of Balb/C mice and did not reduce the formation of CRC tumors in APC^{min} mice (Table 1) [92]. Thus, DIM may have some usefulness as a chemotherapeutic agent, but the lack of chemopreventive activity in APC^{min} mice raises serious questions about its usability in more physiologically relevant models. Also, it shows little use in reducing metastasis. Of note is the fact that I3C must undergo condensation to DIM (and also other polymerization products) under the gastric acidic pH in order to obtain bioactivity. This does not occur in the case of intraperitoneal administration, nor in the case of in vitro cell line treatments. Considering that DIM is effective at much smaller concentrations than I3C, it is possible that a large part of I3C's activity may be due to these polymerization reactions.

3. Organosulfur Compounds Derived from *Allium* Species

Allium is a genus of enormous economic importance, belonging to the *Amaryllidaceae* family. The majority of bioactive organosulfur compounds found in *Allium* vegetables are derived from glutathione, which is formed by a series of reactions between the amino acids L-cysteine, L-glutamic acid, and L-glycine. Glutathione reacts with methacrylic acid, an L-valine derivative, to form S-(2-carboxypropyl)glutathione. This then undergoes a series of reactions to form S-allylcysteine (SAC) [129]. SAC then has a carbonyl group added to its sulfur atom to form S-allylcysteine sulfoxide (SACS), also known as alliin [130]. Upon cellular damage (for example, during garlic crushing before cooking), alliinase is released from the vacuole and comes into contact with alliin. When this occurs, it converts alliin to allylsulfenic acid, which dimerizes to form diallyl thiosulfinate, also known as allicin [131]. Allicin is relatively unstable, so it rapidly rearranges to form numerous derivates. In this review, we will discuss the derivates (Z)-ajoene [132], diallyl sulfide (DAS), diallyl disulfide (DADS), diallyl trisulfide (DATS), and diallyl tetrasulfide [133,134]. Furthermore, during the preparation of aged garlic extract, S-allyl cysteine and diallyl disulfide react to form S-allylmercaptocysteine (SAMC) [135] (Figure 4).

3.1. Allium Vegetable Consumption

Epidemiological studies on *Allium* vegetable consumption and colorectal cancer have been promising, although inconclusive. For example, a study in China with 1666 patients found that consumption of *Allium* vegetables (>16 kg/year) was associated with a lower risk of colon cancer, with the exception of distal colon cancer where no association was found [136]. Moreover, a meta-analysis with a sample of 12,558 patients suggested similar effects from the consumption of the *Allium* genus against CRC [137]. In contrast, a meta-analysis of 5458 CRC patients found no association between *Allium* consumption and CRC risk. It even showed that *Allium* consumption was associated with a 23% increase in CRC incidence in women. The authors attribute this contradictory increase to the consumption of a high concentration of flavonoids present in *Allium* vegetables, which may function as an inhibitor of hormonal metabolism and affect estrogen metabolism [138].

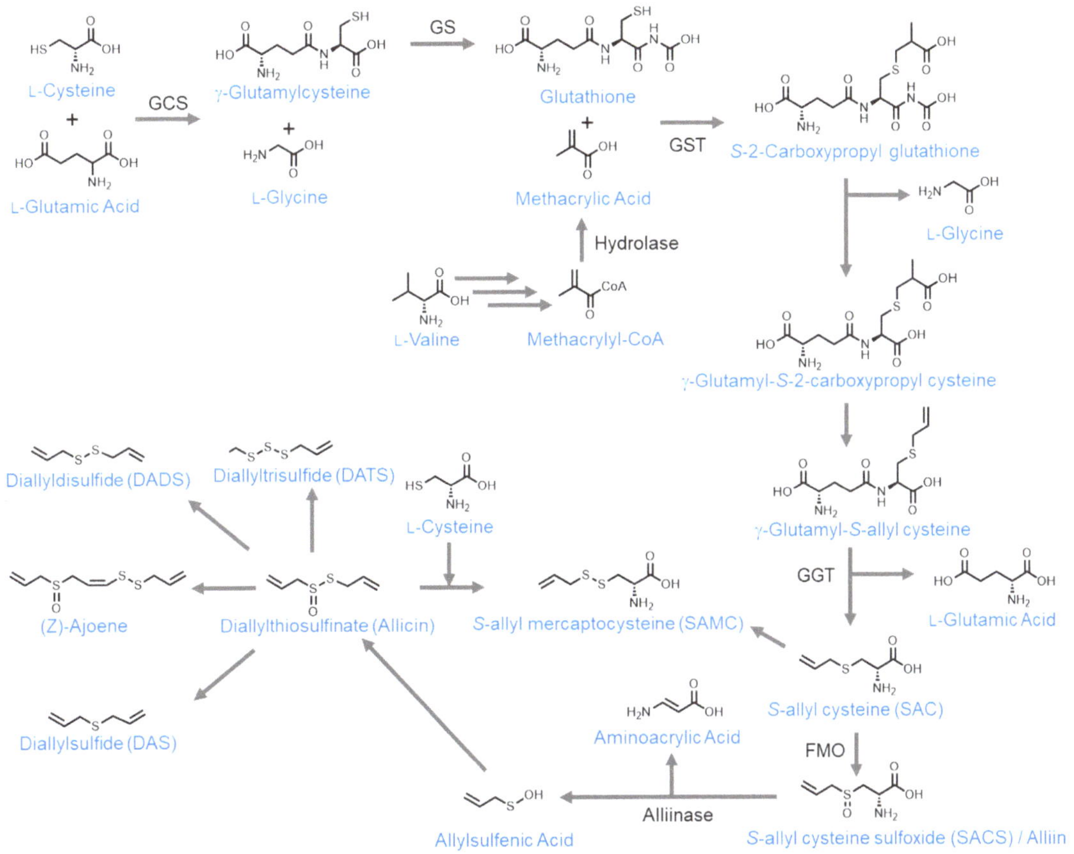

Figure 4. Biosynthesis of organosulfur compounds from allium vegetables. GCS: γ-Glutamylcysteine synthetase, GS: Glutathione synthetase, GST: Glutathione S-transferase, GGT: γ-Glutamyl transferase, FMO: Flavin-containing monooxygenase.

Studies of *Allium* consumption and CRC in humans have many limitations, primarily due to experimental variability. The size of the garlic portions taken in the different studies, the presence of multiple food items, the cooking method, the bioactivity of the different *Allium* vegetables, the length of the studies, the heterogeneity of participants, cultural differences between the different regions studied, and the study design are some prominent examples [139]. Thus, further epidemiological and clinical studies that can minimize these variables are required to draw consistent conclusions.

3.2. Allium Vegetable Extracts

Allium sativum is a vegetable widely used in gastronomy and is commonly known as garlic. The chemical composition of garlic is very broad, highlighting mixtures of soluble organosulfur compounds with different health properties such as antioxidant, antimicrobial, hypoglycemic, antiobesity, or anticancer [140].

Studies conducted by Bagul et al. showed that a crude garlic extract at concentrations of 0.5 μg/mL produced a growth inhibition of greater than 50% in colon cancer cells (Caco-2). They even observed that this effect was enhanced when this cell line was cultured together with a tumor-associated macrophage line (>80% inhibition). This inhibition is due to cell cycle arrest in the G1 phase, and increased apoptosis mediated by caspase-3 activation [141].

Similar results were shown by Su et al. using crude garlic extract at 1 μg/mL in a human colon adenocarcinoma line (Colo205) [142]. However, in a similar study with HT-29 colon adenocarcinoma cells, no cytotoxic effect was found until 1 mg/mL of treatment concentration was used [143]. In this sense, the extraction method as well as the concentration of organosulfur compounds present in these extracts may play a fundamental role in obtaining successful results.

In turn, anticancerogenic effects against CRC have been found in other *Allium* vegetables as well. For example, *Allium subhirsutum* (popularly known as hairy garlic) is known to have concentrations above 100 mg/g of sulfur compounds. Extracts of this *Allium* species have been found to be highly effective in vitro against CRC with an IC_{50} of 71 μg/mL against the colon carcinoma line HCT116 [144]. It has also been shown that *Allium victorialis* var. *platyphyllum* (known as Myung-I in Korea) extract at a concentration of 0.2 mg/mL reduced cell viability by 68% in the HT-29 cell line. In addition, the antimetastatic activity of the extract has even been observed when it is administered intraperitoneally at 1 mg/kg of body weight in a mouse model of CRC metastasis in the lungs [145].

An interesting effect of garlic extracts is the adjuvant effect when administered together with classical chemotherapy treatments such as 5-Fluorouracil or oxaliplatin. In this regard, studies by Ortiz et al. showed a synergistic inhibitory effect over 60% in Caco-2 and HT-29 cell lines when these chemotherapies were used in combination with garlic extract concentrations of up to 200 μg/mL. The use of these combinatorial treatments can reduce the economic cost by 45.3% compared to the use of a single chemotherapy agent, so their implementation could be helpful to healthcare systems [146].

3.2.1. Aged Garlic Extract

Aged garlic extract is a preparation made from fresh garlic that has been preserved for at least 10 months in an ethanol solution at room temperature. Through this process, the garlic components, such as flavonoids, S-allycysteine, pyruvate, benzyl cysteine compounds, lipid-soluble allyl sulfides, nutrient saponins, etc., are concentrated compared to fresh garlic [147]. In vitro studies by Dong et al. showed a cytotoxic effect of aged garlic extract (AGE) on the CRC cell line HT-29. Concentrations of 100 mg/mL of AGE were able to suppress cell growth by up to 64%. This effect is caused by an induction of G0/G1 cell cycle arrest, because of a downregulation of the PI3K/Akt pathway involved in proliferation, migration, and apoptosis [148]. These data were corroborated by Matsura et al. on different CRC lines such as HT-29, Sw480, and Sw620, where concentrations as low as 0.1 mg/mL significantly reduced proliferation. In addition, in their studies using AGE, they observed an anti-angiogenic effect. When AGE was added to endothelial cell lines (ECV304 and TRLECs cells) at a concentration of 10 g/L, an increase in cell adhesion and a reduction in mortality and invasion were observed, preventing the growth and development of cancer cells [149].

Similar results were obtained in studies using CRC rat models with dimethylhydrazine. Katsuki et al. fed the animals with AGE at 4% and observed a reduction in adenomas in the small intestine and a lower number of aberrant crypt foci. Jikihara et al. confirmed these studies with very similar effects when AGE at 3% was administered. In both studies, they found a reduced expression of cell proliferation markers (such as PCNA or MIB-5) and an attenuation of NF-κB activity, concluding that AGE may act as a chemopreventive agent with a suppressive effect on cell proliferation [150,151].

Finally, an interventional study by Tanaka et al. in patients with colorectal adenomas demonstrated promising results. In this study, each patient was administered 2.4 mL of AGE daily for 12 months. Only 37 patients completed this study, but it can be seen that in the group administered with AGE, the number and size of the adenomas was reduced, while in the control group, both values were increased [152]. In this study, the AGE preparation protocol and garlic concentration are not presented.

3.2.2. *Allium roseum* L. var. Grandiflorum Briq. Essential Oil

Considering all the investigations into the anticancer properties of garlic and its extracts, it is surprising that so little has been researched involving garlic oils. Numerous lipophilic molecules identified from garlic have been demonstrated to have antitumoral properties (allicin, DAS, DADS, DATS, etc.) and these molecules individually have received far more attention than intact garlic oil. One study of *Allium roseum* bulb essential oil demonstrated its antiproliferative effects in two CRC cell lines. They found that culturing both HT-29 and Caco-2 cells with 4 and 8 µg/mL of the oil, respectively, was sufficient to reduce cell viability by 50%, while culturing with 20 µg/mL caused 100% cell death in both lines. They also found that the oil had potent antioxidant activities and significantly reduced ROS production in both lines as well [153]. This study should be repeated using healthy colonic cells to ensure that the antiproliferative effects are cancer-cell specific.

3.3. *Bioactive Molecules*
3.3.1. Allicin

Allicin, as one of the primary metabolites of crushed garlic, has tremendous potential as a dietary supplement. Furthermore, its potential in CRC treatment and prevention has been extensively studied. One study of allicin's effects on CRC cells in vitro found that it significantly reduced cell viability in a dose-dependent manner in an HT-29 cell line model, with an IC 50 of 37.5 µM (Table 1). They also found that it reduced cell proliferation and induced both of these effects through a decrease in intracellular glutathione levels, and an increase in the formation of reactive oxygen species (ROS) [98]. A second study confirmed the findings that allicin treatment reduces cell proliferation and viability, with an IC 50 of around 25 µM, but also found that allicin-treated cells had significantly reduced levels of phosphorylated STAT3 [99]. STAT3 is a transcription factor for genes involved in survival and proliferation that must be phosphorylated to initiate its activity. Thus, preventing STAT3 phosphorylation is one of the ways that allicin can reduce tumor cell viability. Incredibly, a third study found that allicin treatment was even more effective in reducing DLD-1 cell line viability than treatment with 5-Fluorouracil, a commonly used chemotherapeutic agent, with an IC 50 near 50 µM (Table 1). Furthermore, they found that allicin was less cytotoxic to healthy cells than 5-Fluorouracil, demonstrating both its effectiveness against CRC cells as well as the reduced likelihood of negative side effects [100]. On top of allicin's antiproliferative activity and its selectivity for tumor tissues, Huang et al. also found that allicin treatment increased the radiosensitivity of HCT116 CRC cells and inhibited their migration ability as tested by a transwell chamber assay at 4 µg/mL [101]. Thus, allicin shows great potential as a pre-radiation therapy chemotherapeutic agent.

Beyond allicin's promise in vitro, several studies have also confirmed its antitumoral effects in vivo. Huang et al. further confirmed its in vitro effects in a BALB/c mouse model with CT-26 cell-derived xenografts. They found that combination therapy of intraperitoneally injected allicin at 5 mg/kg of body weight and radiation significantly reduced the tumor weight compared to allicin treatment or radiation therapy alone [101]. With this said, it should be noted that the allicin and radiation dosage given in the combination treatment was equal to the doses used in the individual treatments. As such, rather than allicin improving the radiosensitivity of the CRC cells, it could be argued that allicin reduces cell viability in a complementary way to radiation therapy and so mixing two complementary treatments is always likely to achieve augmented results. It would be interesting to see if combining reduced doses of allicin and radiation therapy could achieve similar results, as this could reduce the negative side effects associated with chemo and radiation therapies.

Furthermore, allicin has also been demonstrated to function as a chemopreventive agent in an azoxymethane/dextran sodium sulfate (AOM/DSS)-induced CRC mouse model. It was found that dietary allicin supplementation at 48 mg/kg of feed significantly reduced tumor size and incidence compared to controls. The allicin-supplemented mice also recovered significantly more quickly from DSS-induced inflammation than did the un-supplemented controls [99]. This supports findings that associate high garlic intake with a reduced risk of CRC development and indicates that allicin is likely one of the bioactive compounds responsible for this effect. Furthermore, allicin's anti-inflammatory properties could make it particularly useful in the chemoprevention of colitis-associated CRC.

3.3.2. Alliin/S-Allyl-L-Cysteine Sulfoxide (SACS)

Alliin, also known by its chemical name S-Allyl-L-cysteine sulfoxide (SACS), is a major constituent of garlic that becomes rapidly metabolized into allicin when the garlic is crushed or minced. Because of this, it is not regularly consumed in the human diet and has received comparatively less attention than allicin and its derivatives as a potential treatment for CRC. Even so, one study conducted in silico analyses of the binding affinities of various organosulfur compounds toward epithelial growth factor receptor (EGFR) using the CDOCKER software version DS 4.0. They found that, of the compounds used in the investigation, only alliin was predicted to interact with EGFR. They next confirmed the findings in vitro by treating HCT-15 cells with alliin and found that the treatment significantly reduced EGFR gene expression as well as cell viability (IC 50 approximately 100 µg/mL) (Table 1). Furthermore, alliin was found to satisfy Lipinski's rule of five and Veber's protocol for drug-like properties and bioavailability [102].

Numerous in vivo studies have further confirmed alliin's antitumorigenic properties. One study in F344 rats by Hatono et al. found that dietary alliin supplementation at 125 to 250 mg/kg of feed before and during cancer induction significantly reduced the formation of aberrant crypt foci. This effect was not repeated when alliin supplementation began after cancer induction [103], indicating alliin's chemopreventive, rather than chemotherapeutic, potential. This effect was further confirmed in a second study with dietary supplementation at 125 mg/kg of feed (Table 1) [104]. Furthermore, a separate study found that oral administration of alliin by gavage in C57BI/6J mice at a dosage of 200 mg/kg of body weight significantly reduced nuclear damage in colonic mucosal cells. This result was followed up by confirmation that similar alliin administration significantly reduced the frequency of colonic tumors in CF-1 mice. Interestingly, although numerous organosulfur compounds were tested in this study (DAS, DADS, DPS, DPDS, ajoene, SAC, SPC, and SAMC), only alliin (SAC) was shown to be protective of nuclear damage [105].

The mixture of in silico, in vitro, and in vivo results indicates that alliin may have potential as a chemopreventive agent, but not likely as a chemotherapeutic agent. Unfortunately, the lack of in vitro studies limits our knowledge of its mechanism of action. Furthermore, all the in vivo studies that have been conducted on this topic were undertaken before the year 2000, and as such, these results would benefit from further confirmation using more modern techniques.

3.3.3. Diallyl Sulfide (DAS)

Diallyl sulfide (DAS) is one of the most studied lipophilic, garlic-derived molecules. In vitro studies have demonstrated its ability to induce apoptosis and inhibit migration and metastasis in CRC cell lines. One study found that treating a CRC cell line (Colo-320 DM) with 50 µM DAS induced G2/M phase cell cycle arrest, increased ROS, and induced apoptosis via increased caspase-3 expression [106]. A separate study by Lai et al. found that DAS reduced Colo-205 CRC cell proliferation by inducing the downregulation of PI3K, Ras, MEKK3, MKK7, ERK1/2, JNK1/2, and p38. They also found that DAS treatment significantly reduced CRC cells' migration and invasion capabilities via a reduction in matrix metalloproteinase-2 (MMP-2) expression. Although they found that DAS reduced proliferation and inhibited migration and invasion (IC 50 greater than 50 μM), they also found that both diallyl disulfide (DADS) and diallyl trisulfide (DATS) were even more effective in reducing proliferation and inhibiting MMPs (IC 50 25 μM for both) (Table 1) [107]. Interestingly, another study found that DAS treatment of Colo-205 CRC cells caused a decrease in multidrug resistance 1, 3, 4, and 6 proteins (MRP1, 3, 4, 6), indicating that DAS may be particularly useful in co-treatments [108].

In contrast to the positive in vitro results, in vivo studies have yielded more mixed results. The same study that found that DAS treatment decreased MRP expression in cell lines also investigated DAS's effectivity against a Colo-205 CRC cell line-derived xenograft in BALB/C mice via intraperitoneal injections of DAS at 6 mg/kg body weight (Table 1). They found that tumor size trended downward in DAS-treated mice, but that the results were not significant. Furthermore, in the in vivo models, DAS yielded a significant increase in Mdr1 expression, but no significant change in MRP1, 3, 4, or 6 expressions [108]. All of this indicates that the effects of DAS are significantly reduced in vivo. In this study, DAS was injected intraperitoneally, so poor bioavailability or increased liver metabolism could be the cause of the more limited efficacy.

A separate study testing the effects of dietary DAS supplementation in APC^{min} mice found that supplementation with 100 to 300 ppm reduced the incidence of colonic polyps in a dose-dependent manner, although the inhibition was not statistically significant [109]. The authors claim that the lack of statistical significance is likely due to the relatively small number of mice used. DAS may be more useful as a dietary supplement since it readily encounters the colonic mucosa in this way and thus its poor bioavailability is less problematic than what was observed in the xenograft model. Thus, it may have better functionality as a chemopreventive agent than as a chemotherapeutic agent.

3.3.4. Diallyl Disulfide (DADS)

Diallyl disulfide (DADS) is one of the most extensively studied garlic-derived organosulfur compounds in terms of cancer treatment. Its effects in CRC models have been investigated in many in vivo and in vitro studies. Numerous studies have demonstrated DADS' ability to reduce proliferation and induce apoptosis in CRC cell lines (SW620, SW480, HCT116, HT-29, DLD-1), with IC 50s ranging from 5 to 50 μM depending on treatment time and cell line (Table 1). Multiple mechanisms for this effect have been observed including the downregulation of RAC1 and BCL-2, overexpression of BAX, BAD, and TRAIL, reduced NF-κB nuclear localization, and inhibition of the PI3K/Akt pathway [110–112]. A separate study found that DADS treatment induced a release of intracellular calcium stores in SW480 cells. While apoptosis was not measured in this study, the buildup of intracellular calcium would cause endoplasmic reticulum and mitochondrial stress, which could lead to apoptosis via the intrinsic pathway [113].

Beyond its cell viability-reducing effects, DADS has also been found to inhibit migration and invasion. One group found that DADS treatment at 30 to 45 mg/L downregulated Rac1 expression, which reduced PAK1-LIMK1-Cofilins signaling and inhibited the epithelial–mesenchymal transition in SW620 and HT-29 cells, as measured by transwell chamber and wound healing assays (Table 1) [110]. The same study further confirmed these results in vivo by measuring a reduced number of pulmonary metastatic nodules of nude mice that had HT-29 cells injected into their tail veins and received intraperitoneal injections of DADS at 100 mg/kg of body weight. A second study found that DADS treatment at 45 mg/L reduced LIMK1 expression, which reduced destrin and cofilin phosphorylation. This significantly reduced cell migration and invasion as measured by a scratch wound assay [114]. A third study found that DADS treatment at 25 μM induced the downregulation of MMP-2 and 7, as well as RAS, MEKK3, JNK1 and 2, and COX-1 and 2. It also reduced cell migration and invasion more effectively than treatment with diallyl sulfide (DAS) [107]. Also, much like diallyl sulfide (DAS), DADS treatment significantly reduced the expression of MRP1, 3, 4, and 6 in cell culture, indicating its potential in co-treatments [108].

In contrast to diallyl sulfide (DAS), DADS treatment has yielded numerous positive results in in vivo studies. DADS treatment via intraperitoneal injection at 6 mg/kg in BALB/c mice with Colo-205 cell-derived xenografts yielded a significant reduction in xenograft size (Table 1). This is in contrast with DAS, which yielded a downward trend in xenograft size without reaching statistical significance. Furthermore, in contrast to the in vitro studies, gene expression analysis demonstrated that DADS treatment in vivo caused an upregulation of Mdr1, MRP1, MRP4, and MRP6 drug resistance proteins [108]. Thus, while DADS may be more effective than DAS in reducing tumor size, it may not be useful in co-treatments. A second study found that regular DADS treatment, administered intraperitoneally at 100 mg/kg body weight, significantly reduced the size of cell-derived xenografts in BALB/c nude mice. Similarly, to what was found in the in vitro studies, LIMK1 overexpression reduced the effectiveness of the DADS treatment [114]. In contrast to these results, a separate study found that DADS treatment alone did not reduce the size of cell-derived xenografts in BALB/c nude mice. They reported that DADS was only effective against xenografts made of TRAIL overexpressing cells [111]. This study did not specify how the DADS was administered nor its dosage, so it is impossible to compare it with the other studies. Finally, the chemopreventive properties of DADS were investigated in a study where FVB/N mice were fed diets supplemented with 42 ppm DADS and had CRC chemically induced by azoxymethane (AOM) and dextran sodium sulfate (DSS). They found that DADS supplementation significantly reduced tumor incidence, number, and burden when compared with mice fed standard diets. Furthermore, they found that the mice eating the supplemented diet recovered from DSS-induced inflammation much more rapidly than those eating the standard diet [112].

DADS shows tremendous promise as a potential chemotherapeutic agent due to its antiproliferative and antimetastatic properties, as demonstrated both in vitro and in vivo. It also shows great potential as a chemopreventive agent that could be used as a dietary supplement. For these reasons, DADS should be further studied, both in vivo and potentially in clinical human trials. Interestingly, as of writing, no clinical trials of any sort have involved DADS, as per ClinicalTrials.gov; thus, it is likely that further pre-clinical studies are necessary.

A major potential downside to DADS is its hydrophobicity. This gives it low bioavailability and severely limits its use in clinical settings. Several studies have attempted to circumvent this problem and increase DADS' aqueous solubility by encasing it in liposomes [115] or nanoparticles [116]. These techniques have proven fruitful in in vitro studies, with both modifications achieving greater reductions in CRC cell viability and increased intracellular DADS concentrations. As of yet, these studies have not been repeated in vivo.

3.3.5. Diallyl Trisulfide (DATS)

Diallyl trisulfide (DATS) has received comparatively less attention as a potential CRC chemotherapeutic than DADS has, yet it has been proven effective against CRC models both in vitro and in vivo. In vitro, two recent studies have demonstrated that treating SW480 and DLD-1 cell lines, as well as patient-derived primary CRC cells, with DATS induced apoptosis as observed by the increased expression of proapoptotic proteins, DNA condensation, and increased reactive oxygen species with IC 50s ranging from 30 to 40 μM depending on the cell type (Table 1) [117,118]. Furthermore, it has been shown to be effective against CRC stem cells at 40 μM, which are normally particularly difficult to treat. One study found that treating CRC stem cells derived from SW480 and DLD-1 cell lines with DATS inhibited the Wnt/β-catenin pathway and reduced colonosphere formation [117]. This effectiveness against CRC stem cells may be partially because DATS, much like DADS and DAS, significantly reduced the expression of multidrug resistance-associated proteins (MRP 1, 3, 4, 6) in CRC cell culture [108]. Cancer stem cells are known to exhibit increased levels of multidrug resistance-associated proteins [119], and so a decrease in their expression induced by DATS could reduce their chemoresistance. Furthermore, DATS, much like DADS and DAS, has been demonstrated to reduce migration and invasion capabilities at concentrations as low as 10 µM, as measured by transwell chamber and wound healing assays. In contrast to DAS and DADS, DATS was observed to induce the greatest antimetastatic effects [107].

DATS has also shown positive results in numerous rodent CRC models. Two studies of cell-derived xenografts (Colo-205, CT-29) in BALB/c mice have shown that regular treatment with DATS, either by gavage at 50 mg/kg body weight or by intraperitoneal injection at 6 mg/kg body weight, significantly reduced the size and weight of the xenografted tumors when compared to saline-treated controls (Table 1) [108,120]. Similarly, to what was observed with DADS treatment, DATS-treated xenografts showed significantly increased expression of MRP 1, 4, and 6. Once again, this contrasts directly with the results observed in cell lines [108]. Furthermore, these results call into question the validity of the anti-stem cell effects that were observed in cell lines.

DATS has also been demonstrated to be effective as a chemopreventive agent against CRC. One study found that DATS treatment at a concentration of 25 mg/kg body weight, three times per week via an unspecified route, significantly reduced tumor number and incidence in an AOM-induced CRC mouse model (Table 1) [23]. Much like DADS, DATS is also hydrophobic and exhibits low bioavailability, limiting its potential for clinical use. This same study formulated DATS in lipid nanoparticles to increase its bioavailability and found not only its effectiveness in the AOM model but also increased kinetics and greater efficacy at lower doses against RKO and HT-29 cell lines.

Due to its demonstrated efficacy in vitro and in vivo, DATS should be considered a strong candidate for further study and development as a chemotherapeutic agent against CRC. Its main limitations come from its limited bioavailability, but novel techniques such as encapsulation could be effective in circumventing these problems.

3.3.6. Diallyl Tetrasulfide

In contrast to the other diallyl compounds discussed thus far in this review, diallyl tetrasulfide as a CRC chemotherapeutic has only been investigated in one single study. In SW480/620 and HT-29 cell lines, Yagdi et al. discovered that diallyl tetrasulfide is a reversible tubulin binder that induces mitotic arrest and apoptosis via the inhibition of autophagy at 10 to 50 µM (Table 1). p62 protein expression seems to protect the CRC cell lines against diallyl tetrasulfide-induced apoptosis since HT-29 cells contained greater amounts of p62 and were much more resistant than the other cell lines. They further confirmed that the diallyl tetrasulfide treatment of CRC cell lines significantly reduced spheroid formation. This effect was also cell-line-dependent (HT-29 cells were protected) and seemed to be mediated by levels of p62 expression since p62 knockdown sensitized the resistant cells. Furthermore, zebrafish implanted with HT-29 cell-derived xenografts from CRC cell lines experienced significantly less tumor growth when treated with 50 µM dibenzyl tetrasulfide (a synthetic derivative of diallyl tetrasulfide that shows similar effects in vitro) [121].

3.3.7. Methylsulfonylmethane (MSM) ((Dimethyl Sulfone (DMSO2))

Methylsulfonylmethane (MSM) is synthesized by the oxidation of dimethyl sulfoxide (DMSO) and is found in a variety of vegetables [154]. Its activity against a variety of cancers has been investigated, but so far only one study has investigated its function as a chemotherapeutic agent against CRC. Kim et al. found that treating HT-29 cells with MSM reduced viability, induced cell cycle arrest at the G_0/G_1 checkpoint, and induced apoptosis with IC 50s of 250 mM at 24 h and 100 mM at 48 h (Table 1). The cell cycle arrest coincided with the decreased expression of cyclin D and E, CDK4, and Rb proteins, and apoptosis was induced via mitochondrial membrane disruption. Furthermore, MSM reduced the stemness of the CRC cell lines as measured by reduced sphere formation and downregulation of the typical stem cell markers SOX2, NANOG, and OCT4 [24].

Thus, MSM shows some potential as a chemotherapeutic agent against CRC, but the current results are very preliminary. The study discussed above was only conducted in one cell line, and as has been observed with many of the other molecules already discussed in this review, the results are often cell-line-dependent. As such, further experiments in other cell lines are necessary to confirm its usefulness. Furthermore, no in vivo studies have been conducted. Interestingly, MSM supplementation is already somewhat common amongst patients with various ailments, so this molecule may be particularly amenable to investigation in the clinic [154].

3.3.8. (Z)-Ajoene

(Z)-ajoene is an organosulfur compound found in crushed garlic that has only recently begun to be investigated as a potential chemotherapeutic agent against CRC. In the only study to date investigating its impact, Li et al. found that (Z)-ajoene reduced the viability of SW480 CRC cells in a dose-dependent model, with an IC 50 of 30 μM at 72 h (Table 1). Furthermore, they found that (Z)-ajoene significantly reduced nuclear β-catenin protein levels, as well as the levels of proteins whose transcription is activated by the Wnt/β-catenin pathway (c-Myc and cyclin D1). It was further confirmed that (Z)-ajoene increases the phosphorylation of cytosolic β-catenin, which induces its degradation, and thus inhibits the Wnt/β-catenin pathway [122].

3.3.9. S-Allylmercaptocysteine (SAMC)

S-allylmercaptocysteine (SAMC) is a non-volatile and soluble compound present in aged garlic. This compound can be extracted from garlic bulbs or produced by chemical synthesis from L-cysteine and allicin in an aqueous solution at pH 6 [123].

At different SAMC concentrations between 200 and 450 μM, numerous anticarcinogenic properties have been observed. These include the inhibition of cell proliferation via G2-M phase cell cycle arrest, the induction of apoptosis, and the reduction in invasion in different CRC lines such as HT-29, Sw-480, Sw-620, Caco-2, or HT116 [124–127]. The molecular mechanism that explains the induction of apoptosis is via the JNK and p38 pathways that activate a signaling cascade that ultimately leads to the activation of the tumor suppressor genes p53 and Bax [125].

In addition, a very promising effect of SAMC in combinatorial therapy has been observed. Tong et al. observed a synergistic effect on the induction of apoptosis when SAMC at 800 μM was used in combination with MAPK inhibitors. This effect is produced via the activation of the TGF-β pathway, which leads to the expression of apoptotic proteins such as ERK and JNK (Table 1) [127].

A similar effect has been seen by Li et al. when combining SAMC with a chemo-preventive agent such as rapamycin, highlighting the potential adjuvant functions of this compound. In an HCT116 xenograft mouse model, they observed an 80% reduction in tumor growth with this combinatory therapy of oral SAMC administration at 300 mg/kg body weight/day with intraperitoneal injections of rapamycin at 5 mg/kg body weight/day (Table 1). This effect has been shown to be due not only to the induction of apoptosis but also to the activation of autophagy and antioxidant responses via Nrf2 [124].

Finally, it has been shown that SAMC may function as a good agent against CRC metastasis, probably by restoring the expression of E-cadherin levels in Caco-2, SW480, and SW620 cells at a 400 μM concentration [126].

4. Conclusions

Organosulfur compounds show great promise in colorectal cancer prevention, but many questions remain. *Allium* and cruciferous vegetable extracts have shown interesting results in human studies, but complete clinical trials remain to be conducted. For individual molecules, sulforaphane, phenethyl isothiocyanate, 3,3'-diindolylmethane, allicin, and diallyl disulfide have been extensively studied in multiple cell lines and animal models but lack studies in humans. These molecules are effective at physiological doses but require optimized administration. In the case of sulforaphane, intraperitoneal administration to treat xenografts in mice is about 100 times more potent than the tested oral administration (14.1 µg versus 1600 µg per day, respectively). In the case of phenethyl isothiocyanate, 0.8 mg per day per mouse orally reduced the tumor size in a xenograft model, in comparison with 2.4 mg per mouse intraperitoneally as chemoprevention. In the case of 3,3′-diindolylmethane, 1.6 mg per mouse intraperitoneally enhances the xenograft's sensitivity to 5-fluorouracil (but not in an *APC* mouse model), whereas orally, this same dosage caused positive effects in the absence of the drug. In the case of allicin, 0.2 mg per day in mice has chemopreventative properties, in comparison to a mixture of a radiation treatment plus 0.2 mg per day of allicin (intraperitoneal administration) to reduce tumor size. In the case of diallyl disulfide, 0.24 mg per mouse intraperitoneally reduced xenograft size, whereas 0.16 mg per day orally in mice was effective in chemoprevention.

Major limitations for the clinical use of these organosulfur compounds include the difficulty in comparing the achieved bioactivity among the different possible administration routes (gavage, diet, encapsulated, intraperitoneal, in co-therapy, etc.), as well as current data from diverse animal species and lineages. Pharmacokinetic data in humans are unknown for some of these compounds under the desired administration routes, a key limitation for their clinical development as antitumors. These data are needed for selecting the most useful dosage, as well as data from studies including side effects monitoring. Furthermore, the industrial supply for some of these compounds must be secured at a large scale. Finally, clinical trials will be needed to confirm whether any of these molecules are more effective than the current standard of care in humans.

The remaining molecules are not yet candidates for clinical studies due to their lack of investigation in pre-clinical models. Benzyl isothiocyanate has been extensively investigated in cell lines and animal models but requires follow-up studies to address unanswered questions. Other molecules, such as sulforaphene, indole-3-carbinol, diallyl sulfide, and S-allylmercaptocysteine, have been proven effective in cellular models but require confirmation from studies in animal models. The remaining molecules, including alliin, iberin, allyl isothiocyanate, diallyl tetrasulfide, (Z)-ajoene, *Allium roseum* oil, and methylsulfonylmethane, are still in their infancies in terms of CRC investigation. Many of these would benefit from confirmation in a wider variety of cell lines and more in-depth mechanistic studies.

As a final conclusion, in vitro and animal in vivo tests show that the organosulfur compounds with the highest future potential for use at the clinical level are sulforaphane, phenethyl isothiocyanate, 3,3′-diindolylmethane, allicin, and diallyl disulfide. These have been most extensively studied and demonstrated in animal models for colon cancer and show efficacy at reasonable doses.

Funding: Thjs research was funded by the Programa de Ayudas a Grupos de Investigación del Principado de Asturias (IDI/2018/000120), Programa de Ayudas Predoctorales de la Asociación Española Contra el Cáncer (AECC grant PRDAS233973MCAL) (grant to P.L.M.), the research project PID2021-127812OB-I00 from MICINN (Spanish Ministry of Science and Innovation), and the research project MIG-20201012 from CDTI (Spanish Ministry of Science and Innovation-Centro para el Desarrollo Tecnológico e Industrial).

Conflicts of Interest: The authors declare no conflicts of interest.

References

1. Wild, C.; Weiderpass, E.; Stewart, B.W. *World Cancer Report: Cancer Research for Cancer Prevention. Lyon, France: International Agency for Research on Cancer*; Wild, C.P., Weiderpass, E., Stewart, B.W., Eds.; IARC: Lyon, France, 2020.
2. Haque, T.R.; Bradshaw, P.T.; Crockett, S.D. Risk Factors for Serrated Polyps of the Colorectum. *Dig. Dis. Sci.* **2014**, *59*, 2874–2889. [CrossRef] [PubMed]

3. Murphy, N.; Ward, H.A.; Jenab, M.; Rothwell, J.A.; Boutron-Ruault, M.-C.; Carbonnel, F.; Kvaskoff, M.; Kaaks, R.; Kühn, T.; Boeing, H.; et al. Heterogeneity of Colorectal Cancer Risk Factors by Anatomical Subsite in 10 European Countries: A Multinational Cohort Study. *Clin. Gastroenterol. Hepatol.* **2019**, *17*, 1323–1331.e6. [CrossRef] [PubMed]

4. Tabung, F.K.; Brown, L.S.; Fung, T.T. Dietary Patterns and Colorectal Cancer Risk: A Review of 17 Years of Evidence (2000–2016). *Curr. Colorectal. Cancer Rep.* **2017**, *13*, 440–454. [CrossRef] [PubMed]

5. Ahn, J.; Sinha, R.; Pei, Z.; Dominianni, C.; Wu, J.; Shi, J.; Goedert, J.J.; Hayes, R.B.; Yang, L. Human Gut Microbiome and Risk for Colorectal Cancer. *J. Natl. Cancer Inst.* **2013**, *105*, 1907–1911. [CrossRef] [PubMed]

6. Arthur, J.C.; Perez-Chanona, E.; Mühlbauer, M.; Tomkovich, S.; Uronis, J.M.; Fan, T.-J.; Campbell, B.J.; Abujamel, T.; Dogan, B.; Rogers, A.B.; et al. Intestinal Inflammation Targets Cancer-Inducing Activity of the Microbiota. *Science* **2012**, *338*, 120–123. [CrossRef] [PubMed]

7. Fernández, J.; García, L.; Monte, J.; Villar, C.J.; Lombó, F. Functional Anthocyanin-Rich Sausages Diminish Colorectal Cancer in an Animal Model and Reduce pro-Inflammatory Bacteria in the Intestinal Microbiota. *Genes.* **2018**, *9*, 133. [CrossRef] [PubMed]

8. Huang, L.C.; Merchea, A. Dysplasia and Cancer in Inflammatory Bowel Disease. *Surg. Clin. North Am.* **2017**, *97*, 627–639. [CrossRef] [PubMed]

9. Crusz, S.M.; Balkwill, F.R. Inflammation and Cancer: Advances and New Agents. *Nat. Rev. Clin. Oncol.* **2015**, *12*, 584–596. [CrossRef]

10. Fearon, E.R.; Vogelstein, B. A Genetic Model for Colorectal Tumorigenesis. *Cell* **1990**, *61*, 759–767. [CrossRef]

11. Zoratto, F.; Rossi, L.; Verrico, M.; Papa, A.; Basso, E.; Zullo, A.; Tomao, L.; Romiti, A.; Lo Russo, G.; Tomao, S. Focus on Genetic and Epigenetic Events of Colorectal Cancer Pathogenesis: Implications for Molecular Diagnosis. *Tumour Biol.* **2014**, *35*, 6195–6206. [CrossRef]

12. Bettington, M.; Walker, N.; Clouston, A.; Brown, I.; Leggett, B.; Whitehall, V. The Serrated Pathway to Colorectal Carcinoma: Current Concepts and Challenges. *Histopathology* **2013**, *62*, 367–386. [CrossRef]

13. Haque, T.; Greene, K.G.; Crockett, S.D. Serrated Neoplasia of the Colon: What Do We Really Know? *Curr. Gastroenterol. Rep.* **2014**, *16*, 380. [CrossRef]

14. Binefa, G. Colorectal Cancer: From Prevention to Personalized Medicine. *World J. Gastroenterol.* **2014**, *20*, 6786. [CrossRef]

15. Waldecker, M.; Kautenburger, T.; Daumann, H.; Busch, C.; Schrenk, D. Inhibition of Histone-Deacetylase Activity by Short-Chain Fatty Acids and Some Polyphenol Metabolites Formed in the Colon. *J. Nutr. Biochem.* **2008**, *19*, 587–593. [CrossRef]

16. Redondo-Blanco, S.; Fernández, J.; Gutiérrez-del-Río, I.; Villar, C.J.; Lombó, F. New Insights toward Colorectal Cancer Chemotherapy Using Natural Bioactive Compounds. *Front. Pharmacol.* **2017**, *8*, 242711. [CrossRef]

17. Fernández, J.; Moreno, F.J.; Olano, A.; Clemente, A.; Villar, C.J.; Lombó, F. A Galacto-Oligosaccharides Preparation Derived From Lactulose Protects Against Colorectal Cancer Development in an Animal Model. *Front. Microbiol.* **2018**, *9*, 2004. [CrossRef]

18. Fernández, J.; Redondo-Blanco, S.; Miguélez, E.M.; Villar, C.J.; Clemente, A.; Lombó, F. Healthy Effects of Prebiotics and Their Metabolites against Intestinal Diseases and Colorectal Cancer. *AIMS Microbiol.* **2015**, *1*, 48–71. [CrossRef]

19. Fernández, J.; Redondo-Blanco, S.; Gutiérrez-del-Río, I.; Miguélez, E.M.; Villar, C.J.; Lombó, F. Colon Microbiota Fermentation of Dietary Prebiotics towards Short-Chain Fatty Acids and Their Roles as Anti-Inflammatory and Antitumour Agents: A Review. *J. Funct. Foods* **2016**, *25*, 511–522. [CrossRef]

20. Ferreira-Lazarte, A.; Fernández, J.; Gallego-Lobillo, P.; Villar, C.J.; Lombó, F.; Moreno, F.J.; Villamiel, M. Behaviour of Citrus Pectin and Modified Citrus Pectin in an Azoxymethane/Dextran Sodium Sulfate (AOM/DSS)-Induced Rat Colorectal Carcinogenesis Model. *Int. J. Biol. Macromol.* **2021**, *167*, 1349–1360. [CrossRef] [PubMed]

21. Fernández, J.; Ledesma, E.; Monte, J.; Millán, E.; Costa, P.; de la Fuente, V.G.; García, M.T.F.; Martínez-Camblor, P.; Villar, C.J.; Lombó, F. Traditional Processed Meat Products Re-Designed Towards Inulin-Rich Functional Foods Reduce Polyps in Two Colorectal Cancer Animal Models. *Sci. Rep.* **2019**, *9*, 14783. [CrossRef] [PubMed]

22. Pan, M.; Lai, C.; Wu, J.; Ho, C. Molecular Mechanisms for Chemoprevention of Colorectal Cancer by Natural Dietary Compounds. *Mol. Nutr. Food Res.* **2011**, *55*, 32–45. [CrossRef]

23. Alrumaihi, F.; Khan, M.A.; Babiker, A.Y.; Alsaweed, M.; Azam, F.; Allemailem, K.S.; Almatroudi, A.A.; Ahamad, S.R.; Alsugoor, M.H.; Alharbi, K.N.; et al. Lipid-Based Nanoparticle Formulation of Diallyl Trisulfide Chemosensitizes the Growth Inhibitory Activity of Doxorubicin in Colorectal Cancer Model: A Novel In Vitro, In Vivo and In Silico Analysis. *Molecules* **2022**, *27*, 2192. [CrossRef]

24. Kim, D.H.; Kang, D.Y.; Sp, N.; Jo, E.S.; Rugamba, A.; Jang, K.-J.; Yang, Y.M. Methylsulfonylmethane Induces Cell Cycle Arrest and Apoptosis, and Suppresses the Stemness Potential of HT-29 Cells. *Anticancer Res.* **2020**, *40*, 5191–5200. [CrossRef]

25. Wargovich, M.J. Diallylsulfide and Allylmethylsulfide Are Uniquely Effective among Organosulfur Compounds in Inhibiting CYP2E1 Protein in Animal Models. *J. Nutr.* **2006**, *136*, 832S–834S. [CrossRef] [PubMed]

26. Lu, Y.; Zhang, M.; Huang, D. Dietary Organosulfur-Containing Compounds and Their Health-Promotion Mechanisms. *Annu. Rev. Food Sci. Technol.* **2022**, *13*, 287–313. [CrossRef] [PubMed]

27. Petropoulos, S.; Di Gioia, F.; Ntatsi, G. Vegetable Organosulfur Compounds and Their Health Promoting Effects. *Curr. Pharm. Des.* **2017**, *23*, 2850–2875. [CrossRef] [PubMed]

28. Seo, M.-S.; Kim, J.S. Understanding of MYB Transcription Factors Involved in Glucosinolate Biosynthesis in Brassicaceae. *Molecules* **2017**, *22*, 1549. [CrossRef]

29. Mithen, R.F.; Dekker, M.; Verkerk, R.; Rabot, S.; Johnson, I.T. The Nutritional Significance, Biosynthesis and Bioavailability of Glucosinolates in Human Foods. *J. Sci. Food Agric.* **2000**, *80*, 967–984. [CrossRef]
30. Nguyen, V.P.T.; Stewart, J.; Lopez, M.; Ioannou, I.; Allais, F. Glucosinolates: Natural Occurrence, Biosynthesis, Accessibility, Isolation, Structures, and Biological Activities. *Molecules* **2020**, *25*, 4537. [CrossRef]
31. Yagishita, Y.; Fahey, J.W.; Dinkova-Kostova, A.T.; Kensler, T.W. Broccoli or Sulforaphane: Is It the Source or Dose That Matters? *Molecules* **2019**, *24*, 3593. [CrossRef]
32. Zhang, Y. Allyl Isothiocyanate as a Cancer Chemopreventive Phytochemical. *Mol. Nutr. Food Res.* **2010**, *54*, 127. [CrossRef] [PubMed]
33. Zhang, J.; Li, X.; Ge, P.; Zhang, B.; Wen, L.; Gu, C.; Zhou, X. Sulforaphene: Formation, Stability, Separation, Purification, Determination and Biological Activities. *Sep. Purif. Rev.* **2022**, *51*, 330–339. [CrossRef]
34. Al Janobi, A.A.; Mithen, R.F.; Gasper, A.V.; Shaw, P.N.; Middleton, R.J.; Ortori, C.A.; Barrett, D.A. Quantitative Measurement of Sulforaphane, Iberin and Their Mercapturic Acid Pathway Metabolites in Human Plasma and Urine Using Liquid Chromatography-Tandem Electrospray Ionisation Mass Spectrometry. *J. Chromatogr. B Anal. Technol. Biomed. Life Sci.* **2006**, *844*, 223–234. [CrossRef] [PubMed]
35. Liu, F.; Yang, H.; Wang, L.; Yu, B. Biosynthesis of the High-Value Plant Secondary Product Benzyl Isothiocyanate via Functional Expression of Multiple Heterologous Enzymes in Escherichia Coli. *ACS Synth. Biol.* **2016**, *5*, 1557–1565. [CrossRef] [PubMed]
36. Morris, M.E.; Dave, R.A. Pharmacokinetics and Pharmacodynamics of Phenethyl Isothiocyanate: Implications in Breast Cancer Prevention. *AAPS J.* **2014**, *16*, 705–713. [CrossRef] [PubMed]
37. Kokotou, M.G.; Revelou, P.-K.; Pappas, C.; Constantinou-Kokotou, V. High Resolution Mass Spectrometry Studies of Sulforaphane and Indole-3-Carbinol in Broccoli. *Food Chem.* **2017**, *237*, 566–573. [CrossRef]
38. Ciska, E.; Verkerk, R.; Honke, J. Effect of Boiling on the Content of Ascorbigen, Indole-3-Carbinol, Indole-3-Acetonitrile, and 3,3'-Diindolylmethane in Fermented Cabbage. *J. Agric. Food Chem.* **2009**, *57*, 2334–2338. [CrossRef]
39. *Cruciferous Vegetable Consumption and Multiple Health Outcomes: An Umbrella Review of 41 Systematic Reviews and Meta-Analyses of 303 Observational Studies—Food & Function*; RSC Publishing: Tokyo, Japan, 2022.
40. Fang, W.; Qu, X.; Shi, J.; Li, H.; Guo, X.; Wu, X.; Liu, Y.; Li, Z. Cruciferous Vegetables and Colorectal Cancer Risk: A Hospital-Based Matched Case–Control Study in Northeast China. *Eur. J. Clin. Nutr.* **2019**, *73*, 450–457. [CrossRef]
41. Tse, G.; Eslick, G.D. Cruciferous Vegetables and Risk of Colorectal Neoplasms: A Systematic Review and Meta-Analysis. *Nutr. Cancer* **2014**, *66*, 128–139. [CrossRef]
42. Wu, Q.J.; Yang, Y.; Vogtmann, E.; Wang, J.; Han, L.H.; Li, H.L.; Xiang, Y.B. Cruciferous Vegetables Intake and the Risk of Colorectal Cancer: A Meta-Analysis of Observational Studies. *Ann. Oncol.* **2013**, *24*, 1079–1087. [CrossRef]
43. Mori, N.; Sawada, N.; Shimazu, T.; Yamaji, T.; Goto, A.; Takachi, R.; Ishihara, J.; Iwasaki, M.; Inoue, M.; Tsugane, S. Cruciferous Vegetable Intake and Colorectal Cancer Risk: Japan Public Health Center-Based Prospective Study. *Eur. J. Cancer Prev.* **2019**, *28*, 420–427. [CrossRef] [PubMed]
44. Leenders, M.; Siersema, P.D.; Overvad, K.; Tjønneland, A.; Olsen, A.; Boutron-Ruault, M.-C.; Bastide, N.; Fagherazzi, G.; Katzke, V.; Kühn, T.; et al. Subtypes of Fruit and Vegetables, Variety in Consumption and Risk of Colon and Rectal Cancer in the European Prospective Investigation into Cancer and Nutrition. *Int. J. Cancer* **2015**, *137*, 2705–2714. [CrossRef] [PubMed]
45. Kim, S.; Abernathy, B.E.; Trudo, S.P.; Gallaher, D.D. Colon Cancer Risk of a Westernized Diet Is Reduced in Mice by Feeding Cruciferous or Apiaceous Vegetables at a Lower Dose of Carcinogen but Not a Higher Dose. *J. Cancer Prev.* **2020**, *25*, 223–233. [CrossRef] [PubMed]
46. Kim, S.; Trudo, S.P.; Gallaher, D.D. Apiaceous and Cruciferous Vegetables Fed During the Post-Initiation Stage Reduce Colon Cancer Risk Markers in Rats. *J. Nutr.* **2019**, *149*, 249–257. [CrossRef] [PubMed]
47. Lynn, A.; Fuller, Z.; Collins, A.R.; Ratcliffe, B. Comparison of the Effect of Raw and Blanched-Frozen Broccoli on DNA Damage in Colonocytes. *Cell Biochem. Funct.* **2015**, *33*, 266–276. [CrossRef] [PubMed]
48. Yanaka, A.; Suzuki, H.; Mutoh, M.; Kamoshida, T.; Kakinoki, N.; Yoshida, S.; Hirose, M.; Ebihara, T.; Hyodo, I. Chemoprevention against Colon Cancer by Dietary Intake of Sulforaphane. *Funct. Foods Health Dis.* **2019**, *9*, 392–411. [CrossRef]
49. Baenas, N.; Silván, J.M.; Medina, S.; de Pascual-Teresa, S.; García-Viguera, C.; Moreno, D.A. Metabolism and Antiproliferative Effects of Sulforaphane and Broccoli Sprouts in Human Intestinal (Caco-2) and Hepatic (HepG2) Cells. *Phytochem. Rev.* **2015**, *14*, 1035–1044. [CrossRef]
50. Pereira, L.; Silva, P.; Duarte, M.; Rodrigues, L.; Duarte, C.; Albuquerque, C.; Serra, A. Targeting Colorectal Cancer Proliferation, Stemness and Metastatic Potential Using Brassicaceae Extracts Enriched in Isothiocyanates: A 3D Cell Model-Based Study. *Nutrients* **2017**, *9*, 368. [CrossRef]
51. Kassie, F.; Uhl, M.; Rabot, S.; Grasl-Kraupp, B.; Verkerk, R.; Kundi, M.; Chabicovsky, M.; Schulte-Hermann, R.; Knasmüller, S. Chemoprevention of 2-Amino-3-Methylimidazo[4, 5-f]Quinoline (IQ)-Induced Colonic and Hepatic Preneoplastic Lesions in the F344 Rat by Cruciferous Vegetables Administered Simultaneously with the Carcinogen. *Carcinogenesis* **2003**, *24*, 255–261. [CrossRef]
52. Uhl, M.; Kassie, F.; Rabot, S.; Grasl-Kraupp, B.; Chakraborty, A.; Laky, B.; Kundi, M.; Knasmüller, S. Effect of Common Brassica Vegetables (Brussels Sprouts and Red Cabbage) on the Development of Preneoplastic Lesions Induced by 2-Amino-3-Methylimidazo[4,5-f]Quinoline (IQ) in Liver and Colon of Fischer 344 Rats. *J. Chromatogr. B* **2004**, *802*, 225–230. [CrossRef]

53. Rajendran, P.; Dashwood, W.-M.; Li, L.; Kang, Y.; Kim, E.; Johnson, G.; Fischer, K.A.; Löhr, C.V.; Williams, D.E.; Ho, E.; et al. Nrf2 Status Affects Tumor Growth, HDAC3 Gene Promoter Associations, and the Response to Sulforaphane in the Colon. *Clin. Epigenetics* **2015**, *7*, 102. [CrossRef] [PubMed]
54. Wilson, A.J.; Byun, D.-S.; Popova, N.; Murray, L.B.; L'Italien, K.; Sowa, Y.; Arango, D.; Velcich, A.; Augenlicht, L.H.; Mariadason, J.M. Histone Deacetylase 3 (HDAC3) and Other Class I HDACs Regulate Colon Cell Maturation and P21 Expression and Are Deregulated in Human Colon Cancer*. *J. Biol. Chem.* **2006**, *281*, 13548–13558. [CrossRef]
55. Zhu, Y.; Hu, H.; Huang, K.; Sun, X.; Li, M. Sulforaphane Inhibits Proliferation and Apoptosis of Colorectal Cancer Cells by Down-Regulating the Cyclooxygenase-2/Protein Kinase B/Glycogen Synthase Kinase-3 Beta Signaling Pathway. *Indian. J. Pharm. Sci.* **2022**, *84*, 219–223. [CrossRef]
56. Hao, Q.; Wang, M.; Sun, N.-X.; Zhu, C.; Lin, Y.-M.; Li, C.; Liu, F.; Zhu, W.-W. Sulforaphane Suppresses Carcinogenesis of Colorectal Cancer through the ERK/Nrf2 UDP Glucuronosyltransferase 1A Metabolic Axis Activation. *Oncol. Rep.* **2020**, *43*, 1067–1080. [CrossRef]
57. Zhou, J.-W.; Wang, M.; Sun, N.-X.; Qing, Y.; Yin, T.-F.; Li, C.; Wu, D. Sulforaphane induced Epigenetic Regulation of Nrf2 Expression by DNA Methyltransferase in Human Caco 2 Cells. *Oncol. Lett.* **2019**, *18*, 2639–2647. [CrossRef]
58. Martin, S.L.; Kala, R.; Tollefsbol, T.O. Mechanisms for the Inhibition of Colon Cancer Cells by Sulforaphane through Epigenetic Modulation of MicroRNA-21 and Human Telomerase Reverse Transcriptase (HTERT) Down-Regulation. *Curr. Cancer Drug Targets* **2018**, *18*, 97–106. [CrossRef] [PubMed]
59. Okonkwo, A.; Mitra, J.; Johnson, G.S.; Li, L.; Dashwood, W.M.; Hegde, M.L.; Yue, C.; Dashwood, R.H.; Rajendran, P. Heterocyclic Analogs of Sulforaphane Trigger DNA Damage and Impede DNA Repair in Colon Cancer Cells: Interplay of HATs and HDACs. *Mol. Nutr. Food Res.* **2018**, *62*, 1800228. [CrossRef] [PubMed]
60. Tafakh, M.S.; Saidijam, M.; Ranjbarnejad, T.; Malih, S.; Mirzamohammadi, S.; Najafi, R. Sulforaphane, a Chemopreventive Compound, Inhibits Cyclooxygenase-2 and Microsomal Prostaglandin E Synthase-1 Expression in Human HT-29 Colon Cancer Cells. *Cells Tissues Organs* **2018**, *206*, 46–53. [CrossRef]
61. Liu, K.-C.; Shih, T.-Y.; Kuo, C.-L.; Ma, Y.-S.; Yang, J.-L.; Wu, P.-P.; Huang, Y.-P.; Lai, K.-C.; Chung, J.-G. Sulforaphane Induces Cell Death Through G2/M Phase Arrest and Triggers Apoptosis in HCT 116 Human Colon Cancer Cells. *Am. J. Chin. Med.* **2016**, *44*, 1289–1310. [CrossRef]
62. Chung, Y.-K.; Chi-Hung Or, R.; Lu, C.-H.; Ouyang, W.-T.; Yang, S.-Y.; Chang, C.-C. Sulforaphane Down-Regulates SKP2 to Stabilize P27KIP1 for Inducing Antiproliferation in Human Colon Adenocarcinoma Cells. *J. Biosci. Bioeng.* **2015**, *119*, 35–42. [CrossRef]
63. Rajendran, P.; Kidane, A.I.; Yu, T.-W.; Dashwood, W.-M.; Bisson, W.H.; Löhr, C.V.; Ho, E.; Williams, D.E.; Dashwood, R.H. HDAC Turnover, CtIP Acetylation and Dysregulated DNA Damage Signaling in Colon Cancer Cells Treated with Sulforaphane and Related Dietary Isothiocyanates. *Epigenetics* **2013**, *8*, 612–623. [CrossRef]
64. Chen, M.-J.; Tang, W.-Y.; Hsu, C.-W.; Tsai, Y.-T.; Wu, J.-F.; Lin, C.-W.; Cheng, Y.-M.; Hsu, Y.-C. Apoptosis Induction in Primary Human Colorectal Cancer Cell Lines and Retarded Tumor Growth in SCID Mice by Sulforaphane. *Evid. -Based Complement. Altern. Med.* **2011**, *2012*, e415231. [CrossRef] [PubMed]
65. Wang, M.; Chen, S.; Wang, S.; Sun, D.-F.; Chen, J.; Li, Y.-Q.; Han, W.; Yang, X.-Y.; Gao, H.-Q. Effects of Phytochemicals Sulforaphane on Uridine Diphosphate-Glucuronosyltransferase Expression as Well as Cell-Cycle Arrest and Apoptosis in Human Colon Cancer Caco-2 Cells. *Chin. J. Physiol.* **2012**, *55*, 134–144. [CrossRef] [PubMed]
66. Yasuda, S.; Horinaka, M.; Sakai, T. Sulforaphane Enhances Apoptosis Induced by Lactobacillus Pentosus Strain S PT84 via the TNFα Pathway in Human Colon Cancer Cells. *Oncol. Lett.* **2019**, *18*, 4253–4261. [CrossRef]
67. Bessler, H.; Djaldetti, M. Broccoli and Human Health: Immunomodulatory Effect of Sulforaphane in a Model of Colon Cancer. *Int. J. Food Sci. Nutr.* **2018**, *69*, 946–953. [CrossRef] [PubMed]
68. Kim, D.H.; Sung, B.; Kang, Y.J.; Hwang, S.Y.; Kim, M.J.; Yoon, J.-H.; Im, E.; Kim, N.D. Sulforaphane Inhibits Hypoxia-Induced HIF-1α and VEGF Expression and Migration of Human Colon Cancer Cells. *Int. J. Oncol.* **2015**, *47*, 2226–2232. [CrossRef] [PubMed]
69. Milczarek, M.; Pogorzelska, A.; Wiktorska, K. Synergistic Interaction between 5-FU and an Analog of Sulforaphane—2-Oxohexyl Isothiocyanate—In an In Vitro Colon Cancer Model. *Molecules* **2021**, *26*, 3019. [CrossRef] [PubMed]
70. Liu, F.; Lv, R.-B.; Liu, Y.; Hao, Q.; Liu, S.-J.; Zheng, Y.-Y.; Li, C.; Zhu, C.; Wang, M. Salinomycin and Sulforaphane Exerted Synergistic Antiproliferative and Proapoptotic Effects on Colorectal Cancer Cells by Inhibiting the PI3K/Akt Signaling Pathway in Vitro and in Vivo. *Onco Targets Ther.* **2020**, *13*, 4957–4969. [CrossRef] [PubMed]
71. Erzinger, M.M.; Bovet, C.; Hecht, K.M.; Senger, S.; Winiker, P.; Sobotzki, N.; Cristea, S.; Beerenwinkel, N.; Shay, J.W.; Marra, G.; et al. Sulforaphane Preconditioning Sensitizes Human Colon Cancer Cells towards the Bioreductive Anticancer Prodrug PR-104A. *PLoS ONE* **2016**, *11*, e0150219. [CrossRef]
72. Čižauskaitė, A.; Šimčikas, D.; Schultze, D.; Kallifatidis, G.; Bruns, H.; Čekauskas, A.; Herr, I.; Baušys, A.; Strupas, K.; Schemmer, P. Sulforaphane Has an Additive Anticancer Effect to FOLFOX in Highly Metastatic Human Colon Carcinoma Cells. *Oncol. Rep.* **2022**, *48*, 205. [CrossRef]
73. Hu, R. Cancer Chemoprevention of Intestinal Polyposis in ApcMin/+ Mice by Sulforaphane, a Natural Product Derived from Cruciferous Vegetable. *Carcinogenesis* **2006**, *27*, 2038–2046. [CrossRef] [PubMed]

74. Byun, S.; Shin, S.H.; Park, J.; Lim, S.; Lee, E.; Lee, C.; Sung, D.; Farrand, L.; Lee, S.R.; Kim, K.H.; et al. Sulforaphene Suppresses Growth of Colon Cancer-Derived Tumors via Induction of Glutathione Depletion and Microtubule Depolymerization. *Mol. Nutr. Food Res.* **2016**, *60*, 1068–1078. [CrossRef] [PubMed]

75. Gao, L.; Du, F.; Wang, J.; Zhao, Y.; Liu, J.; Cai, D.; Zhang, X.; Wang, Y.; Zhang, S. Examination of the Differences between Sulforaphane and Sulforaphene in Colon Cancer: A Study Based on Next generation Sequencing. *Oncol. Lett.* **2021**, *22*, 690. [CrossRef]

76. Chiang, J.-H.; Tsai, F.-J.; Hsu, Y.-M.; Yin, M.-C.; Chiu, H.-Y.; Yang, J.-S. Sensitivity of Allyl Isothiocyanate to Induce Apoptosis via ER Stress and the Mitochondrial Pathway upon ROS Production in Colorectal Adenocarcinoma Cells. *Oncol. Rep.* **2020**, *44*, 1415–1424. [CrossRef] [PubMed]

77. Lai, K.-C.; Lu, C.-C.; Tang, Y.-J.; Chiang, J.-H.; Kuo, D.-H.; Chen, F.-A.; Chen, I.-L.; Yang, J.-S. Allyl Isothiocyanate Inhibits Cell Metastasis through Suppression of the MAPK Pathways in Epidermal Growth Factor stimulated HT29 Human Colorectal Adenocarcinoma Cells. *Oncol. Rep.* **2014**, *31*, 189–196. [CrossRef]

78. Chen, Y.; Li, Y.; Wang, X.; Meng, Y.; Zhang, Q.; Zhu, J.; Chen, J.; Cao, W.; Wang, X.; Xie, C.; et al. Phenethyl Isothiocyanate Inhibits Colorectal Cancer Stem Cells by Suppressing Wnt/β-Catenin Pathway. *Phytother. Res.* **2018**, *32*, 2447–2455. [CrossRef]

79. Gupta, R.; Bhatt, L.K.; Momin, M. Potent Antitumor Activity of Laccaic Acid and Phenethyl Isothiocyanate Combination in Colorectal Cancer via Dual Inhibition of DNA Methyltransferase-1 and Histone Deacetylase-1. *Toxicol. Appl. Pharmacol.* **2019**, *377*, 114631. [CrossRef]

80. Shin, J.M.; Lim, E.; Cho, Y.S.; Nho, C.W. Cancer-Preventive Effect of Phenethyl Isothiocyanate through Tumor Microenvironment Regulation in a Colorectal Cancer Stem Cell Xenograft Model. *Phytomedicine* **2021**, *84*, 153493. [CrossRef]

81. Liu, X.; Takano, C.; Shimizu, T.; Yokobe, S.; Abe-Kanoh, N.; Zhu, B.; Nakamura, T.; Munemasa, S.; Murata, Y.; Nakamura, Y. Inhibition of Phosphatidylinositide 3-Kinase Ameliorates Antiproliferation by Benzyl Isothiocyanate in Human Colon Cancer Cells. *Biochem. Biophys. Res. Commun.* **2017**, *491*, 209–216. [CrossRef]

82. Abe, N.; Hou, D.-X.; Munemasa, S.; Murata, Y.; Nakamura, Y. Nuclear Factor-KappaB Sensitizes to Benzyl Isothiocyanate-Induced Antiproliferation in P53-Deficient Colorectal Cancer Cells. *Cell Death Dis.* **2014**, *5*, e1534. [CrossRef]

83. Wargovich, M.J.; Chen, C.D.; Jimenez, A.; Steele, V.E.; Velasco, M.; Stephens, L.C.; Price, R.; Gray, K.; Kelloff, G.J. Aberrant Crypts as a Biomarker for Colon Cancer: Evaluation of Potential Chemopreventive Agents in the Rat. *Cancer Epidemiol. Biomark. Prev.* **1996**, *5*, 355–360.

84. Barrera, L.N.; Johnson, I.T.; Bao, Y.; Cassidy, A.; Belshaw, N.J. Colorectal Cancer Cells Caco-2 and HCT116 Resist Epigenetic Effects of Isothiocyanates and Selenium in Vitro. *Eur. J. Nutr.* **2013**, *52*, 1327–1341. [CrossRef] [PubMed]

85. Slaby, O.; Sachlova, M.; Brezkova, V.; Hezova, R.; Kovarikova, A.; Bischofová, S.; Sevcikova, S.; Bienertova-Vasku, J.; Vasku, A.; Svoboda, M.; et al. Identification of MicroRNAs Regulated by Isothiocyanates and Association of Polymorphisms Inside Their Target Sites with Risk of Sporadic Colorectal Cancer. *Nutr. Cancer* **2013**, *65*, 247–254. [CrossRef] [PubMed]

86. Lee, J.Y.; Lim, H.M.; Lee, C.M.; Park, S.-H.; Nam, M.J. Indole-3-Carbinol Inhibits the Proliferation of Colorectal Carcinoma LoVo Cells through Activation of the Apoptotic Signaling Pathway. *Hum. Exp. Toxicol.* **2021**, *40*, 2099–2112. [CrossRef]

87. Megna, B.W.; Carney, P.R.; Nukaya, M.; Geiger, P.; Kennedy, G.D. Indole-3-Carbinol Induces Tumor Cell Death: Function Follows Form. *J. Surg. Res.* **2016**, *204*, 47–54. [CrossRef]

88. Fadlalla, K.; Elgendy, R.; Gilbreath, E.; Pondugula, S.R.; Yehualaeshet, T.; Mansour, M.; Serbessa, T.; Manne, U.; Samuel, T. 3-(2-Bromoethyl)-Indole Inhibits the Growth of Cancer Cells and NF-KB Activation. *Oncol. Rep.* **2015**, *34*, 495–503. [CrossRef]

89. de Moura, N.A.; Caetano, B.F.R.; de Moraes, L.N.; Carvalho, R.F.; Rodrigues, M.A.M.; Barbisan, L.F. Enhancement of Colon Carcinogenesis by the Combination of Indole-3 Carbinol and Synbiotics in Hemin-Fed Rats. *Food Chem. Toxicol.* **2018**, *112*, 11–18. [CrossRef]

90. de Moura, N.A.; Caetano, B.F.R.; Bidinotto, L.T.; Rodrigues, M.A.M.; Barbisan, L.F. Synbiotic Supplementation Attenuates the Promoting Effect of Indole-3-Carbinol on Colon Tumorigenesis. *Benef. Microbes* **2021**, *12*, 493–501. [CrossRef]

91. Wu, Y.; He, Q.; Yu, L.; Pham, Q.; Cheung, L.; Kim, Y.S.; Wang, T.T.Y.; Smith, A.D. Indole-3-Carbinol Inhibits Citrobacter Rodentium Infection through Multiple Pathways Including Reduction of Bacterial Adhesion and Enhancement of Cytotoxic T Cell Activity. *Nutrients* **2020**, *12*, 917. [CrossRef]

92. Tian, X.; Liu, K.; Zu, X.; Ma, F.; Li, Z.; Lee, M.; Chen, H.; Li, Y.; Zhao, Y.; Liu, F.; et al. 3,3′-Diindolylmethane Inhibits Patient-Derived Xenograft Colon Tumor Growth by Targeting COX1/2 and ERK1/2. *Cancer Lett.* **2019**, *448*, 20–30. [CrossRef]

93. Zhang, X.; Sukamporn, P.; Zhang, S.; Min, K.-W.; Baek, S.J. 3,3′-Diindolylmethane Downregulates Cyclin D1 through Triggering Endoplasmic Reticulum Stress in Colorectal Cancer Cells. *Oncol. Rep.* **2017**, *38*, 569–574. [CrossRef] [PubMed]

94. Lerner, A.; Grafi-Cohen, M.; Napso, T.; Azzam, N.; Fares, F. The Indolic Diet-Derivative, 3,3′-Diindolylmethane, Induced Apoptosis in Human Colon Cancer Cells through Upregulation of NDRG1. *BioMed. Res. Int.* **2011**, *2012*, e256178. [CrossRef] [PubMed]

95. Lee, S.-H.; Min, K.-W.; Zhang, X.; Baek, S.J. 3,3′-Diindolylmethane Induces Activating Transcription Factor 3 (ATF3) via ATF4 in Human Colorectal Cancer Cells. *J. Nutr. Biochem.* **2013**, *24*, 664–671. [CrossRef]

96. Wang, D.; Neupane, P.; Ragnarsson, L.; Capon, R.J.; Lewis, R.J. Diindolylmethane Derivatives: New Selective Blockers for T-Type Calcium Channels. *Membranes* **2022**, *12*, 749. [CrossRef] [PubMed]

97. Zhang, J.; Zou, S.; Zhang, Y.; Yang, Z.; Wang, W.; Meng, M.; Feng, J.; Zhang, P.; Xiao, L.; Lee, M.-H.; et al. 3,3′-Diindolylmethane Enhances Fluorouracil Sensitivity via Inhibition of Pyrimidine Metabolism in Colorectal Cancer. *Metabolites* **2022**, *12*, 410. [CrossRef] [PubMed]

98. Gruhlke, M.C.H.; Nicco, C.; Batteux, F.; Slusarenko, A.J. The Effects of Allicin, a Reactive Sulfur Species from Garlic, on a Selection of Mammalian Cell Lines. *Antioxidants* **2017**, *6*, 1. [CrossRef] [PubMed]

99. Li, X.; Ni, J.; Tang, Y.; Wang, X.; Tang, H.; Li, H.; Zhang, S.; Shen, X. Allicin Inhibits Mouse Colorectal Tumorigenesis through Suppressing the Activation of STAT3 Signaling Pathway. *Nat. Prod. Res.* **2019**, *33*, 2722–2725. [CrossRef]

100. Ţigu, A.B.; Toma, V.-A.; Moţ, A.C.; Jurj, A.; Moldovan, C.S.; Fischer-Fodor, E.; Berindan-Neagoe, I.; Pârvu, M. The Synergistic Antitumor Effect of 5-Fluorouracil Combined with Allicin against Lung and Colorectal Carcinoma Cells. *Molecules* **2020**, *25*, 1947. [CrossRef]

101. Huang, W.; Wu, S.; Xu, S.; Ma, Y.; Wang, R.; Jin, S.; Zhou, S. Allicin Enhances the Radiosensitivity of Colorectal Cancer Cells via Inhibition of NF-KB Signaling Pathway. *J. Food Sci.* **2020**, *85*, 1924–1931. [CrossRef]

102. Roy, N.; Nazeem, P.A.; Babu, T.D.; Abida, P.S.; Narayanankutty, A.; Valsalan, R.; Valsala, P.A.; Raghavamenon, A.C. EGFR Gene Regulation in Colorectal Cancer Cells by Garlic Phytocompounds with Special Emphasis on S-Allyl-L-Cysteine Sulfoxide. *Interdiscip. Sci.* **2018**, *10*, 686–693. [CrossRef]

103. Hatono, S.; Jimenez, A.; Wargovich, M.J. Chemopreventive Effect of S-Allylcysteine and Its Relationship to the Detoxification Enzyme Glutathione S-Transferase. *Carcinogenesis* **1996**, *17*, 1041–1044. [CrossRef]

104. Wargovich, M.J.; Jimenez, A.; McKee, K.; Steele, V.E.; Velasco, M.; Woods, J.; Price, R.; Gray, K.; Kelloff, G.J. Efficacy of Potential Chemopreventive Agents on Rat Colon Aberrant Crypt Formation and Progression. *Carcinogenesis* **2000**, *21*, 1149–1155. [CrossRef]

105. Sumiyoshi, H.; Wargovich, M.J. Chemoprevention of 1,2-Dimethylhydrazine-Induced Colon Cancer in Mice by Naturally Occurring Organosulfur Compounds1. *Cancer Res.* **1990**, *50*, 5084–5087.

106. Sriram, N.; Kalayarasan, S.; Ashokkumar, P.; Sureshkumar, A.; Sudhandiran, G. Diallyl Sulfide Induces Apoptosis in Colo 320 DM Human Colon Cancer Cells: Involvement of Caspase-3, NF-KB, and ERK-2. *Mol. Cell Biochem.* **2008**, *311*, 157–165. [CrossRef]

107. Lai, K.-C.; Hsu, S.-C.; Kuo, C.-L.; Yang, J.-S.; Ma, C.-Y.; Lu, H.-F.; Tang, N.-Y.; Hsia, T.-C.; Ho, H.-C.; Chung, J.-G. Diallyl Sulfide, Diallyl Disulfide, and Diallyl Trisulfide Inhibit Migration and Invasion in Human Colon Cancer Colo 205 Cells through the Inhibition of Matrix Metalloproteinase-2, -7, and -9 Expressions. *Env. Toxicol.* **2013**, *28*, 479–488. [CrossRef]

108. Lai, K.-C.; Kuo, C.-L.; Ho, H.-C.; Yang, J.-S.; Ma, C.-Y.; Lu, H.-F.; Huang, H.-Y.; Chueh, F.-S.; Yu, C.-C.; Chung, J.-G. Diallyl Sulfide, Diallyl Disulfide and Diallyl Trisulfide Affect Drug Resistant Gene Expression in Colo 205 Human Colon Cancer Cells in Vitro and in Vivo. *Phytomedicine* **2012**, *19*, 625–630. [CrossRef]

109. Kang, J.-S.; Kim, T.-M.; Shim, T.-J.; Salim, E.I.; Han, B.-S.; Kim, D.-J. Modifying Effect of Diallyl Sulfide on Colon Carcinogenesis in C57BL/6J-Apc^Min/+ Mice. *Asian Pac. J. Cancer Prev.* **2012**, *13*, 1115–1118. [CrossRef] [PubMed]

110. Xia, L.; Lin, J.; Su, J.; Oyang, L.; Wang, H.; Tan, S.; Tang, Y.; Chen, X.; Liu, W.; Luo, X.; et al. Diallyl Disulfide Inhibits Colon Cancer Metastasis by Suppressing Rac1-Mediated Epithelial-Mesenchymal Transition. *Onco Targets Ther.* **2019**, *12*, 5713–5728. [CrossRef] [PubMed]

111. Kim, H.J.; Kang, S.; Kim, D.Y.; You, S.; Park, D.; Oh, S.C.; Lee, D.-H. Diallyl Disulfide (DADS) Boosts TRAIL-Mediated Apoptosis in Colorectal Cancer Cells by Inhibiting Bcl-2. *Food Chem. Toxicol.* **2019**, *125*, 354–360. [CrossRef] [PubMed]

112. Saud, S.M.; Li, W.; Gray, Z.; Matter, M.S.; Colburn, N.H.; Young, M.R.; Kim, Y.S. Diallyl Disulfide (DADS), a Constituent of Garlic, Inactivates NF-KB and Prevents Colitis-Induced Colorectal Cancer by Inhibiting GSK-3β. *Cancer Prev. Res.* **2016**, *9*, 607–615. [CrossRef] [PubMed]

113. Chen, C.-Y.; Huang, C.-F.; Tseng, Y.-T.; Kuo, S.-Y. Diallyl Disulfide Induces Ca2+ Mobilization in Human Colon Cancer Cell Line SW480. *Arch. Toxicol.* **2012**, *86*, 231–238. [CrossRef]

114. Su, J.; Zhou, Y.; Pan, Z.; Shi, L.; Yang, J.; Liao, A.; Liao, Q.; Su, Q. Downregulation of LIMK1–ADF/Cofilin by DADS Inhibits the Migration and Invasion of Colon Cancer. *Sci. Rep.* **2017**, *7*, 45624. [CrossRef]

115. Alrumaihi, F.; Khan, M.A.; Babiker, A.Y.; Alsaweed, M.; Azam, F.; Allemailem, K.S.; Almatroudi, A.A.; Ahamad, S.R.; AlSuhaymi, N.; Alsugoor, M.H.; et al. The Effect of Liposomal Diallyl Disulfide and Oxaliplatin on Proliferation of Colorectal Cancer Cells: In Vitro and In Silico Analysis. *Pharmaceutics* **2022**, *14*, 236. [CrossRef]

116. Saraf, A.; Dubey, N.; Dubey, N.; Sharma, M. Enhancement of Cytotoxicty of Diallyl Disulfide toward Colon Cancer by Eudragit S100/PLGA Nanoparticles. *J. Drug Deliv. Sci. Technol.* **2021**, *64*, 102580. [CrossRef]

117. Zhang, Q.; Li, X.-T.; Chen, Y.; Chen, J.-Q.; Zhu, J.-Y.; Meng, Y.; Wang, X.-Q.; Li, Y.; Geng, S.-S.; Xie, C.-F.; et al. Wnt/β-Catenin Signaling Mediates the Suppressive Effects of Diallyl Trisulfide on Colorectal Cancer Stem Cells. *Cancer Chemother. Pharmacol.* **2018**, *81*, 969–977. [CrossRef]

118. Yu, C.-S.; Huang, A.-C.; Lai, K.-C.; Huang, Y.-P.; Lin, M.-W.; Yang, J.-S.; Chung, J.-G. Diallyl Trisulfide Induces Apoptosis in Human Primary Colorectal Cancer Cells. *Oncol. Rep.* **2012**, *28*, 949–954. [CrossRef]

119. Garcia-Mayea, Y.; Mir, C.; Masson, F.; Paciucci, R.; LLeonart, M.E. Insights into New Mechanisms and Models of Cancer Stem Cell Multidrug Resistance. *Semin. Cancer Biol.* **2020**, *60*, 166–180. [CrossRef]

120. Wu, P.-P.; Liu, K.-C.; Huang, W.-W.; Chueh, F.-S.; Ko, Y.-C.; Chiu, T.-H.; Lin, J.-P.; Kuo, J.-H.; Yang, J.-S.; Chung, J.-G. Diallyl Trisulfide (DATS) Inhibits Mouse Colon Tumor in Mouse CT-26 Cells Allograft Model in Vivo. *Phytomedicine* **2011**, *18*, 672–676. [CrossRef] [PubMed]

121. Yagdi, E.E.; Mazumder, A.; Lee, J.-Y.; Gaigneaux, A.; Radogna, F.; Nasim, M.J.; Christov, C.; Jacob, C.; Kim, K.-W.; Dicato, M.; et al. Tubulin-Binding Anticancer Polysulfides Induce Cell Death via Mitotic Arrest and Autophagic Interference in Colorectal Cancer. *Cancer Lett.* **2017**, *410*, 139–157. [CrossRef] [PubMed]

122. Li, H.; Jeong, J.H.; Kwon, S.W.; Lee, S.K.; Lee, H.J.; Ryu, J.-H. Z-Ajoene Inhibits Growth of Colon Cancer by Promotion of CK1α Dependent β-Catenin Phosphorylation. *Molecules* **2020**, *25*, 703. [CrossRef] [PubMed]

123. Lv, Y.; So, K.F.; Wong, N.K.; Xiao, J. Anti-Cancer Activities of S-Allylmercaptocysteine from Aged Garlic. *Chin. J. Nat. Med.* **2019**, *17*, 43–49. [CrossRef]

124. Li, S.; Guang, Y.; Zhu, X.; Lin, C.; Sun, Y.; Zhao, Z. Combination of Rapamycin and Garlic-Derived S-Allylmercaptocysteine Induces Colon Cancer Cell Apoptosis and Suppresses Tumor Growth in Xenograft Nude Mice through Autophagy/P62/Nrf2 Pathway. *Oncol. Rep.* **2017**, *38*, 1637–1644. [CrossRef]

125. Zhang, Y.; Li, H.Y.; Zhang, Z.H.; Bian, H.L.; Lin, G. Garlic-Derived Compound S-Allylmercaptocysteine Inhibits Cell Growth and Induces Apoptosis via the JNK and P38 Pathways in Human Colorectal Carcinoma Cells. *Oncol. Lett.* **2014**, *8*, 2591–2596. [CrossRef]

126. Liang, D.; Qin, Y.; Zhao, W.; Zhai, X.; Guo, Z.; Wang, R.; Tong, L.; Lin, L.; Chen, H.; Wong, Y.C.; et al. S-Allylmercaptocysteine Effectively Inhibits the Proliferation of Colorectal Cancer Cells under in Vitro and in Vivo Conditions. *Cancer Lett.* **2011**, *310*, 69–76. [CrossRef]

127. Tong, D.; Qu, H.; Meng, X.; Jiang, Y.; Liu, D.; Ye, S.; Chen, H.; Jin, Y.; Fu, S.; Geng, J. S-Allylmercaptocysteine Promotes MAPK Inhibitor-Induced Apoptosis by Activating the TGF-β Signaling Pathway in Cancer Cells. *Oncol. Rep.* **2014**, *32*, 1124–1132. [CrossRef] [PubMed]

128. Laukens, D.; Brinkman, B.M.; Raes, J.; De Vos, M.; Vandenabeele, P. Heterogeneity of the Gut Microbiome in Mice: Guidelines for Optimizing Experimental Design. *FEMS Microbiol. Rev.* **2016**, *40*, 117–132. [CrossRef]

129. Kodera, Y.; Suzuki, A.; Imada, O.; Kasuga, S.; Sumioka, I.; Kanezawa, A.; Taru, N.; Fujikawa, M.; Nagae, S.; Masamoto, K.; et al. Physical, Chemical, and Biological Properties of s-Allylcysteine, an Amino Acid Derived from Garlic. *J. Agric. Food Chem.* **2002**, *50*, 622–632. [CrossRef] [PubMed]

130. Yoshimoto, N.; Saito, K. S-Alk(En)Ylcysteine Sulfoxides in the Genus Allium: Proposed Biosynthesis, Chemical Conversion, and Bioactivities. *J. Exp. Bot.* **2019**, *70*, 4123–4137. [CrossRef]

131. Borlinghaus, J.; Albrecht, F.; Gruhlke, M.C.H.; Nwachukwu, I.D.; Slusarenko, A.J. Allicin: Chemistry and Biological Properties. *Molecules* **2014**, *19*, 12591–12618. [CrossRef]

132. Yamaguchi, Y.; Kumagai, H. Characteristics, Biosynthesis, Decomposition, Metabolism and Functions of the Garlic Odour Precursor, S-Allyl-L-Cysteine Sulfoxide. *Exp. Ther. Med.* **2020**, *19*, 1528–1535. [CrossRef] [PubMed]

133. Rao, P.S.S.; Midde, N.M.; Miller, D.D.; Chauhan, S.; Kumar, A.; Kumar, S. Diallyl Sulfide: Potential Use in Novel Therapeutic Interventions in Alcohol, Drugs, and Disease Mediated Cellular Toxicity by Targeting Cytochrome P450 2E1. *Curr. Drug Metab.* **2015**, *16*, 486–503. [CrossRef]

134. Puccinelli, M.T.; Stan, S.D. Dietary Bioactive Diallyl Trisulfide in Cancer Prevention and Treatment. *Int. J. Mol. Sci.* **2017**, *18*, 1645. [CrossRef] [PubMed]

135. Matsutomo, T.; Kodera, Y. Development of an Analytic Method for Sulfur Compounds in Aged Garlic Extract with the Use of a Postcolumn High Performance Liquid Chromatography Method with Sulfur-Specific Detection. *J. Nutr.* **2016**, *146*, 450S–455S. [CrossRef] [PubMed]

136. Wu, X.; Shi, J.; Fang, W.X.; Guo, X.Y.; Zhang, L.Y.; Liu, Y.P.; Li, Z. Allium Vegetables Are Associated with Reduced Risk of Colorectal Cancer: A Hospital-Based Matched Case-Control Study in China. *Asia Pac. J. Clin. Oncol.* **2019**, *15*, e132–e141. [CrossRef]

137. Zhou, X.; Qian, H.; Zhang, D.; Zeng, L. Garlic Intake and the Risk of Colorectal Cancer: A Meta-Analysis. *Medicine* **2020**, *99*, e18575. [CrossRef]

138. Zhu, B.; Zou, L.; Qi, L.; Zhong, R.; Miao, X. Allium Vegetables and Garlic Supplements Do Not Reduce Risk of Colorectal Cancer, Based on Meta-Analysis of Prospective Studies. *Clin. Gastroenterol. Hepatol.* **2014**, *12*, 1991–2001.e4. [CrossRef]

139. Satia, J.A.; Littman, A.; Slatore, C.G.; Galanko, J.A.; White, E. Associations of Herbal and Specialty Supplements with Lung and Colorectal Cancer Risk in the VITamins and Lifestyle Study. *Cancer Epidemiol. Biomark. Prev.* **2009**, *18*, 1419–1428. [CrossRef]

140. Cascajosa-Lira, A.; Andreo-Martínez, P.; Prieto, A.I.; Baños, A.; Guillamón, E.; Jos, A.; Cameán, A.M. In Vitro Toxicity Studies of Bioactive Organosulfur Compounds from *Allium* spp. with Potential Application in the Agri-Food Industry: A Review. *Foods* **2022**, *11*, 2620. [CrossRef] [PubMed]

141. Bagul, M.; Kakumanu, S.; Wilson, T.A. Crude Garlic Extract Inhibits Cell Proliferation and Induces Cell Cycle Arrest and Apoptosis of Cancer Cells in Vitro. *J. Med. Food* **2015**, *18*, 731–737. [CrossRef]

142. Su, C.-C.; Chen, G.W.; Tan, T.-W.; Lin, J.-G.; Chung, J.-G. Crude Extract of Garlic Induced Caspase-3 Gene Expression Leading to Apoptosis in Human Colon Cancer Cells. *Int. Inst. Anticancer Res.* **2006**, *20*, 85–90.

143. Ghazanfari, T.; Yaraee, R.; Rahmati, B.; Hakimzadeh, H.; Shams, J.; Jalali-Nadoushan, M.R. In Vitro Cytotoxic Effect of Garlic Extract on Malignant and Nonmalignant Cell Lines. *Immunopharmacol. Immunotoxicol.* **2011**, *33*, 603–608. [CrossRef]

144. Sut, S.; Maggi, F.; Bruno, S.; Badalamenti, N.; Quassinti, L.; Bramucci, M.; Beghelli, D.; Lupidi, G.; Dall'Acqua, S. Hairy Garlic (*Allium subhirsutum*) from Sicily (Italy): LC-DAD-MSn Analysis of Secondary Metabolites and in Vitro Biological Properties. *Molecules* **2020**, *25*, 2837. [CrossRef]

145. Kim, H.-J.; Park, M.J.; Park, H.-J.; Chung, W.-Y.; Kim, K.-R.; Park, K.-K. Chemopreventive and Anticancer Activities of *Allium victorialis* var. *platyphyllum* Extracts. *J. Cancer Prev.* **2014**, *19*, 179–186. [CrossRef]
146. Perez-Ortiz, J.M.; Galan-Moya, E.M.; de la Cruz-Morcillo, M.A.; Rodriguez, J.F.; Gracia, I.; Garcia, M.T.; Redondo-Calvo, F.J. Cost Effective Use of a Thiosulfinate-Enriched *Allium sativum* Extract in Combination with Chemotherapy in Colon Cancer. *Int. J. Mol. Sci.* **2020**, *21*, 2766. [CrossRef] [PubMed]
147. Miraghajani, M.; Rafie, N.; Hajianfar, H.; Larijani, B.; Azadbakht, L. Aged Garlic and Cancer: A Systematic Review. *Int. J. Prev. Med.* **2018**, *9*, 84. [PubMed]
148. Dong, M.; Yang, G.; Liu, H.; Liu, X.; Lin, S.; Sun, D.; Wang, Y. Aged Black Garlic Extract Inhibits HT29 Colon Cancer Cell Growth via the PI3K/Akt Signaling Pathway. *Biomed. Rep.* **2014**, *2*, 250–254. [CrossRef] [PubMed]
149. Matsuura, N.; Miyamae, Y.; Yamane, K.; Nagao, Y.; Hamada, Y.; Kawaguchi, N.; Katsuki, T.; Hirata, K.; Sumi, S.-I.; Ishikawa, H. Significance of Garlic and Its Constituents in Cancer and Cardiovascular Disease. *J. Nutr.* **2006**, *136*, 716–725.
150. Katsuki, T.; Hirata, K.; Ishikawa, H.; Matsuura, N.; Sumi, S.-I.; Itoh, H. Significance of Garlic and Its Constituents in Cancer and Cardiovascular Disease Aged Garlic Extract Has Chemopreventative Effects On 1,2-Dimethylhydrazine-Induced Colon Tumors in Rats. *J. Nutr.* **2006**, *136*, 847S–851S. [CrossRef] [PubMed]
151. Jikihara, H.; Qi, G.; Nozoe, K.; Hirokawa, M.; Sato, H.; Sugihara, Y.; Shimamoto, F. Aged Garlic Extract Inhibits 1,2-Dimethylhydrazine-Induced Colon Tumor Development by Suppressing Cell Proliferation. *Oncol. Rep.* **2015**, *33*, 1131–1140. [CrossRef] [PubMed]
152. Tanaka, S.; Haruma, K.; Yoshihara, M.; Kajiyama, G.; Kira, K.; Amagase, H.; Chayama, K. Significance of Garlic and Its Constituents in Cancer and Cardiovascular Disease Aged Garlic Extract Has Potential Suppressive Effect on Colorectal Adenomas in Humans. *J. Nutr.* **2006**, *136*, 821S–826S. [CrossRef]
153. Touihri, I.; Boukhris, M.; Marrakchi, N.; Luis, J.; Hanchi, B.; Kallech-Ziri, O. Chemical Composition and Biological Activities of *Allium roseum* L. var. Grandiflorum Briq. Essential Oil. *J. Oleo Sci.* **2015**, *64*, ess15055. [CrossRef]
154. Butawan, M.; Benjamin, R.L.; Bloomer, R.J. Methylsulfonylmethane: Applications and Safety of a Novel Dietary Supplement. *Nutrients* **2017**, *9*, 290. [CrossRef] [PubMed]

Review

Therapeutic Effects of Natural Products on Liver Cancer and Their Potential Mechanisms

Jinhong Guo [1], Wenjie Yan [1], Hao Duan [1], Diandian Wang [1], Yaxi Zhou [1], Duo Feng [2], Yue Zheng [3], Shiqi Zhou [1], Gaigai Liu [1] and Xia Qin [4,*]

[1] Beijing Key Laboratory of Bioactive Substances and Functional Food, Beijing Union University, Beijing 100023, China; gjh121133@163.com (J.G.); meyanwenjie@buu.edu.cn (W.Y.); dhuanao@163.com (H.D.); spwdd2018@163.com (D.W.); 15239407080@163.com (Y.Z.); shiqizhougood@163.com (S.Z.); lgg010403@163.com (G.L.)

[2] Institute of Food and Nutrition Development, Ministry of Agriculture and Rural Affairs, Beijing 100081, China; 15525926785@163.com

[3] College of Food Science and Engineering, Northwest A&F University, Yangling 712100, China; zheng_yue21@163.com

[4] Graduate Department, Beijing Union University, Beijing 100101, China

* Correspondence: qinxia@buu.edu.cn; Tel.: +86-15901251686

Abstract: Liver cancer ranks third globally among causes of cancer-related deaths, posing a significant public health challenge. However, current treatments are inadequate, prompting a growing demand for novel, safe, and effective therapies. Natural products (NPs) have emerged as promising candidates in drug development due to their diverse biological activities, low toxicity, and minimal side effects. This paper begins by reviewing existing treatment methods and drugs for liver cancer. It then summarizes the therapeutic effects of NPs sourced from various origins on liver cancer. Finally, we analyze the potential mechanisms of NPs in treating liver cancer, including inhibition of angiogenesis, migration, and invasion; regulation of the cell cycle; induction of apoptosis, autophagy, pyroptosis, and ferroptosis; influence on tumor metabolism; immune regulation; regulation of intestinal function; and regulation of key signaling pathways. This systematic review aims to provide a comprehensive overview of NPs research in liver cancer treatment, offering a foundation for further development and application in pharmaceuticals and functional foods.

Keywords: natural products; liver cancer; therapeutic drugs; mechanisms of action

Citation: Guo, J.; Yan, W.; Duan, H.; Wang, D.; Zhou, Y.; Feng, D.; Zheng, Y.; Zhou, S.; Liu, G.; Qin, X. Therapeutic Effects of Natural Products on Liver Cancer and Their Potential Mechanisms. *Nutrients* **2024**, *16*, 1642. https://doi.org/10.3390/nu16111642

Academic Editor: Alessandro Granito

Received: 23 April 2024
Revised: 22 May 2024
Accepted: 24 May 2024
Published: 27 May 2024

1. Introduction

Liver cancer is the third most common cause of cancer-related deaths worldwide, and is categorized into primary and secondary liver cancer [1].

Primary liver cancer encompasses hepatocellular carcinoma, cholangiocarcinoma, and mixed hepatocellular-cholangiocarcinoma, whereas secondary liver cancer, also known as hepatoblastoma, is comparatively less common. Projections suggest that by 2025, liver cancer incidence will surpass one million cases, with hepatocellular carcinoma (HCC) comprising 90% of diagnoses [2]. This alarming trend presents substantial economic and public health challenges. Consequently, there is a pressing demand to investigate more effective and cost-efficient treatment strategies.

Treatment options for liver cancer are varied, with surgical intervention remaining the primary approach. However, even after curative surgery, challenges such as cancer metastasis, high recurrence rates, and poor prognosis persist [3]. For patients with advanced HCC who are ineligible for transplantation or have failed local and regional therapies, chemotherapy drugs like sorafenib are often employed. However, their effectiveness is hindered by drug resistance and accompanied by numerous side effects [4,5]. Consequently, there is an urgent need to develop anticancer drugs that are more cost-effective, safer, and efficacious to address the limitations of current treatment modalities and medications.

Natural products (NPs) exhibit diverse biological activities, including anti-inflammatory and antioxidant properties, rendering them promising sources for developing novel therapies for liver cancer [6,7]. Extensive in vitro and in vivo studies have unveiled the anti-tumor potential of NPs against liver cancer. Examples of such NPs encompass natural polysaccharides, flavonoids, terpenes, alkaloids, polyphenols, quinones, and essential oils. These NPs hold the potential to ameliorate and combat liver cancer through multiple pathways.

This article commences with an overview of current treatment modalities and medications for liver cancer. Additionally, we conducted a comprehensive search across "Google Scholar", "Web of Science", and "X-Mol", yielding 181 articles pertaining to liver cancer and natural compounds from 2018 to 2024. The search utilized keyword combinations such as "liver cancer AND natural compounds" and "hepatic cancer AND natural compounds". Included studies met specific criteria: (1) animal in vivo experiments showcasing the anti-cancer effects of natural compounds on liver cancer; (2) cell-based in vitro experiments demonstrating the anti-cancer effects of natural compounds on liver cancer; (3) original research articles in English. Exclusions comprised retrospective studies and meta-analyses. Subsequently, the article evaluates the efficacy and mechanisms of action of natural products sourced from various origins against liver cancer. Its objective is to furnish a systematic synthesis of research on natural products for liver cancer treatment, furnishing a theoretical foundation for the development and application of such products in pharmaceuticals and functional foods.

2. Therapeutic Drugs and Methods of Liver Cancer Treatment

Liver cancer treatment has evolved into a multidisciplinary approach, typically customized to individual patients based on tumor staging, the intricate interaction with underlying liver disease, and the patient's overall health status [8]. Various treatment modalities are available depending on the stage of the tumor. Early-stage liver cancer patients are often candidates for surgical resection, transplantation, or local ablation. Intermediate-stage liver cancer, characterized by multifocal tumors, may be managed with transarterial chemoembolization (TACE). Late-stage liver cancer diagnoses are more common, offering a broader array of treatment options, including chemotherapy, immunotherapy, targeted drug therapy, and various other interventions [9,10].

2.1. Surgical Treatment

Surgical intervention, comprising hepatic resection and liver transplantation, has historically been the cornerstone of curative treatment for liver cancer, offering the most favorable outcomes with a 5-year survival rate of around 70–80%. Typically, candidates eligible for resection must have tumors situated in surgically accessible locations, adequate hepatic reserve, a sufficient volume of remaining liver, and undergo evaluation based on clinical and biochemical indicators or liver blood volume determination [11].

Patients with cirrhosis and limited tumor burden may opt for liver transplantation, which offers 5-year and 10-year survival rates of 70% and 50%, respectively, with a 5-year recurrence rate of 10–15% [12,13]. Liver transplantation's long-term outcomes are generally deemed superior to those of resection. However, the scarcity of liver donors poses a challenge, as patients awaiting transplantation also confront the risk of tumor progression.

2.2. Local (Ablative) Treatment

Local ablation therapy serves as a curative option for solitary liver tumors with a maximum diameter of ≤ 5 cm, as well as for multiple liver tumors with a total count of ≤ 3 and a maximum diameter of ≤ 3 cm. Ethanol injection, microwave ablation, radiofrequency ablation, and cryotherapy are among the common local ablation therapies employed for early to intermediate-stage liver cancer patients. Numerous studies have validated that the 5-year survival rate of early-stage small liver cancer treated with local ablation approaches is approximately 60%, exhibiting no statistical variance compared to surgical resection [14,15]. Hence, local ablation therapy is acknowledged as another treatment modality with the potential for localized cure of hepatocellular carcinoma (HCC), following surgical resection (inclusive of liver transplantation) [16].

2.3. Hepatic Artery Therapy

Transarterial chemoembolization (TACE) has gained widespread adoption globally as the standard treatment for intermediate-stage liver cancer patients [17,18]. TACE is capable of addressing a wider spectrum of tumors, ranging from individual small liver cancers that cannot be treated via ablation due to technical limitations to intermediate-sized nodular liver cancers. Presently, there are two primary types of TACE: conventional TACE (cTACE) and drug-eluting bead TACE (DEB-TACE). Radioembolization, also referred to as selective internal radiation therapy (SIRT), serves as an alternative treatment for intermediate-stage liver cancer patients who either do not respond to TACE or have contraindications. It effectively reduces large liver cancers beyond Milan criteria and can act as a bridge to transplantation [19].

2.4. Systemic Therapy-Chemotherapy

Systemic chemotherapy plays a pivotal role in liver cancer treatment, especially for patients ineligible for surgery, particularly those in advanced stages. Sorafenib and lenvatinib, both tyrosine kinase inhibitors, are commonly used as first-line treatment drugs for advanced liver cancer patients [20,21]. Second-line treatment options for advanced-stage liver cancer encompass regorafenib, cabozantinib, and ramucirumab [22–24]. However, due to the considerable resistance of hepatocellular carcinoma (HCC) and the potential hepatotoxicity of drugs in patients with underlying liver disease, there exists no current consensus on systemic chemotherapy regimens [25]. Medical oncologists need to consider factors such as liver reserve, treatment objectives, treatment availability, and eligibility for clinical trials when deliberating systemic chemotherapy. Moreover, patients with liver cancer frequently develop resistance to drug therapy, resulting in a poor prognosis and high rates of recurrence, with an overall recurrence rate as elevated as 70% [26].

2.5. Radiotherapy

Radiotherapy, commonly referred to as radiation therapy, is a local treatment method for tumors using radiation beams or radioactive isotopes, alongside surgery and chemotherapy, constituting the three main traditional cancer treatment methods. It offers a better option for treating liver cancer in intermediate- to advanced-stage patients either alone or in combination with other techniques, and for patients in the early stages who are not suitable for surgical resection. Currently, the primary modality of external radiotherapy is stereotactic body radiotherapy (SBRT) [27].

2.6. Immunotherapy

Immune therapy for liver cancer is an emerging treatment method that works by activating or enhancing the patient's own immune system to attack cancer cells. It mainly includes immune checkpoint inhibitors, adoptive cell therapy, tumor vaccines, oncolytic viruses, and non-specific immunotherapy [28]. Presently, immune checkpoint inhibitors are at the forefront of immune therapy, targeting critical checkpoints like programmed cell death protein 1 (PD-1) and its ligand PD-L1, along with cytotoxic T-lymphocyte-associated antigen 4 (CTLA-4). Moreover, combining immune therapy with targeted drugs used in liver cancer treatment can elicit a synergistic effect [26].

2.7. TCM (Traditional Chinese Medicine) Therapy

Currently, in clinical practice, anti-tumor drugs have shown limited efficacy in extending patient survival, and the outcomes of single-targeted therapy are also unsatisfactory. According to traditional Chinese medicine (TCM) theory, tumor diseases have diverse pathogenic factors, mainly characterized by deficiency syndrome with excess syndrome. TCM treatment emphasizes multi-targeting and holistic therapy, which can positively impact alleviating adverse reactions to chemotherapy, preventing tumor recurrence, and prolonging life [29]. In recent years, traditional Chinese medicine has demonstrated unique advantages in the prevention and treatment of primary liver cancer, significantly improving the quality of life, enhancing immunity, and prolonging life for liver cancer patients [30]. Figure 1 depicts the drugs and methods used to treat liver cancer and their limitations.

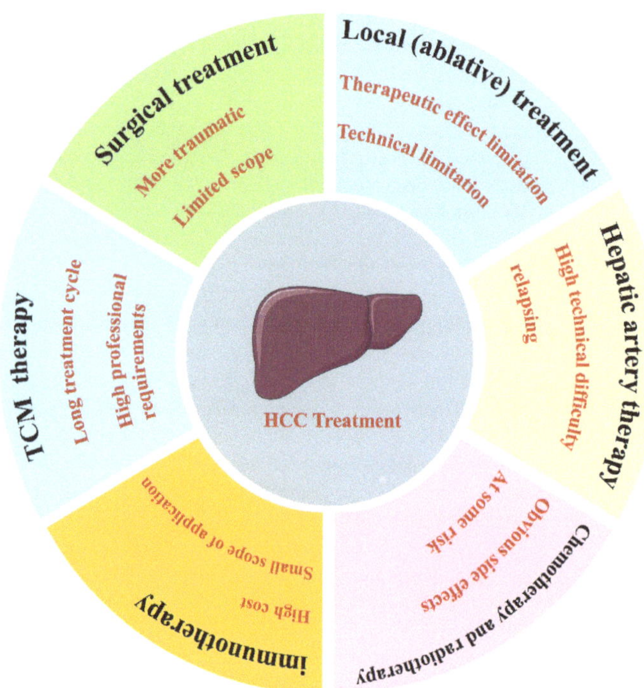

Figure 1. Current drugs and methods of treatment for liver cancer. Highlighting in red font indicates the limitations of the approach.

3. NPs with Therapeutic Activity for Liver Cancer

Based on the chemical structure and categories of compounds, we roughly classify NPs with therapeutic activity against liver cancer into seven specific categories: polysaccharides, flavonoids, terpenoids, alkaloids, polyphenols, quinones, and volatile oils. The therapeutic effects of each type of natural product on liver cancer, as well as their sources, doses, and mechanisms of action, will be detailed in the following sections.

3.1. Polysaccharides

Polysaccharides are biologically active macromolecules composed of 10 or more monomers linked by glycosidic bonds. Extensive research has highlighted the therapeutic and ameliorative effects of natural polysaccharides on liver cancer, sourced from plants, animals, marine organisms, and fungi [7]. For instance, polysaccharides purified from Panax ginseng (PHP-1) and active polysaccharides from Ganoderma lucidum (GLPS) function as tumor immunotherapy modulators. Both are capable of polarizing M2 macrophages into the M1 type by activating NF-κB and MAPK signaling pathways, thereby exerting anti-tumor activity [31,32].

Furthermore, high molecular weight polysaccharides extracted from Cordyceps sinensis not only enhance the immune function of mice but also induce tumor cell apoptosis through the Cytochrome-c/Caspase8/3 and IL-10/STAT3/Bcl2 pathways, thereby exerting anti-tumor effects [33]. In H22 hepatocellular carcinoma tumor-bearing mouse models,

Ulva lactuca polysaccharide (UIP) extracted from green algae inhibits tumor growth by modulating gut microbiota and metabolite composition. Additionally, UIP has been found to suppress tumors in HepG2 and H22 cells by promoting ROS production [34]. Researchers, utilizing metabolomics, have demonstrated that Black fungus polysaccharide (BFP) inhibits tumor cell proliferation in H22 tumor-bearing mice by promoting DNA damage, weakening DNA damage repair, and inhibiting DNA synthesis [35].

Based on a large number of studies on the improvement of liver cancer by natural polysaccharides, natural polysaccharides have shown very good effects on immune regulation. On one hand, polysaccharides directly inhibit tumor growth through pathways such as blocking the cell cycle, inducing cell apoptosis, and inhibiting angiogenesis. On the other hand, polysaccharides can also regulate the host immune system, indirectly exerting anti-tumor effects by stimulating non-specific and specific immune responses [36]. Through the aforementioned review, it is evident that natural polysaccharides primarily exert their beneficial effects on liver cancer by inducing cell apoptosis, inducing cell cycle arrest, inhibiting angiogenesis, suppressing cancer cell invasion and metastasis, and modulating immune responses. In Table 1, we have compiled findings from 18 recently published studies elucidating the therapeutic effects of natural polysaccharides on liver cancer and their potential mechanisms of action.

Table 1. Therapeutic effects and mechanisms of natural polysaccharides on liver cancer.

No.	Name	Origin	In Vitro (a) or In Vivo (b)	Optimal Doses (/kg Body Weight)	Model	Potential Mechanism	References
1	Wild Cordyceps polysaccharides	*Cordyceps sinensis*	b	100 mg	H22 tumor-bearing BALB/c mice	Modulates immunity. Modulation of IL-10/STAT3/Bcl2 and Cytoc/Caspase8/3 signaling pathways promotes apoptosis.	[33]
2	Low molecular weight polysaccharide from Grifola frondosa	*Grifola frondosa*	b	200 mg	H22 tumor-bearing BALB/c mice	Modulates immune activity; promotes tumor cell apoptosis via the mitochondrial apoptotic pathway.	[37]
3	G. frondosa polysaccharide	*G. frondosa*	b	200 mg	H22 tumor-bearing mice	Enhances immunity and induces cell cycle arrest at G0/G1 phase.	[37]
4	Pleurotus citrinopileatus polysaccharides	*Pleurotus citrinopileatus*	b	300 mg	H22 tumor-bearing mice	Enhances immunity and induces cell cycle arrest in S phase.	[38]
5	Ganoderma lucidum polysaccharide	*Ganoderma lucidum*	a + b	200 mg	Hepatic carcinoma Hepa1-6 allograft mice; RAW 264.7 and Hepa1-6 co-culture	Regulates macrophage polarization through activating MAPK/NF-κB signaling.	[39]

Table 1. *Cont.*

No.	Name	Origin	In Vitro (a) or In Vivo (b)	Optimal Doses (/kg Body Weight)	Model	Potential Mechanism	References
6	Bletilla striata polysaccharides	*Bletilla striata*	b	200 mg	H22 tumor-bearing BALB/c mice	Modulates immunity and induces cell cycle arrest at G1 phase.	[40]
7	Acetylaminoglucan	/	b	50 mg	H22 tumor-bearing BALB/c mice	Modulates immunity and induces cell cycle arrest in S phase.	[41]
8	Bupleurum chinense DC root polysaccharides	*Bupleurum chinense* DC root	b	300 mg	H22 tumor-bearing Kunming mice	Induces S-phase block of the cell cycle and activates the mitochondrial pathway to induce apoptosis.	[42]
9	Angelica dahurica polysaccharide	*Angelica dahurica*	b	300 mg	H22 tumor-bearing BALB/c mice	Induces apoptosis by cell cycle arrest in G0/G1 phase and reduction of cellular mitochondrial membrane potential.	[43]
10	Rhodiola rosea L. root polysaccharide	*Rhodiola rosea* L. root	b	300 mg	H22 tumor-bearing mice	Induces cell cycle S-phase block by disrupting mitochondrial membrane potential and inducing apoptosis in tumor cells.	[44]
11	Dandelion polysaccharides	*Dandelion*	a + b	400 mg	H22 tumor-bearing BALB/c mice; HepG2, Huh7, and Hepa1-6 cells	Down-regulates PI3K/AKT/mTOR pathway, inhibits cell proliferation and apoptosis, and enhances immune response.	[45]
12	Dandelion polysaccharide	*Dandelion*	a + b	200 mg	H22 tumor-bearing BALB/c mice; HepG2, Huh7, Hepa1-6, and H22 cells	Inhibits IL-6-activated JAK-STAT signaling pathway; reduces hepcidin.	[46]
13	Eucommia folium polysaccharide	*Eucommia folium*	b	300 mg	H22 tumor-bearing Kunming mice	Induces S-phase block of the cell cycle and apoptosis via the mitochondrial pathway.	[47]
14	Pseudostellaria heterophylla polysaccharide	*Pseudostellaria heterophylla*	a + b	50 mg	H22 tumor-bearing C57BL/6 mice; RAW264.7 and Huh-7 cells	Regulates macrophage polarization through activating MAPK/NF-κB signaling.	[31]

Table 1. *Cont.*

No.	Name	Origin	In Vitro (a) or In Vivo (b)	Optimal Doses (/kg Body Weight)	Model	Potential Mechanism	References
15	Black fungus polysaccharide	*Black fungus*	a + b	5 mg	HCCLM3 tumor-bearing BALB/c mice; HepG2, HCCLM3, and SK-Hep1 cells	Suppresses tumor cell proliferation by promoting DNA damage, attenuating DNA damage repair, and inhibiting DNA synthesis.	[35]
16	Fucoidan	*Brown algae*	a + b	15 mg	MHCI297H tumor-bearing BALB/c mice; MHCC-97H and Hep3B cells	Causes lncRNA Linc00261 overexpression.	[48]
17	Ulva lactuca polysaccharide	*Ulva lactuca*	a + b	300 mg	H22 liver cancer tumor-bearing mice; HepG2 and H22 cells	Inhibits tumor growth through modulation of gut microbial community and metabolism and modulation of miR-98-5p/ROS signaling pathway.	[34]
18	Sipunculus nudus polysaccharide	*Sipunculus nudus*	b	200 mg	HepG2 tumor-bearing athymic nu/nu mice	Enhances immunity and induces apoptosis in tumor cells via the mitochondrial apoptotic pathway.	[49]

MAPK, mitogen-activated protein kinase; NF-κB, nuclear factor kappa-B; PI3K, phosphatidylinositide 3-kinase; AKT, protein kinase B; mTOR, mammalian target of rapamycin; IL-6, interleukin 6; DNA, deoxyribonucleic acid; lncRNA, long non-coding RNAs.

3.2. Flavonoids

Flavonoid compounds are commonly found in various dietary components, including vegetables, fruits, wine, flowers, tea, and other plant-based food sources. They are celebrated for their myriad health benefits, encompassing anti-cancer, cardio-protective, anti-inflammatory, antimicrobial, antioxidant, and hepatoprotective properties. Extensive research underscores the therapeutic and preventive potential of plant-derived flavonoids against liver cancer. For instance, flavonoids isolated from Resina Draconis inhibit the progression of liver cancer by upregulating p21 expression driven by DNA damage, inducing cell apoptosis, and causing cell cycle G2/M phase arrest in human liver cancer cells [50]. Neobavaisoflavone (NBIF), a natural active ingredient derived from psoralen, exhibits anti-inflammatory, anti-cancer, and antioxidant properties. Researchers have found that NBIF induces mitochondrial outer membrane protein Tom20 alteration through ROS both in vivo and in vitro, thereby enhancing Bax recruitment to mitochondria and activating the calpain I-GSDME pathway in liver cancer cells to induce pyroptosis. Sinensetin (SIN), a polymethoxylated flavone found in citrus fruits, inhibits angiogenesis in liver cancer by targeting the VEGF/VEGFR2/AKT signaling pathway in a murine xenograft tumor model [51]. Studies indicate that extramedullary hematopoiesis (EMH) is a crucial mechanism for myeloid-derived suppressor cells (MDSCs) accumulation in tumor tissues, contributing to disease progression. Icaritin dampens tumoral immunosuppression, eliciting anti-tumor immune responses by preventing MDSC generation via EMH attenuation [52].

In summary, flavonoid compounds inhibit the progression of liver cancer by inducing cell apoptosis, blocking the cell cycle, inhibiting angiogenesis, and exerting antioxidant effects, consistent with previous research findings [53]. Additionally, flavonoid compounds also exhibit anti-cancer efficacy by inducing cell pyroptosis and modulating immune responses. This article summarizes 18 flavonoid compounds with therapeutic effects on liver cancer from the recent literature. Details regarding their sources, doses, and mechanisms of action are provided in Table 2.

Table 2. Therapeutic effects and mechanisms of flavonoids on liver cancer.

No.	Name	Origin	In Vitro (a) or In Vivo (b)	Optimal Doses (/kg Body Weight)	Model	Potential Mechanism	References
1	(R)-7,3′-dihydroxy-4′-methoxy-8-methylflavane	*Resina Draconis*	a + b	20 mg	HepG2 tumor-bearing BALB/c mice and H22 tumor-bearing Kunming mice; HepG2 and SK-HEP-1 cells	Induction of apoptosis and G2/M phase block by DNA damage-driven upregulation of p21 expression in human hepatocellular carcinoma cells.	[50]
2	Neobavaisoflavone	*Psoralea*	a + b	30 mg	HCCLM3 tumor-bearing BALB/c mice; HepG2 and HCCLM3 cells	Induction of cellular pyroptosis via the tom20/bax/caspase3/GSDME pathway.	[54]
3	Isorhamnetin	/	b	100 mg	DEN + CCl4-induced HCC mice	Suppression of inflammation; regulates Akt and MAPKs to inhibit Nrf2 signaling; activates PPAR-γ and autophagy.	[55]
4	Sinensetin	citrus fruits.	b	40 mg	HepG2/C3A tumor-bearing BALB/c nude mice	Inhibition of angiogenesis in hepatocellular carcinoma by regulating VEGF/VEGFR2/AKT signaling.	[51]
5	Icaritin	plants of the Epimedium genus	b	70 mg	H22 tumor-bearing BALB/c mice	Prevention of MDSC generation via the attenuation of EMH.	[52]
6	Apigenin	parsley and chamomile	a + b	50 mg	Huh7 tumor-bearing BALB/c mice; Hep3B cells	Downregulation of H19 induces inactivation of the Wnt/β-catenin signaling pathway.	[56]

Table 2. *Cont.*

No.	Name	Origin	In Vitro (a) or In Vivo (b)	Optimal Doses (/kg Body Weight)	Model	Potential Mechanism	References
7	Lysionotin	*Lysionotus pauciflorus* Maxim	a + b	20 mg	HepG2 nude mice; HepG2 and SMMC-7721 cells	Induction of apoptosis in hepatocellular carcinoma cells via caspase-3-mediated mitochondrial pathway.	[57]
8	Kaempferide	*Mountain apple* root	a + b	25 mg	N1S1 orthotopically injected SD rats; HepG2, Huh7, and N1S1 cells	Induction of apoptosis.	[58]
9	Wogonin	*Scutellaria baicalensis*	a + b	50 mg	Orthotopically HCC-implantation mice; HepG2 cells	Inhibition of cell cycle progression by activating the glycogen synthase kinase-3 beta.	[59]
10	Total flavonoids	*Oldenlandia diffusa*	a + b	0.4 mg	H22 tumor-bearing BALB/c mice; HepG2, Hep3B, and HCCLM3 H22 cells	Induction of apoptosis and autophagy of HCC cells by inducing endoplasmic reticulum (ER) stress response and activating PERK-eIF2α-ATF4 signaling pathway.	[60]
11	Isoliquiritigenin	roots of plants belonging to licorice	a + b	50 mg	SMMC7721 tumor-bearing BALB/c mice; MHCC97-H and SMMC7721 cells	Downregulation of PI3K/AKT/mTOR pathway induces apoptosis and autophagy in hepatocellular carcinoma cells.	[61]
12	Prunetin	/	b	100 μM	DEN-induced HCC mice	Regulation of the NF-κB/PI3K/AKT signaling pathway.	[62]
13	Baicalein and baicalin	*Scutellaria baicalensis* Georgi	a + b	50 mg 80 mg	H22 tumor-bearing BALB/c mice; SMMC-7721 and HepG2 cells	Promotion of anti-tumor immunity by inhibiting PD-L1 expression.	[63]
14	Hesperidin	Citrus	b	200 mg	DEN-induced HCC mice	Downregulation of the PI3K/Akt signaling pathway.	[64]
15	Cardamonin	cardamom	b	50 mg	HepG2 tumor-bearing nude mice	Inhibition of NF-κB signaling pathway.	[65]

Table 2. *Cont.*

No.	Name	Origin	In Vitro (a) or In Vivo (b)	Optimal Doses (/kg Body Weight)	Model	Potential Mechanism	References
16	Safflower yellow	*Carthamus tinctorius*	a + b	5 mg	DEN-induced HCC mice; Hepa1-6 cells	Inhibition of inflammatory response; promotes collagen degradation and modulates gut microbiota to improve immune microenvironment.	[66]
17	Daidzin	soybean	a + b	100 mg	Hep3B tumor-bearing nude mice; HCCLM3 and Hep3B cells	Interference with hepatocellular carcinoma survival and migration via TPI1 and gluconeogenic pathways.	[67]
18	Salvigenin	*Scutellariae barbatae* Herba	a + b	10 μg	Huh7 tumor-bearing BALB/c mice; Human HCC cell lines	Blocking of aerobic glycolysis in HCC cells by inhibiting the PI3K/AKT/GSK-3β pathway.	[68]

DNA, deoxyribonucleic acid; DEN, diethylnitrosamine; CCL4, carbon tetrachloride, HCC, hepatocellular carcinoma, Akt, protein kinase B; MAPK, mitogen activated protein kinase, VEGF, vascular endothelial growth factor; MDSC, myeloid-derived suppressor cells; EMH, extramedullary hematopoiesis; H19, long non-coding RNAs H19; SD, sprague dawley; PI3K, phosphatidylinositide 3-kinase; NF-κB, nuclear factor kappa-B; "/" indicates that the reference is not mentioned or is unclear.

3.3. Terpenoids

Terpenoids, a class of naturally occurring hydrocarbon compounds, are abundantly present in plants, especially coniferous trees. They are classified into various types based on the number of isoprene units in their structure, including monoterpenes, sesquiterpenes, diterpenes, triterpenes, tetraterpenes, and polyterpenes. Many terpenoids showcase notable biological activities, encompassing anti-inflammatory, anticancer, antimicrobial, and antiviral properties. As a result, they serve as invaluable resources for investigating natural products and advancing drug development initiatives [69]. A recent review has succinctly outlined the anticancer attributes of terpenoids, emphasizing their potent efficacy against tumors. This underscores their potential as additional avenues for the development of anticancer medications [70–72]. For liver cancer, some terpenoids have also demonstrated promising therapeutic effects.

Ginsenosides, vital active constituents of ginseng, exhibit a spectrum of therapeutic effects, including antioxidant, anti-inflammatory, vasodilatory, anti-allergic, and anticancer properties [73]. Recent research has elucidated the therapeutic mechanisms through which ginsenosides combat liver cancer. For instance, ginsenoside Rh4 impedes the progression of inflammation-related hepatocellular carcinoma by targeting the HDAC4/IL-6/STAT3 signaling pathway [74]. In a DEN-induced SD rat model, ginsenoside CK may suppress liver cancer cell proliferation by downregulating Bclaf1 expression, thereby inhibiting the HIF-1α-mediated glycolytic pathway and proliferation [75]. Both in vitro and in vivo studies demonstrate that ginsenoside Rk3 not only curtails cell proliferation and induces cell cycle arrest but also fosters HCC cell autophagy and apoptosis via the PI3K/AKT pathway [76]. Ginsenoside Rh2 and its octyl ester derivative (Rh2-O) suppress the invasion and metastasis of liver cancer through the c-Jun/COX2/PGE2 pathway [77]. Rh2-O also exhibits immune-regulatory effects on splenic lymphocytes in H22 tumor-bearing mice [78]. Betulinic acid inhibits the proliferation of liver cancer HUH7 and HCCLM3 cells by activating ferritinophagy in cancer cells, thereby inducing ferroptosis through the NCOA4/FTH1/LC3II signaling pathway [79]. Additionally, ganoderma triterpenoids regulate the gut microbiota to inhibit the progression of liver cancer in S180 and H22 tumor-bearing mice [80].

Therefore, terpenoids demonstrate anticancer effects through various mechanisms, including apoptosis induction, autophagy stimulation, ferroptosis initiation, cell cycle arrest, inhibition of cell migration and invasion, modulation of immunity, metabolism, and gut microbiota. Table 3 provides a summary of the therapeutic effects and potential mechanisms of terpenoids on liver cancer.

Table 3. Therapeutic effects of terpenoids on liver cancer and their potential mechanisms.

No.	Name	Origin	In Vitro (a) or In Vivo (b)	Optimal Doses (/kg Body Weight)	Model	Potential Mechanism	References
1	Ginsenoside Rh4	ginseng	a + b	/	HCC tumor-bearing BALB/c mice; HUH7 and LM3 cells	Inhibition of inflammation-associated HCC progression by targeting HDAC4/IL-6/STAT3 signaling.	[74]
2	Octyl ester derivative of ginsenoside Rh2 (Rh2-O)	ginseng	b	10 mg	H22 tumor-bearing Tlr4$^{-/-}$ mice	Anti-hepatocellular carcinoma through TLR4-mediated immunomodulation of lymphocytes.	[78]
3	Ginsenoside Compound K	ginseng	a + b	5 mg	DEN-induced SD mice; Bel-7404 and Huh7 cells	Regulation of HIF-1α-mediated glycolysis by Bclaf1 inhibits cell proliferation.	[75]
4	Ginsenoside Rk3	ginseng	a + b	100 mg	HCC-LM3 tumor-bearing BALB/c mice; HepG2 and HCC-LM3 cells	Inhibition of cell proliferation and induction of cell cycle arrest. Promotion of cell autophagy and apoptosis via PI3K/AKT.	[76]
5	Ginsenoside Rh2	ginseng	a + b	30 μmol	H22 tumor-bearing C57BL/6 mice; Huh-7 and H22 cells	Suppression of hepatocellular carcinoma invasion and metastasis by inhibition of c-Jun/COX2/PGE2 pathway-mediated EMT.	[77]

Table 3. *Cont.*

No.	Name	Origin	In Vitro (a) or In Vivo (b)	Optimal Doses (/kg Body Weight)	Model	Potential Mechanism	References
6	Ginsenoside CK	ginseng	a + b	60 mg	HCC-LM3 tumor-bearing BALB/c mice; HepG2, SMMC-7721, HCC-LM3, and Huh-7 cells	Inhibition of hypoxia-induced epithelial mesenchymal transition in hepatocellular carcinoma through the HIF-1α/NF-κB feedback pathway.	[81]
7	Saikosaponin-b2 (SS-b2)	*Radix bupleuri*	a + b	20 mg	H22 tumor-bearing C57BL/6 mice; HepG2 and HUVECs cells	Inhibition of angiogenesis through inhibition of the VEGF/ERK/HIF-1α signaling pathway.	[82]
8	Saikosaponin-b2 (SS-b2)	*Radix bupleuri*	a + b	6 mg	DEN-induced BALB/c mice; RAW 264.7 macrophages	Upregulation of STK4 inhibits IRAK1/NF-κB signaling axis effectively suppresses PLCs.	[83]
9	Astragaloside IV (AS-IV)	*Astragalus membranaceus*	a + b	40 mg; 20 μmol/L	HT, genotype: pSmad3C$^{+/-}$; HO, genotype: Nrf2$^{-/-}$; lentivirus-transfected HepG2 cells	Amelioration of hepatocellular carcinoma by Nrf2-mediated transformation of pSmad3C/3L.	[84]
10	Dihydrotanshinon	*Salvia miltiorrhiza Bunge*	a + b	15 mg; 4 μg/mL	SMMC7721 tumor-bearing Balb/c mice; HCCLM3, SMMC7721, Hep3B. and HepG2 cells	Promotion of apoptosis by blocking the JAK2/STAT3 pathway.	[85]
11	Betulinic acid	bark of several plants	a + b	40 mg	HUH7 tumor-bearing BALB/c mice; HUH7 and HCCLM3 cells	Inhibition of hepatocellular carcinoma cell growth through activation of NCOA4-mediated ferritin phagocytosis pathway.	[79]
12	2α, 3α, 23-trihydroxy-urs-12-en-28-oic acid,	*Ganoderma lucidum*	b	400 mg	S180 and H22 tumor-bearing Kunming mice	Amelioration of lipid peroxidation. Down-regulation of Bcl-2 and up-regulation of Bax. Increase in intestinal flora richness and structure.	[80]

Table 3. *Cont.*

No.	Name	Origin	In Vitro (a) or In Vivo (b)	Optimal Doses (/kg Body Weight)	Model	Potential Mechanism	References
13	Pseudolaric acid B	*Pseudolarix kaempferi*	a + b	20 mg	Hepa1–6 tumor-bearing C57BL/6 mice; Hepa1–6 cells	Triggering apoptosis through activation of the AMPK/JNK/dr P1/mitochondrial fission pathway.	[86]
14	Mallotucin D	*Croton crassifolius*	a	20 µM	HepG2 cells	Induction of cellular pyroptosis. Inhibition of the PI3K/AKT/mTOR pathway activates mitochondrial autophagy.	[87]
15	Heteronemin	*Hippospongia* sp.	a	20 µM	HA22T and HA59T cells	Induction of apoptosis via the caspase pathway. Induction of ferroptosis by down-regulation of GPX4.	[88]

HCC, hepatocellular carcinoma; HDAC4, histone deacetylase 4; IL-6, interleukin 6; STAT3, signal transducer and activator of transcription 3; DEN, diethylnitrosamine; SD, sprague-dawley; PI3K, phosphatidylinositide 3-kinase; AKT, protein kinase B; PGF2, prostaglandin F2; EMT, epithelial-mesenchymal transition; HIF-1, hypoxia inducible factor-1; NF-κB, nuclear factor kappa-B; VEGF, vascular endothelial growth factor; ERK, extracellular regulated protein kinases; PLC: primary liver cancer; HT genotype, pSmad3C$^{+/-}$: Smad3 C-terminal phosphorylation site-heterozygous mutant mice model; HO genotype, Nrf2$^{-/-}$: Nrf2 conventional knockout homozygous mice; JAK2, janus kinase-2; Bcl-2, B-cell lymphoma-2; AMPK, AMP-activated protein kinase; JNK, c-Jun N-terminal kinase; GPX4, glutathione peroxidase 4; "/" indicates that the reference is not mentioned or is unclear.

3.4. Alkaloids

Alkaloids are organic compounds containing nitrogen, derived from amino acids and found in plants, animals, and microorganisms. They are classified based on their chemical structure into various types, including pyridine alkaloids, isoquinoline alkaloids, indole alkaloids, terpenoid alkaloids, steroidal alkaloids, quinoline alkaloids, and others. Alkaloids of natural origin exhibit diverse pharmacological activities such as anti-inflammatory, immunomodulatory, and anticancer properties. Numerous studies have highlighted the therapeutic effects of natural alkaloids from various sources on HCC [89]. Matrine inhibits HCC invasion and migration by PTEN/Akt-dependent suppression of epithelial-mesenchymal transition (EMT) [90]. Anisodamine (ANI), an alkaloid extracted from Anisodus, exhibits effective therapeutic activity against HCC in xenograft mouse models. At a dose of 200 mg/kg/day, ANI inhibits the growth of HCC cells by suppressing the activation of the NLRP3 inflammasome, inducing cell apoptosis, and modulating levels of inflammatory factors [91]. Both in vitro and in vivo studies have shown the anti-HCC effects of epipolythiodioxopiperazine (ETP) alkaloids such as Chaetocochin J, primarily attributed to the inhibition of the PI3K/Akt/mTOR/p70S6K/4EBP1 pathway and downregulation of HIF-1α expression under hypoxic conditions, disrupting the binding of HIF-1α/p300. In a patient-derived xenograft model, Stachydrine hydrochloride (SH) regulates autophagy and cell cycle arrest via the LIF/AMPK axis, inducing cell senescence, thereby suppressing tumor initiation and progression in HCC [92]. Additionally, Cepharanthine extracted from Stephania cepharantha Hayata inhibits HCC cell growth and proliferation by modulating amino acid metabolism and suppressing tumor formation in vivo [93].

A recent review highlights the significant therapeutic potential of natural alkaloids for HCC [89]. In summary, the potential mechanisms of natural alkaloids in treating liver cancer are closely associated with inhibiting cell proliferation, migration, and invasion, blocking the cell cycle, inducing apoptosis and autophagy, regulating metabolism, and modulating immune function. Table 4 summarizes the therapeutic effects and potential mechanisms of alkaloids on liver cancer.

Table 4. Therapeutic effects and potential mechanisms of alkaloids on liver cancer.

No.	Name	Origin	In Vitro (a) or In Vivo (b)	Optimal Dose (/kg Body Weight) or Concentration	Model	Potential Mechanism	References
1	Matrine	*Sophora flavescens* Aition	b	100 mg/kg matrine and 2 mg/kg cisplatin.	HepG2 tumor-bearing BALB/c mice	Promotion of apoptosis via suppression of survivin and activation of the caspase pathway.	[94]
2	Cepharanthine	*Stephania cephalantha* Hayata	a + b	20 mg; 60 μM	Huh7 tumor-bearing BALB/c mice; Huh7 and HepG2 cells	Inhibition of the Wnt/β-catenin/Hedgehog signaling pathway.	[95]
3	Veratramine	*Veratrum nigrum* L.	a + b	2 mg; 10, 20 and 40 μM	HepG2 tumor-bearing BALB/c mice; HepG2 cells	Activation of autophagy-mediated apoptosis through inhibition of the PI3K/Akt/mTOR signaling pathway.	[96]
4	Anisodamine	*Anisodus*	b	200 mg	HepG2 tumor-bearing BALB/c mice	Inhibition of NLRP3 inflammatory vesicles; induction of apoptosis.	[91]
5	Chaetocochin J	*Chaetomium* sp.	a + b	0.5 mg	HepG2 tumor-bearing nude mice; HepG2 and Hep3B cells	Inhibition of the PI3K/Akt/mTOR/p70S6K/4EBP1 pathway. Disruption of HIF-1α/p300 binding.	[97]
6	Sophoridine	*Sophora alopecuroides* L.	a + b	50 mg	HepG2 LR tumor-bearing BALB/c mice; HepG2 and Huh7 cells	Inhibition of the RAS/MEK/ERK axis by decreasing VEGFR2 expression.	[98]
7	Stachydrine hydrochloride	*Panzeria alaschanica* Kupr	a + b	30 mg	Patient-derived xenograft model; SMMC-7721 and HepG2 cells	Regulation of LIF/AMPK induced autophagy and senescence.	[92]

Table 4. *Cont.*

No.	Name	Origin	In Vitro (a) or In Vivo (b)	Optimal Dose (/kg Body Weight) or Concentration	Model	Potential Mechanism	References
8	Cyclovirobuxine D	*Buxus microphylla*	a + b	10 mg	HepG2 tumor-bearing BALB/c mice; HepG2 and HCCLM3 cells	Suppression of cell proliferation, migration, and invasion through inhibition of the EGFR-FAK-AKT/ERK1/2-Slug signaling pathway.	[99]
9	Piperlongumine	*Piper longum* L.	a + b	10 mg/kg PL and 5 mg/kg sorafenib	HCCLM3 tumor-bearing BALB/c mice; HCCLM3 and SMMC7721 cells	Mediation of ROS-AMPK activation and targeting of CPSF7.	[100]
10	Bufothionine	cinobufacini injection	a + b	343.35 µg/kg	H22-tumor-bearing Kunming mice; SMMC7721 cells	Induction of autophagy in HCC by inhibiting JAK2/STAT3 pathway.	[101]
11	Abrine	*Abrus cantonments*	a + b	15 mg	HepG2 tumor-bearing BALB/c mice; HepG2 and Huh7 cells	Regulation of hepatocellular carcinoma cell growth and apoptosis through the KAT5/PD-L1 axis and regulation of T cell proliferation and activation.	[102]
12	Cepharanthine	*Stephania cepharantha* Hayata	a + b	20 mg	Hep3B tumor-bearing nude mice; Hep3B and HCCLM3 cells	Inhibition of HCC cell proliferation by regulating amino acid metabolism and cholesterol metabolism; promotion of apoptosis and necrosis.	[93]
13	Ventilagolin	*Ventilago leiocarpa* Benth	b	12 mg	SMMC-7721 tumor-bearing BALB/c-nu nude mice	Inhibition of HCC cell growth, migration, and invasion by regulating Pim-1 expression and EMT markers.	[103]

PI3K, phosphatidylinositide 3-kinase; AKT, protein kinase B; mTOR, mammalian target of rapamycin; NLRP3, NOD-like receptor thermal protein domain associated protein 3; RAS, renin-angiotensin system; MEK, mitogen-activated extracellular signal-regulated kinase; ERK, extracellular regulated protein kinases; LIF, leukemia inhibitory factor; AMPK, AMP-activated protein kinase; EGFR, epidermal growth factor receptor; FAK, focal adhesion kinase; ROS, reactive oxygen species; HCC, hepatocellular carcinoma; JAK2, just another kinase2; STAT3, signal transducer and activator of transcription 3; PD-L1, programmed cell death 1 ligand 1; EMT, epithelial–mesenchymal transition.

3.5. Polyphenol Compounds

Polyphenols, typically abundant in vegetables and fruits, serve as secondary metabolites in many plants and represent the richest natural source of antioxidants in the human diet. Numerous studies suggest that consuming polyphenol-rich foods can help prevent and treat common chronic diseases [104]. Due to their potent antioxidant and anti-inflammatory properties, polyphenols exhibit a significant therapeutic effect on liver cancer [105]. The fruits of Terminalia bellirica possess various pharmacological activities. Researchers purified crude extracts of TB to produce total tannin fractions (TB-TF) and found that TB-TF inhibited tumor growth in H22 tumor-bearing mice by inducing apoptosis, reducing angiogenesis, and modulating immune suppression in the tumor microenvironment [106]. Catechin acts on Hep-G2 and Huh-7 cells, inhibiting the proliferation of human liver cancer cells by suppressing the NF-κB signaling pathway and triggering mitochondrial apoptosis [107]. In vitro experiments demonstrate that curcumin inhibits the proliferation of liver cancer cells by reducing DJ-1 expression and inhibiting the PTEN/PI3K/AKT signaling pathway [108].In vivo studies suggest that curcumin improves diethylnitrosamine-induced liver cancer by regulating oxidative stress, inflammation, and gut microbiota [109]. Resveratrol induces apoptosis and inhibits proliferation, migration, and invasion of liver cancer cell lines (HepG2 and Hep3B), and suppresses HCC progression by downregulating MARCH1 expression [110].

In summary, polyphenolic compounds primarily inhibit the progression of liver cancer by inducing cell apoptosis, suppressing migration and invasion, regulating immunity, and improving gut microbiota abundance. Table 5 summarizes the effects and potential mechanisms of polyphenolic compounds in treating liver cancer.

Table 5. Therapeutic effects and potential mechanisms of polyphenols on liver cancer.

No.	Name	Origin	In Vitro (a) or In Vivo (b)	Optimal Dose (/kg Body Weight) or Concentration	Model	Potential Mechanism	References
1	Tannins	*Terminalia bellirica*	b	2 g	HepG2 tumor-bearing ICR mice	Regulation of the EGFR signaling pathway and modulation of the immunosuppressive tumor microenvironment.	[106]
2	Proanthocyanidin-B2	peanut skin	a + b	300 mg	DEN + CCl4-induced HCC mice; Huh7 and Smmc-7721 cells	Inhibition of AKT leads to cell cycle arrest and tumor cell metabolic suppression.	[111]
3	Procyanidin B1	grape seed	b	15 mg	HepG2 tumor-bearing Balb/c mice	Inhibition of tumor growth in mice by inhibiting Kv10.1 current.	[112]
4	Chlorogenic acid	*Eucommia ulmoides* Oliver	a	0, 30 and 300 μM	HepG2 and Huh-7cells	Inhibition of the NF-κB signaling pathway and triggering of mitochondrial apoptosis.	[107]

Table 5. *Cont.*

No.	Name	Origin	In Vitro (a) or In Vivo (b)	Optimal Dose (/kg Body Weight) or Concentration	Model	Potential Mechanism	References
5	Curcumae	turmeric	b	10 mg	DEN-induced albino Wistar rats	Regulation of oxidative stress, inflammatory response, and gut microbiota.	[109]
6	Curcumin	turmeric	a	0, 5, 10 μM	SMMC-7721 and HepG2 cells	Down-regulation of DJ-1 inhibits cell proliferation.	[108]
7	Resveratrol		a + b	100 mg	HepG2 tumor-bearing BALB/c nude mice; HepG2 and Hep3B cells	Down-regulation of MARCH1 regulates PTEN/AKT signaling.	[110]

DEN, diethylnitrosamine; HCC, hepatocellular carcinoma; EGFR, epidermal growth factor receptor; AKT, protein kinase B; NF-κB, nuclear factor kappa-B; MARCH, membrane associated RING-CH; PTEN, phosphatase and tensin homolog.

3.6. Quinone

Quinone compounds constitute a class of secondary metabolites in plants. Based on the number of benzene rings, quinone compounds can be divided into anthraquinones, naphthoquinones, benzoquinones, and phenanthraquinones [113]. Several studies have demonstrated the therapeutic activity of quinones against liver cancer. Upon acting on HepG2 cells, emodin induces autophagy and suppresses epithelial–mesenchymal transition (EMT) by inhibiting the PI3K/AKT/mTOR and Wnt/β-catenin pathways [114]. Additionally, a separate study found that a structural derivative of emodin, emodin succinyl ester (ESE), inhibits malignant proliferation and migration of hepatocellular carcinoma by suppressing the interaction between androgen receptor (AR) and enhancer of zeste homolog 2 (EZH2) [115]. Thymoquinone (TQ), the main bioactive component of black seed, has been shown to suppress tumor angiogenesis by regulating miR-1-3p [116]. Furthermore, plumbagin has been found to induce apoptosis by down-regulating GPX4 in a subcutaneous xenograft tumor model [117].

In summary, quinone compounds exert anticancer effects against liver cancer through the induction of apoptosis, autophagy, inhibition of migration and invasion, and suppression of angiogenesis. However, research on the role of quinone compounds in liver cancer is relatively limited, leaving room for exploration of their anti-tumor mechanisms and targets. Researchers can employ bioinformatics, computer simulation techniques, and molecular docking to analyze the molecular mechanisms and targets of quinone compounds in anti-tumor effects, providing a theoretical basis for subsequent experimental studies. Table 6 summarizes the anticancer effects and potential mechanisms of quinone compounds against liver cancer.

Table 6. Therapeutic effects and potential mechanisms of quinones on liver cancer.

No.	Name	Origin	In Vitro (a) or In Vivo (b)	Optimal Dose (/kg Body Weight) or Concentration	Model	Potential Mechanism	Reference
1	Emodin	*Rheum palmatum* L.	a	0, 25, 50, 75, and 100 μM	HepG2 cells	Induction of autophagy and inhibited EMT through inhibition of PI3K/AKT/mTOR and Wnt/β-catenin pathways.	[114]
2	Emodin succinyl ester	*Rheum palmatum*	b	200 mg	DEN-induced HCC mice	Inhibition of cell proliferation and migration through inhibition of AR interaction with EZH2.	[115]
3	Thymoquinone	*Nigella sativa*	b	5 mg	DEN-induced HCC mice	Inhibition of angiogenesis by targeting miR-1-3p.	[116]
4	Plumbagin	*Plumbagin zeylanica* L	b	2 mg	HepG2 tumor-bearing Balb/c mice	Inhibition of USP31 activity leads to GPX4 protein degradation and apoptosis.	[117]

EMT, epithelial–mesenchymal transition; PI3K, phosphatidylinositide 3-kinase; AKT, protein kinase B; mTOR, mammalian target of rapamycin; DEN, dimethylnitrosamine; HCC, hepatocellular carcinoma; EZH2, enhancer of zeste homolog; GPX4, glutathione peroxidase 4.

3.7. Volatile Oils

Volatile oils are oily liquids found within plants, and research has highlighted their anti-tumor effects. In vitro and in vivo studies indicate that essential oil extracted from Artemisia argyi downregulates the mRNA and protein expression of DEPDC1, weakens Wnt/β-catenin signaling by reducing the production of Wnt1 and β-catenin, and prevents epithelial–mesenchymal transition (EMT) by downregulating vimentin and upregulating E-cadherin [118]. Additionally, essential oils extracted from the leaves of *Conobea scoparioides* Benth, Duguetia pycnastera Sandwith, and the bark of Aniba parviflora Mez inhibit the proliferation of HepG2 cells and suppress the growth of xenograft tumors in mice [119–121]. Pogostemon cablin essential oils, essential oil of lemon myrtle, and *Cyperus articulatus* L. rhizome essential oil exert anti-tumor effects by blocking the cell cycle and inducing apoptosis in liver cancer cells [122,123]. Table 7 summarizes the effects and potential mechanisms of volatile oils in treating liver cancer.

Table 7. Therapeutic effects and potential mechanisms of volatile oil on liver.

No.	Name	Origin	In Vitro (a) or In Vivo (b)	Optimal Dose (/kg Body Weight) or Concentration	Model	Potential Mechanism	Reference
1	Artemisia argyi essential oil	*Artemisia argyi*	a + b	230 mg	HepG2-Luc tumor-bearing BALB/c mice; HepG2 and SMMC-7721 cells	Inhibition of the Wnt/β-catenin signaling pathway suppresses metastasis of hepatocellular carcinoma.	[118]
2	D. pycnastera leaf essential oil	*Duguetia pycnastera* Sandwith	a + b	40 mg	HepG2 tumor-bearing C.B-17 SCID mice; HepG2 cell	Inhibition of tumor growth.	[119]
3	Conobea scoparioides essential oil	*Conobea scoparioides*	a + b	80 mg	HepG2 tumor-bearing C.B-17 SCID mice; HepG2 cell	Induction of apoptosis and inhibition of tumor growth.	[120]
4	Aniba parviflora essential oil	*Aniba parviflora*	a + b	80 mg	HepG2 tumor-bearing C.B-17 SCID mice	Inhibition of tumor cell proliferation and inhibition of tumor growth.	[121]
5	C. articulatus rhizome essential oil	*Cyperus articulatus* L	b	80 mg	HepG2 tumor-bearing C.B-17 SCID mice	Blocking the cell cycle and inducing apoptosis.	[122]
6	Essential oil of lemon myrtle	*Lemon myrtle*	a	40.90 µM	HepG2 cells	Blocking the cell cycle and inducing apoptosis	[123]
7	Pogostemon cablin essential oils	*Pogostemon cablin*	a + b	200 mg	HepG2 tumor-bearing Balb/c nude mice; HCC, SVEC, MDCK, and BNL CL.2 cells	Blocking the cell cycle and inducing apoptosis.	[124]

SCID, severe combined immunodeficient.

4. The Main Action Pathway and Potential Mechanism of Natural Compounds in the Treatment of Liver Cancer

The mechanism of action of NPs for treating liver cancer is currently under exploration in numerous studies. Here, we will detail the main pathways of action of natural compounds for liver cancer treatment. Utilizing multiple types of liver cancer cell lines in vitro and various animal models in vivo, studies have shown that the ameliorative and therapeutic effects of NPs on liver cancer primarily occur through eleven pathways.

Figure 2 illustrates the mechanism of action of NPs in liver cancer treatment, while Figure 3 shows the diagram of signaling pathways associated with NPs treatment of liver cancer.

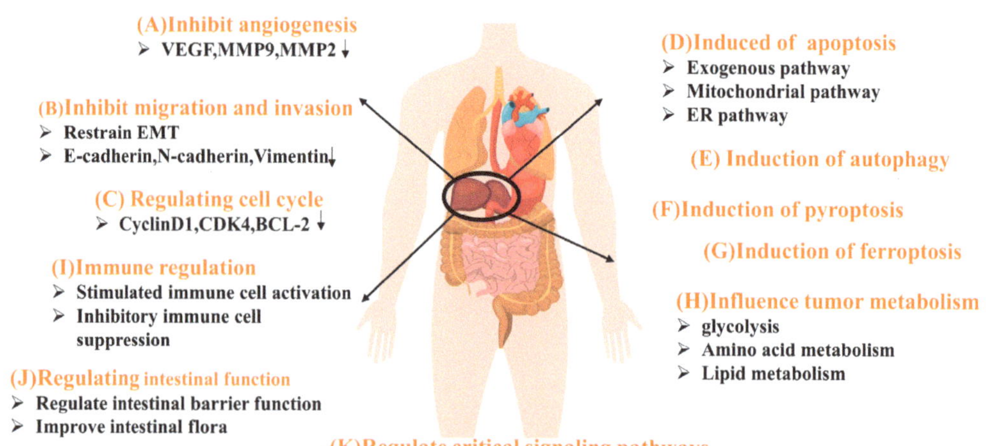

Figure 2. Mechanisms of NPs in treating liver cancer. A, B, C, D, E, F, G, H, I, J represent the primary pathways through which NPs treat liver cancer; K denotes the key signaling pathway through which NPs improve liver cancer.

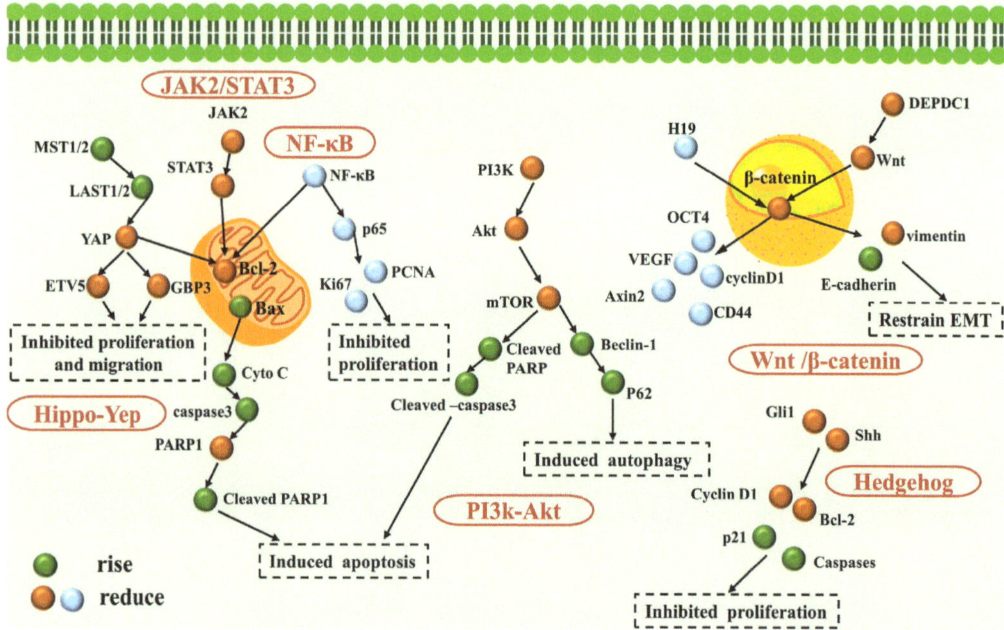

Figure 3. Mechanistic map of signaling pathways associated with NPs for liver cancer treatment. H19, long non-coding RNA H19; EMT, epithelial–mesenchymal transition.

4.1. Inhibition of Angiogenesis

Cancer is characterized by highly vascularized solid tumors, where the continuous growth of new blood vessels supplies oxygen and nutrients to tumor cells, consequently fueling cancer progression [125]. Hence, anti-angiogenesis emerges as a promising approach for treating aggressively prognosed cancers [126,127].

The terpenoid compound saikosaponin b2 (SSb2), traditionally used in Chinese medicine for fever reduction and liver protection, has demonstrated anti-angiogenic effects both in vivo and in vitro. Its mechanism involves the inhibition of the VEGF/ERK/HIF-1α signaling pathway [82]. Thymoquinone (TQ), a quinone compound, intervened in DEN-induced hepatocellular carcinoma (HCC) in rats. The results revealed decreased expression of MMP2, MMP9, and VEGF in the rat liver, accompanied by increased levels of TIMP3 and miR-1-3p expression, suggesting that the anti-angiogenic effect of TQ in HCC is mediated through miR-1-3p regulation [116]. Natural polysaccharides like Asparagus polysaccharide, flavonoids like sinensetin, and polyphenols like tannins have also been found to exert anti-hepatocellular carcinoma effects by inhibiting tumor angiogenesis [51,106,128].

4.2. Inhibition of Migration and Invasion

Malignant cell invasion and migration significantly contribute to mortality in liver cancer patients. The process of tumor migration and invasion is intricate, involving multiple steps and factors, and regulated by various signaling pathways. Therefore, targeting these signaling molecules to disrupt tumor metastasis emerges as one of the strategies for treating liver cancer using natural compounds.

Epithelial–mesenchymal transition (EMT) stands as a pivotal process driving cancer cell migration. EMT is characterized by decreased cell adhesion and apical–basal polarity, ultimately fostering cell movement and invasion. Research findings suggest that ginsenoside Rh2 and its octyl ester derivative inhibit the invasion and metastasis of hepatocellular carcinoma via the c-Jun/COX2/PGE2 pathway [77]. Both in vivo and in vitro studies indicate that ginsenoside CK inhibits hypoxia-induced epithelial–mesenchymal transition in liver cancer cells through the HIF-1α/NF-κB feedback pathway [81]. The alkaloid CVB-D suppresses proliferation, migration, invasion, and EMT of HCC cell lines by inhibiting the EGFR-FAK-AKT/ERK1/2-Slug signaling pathway in human HCC [99]. Both in vivo and in vitro studies demonstrate that the alkaloid ventilagolin downregulates the expression of Pim-1, N-cadherin, and vimentin, upregulates the expression of E-cadherin, and inhibits the migration, invasion, and EMT of HCC cells [103]. These studies collectively confirm that NPs can exert anti-tumor effects by inhibiting migration and invasion.

4.3. Regulating Cell Cycle

The cell cycle is primarily regulated by cell cycle proteins, protein kinases, and kinase inhibitors. Dysregulation of these components can lead to aberrant cell cycle regulation, resulting in abnormal cell proliferation. Recent studies highlight that targeting the cell cycle is a key strategy in combating liver cancer [129]. Scutellaria barbata polysaccharides (SBP) disrupt the cell cycle at the G0/G1 phase by downregulating the expression levels of CyclinD1 and CDK4 in H22 tumor cells. Additionally, they upregulate the expression levels of p53 and Bax/Bcl-2, thereby inducing apoptosis in H22 tumor cells [130]. In vivo and in vitro studies have demonstrated that the flavonoid compound wogonin induces the degradation of cell cycle protein D1 by activating glycogen synthase kinase-3 beta (GSK3β), effectively blocking the cell cycle in H22 tumor cells [59]. Researchers have also discovered that the triterpenoid compound Cucurbitacin B (CuB) effectively impedes HCC progression by inducing DNA damage-dependent cell cycle arrest. This effect is mediated through the ataxia telangiectasia mutated (ATM)-dependent p53-p21-CDK1 and checkpoint kinase 1 (CHK1)-CDC25C signaling pathways [131]. Additionally, the alkaloid Stachydrine hydrochloride, volatile oils extracted from plants, and the polyphenolic compound Proanthocyanidin-B2 all exert anti-cancer effects by regulating the cell cycle [92,111,123].

4.4. Induction of Apoptosis

Apoptosis, also known as programmed cell death, is a self-regulatory process that cells undergo during development and aging to maintain dynamic equilibrium [132]. Three different pathways can initiate apoptosis: the extrinsic or cell surface death receptor pathway, the mitochondrial pathway, and the endoplasmic reticulum (ER) pathway [133]. Cancer cells tend to develop the ability to evade apoptosis, thereby resisting the effects of drug therapy [134]. Therefore, targeting apoptosis-related signals to induce cancer cell death is a critical mechanism of action for natural product anticancer activity. The aforementioned seven natural compounds all exert their anti-liver cancer effects by inducing apoptosis. Researchers have discovered that water-soluble polysaccharides extracted from Eucommia ulmoides leaves induce apoptosis via the mitochondrial pathway in H22 tumor-bearing mice [47]. The flavonoid compound lysionotin, upon acting on liver cancer cells and xenograft mouse models, enhances the expression levels of PARP, FasL, Bax, Bad, and cleaved caspases, while reducing the expression level of Bcl-xL. This leads to the induction of apoptosis through the caspase-3 mediated mitochondrial pathway [57]. The diterpenoid compound pseudolaric acid B triggers apoptosis in hepatocellular carcinoma cells by activating the AMPK/JNK/DRP1/mitochondrial fission pathway [86]. Additionally, the alkaloid protopine inhibits the viability of liver cancer cells, induces caspase-dependent apoptosis via an intrinsic pathway, and induces ROS, further blocking the PI3K/Akt signaling pathway [135].

4.5. Induction of Autophagy

Autophagy refers to the process, mediated by autophagy-related genes, of using lysosomes to degrade damaged organelles and certain macromolecules, playing a unique role in maintaining cellular homeostasis, and is considered a form of programmed cell death [136]. Aberrations in autophagy can lead to various pathological conditions, including cancer. While autophagy is generally seen as a survival mechanism, excessive activation can result in non-apoptotic cell death. Therefore, regulating autophagy presents an opportunity to improve the cure rate of liver cancer, offering new treatment targets and directions [137]. Studies have shown that the alkaloid veratramine significantly inhibits the proliferation, migration, and invasion of HepG2 cells and induces autophagy-mediated cell death by inhibiting the PI3K/Akt/mTOR signaling pathway [96]. Both in vitro and in vivo studies have demonstrated that the main active component, total flavonoids of Hedyotis diffusa Willd, induces endoplasmic reticulum stress (ER stress) and activates the PERK/EIF2α/ATF4 signaling pathway, inducing apoptosis and activating autophagy in HCC cells [60].

4.6. Induction of Pyroptosis

Pyroptosis, also known as inflammatory necrosis of cells, is a novel form of programmed cell death characterized by cell swelling and the formation of pores on the plasma membrane surface [138]. Gasdermin D (GSDMD) and gasdermin E (GSDME), members of the gasdermin family, serve as hallmarks for inducing pyroptosis through proteolytic fragmentation catalyzed by caspases 1/4/5/11 and caspase 3, respectively [138,139]. Research indicates that the activation of pyroptosis can promote the death of liver cancer cells, thereby exhibiting its anti-cancer properties [139]. Consequently, promoting pyroptosis in liver cancer cells may serve as a promising new therapeutic target for treating liver cancer. In vitro and in vivo studies have shown that flavonoid compound Neobavaisoflavone (NBIF) promotes ROS production in HCC cells, affecting Tom20 protein expression. This facilitates Bax recruitment to mitochondria, activating caspase-3, cleaving GSDME, and inducing pyroptosis [54]. The terpenoid compound Mallotucin D promotes cytochrome C release from mitochondria into the cytoplasm, leading to caspase-9 and caspase-3 cleavage, inducing GSDMD-related pyroptosis [87]. The alkaloid ajmalicine activates ROS-induced pyroptosis, exerting anti-tumor effects through the non-canonical caspase-3-GSDME pyroptosis pathway [140].

4.7. Induction of Ferroptosis

Iron death is a novel regulatory mechanism of cell demise, distinct from previously known forms like necrosis, apoptosis, and autophagy. It involves disruptions in cellular iron metabolism and accumulation of peroxides. Iron death plays a crucial role in preventing and treating HCC. Activating ferroptosis in liver cancer cells and enhancing their sensitivity to ferroptosis can induce cell death, thereby preserving normal liver function [141,142].

Hepcidin, a key regulator of iron metabolism, is expressed at higher levels in HCC tumor tissues. Dandelion polysaccharide (DP) regulates iron overload in HCC cells and transplant tumors by inhibiting interleukin-6 (IL-6)-induced JAK-STAT signaling, exerting anti-tumor effects [46]. The terpenoid compound heteronemin induces intracellular ROS formation and inhibits cell apoptosis via the p38/JNK MAPK signaling pathway. Additionally, it suppresses glutathione peroxidase 4 (GPX4) expression, a critical regulator of lipid peroxidation and iron metabolism, leading to lipid peroxide accumulation and iron death, thereby inhibiting HCC progression [88].

4.8. Influence Tumor Metabolism

Metabolic abnormalities play a crucial role in tumor survival and progression [143]. Abnormal tumor metabolism can affect glycolysis [144], amino acid metabolism [145], and lipid metabolism [146]. Targeting metabolic dysregulation within the organism can prevent and combat the development of liver cancer, which holds significant implications for identifying novel cancer management and treatment strategies.

Daidzin (DDZ), a natural brassin-like compound derived from soybeans, has been shown in in vitro and in vivo studies to interfere with the survival and migration of hepatocellular carcinoma (HCC) cells by downregulating the expression of TPI1, a gene involved in the glycolysis pathway and prognostically relevant for HCC [67].

Cepharanthine (CEP), an alkaloid extracted from Stephania cepharantha Hayata, has shown efficacy in inhibiting the proliferation of HCC cells by modulating cellular metabolism, particularly amino acid and cholesterol metabolism [93]. Moreover, through LC-MS metabolomics and lipidomics analyses, researchers have identified that the active components of cepharanthine injection impede the progression of HepG2 cells by disrupting lipid, energy, and amino acid metabolism [147].

4.9. Immune Regulation

In recent years, immunotherapy has gained traction for improving liver cancer treatment by countering the liver's immune-suppressive environment, which enables cancer cells to evade therapy. NPs offer promising immunomodulatory effects, revitalizing immune surveillance and bolstering responses against liver cancer. They achieve this by targeting multiple immune pathways, either by boosting immune activators or by inhibiting suppressive factors, thus exerting anti-tumor effects [137].

4.9.1. Activation of Stimulatory Immune Cells

The tumor microenvironment (TME) is pivotal in driving tumor initiation and progression, comprising a complex ecosystem of tumor cells, immune cells, and other surrounding cells [148]. In HCC, immune cells within the TME can be broadly categorized into immunostimulatory lymphoid cells and immunosuppressive myeloid cells and lymphocytes based on their differentiation and function.

Representative stimulatory immune cells include cytotoxic CD8+ T cells, CD4+ T cells with a type 1 helper phenotype, and natural killer (NK) cells. These cells play crucial roles in anti-tumor immunity [137].

Activation of stimulatory immune cells to promote the release of anti-tumor pro-inflammatory cytokines is one of the mechanisms through which natural products inhibit tumor growth in HCC immunotherapy. One study found that the polysaccharide from wild Cordyceps significantly inhibited tumor growth and metastasis in H22 tumor-bearing mice, improved blood routine indicators, increased the thymus index, decreased the spleen index, and increased the proportion of CD4+ T cells, CD8+ T cells, and macrophages. It exhibited significant anti-tumor activity by enhancing host immune function [33]. Furthermore, a polysaccharide (ESPS) purified from Eupolyphaga sinensis Walker promoted lymphocyte proliferation and inhibited liver cancer cell growth by enhancing lymphocyte activity in vitro, primarily natural killer (NK) cells. Additionally, ESPS markedly enhanced immunity in H22-bearing mice by increasing the spleen and thymus indices and effectively inhibiting H22 cell growth in vivo [149]. Safflower yellow (SY) is the primary active ingredient isolated from the traditional Chinese medicine Carthamus tinctorius. Studies have shown that SY degrades collagens via the TGF-β/Smad signaling pathway to promote infiltration of CD8+ T cells and Gr-1+ macrophages. Additionally, it can regulate the gut microbiota to enhance hepatic immune infiltration, thereby inhibiting the development of HCC [66].

4.9.2. Suppression of Inhibitory Immune Cells

In addition to stimulatory immune cells, the TME also hosts inhibitory immune cells, notably regulatory T (Treg) cells, myeloid-derived suppressor cells (MDSCs), and tumor-associated macrophages (TAMs). These cells secrete diverse pro-inflammatory cytokines, which not only fuel cancer cell proliferation but also dampen the activity of stimulatory immune cells. In specific scenarios, natural products (NPs) demonstrate anti-tumor efficacy by thwarting the activation of inhibitory immune cells and curtailing the release of pro-tumorigenic cytokines [137].

Treg cells play a significant role as inhibitory immune cells in the HCC microenvironment. Studies have shown that resveratrol can inhibit tumor growth in HCC mouse models by reducing the number of CD8+CD122+ Tregs, downregulating immunosuppressive cytokines TGF-β1 and interleukin-10, and upregulating anti-tumor cytokines TNF-α and IFN-γ, thus reversing the immunosuppressive tumor microenvironment [150].

Myeloid-derived suppressor cells (MDSCs) are a group of immature immunosuppressive cells that help establish an immune-suppressive microenvironment in hepatocellular carcinoma (HCC). Recent studies have demonstrated that splenic extramedullary hematopoiesis (EMH) is an important mechanism for the accumulation of MDSCs in tumor tissues, thus contributing to disease progression. Icaritin, a prenylflavonoid derivative from plants of the Epimedium genus, has been shown to suppress tumor-associated splenic EMH, thereby inhibiting the production and activation of MDSCs and increasing the number and activity of cytotoxic T cells. As a result, it suppressed tumor progression and significantly prolonged the survival of mice bearing orthotopic and subcutaneous HCC tumors [52].

Tumor-associated macrophages (TAMs) play a crucial role in suppressing the anti-tumor activity of T cells or other immune-stimulating cells. Numerous studies have indicated that TAMs are typical pro-tumor macrophages responsible for the release of immunosuppressive cytokines [151]. Generally, macrophages can polarize into either the classically activated (pro-inflammatory) M1 state or the alternatively activated (anti-inflammatory) M2 state [152]. Polysaccharides from Radix Codonopsis Pilosulae (PHP-1) and Ganoderma lucidum polysaccharides (GLPS) can trigger the NF-κB and MAPK signaling pathways in the HCC microenvironment, shifting the M2 phenotype towards the M1 phenotype and alleviating immune suppression [31,32]. Research has found that Cinobufacini injection inhibits cholesterol metabolism via the AMPK/SREBP1/FASN pathway, affecting macrophage polarization, weakening hepatocellular carcinoma growth and migration, and promoting apoptosis [153].

Among the seven classes of compounds summarized in this article, natural polysaccharides have been extensively studied for their immunomodulatory effects in improving liver cancer progression, highlighting the significant potential of polysaccharides in immune regulation. Future research should explore the relationship between other types of compounds and their roles in anti-tumor activity and immune modulation, providing robust evidence for the use of immunotherapy in cancer treatment.

4.10. Regulates Intestinal Function

The bidirectional communication between the intestine and liver via the portal vein, bile ducts, and systemic circulation forms the gut–liver axis, a pivotal component in HCC pathogenesis. Dysfunction of intestinal barrier function and dysbiosis of gut microbiota are significant contributors to HCC development. In recent years, natural products have gained increasing attention in cancer therapy owing to their potent biological activity and minimal side effects. Leveraging the concept of the gut–liver axis, active ingredients in natural products play a crucial role in preventing and treating HCC by effectively intercepting intestinal–liver signaling pathways [154].

NPs inhibit the growth of HCC by modulating intestinal barrier function. In vivo studies have shown that Ganoderma lucidum polysaccharides can stimulate immune modulation and anti-tumor effects by regulating intestinal mucosal immune function [155]. In interventions with HCC mice established by HepG2 cells, ginsenoside Rg3 inhibits the overproduction of inflammatory factors and modulates the pathway of phosphoinositide 3-kinase, suppressing the growth of liver cancer cells [76,156]. This suggests that ginseng can inhibit the growth of HCC by suppressing inflammatory factors, increasing tight junction protein contacts, and repairing the intestinal barrier.

NPs impede the growth of HCC by enhancing gut microbiota. Recent findings have revealed a close association between gut microbiota and the occurrence, progression, and treatment of HCC, with microbial communities even discernible in late-stage tumor tissues [157]. The gut microbiota consists of probiotics and opportunistic pathogens, which promote T cell differentiation, NK cell activation, and dendritic cell maturation, actively regulating the immune system, surveilling for mutated cells, and thwarting tumor formation [158]. Studies have shown that intervention with stigmasterol in H22 tumor-bearing mice can alter the α and β diversity of the gut microbiota, significantly increasing the abundance of Lactobacillus johnsonii, Lactobacillus gasseri, and Lactobacillus vaginalis, leading to a decrease in the ratio of regulatory T cells (Tregs) to CD8+ T cells in both intestinal and tumor tissues. This enhances the immune response in the host tumor microenvironment (TME), exerting anti-tumor effects [159]. Ulva lactuca polysaccharide (ULP) inhibits tumor growth in H22 tumor-bearing mice by modulating the composition and metabolites of the gut microbiota [34]. Additionally, the Grifola frondosa polysaccharide-protein complex extracted from maitake mushrooms altered the gut microbiota composition and abundance in H22 tumor-bearing mice. This complex enriched norank_f__Muribaculaceae, Bacillus, and Bacteroides while reducing the abundance of Lactobacillus, thereby exerting anti HCC effects through gut microbiota modulation [160].

Although the interactions between the intestine and liver are not yet fully understood, intervention with NPs in liver cancer presents several advantages, including multi-target effects, comprehensive actions, multiple components, and minimal adverse reactions. However, the pathogenesis of liver cancer is exceedingly complex, requiring further elucidation of its progression and its correlation with intestinal barrier function and the gut microbiota. Moreover, the intricate nature of natural compound constituents and the diversity of therapeutic targets suggest that the specific mechanisms by which natural compounds regulate intestinal dysfunction to achieve preventive and therapeutic effects against liver cancer are still under exploration.

4.11. Regulation of Key Signaling Pathways

4.11.1. Wnt/β-Catenin Signaling Pathway

The Wnt/β-catenin signaling pathway regulates the occurrence and development of HCC through mechanisms such as the expression of cancer-related genes, activation of hepatic stellate cells, modulation of hepatic stem cell behavior, and promotion of HCC cell invasion and metastasis [161].

Both in vivo and in vitro studies have shown that the flavonoid compound apigenin can inhibit tumor growth in HCC by modulating the Wnt/β-catenin signaling axis mediated by the long non-coding RNA H19 [56]. DEPDC1, an upregulated novel tumor antigen in HCC, is considered a molecular target for novel therapeutic drugs. Research has found that Artemisia argyi essential oil (AAEO) effectively inhibits HCC metastasis by suppressing DEPDC1 expression, weakening Wnt/β-catenin signal transduction, and inhibiting epithelial–mesenchymal transition (EMT) [118].

4.11.2. PI3K/Akt Signaling Pathway

Research indicates abnormal regulation of the PI3K/Akt signaling pathway in HCC. This pathway is widely present in various biological cells and participates in regulating processes such as cell growth, survival, migration, tumor formation, and angiogenesis, all of which are closely associated with the growth, apoptosis, metastasis, and invasion of liver cancer cells [162]. Isoliquiritigenin (ISL), a flavonoid compound, induces both cell apoptosis and autophagy by upregulating the expression levels of LC3-II and cleaved-caspase-3, thereby inhibiting the PI3K/Akt/mTOR pathway [61]. Veratramine, an alkaloid, induces autophagic cell death by inhibiting the PI3K/Akt/mTOR signaling pathway, effectively suppressing the growth of HepG2 liver cancer cells both in vitro and in vivo [96]. Dandelion polysaccharides notably downregulate the protein levels of p-PI3K, p-Akt, and p-mTOR in HCC cells, consequently hindering HCC cell proliferation and inducing cell apoptosis through the suppression of the PI3K/Akt/mTOR pathway [45].

4.11.3. JAK2/STAT3 Signaling Pathway

The transcription factor STAT3, which stands for Signal Transducer and Activator of Transcription 3, undergoes transient activation in normal cells but exhibits heightened activity in many tumor cells and adjacent cancer tissues [163]. Aberrant activation of the JAK2/STAT3 pathway is a common mechanism leading to the occurrence of liver cancer, where overexpression of STAT3 is often observed [164]. STAT3, acting as a driving factor, plays a critical role in the initiation, progression, metastasis, and immune evasion of liver cancer and is associated with poor prognosis.

The quinone compound dihydroartemisinin inhibits the activation of the JAK2/STAT3 pathway, effectively suppressing cell proliferation and promoting apoptosis, thereby inhibiting the progression of HCC [85]. The bufadienolide compound bufothionine abundant in Huachansu injection induces autophagy by upregulating the expression of autophagy-related proteins Atg5, Atg7, Beclin1, and LC3II. This is primarily achieved by inhibiting the IL-6/JAK/STAT3 pathway [101].

4.11.4. Hippo–YAP Signaling Pathway

The Hippo–YAP-associated protein (YAP) signaling pathway is a recently identified pathway that holds significant importance in regulating HCC formation. It plays a critical role in controlling malignant behaviors like proliferation, apoptosis, invasion, and metastasis of liver cancer cells. Studies suggest that the alkaloid evodiamine inhibits proliferation and induces apoptosis of hepatocellular carcinoma cells through modulation of the Hippo–YAP signaling pathway [165].

Furthermore, cordycepin extracted from Cordyceps militaris promotes the expression of MST1 and LAST1 in HepG-2 cells, thereby inhibiting the expression of YAP1. Cordycepin regulates the expression of GBP3 and ETV5 by modulating the Hippo signaling pathway, thereby suppressing the proliferation and migration of HepG-2 liver cancer cells. Additionally, cordycepin regulates the expression of Bax and Bcl-2 through the Hippo signaling pathway, activating the mitochondrial apoptosis pathway. After cytochrome c is released into the cytoplasm, it activates Caspase 3, ultimately leading to cell apoptosis [166].

4.11.5. NF-κB Signaling Pathway

The Nuclear Factor-κB (NF-κB) signaling pathway, a crucial transcription regulatory factor, exerts its biological effects by modulating the transcriptional expression of various genes. NF-κB plays a significant role in regulating the growth and proliferation of liver cancer cells and can contribute to spontaneous liver damage, fibrosis, and the development of liver cancer [167]. Studies have demonstrated that intervention with Cardamonin (CADMN) in xenografted nude mice results in the downregulation of proteins such as PCNA, Ki-67, and Bcl-2, alongside upregulation of the Bax protein. Furthermore, NF-κB-p65 and Ikkβ proteins are also downregulated. Thus, CADMN exerts anti-tumor effects on human liver cancer xenografts in nude mice by inhibiting the NF-κB pathway [65].

4.11.6. Hedgehog Signaling Pathway

The Hedgehog signaling pathway, known for its high conservation, plays a crucial role in regulating fundamental processes like cell proliferation, differentiation, migration, and adhesion [168]. Moreover, it has been closely linked with the development of liver cancer [169]. Cepharanthine (CH), a biscoclaurine alkaloid derived from the roots of Stephania cephalantha Hayata, exhibits promising anti-tumor activity against various cancer types. Both in vivo and in vitro studies have highlighted CH's ability to inhibit the Hedgehog/Gli1 signaling pathway by suppressing Gli1 transcription and its transcriptional activity. Additionally, CH inhibits Wnt/β-catenin signal transduction, an upstream regulatory factor of Hedgehog signaling in cancer cells treated with CH [95].

5. Summary and Outlook

The biological activity of NPs has ignited significant interest among researchers across pharmaceuticals, health foods, and cosmetics industries [170]. NPs, prized for their minimal side effects, high safety profile, and potent efficacy against liver cancer, are emerging as promising sources for preventive and therapeutic interventions in liver cancer.

This paper outlines seven categories of natural compounds, each demonstrating therapeutic and ameliorative effects on liver cancer. They operate through diverse mechanisms, including angiogenesis inhibition, migration and invasion suppression, cell cycle regulation, apoptosis induction, autophagy induction, ferroptosis induction, modulation of tumor metabolism, immunity regulation, intestinal function regulation, and modulation of key signaling pathways.

While considerable progress has been achieved in researching the therapeutic effects of NPs on liver cancer, several unresolved issues still demand attention.

Primarily, there is a need to augment in vivo activity experiments utilizing animal models closely resembling clinical scenarios. Concurrently, combined in vivo and in vitro studies should be conducted to delve deeply into the anticancer effects and mechanisms of action of NPs in anti-liver cancer drugs.

Furthermore, there is a call for further elucidation of the active components and structures of certain NPs. For instance, understanding the composition of natural polysaccharide monosaccharides necessitates additional steps such as purification, characterization, and modification to enhance their activity. It is imperative to systematically evaluate the toxicity and safety profiles of natural products, including supplementary studies on toxicity and relevant dosages, thus laying a robust foundation for clinical validation.

Additionally, the treatment of liver cancer necessitates targeting multiple pathways to establish a comprehensive therapeutic strategy. Some recently identified therapeutic targets are yet to be validated and translated into clinical practice. For instance, research on the tumor immune microenvironment and immunotherapy for liver cancer remains in need of extensive preclinical studies. In forthcoming research endeavors, it is not only important to explore novel treatment approaches but also to investigate the synergistic effects of combining multiple therapeutic modalities, thereby offering new avenues for the development of anti-liver cancer agents. Sustaining the quest for highly effective and low side-effect natural products for liver cancer treatment holds paramount importance. Combining different NPs might amplify the therapeutic effects against liver cancer, underscoring the significance of researching combination therapy involving NPs.

Conventional oral NPs face challenges such as low bioavailability, rapid metabolism, poor absorption, and low solubility. However, nanoparticles offer a solution by substantially enhancing the bioavailability of NPs during drug delivery, mitigating toxicity, and targeting tumor sites effectively. The nanoparticles commonly loaded with natural compounds for targeting liver cancer can be classified into organic and inorganic nanoparticles. Organic nanoparticles, primarily composed of liposomes, polymers, and polymer micelles, offer significant advantages in terms of biocompatibility, stability, and resistance to degradation by digestive enzymes. They are extensively utilized in the prevention and treatment of liver cancer [171]. For instance, the terpenoid Triptolide (TPL) presents limitations in liver cancer therapy due to its severe systemic toxicity and poor water solubility. Researchers addressed these challenges by preparing TPL-loaded membrane protein-embedded liposomes through thin-layer evaporation. Injecting these liposomes significantly inhibited hepatocellular carcinoma growth in mice with minimal damage observed in vital organs such as the liver, kidney, and others [172]. Curcumin (Cur), a well-known anticancer polyphenolic compound, faces challenges due to its poor water solubility and low bioavailability [173]. To overcome these limitations, researchers utilized d-alpha-tocopheryl polyethylene glycol 1000 succinate (TPGS) as an emulsifier to prepare Cur-loaded poly(lactic-co-glycolic acid) (PLGA) nanoparticles. This approach notably increased the concentration of Cur at the liver cancer site in mice, resulting in a reduction in tumor volume [174]. Cantharidin (CTD) has long been used in liver cancer prevention and treatment, but its direct administration entails high toxicity, and its standalone usage suffers from low bioavailability [175]. Researchers found that delivering CTD using methoxy polyethylene glycol-polylactic acid (mPEG-PLGA) micelles significantly enhances CTD's pro-apoptotic effect on HepG2 cells. Injection of mPEG-PLGA-CTD in liver cancer mice prolongs CTD's half-life in vivo, leading to increased accumulation at the liver cancer site [176].

Inorganic nanoparticles, primarily composed of metals and metalloids such as gold (Au) and silicon (Si), exhibit unique physical and chemical properties. Inorganic nanoparticles containing anti-tumor bioactive components show promising potential in the diagnosis and treatment of liver cancer [171]. The Chinese herbal medicine Marsdenia tenacissima (MT) possesses anticancer and hepatoprotective properties, with MT extracts demonstrating anti-hepatocellular carcinoma (HCC) activity [177–179]. However, due to the lack of specific targeting capability, MT extracts suffer from inadequate concentrations at tumor sites. To overcome this limitation, researchers have developed gold nanoparticles loaded with MT extracts, capable of inducing apoptosis in HepG2 cells. Furthermore, MT-gold nanoparticles exhibit superior targeting and inhibitory effects on HepG2 cells compared to MT extracts alone [180]. Due to poor water solubility and high toxicity, the clinical application of the alkaloid Hydroxycamptothecin (HCPT) is limited. Researchers have employed carboxymethyl chitosan (CMC) and hyaluronic acid (HA) modified graphene oxide (GO) as nano-carrier materials to deliver HCPT. This carrier induces cell apoptosis in vitro and demonstrates superior therapeutic efficacy against liver cancer in vivo compared to HCPT alone, suggesting its potential for targeted drug delivery in liver cancer treatment [181].

In conclusion, we aim for this review to offer a systematic and reliable summary, aiding researchers in comprehending the treatment and ameliorative effects of NPs on liver cancer. This synthesis aims to furnish a theoretical foundation for the development and application of NPs in pharmaceuticals and functional foods.

Author Contributions: Conceptualization, J.G.; methodology, H.D.; software, W.Y.; validation, W.Y.; formal analysis, D.W.; investigation, D.F.; resources, Y.Z. (Yaxi Zhou); data curation, Y.Z. (Yue Zheng); writing—original draft preparation, J.G.; writing—review and editing, X.Q.; visualization, S.Z.; supervision, G.L.; project administration, X.Q.; funding acquisition, W.Y. All authors have read and agreed to the published version of the manuscript.

Funding: This work was supported by the National Key Research and Development Program of China [2023YFF1103802].

Institutional Review Board Statement: Not applicable.

Informed Consent Statement: Not applicable.

Data Availability Statement: Not applicable.

Acknowledgments: The authors would like to thank Wenjie Yan for his guidance and financial help.

Conflicts of Interest: The authors declare no conflicts of interest.

References

1. Sällberg, M.; Pasetto, A. Liver, Tumor and Viral Hepatitis: Key Players in the Complex Balance Between Tolerance and Immune Activation. *Front. Immunol.* **2020**, *11*, 552. [CrossRef] [PubMed]
2. Balogh, J.; Victor Iii, D.; Asham, E.H.; Burroughs, S.G.; Boktour, M.; Saharia, A.; Li, X.; Ghobrial, R.M.; Monsour, H.P., Jr. Hepatocellular carcinoma: A review. *J. Hepatocell. Carcinoma* **2016**, *3*, 41–53. [CrossRef] [PubMed]
3. Man, S.; Luo, C.; Yan, M.; Zhao, G.; Ma, L.; Gao, W. Treatment for liver cancer: From sorafenib to natural products. *Eur. J. Med. Chem.* **2021**, *224*, 113690. [CrossRef]
4. Pinter, M.; Sieghart, W.; Graziadei, I.; Vogel, W.; Maieron, A.; Königsberg, R.; Weissmann, A.; Kornek, G.; Plank, C.; Peck-Radosavljevic, M. Sorafenib in unresectable hepatocellular carcinoma from mild to advanced stage liver cirrhosis. *Oncol.* **2009**, *14*, 70–76. [CrossRef]
5. Dika, I.E.; Abou-Alfa, G.K. Treatment options after sorafenib failure in patients with hepatocellular carcinoma. *Clin. Mol. Hepatol.* **2017**, *23*, 273–279. [CrossRef]
6. Zheng, Y.; Zhang, W.; Xu, L.; Zhou, H.; Yuan, M.; Xu, H. Recent Progress in Understanding the Action of Natural Compounds at Novel Therapeutic Drug Targets for the Treatment of Liver Cancer. *Front. Oncol.* **2022**, *11*, 795548. [CrossRef]
7. Gao, S.; Jiang, X.; Wang, L.; Jiang, S.; Luo, H.; Chen, Y.; Peng, C. The pathogenesis of liver cancer and the therapeutic potential of bioactive substances. *Front. Pharmacol.* **2022**, *13*, 1029601. [CrossRef]
8. Kamarajah, S.K.; Frankel, T.L.; Sonnenday, C.; Cho, C.S.; Nathan, H. Critical evaluation of the American Joint Commission on Cancer (AJCC) 8th edition staging system for patients with Hepatocellular Carcinoma (HCC): A Surveillance, Epidemiology, End Results (SEER) analysis. *J. Surg. Oncol.* **2018**, *117*, 644–650. [CrossRef] [PubMed]
9. Edge, S.B.; Compton, C.C. The American Joint Committee on Cancer: The 7th edition of the AJCC cancer staging manual and the future of TNM. *Ann. Surg. Oncol.* **2010**, *17*, 1471–1474. [CrossRef]
10. Vauthey, J.N. Simplified Staging for Hepatocellular Carcinoma. *J. Clin. Oncol.* **2002**, *20*, 1527–1536. [CrossRef]
11. Hemming, A.W.; Scudamore, C.H.; Shackleton, C.R.; Pudek, M.; Erb, S.R. Indocyanine green clearance as a predictor of successful hepatic resection in cirrhotic patients. *Am. J. Surg.* **1992**, *163*, 515–518. [CrossRef] [PubMed]
12. Kamo, N.; Kaido, T.; Yagi, S.; Okajima, H.; Uemoto, S. Liver transplantation for small hepatocellular carcinoma. *Hepatobiliary Surg. Nutr.* **2016**, *5*, 391–398. [CrossRef]
13. Tabrizian, P.; Holzner, M.; Halazun, K.; Agopian, V.G.; Busuttil, R.W.; Yao, F.; Roberts, J.; Florman, S.S.; Schwartz, M.E.; Emond, J.C. A US multicenter analysis of 2529 HCC patients undergoing liver transplantation: 10-year outcome assessing the role of down-staging to within Milan criteria. *Hepatology* **2019**, *70*, 10–11.
14. Zhou, M.; Wang, H.; Zeng, X.; Yin, P.; Zhu, J.; Chen, W.; Li, X.; Wang, L.; Wang, L.; Liu, Y.; et al. Mortality, morbidity, and risk factors in China and its provinces, 1990-2017: A systematic analysis for the Global Burden of Disease Study 2017. *Lancet* **2019**, *394*, 1145–1158. [CrossRef] [PubMed]
15. Takayama, T.; Hasegawa, K.; Izumi, N.; Kudo, M.; Shimada, M.; Yamanaka, N.; Inomata, M.; Kaneko, S.; Nakayama, H.; Kawaguchi, Y.; et al. Surgery versus Radiofrequency Ablation for Small Hepatocellular Carcinoma: A Randomized Controlled Trial (SURF Trial). *Liver Cancer* **2021**, *11*, 209–218. [CrossRef] [PubMed]
16. Li, W.; Ni, C.-F. Current status of the combination therapy of transarterial chemoembolization and local ablation for hepatocellular carcinoma. *Abdom. Radiol.* **2019**, *44*, 2268–2275. [CrossRef] [PubMed]
17. Heimbach, J.K.; Kulik, L.M.; Finn, R.S.; Sirlin, C.B.; Abecassis, M.M.; Roberts, L.R.; Zhu, A.X.; Murad, M.H.; Marrero, J.A. Aasld guidelines for the treatment of hepatocellular carcinoma. *Hepatology* **2018**, *67*, 358–380. [CrossRef]
18. Vogel, A.; Cervantes, A.; Chau, I.; Daniele, B.; Llovet, J.M.; Meyer, T.; Nault, J.C.; Neumann, U.; Ricke, J.; Sangro, B.; et al. Hepatocellular carcinoma: ESMO Clinical Practice Guidelines for diagnosis, treatment and follow-up. *Ann. Oncol.* **2019**, *30*, 871–873. [CrossRef]
19. Kudo, M.; Kawamura, Y.; Hasegawa, K.; Tateishi, R.; Kariyama, K.; Shiina, S.; Toyoda, H.; Imai, Y.; Hiraoka, A.; Ikeda, M.; et al. Management of Hepatocellular Carcinoma in Japan: JSH Consensus Statements and Recommendations 2021 Update. *Liver Cancer* **2021**, *10*, 181–223. [CrossRef]
20. Kudo, M.; Finn, R.S.; Qin, S.; Han, K.-H.; Ikeda, K.; Piscaglia, F.; Baron, A.; Park, J.-W.; Han, G.; Jassem, J.; et al. Lenvatinib versus sorafenib in first-line treatment of patients with unresectable hepatocellular carcinoma: A randomised phase 3 non-inferiority trial. *Lancet* **2018**, *391*, 1163–1173. [CrossRef]
21. Huang, A.; Yang, X.-R.; Chung, W.-Y.; Dennison, A.R.; Zhou, J. Targeted therapy for hepatocellular carcinoma. *Signal Transduct. Target. Ther.* **2020**, *5*, 146. [CrossRef] [PubMed]
22. Abou-Alfa, G.K.; Meyer, T.; Cheng, A.-L.; El-Khoueiry, A.B.; Rimassa, L.; Ryoo, B.-Y.; Cicin, I.; Merle, P.; Chen, Y.; Park, J.-W.; et al. Cabozantinib in Patients with Advanced and Progressing Hepatocellular Carcinoma. *N. Engl. J. Med.* **2018**, *379*, 54–63. [CrossRef]
23. Bruix, J.; Qin, S.; Merle, P.; Granito, A.; Huang, Y.-H.; Bodoky, G.; Pracht, M.; Yokosuka, O.; Rosmorduc, O.; Breder, V.; et al. Regorafenib for patients with hepatocellular carcinoma who progressed on sorafenib treatment (RESORCE): A randomised, double-blind, placebo-controlled, phase 3 trial. *Lancet* **2017**, *389*, 56–66. [CrossRef] [PubMed]

24. Zhu, X.; Bian, H.; Wang, L.; Sun, X.; Xu, X.; Yan, H.; Xia, M.; Chang, X.; Lu, Y.; Li, Y.; et al. Berberine attenuates nonalcoholic hepatic steatosis through the AMPK-SREBP-1c-SCD1 pathway. *Free Radic. Biol. Med.* **2019**, *141*, 192–204. [CrossRef]

25. Chen, K.-W. Current systemic treatment of hepatocellular carcinoma: A review of the literature. *World J. Hepatol.* **2015**, *7*, 1412–1420. [CrossRef]

26. Wang, A.; Wu, L.; Lin, J.; Han, L.; Bian, J.; Wu, Y.; Robson, S.C.; Xue, L.; Ge, Y.; Sang, X.; et al. Whole-exome sequencing reveals the origin and evolution of hepato-cholangiocarcinoma. *Nat. Commun.* **2018**, *9*, 894. [CrossRef]

27. Chen, L.; Chen, S.; Zheng, C. Research progress in radiotherapy for liver cancers. *J. Interv. Radiol.* **2020**, *29*, 631–635. [CrossRef]

28. Zheng, Y.; Li, Y.; Feng, J.; Li, J.; Ji, J.; Wu, L.; Yu, Q.; Dai, W.; Wu, J.; Zhou, Y.; et al. Cellular based immunotherapy for primary liver cancer. *J. Exp. Clin. Cancer Res.* **2021**, *40*, 250. [CrossRef]

29. Liu, Y.; Fang, C.; Luo, J.; Gong, C.; Wang, L.; Zhu, S. Traditional Chinese Medicine for Cancer Treatment. *Am. J. Chin. Med.* **2024**, 1–22. [CrossRef]

30. Peng, Y.; Wu, X.; Zhang, Y.; Yin, Y.; Chen, X.; Zheng, D.; Wang, J. An Overview of Traditional Chinese Medicine in the Treatment After Radical Resection of Hepatocellular Carcinoma. *J. Hepatocell. Carcinoma* **2023**, *10*, 2305–2321. [CrossRef]

31. Pu, Y.; Zhu, J.; Xu, J.; Zhang, S.; Bao, Y. Antitumor effect of a polysaccharide from Pseudostellaria heterophylla through reversing tumor-associated macrophages phenotype. *Int. J. Biol. Macromol.* **2022**, *220*, 816–826. [CrossRef] [PubMed]

32. Liu, Z.; Xing, J.; Huang, Y.; Bo, R.; Zheng, S.; Luo, L.; Niu, Y.; Zhang, Y.; Hu, Y.; Liu, J.; et al. Activation effect of Ganoderma lucidum polysaccharides liposomes on murine peritoneal macrophages. *Int. J. Biol. Macromol.* **2015**, *82*, 973–978. [CrossRef] [PubMed]

33. Tan, L.; Liu, S.; Li, X.; He, J.; He, L.; Li, Y.; Yang, C.; Li, Y.; Hua, Y.; Guo, J. The Large Molecular Weight Polysaccharide from Wild Cordyceps and Its Antitumor Activity on H22 Tumor-Bearing Mice. *Molecules* **2023**, *28*, 3351. [CrossRef] [PubMed]

34. Qiu, Y.; Xu, J.; Liao, W.; Wen, Y.; Jiang, S.; Wen, J.; Zhao, C. Suppression of hepatocellular carcinoma by Ulva lactuca ulvan via gut microbiota and metabolite interactions. *J. Adv. Res.* **2023**, *52*, 103–117. [CrossRef] [PubMed]

35. Cai, L.; Zhou, S.; Wang, Y.; Xu, X.; Zhang, L.; Cai, Z. New insights into the anti- hepatoma mechanism of triple-helix β- glucan by metabolomics profiling. *Carbohydr. Polym.* **2021**, *269*, 118289. [CrossRef] [PubMed]

36. Ying, Y.; Hao, W. Immunomodulatory function and anti-tumor mechanism of natural polysaccharides: A review. *Front. Immunol.* **2023**, *14*, 1147641. [CrossRef] [PubMed]

37. Yu, J.; Ji, H.-Y.; Liu, C.; Liu, A.-J. The structural characteristics of an acid-soluble polysaccharide from Grifola frondosa and its antitumor effects on H22-bearing mice. *Int. J. Biol. Macromol.* **2020**, *158*, 1288–1298. [CrossRef] [PubMed]

38. Wang, Q.; Niu, L.-L.; Liu, H.-P.; Wu, Y.-R.; Li, M.-Y.; Jia, Q. Structural characterization of a novel polysaccharide from Pleurotus citrinopileatus and its antitumor activity on H22 tumor-bearing mice. *Int. J. Biol. Macromol.* **2020**, *168*, 251–260. [CrossRef] [PubMed]

39. Li, G.-L.; Tang, J.-F.; Tan, W.-L.; Zhang, T.; Zeng, D.; Zhao, S.; Ran, J.-H.; Li, J.; Wang, Y.-P.; Chen, D.-L. The anti-hepatocellular carcinoma effects of polysaccharides from *Ganoderma lucidum* by regulating macrophage polarization via the MAPK/NF-κB signaling pathway. *Food Funct.* **2023**, *14*, 3155–3168. [CrossRef]

40. Liu, C.; Dai, K.-Y.; Ji, H.-Y.; Jia, X.-Y.; Liu, A.-J. Structural characterization of a low molecular weight *Bletilla striata* polysaccharide and antitumor activity on H22 tumor-bearing mice. *Int. J. Biol. Macromol.* **2022**, *205*, 553–562. [CrossRef]

41. Zhang, J.; Dai, K.; Li, M. Preparation of Water-Soluble Acetylaminoglucan with Low Molecular Weight and Its Anti-Tumor Activity on H22 Tumor-Bearing Mice. *Molecules* **2022**, *27*, 7273. [CrossRef] [PubMed]

42. Shi, S.; Chang, M.; Liu, H.; Ding, S.; Yan, Z.; Si, K.; Gong, T. The Structural Characteristics of an Acidic Water-Soluble Polysaccharide from Bupleurum chinense DC and Its In Vivo Anti-Tumor Activity on H22 Tumor-Bearing Mice. *Polymers* **2022**, *14*, 1119. [CrossRef] [PubMed]

43. Zhang, X.; Li, M.; Liu, H.; Zhang, S.; Wang, B.; Liu, Y.; Zhang, X.; Zhang, M. Polysaccharide of *Asparagus cochinchinensis* (Lour.) Merr: Preliminary characterization and antitumor activity in vivo. *Food Biosci.* **2023**, *56*, 103387. [CrossRef]

44. Wu, Y.; Wang, Q.; Liu, H.; Niu, L.; Li, M.; Jia, Q. A heteropolysaccharide from *Rhodiola rosea* L.: Preparation, purification and anti-tumor activities in H22-bearing mice. *Food Sci. Hum. Wellness* **2023**, *12*, 536–545. [CrossRef]

45. Ren, F.; Li, J.; Yuan, X.; Wang, Y.; Wu, K.; Kang, L.; Luo, Y.; Zhang, H.; Yuan, Z. Dandelion polysaccharides exert anticancer effect on Hepatocellular carcinoma by inhibiting PI3K/AKT/mTOR pathway and enhancing immune response. *J. Funct. Foods* **2019**, *55*, 263–274. [CrossRef]

46. Ren, F.; Yang, Y.; Wu, K.; Zhao, T.; Shi, Y.; Song, M.; Li, J. The Effects of Dandelion Polysaccharides on Iron Metabolism by Regulating Hepcidin via JAK/STAT Signaling Pathway. *Oxidative Med. Cell. Longev.* **2021**, *2021*, 7184760. [CrossRef] [PubMed]

47. Yan, Z.-Q.; Ding, S.-Y.; Chen, P.; Liu, H.-P.; Chang, M.-L.; Shi, S.-Y. A water-soluble polysaccharide from Eucommia folium: The structural characterization and anti-tumor activity in vivo. *Glycoconj. J.* **2022**, *39*, 759–772. [CrossRef] [PubMed]

48. Ma, D.; Wei, J.; Chen, S.; Wang, H.; Ning, L.; Luo, S.-H.; Liu, C.-L.; Song, G.; Yao, Q. Fucoidan Inhibits the Progression of Hepatocellular Carcinoma via Causing lncRNA LINC00261 Overexpression. *Front. Oncol.* **2021**, *11*, 653902. [CrossRef]

49. Su, J.; Liao, D.; Su, Y.; Liu, S.; Jiang, L.; Wu, J.; Liu, Z.; Wu, Y. Novel polysaccharide extracted from *Sipunculus nudus* inhibits HepG2 tumour growth in vivo by enhancing immune function and inducing tumour cell apoptosis. *J. Cell. Mol. Med.* **2021**, *25*, 8338–8351. [CrossRef]

50. Tian, Y.; Wang, L.; Chen, X.; Zhao, Y.; Yang, A.; Huang, H.; Ouyang, L.; Pang, D.; Xie, J.; Liu, D.; et al. DHMMF, a natural flavonoid from *Resina Draconis*, inhibits hepatocellular carcinoma progression via inducing apoptosis and G2/M phase arrest mediated by DNA damage-driven upregulation of p21. *Biochem. Pharmacol.* **2023**, *211*, 115518. [CrossRef]

51. Li, X.; Li, Y.; Wang, Y.; Liu, F.; Liu, Y.; Liang, J.; Zhan, R.; Wu, Y.; Ren, H.; Zhang, X.; et al. Sinensetin suppresses angiogenesis in liver cancer by targeting the VEGF/VEGFR2/AKT signaling pathway. *Exp. Ther. Med.* **2022**, *23*, 360. [CrossRef] [PubMed]

52. Tao, H.; Liu, M.; Wang, Y.; Luo, S.; Xu, Y.; Ye, B.; Zheng, L.; Meng, K.; Li, L. Icaritin Induces Anti-tumor Immune Responses in Hepatocellular Carcinoma by Inhibiting Splenic Myeloid-Derived Suppressor Cell Generation. *Front. Immunol.* **2021**, *12*, 609295. [CrossRef] [PubMed]

53. Baby, J.; Devan, A.R.; Kumar, A.R.; Gorantla, J.N.; Nair, B.; Aishwarya, T.S.; Nath, L.R. Cogent role of flavonoids as key orchestrators of chemoprevention of hepatocellular carcinoma: A review. *J. Food Biochem.* **2021**, *45*, e13761. [CrossRef] [PubMed]

54. Li, Y.; Zhao, R.; Xiu, Z.; Yang, X.; Zhu, Y.; Han, J.; Li, S.; Li, Y.; Sun, L.; Li, X.; et al. Neobavaisoflavone induces pyroptosis of liver cancer cells via Tom20 sensing the activated ROS signal. *Phytomedicine* **2023**, *116*, 154869. [CrossRef] [PubMed]

55. Sarkar, S.; Das, A.K.; Bhattacharya, S.; Gachhui, R.; Sil, P.C. Isorhamnetin exerts anti-tumor activity in DEN + CCl$_4$-induced HCC mice. *Med. Oncol.* **2023**, *40*, 188. [CrossRef] [PubMed]

56. Pan, F.-f.; Zheng, Y.-B.; Shi, C.-J.; Zhang, F.-w.; Zhang, J.-f.; Fu, W.-m. H19-Wnt/β-catenin regulatory axis mediates the suppressive effects of apigenin on tumor growth in hepatocellular carcinoma. *Eur. J. Pharmacol.* **2021**, *893*, 173810. [CrossRef] [PubMed]

57. Yang, A.; Zhang, P.; Sun, Z.; Liu, X.; Zhang, X.; Liu, X.; Wang, D.; Meng, Z. Lysionotin induces apoptosis of hepatocellular carcinoma cells via caspase-3 mediated mitochondrial pathway. *Chem. -Biol. Interact.* **2021**, *344*, 109500. [CrossRef] [PubMed]

58. Chandrababu, G.; Varkey, M.; Devan, A.R.; Anjaly, M.V.; Unni, A.R.; Nath, L.R. Kaempferide exhibits an anticancer effect against hepatocellular carcinoma in vitro and in vivo. *Naunyn-Schmiedeberg's Arch. Pharmacol.* **2023**, *396*, 2461–2467. [CrossRef] [PubMed]

59. Hong, M.; Almutairi, M.M.; Li, S.; Li, J. Wogonin inhibits cell cycle progression by activating the glycogen synthase kinase-3 beta in hepatocellular carcinoma. *Phytomedicine* **2020**, *68*, 153174. [CrossRef]

60. Chen, H.; Shang, X.; Yuan, H.; Niu, Q.; Chen, J.; Luo, S.; Li, W.; Li, X. Total flavonoids of *Oldenlandia diffusa* (Willd.) Roxb. suppresses the growth of hepatocellular carcinoma through endoplasmic reticulum stress-mediated autophagy and apoptosis. *Front. Pharmacol.* **2022**, *13*, 1019670. [CrossRef]

61. Song, L.; Luo, Y.; Li, S.; Hong, M.; Wang, Q.; Chi, X.; Yang, C. ISL Induces Apoptosis and Autophagy in Hepatocellular Carcinoma via Downregulation of PI3K/AKT/mTOR Pathway in vivo and in vitro. *Drug Des. Dev. Ther.* **2020**, *14*, 4363–4376. [CrossRef]

62. Li, G.; Qi, L.; Chen, H.; Tian, G. Involvement of NF-κB/PI3K/AKT signaling pathway in the protective effect of prunetin against a diethylnitrosamine induced hepatocellular carcinogenesis in rats. *J. Biochem. Mol. Toxicol.* **2022**, *36*, e23016. [CrossRef] [PubMed]

63. Ke, M.; Zhang, Z.; Xu, B.; Zhao, S.; Ding, Y.; Wu, X.; Wu, R.; Lv, Y.; Dong, J. Baicalein and baicalin promote antitumor immunity by suppressing PD-L1 expression in hepatocellular carcinoma cells. *Int. Immunopharmacol.* **2019**, *75*, 105824. [CrossRef] [PubMed]

64. Mo'men, Y.S.; Hussein, R.M.; Kandeil, M.A. Involvement of PI3K/Akt pathway in the protective effect of hesperidin against a chemically induced liver cancer in rats. *J. Biochem. Mol. Toxicol.* **2019**, *33*, e22305. [CrossRef] [PubMed]

65. Badroon, N.; Abdul Majid, N.; Al-Suede, F.; Nazari, V.M.; Giribabu, N.; Abdul Majid, A.; Eid, E.; Alshawsh, M. Cardamonin Exerts Antitumor Effect on Human Hepatocellular Carcinoma Xenografts in Athymic Nude Mice through Inhibiting NF-κβ Pathway. *Biomedicines* **2020**, *8*, 586. [CrossRef]

66. Fu, H.; Liu, X.; Jin, L.; Lang, J.; Hu, Z.; Mao, W.; Cheng, C.; Shou, Q. Safflower yellow reduces DEN-induced hepatocellular carcinoma by enhancing liver immune infiltration through promotion of collagen degradation and modulation of gut microbiota. *Food Funct.* **2021**, *12*, 10632–10643. [CrossRef]

67. Li, L.; Xu, H.; Qu, L.; Xu, K.; Liu, X. Daidzin inhibits hepatocellular carcinoma survival by interfering with the glycolytic/gluconeogenic pathway through downregulation of TPI1. *Biofactors* **2022**, *48*, 883–896. [CrossRef]

68. Shao, H.; Chen, J.; Li, A.; Ma, L.; Tang, Y.; Chen, H.; Chen, Y.; Liu, J. Salvigenin Suppresses Hepatocellular Carcinoma Glycolysis and Chemoresistance Through Inactivating the PI3K/AKT/GSK-3β Pathway. *Appl. Biochem. Biotechnol.* **2023**, *195*, 5217–5237. [CrossRef]

69. Yang, W.; Chen, X.; Li, Y.; Guo, S.; Wang, Z.; Yu, X. Advances in Pharmacological Activities of Terpenoids. *Nat. Prod. Commun.* **2020**, *15*, 1934578X20903555. [CrossRef]

70. Wróblewska-Łuczka, P.; Cabaj, J.; Bargieł, J.; Łuszczki, J.J. Anticancer effect of terpenes: Focus on malignant melanoma. *Pharmacol. Rep.* **2023**, *75*, 1115–1125. [CrossRef]

71. de Vasconcelos Cerqueira Braz, J.; Carvalho Nascimento Júnior, J.A.; Serafini, M.R. Terpenes with Antitumor Activity: A Patent Review. *Recent Pat. Anti-Cancer Drug Discov.* **2020**, *15*, 321–328. [CrossRef] [PubMed]

72. Xin, Y.; Huo, R.; Su, W.; Xu, W.; Qiu, Z.; Wang, W.; Qiu, Y. The anticancer mechanism of Terpenoid-based drug candi-dates: Focus on tumor microenvironment. *Anti-Cancer Agents Med. Chem.* 2023. *ahead of print*. [CrossRef] [PubMed]

73. Kim, J.-H. Pharmacological and medical applications of Panax ginseng and ginsenosides: A review for use in cardiovascular diseases. *J. Ginseng Res.* **2018**, *42*, 264–269. [CrossRef]

74. Jiang, R.; Luo, S.; Zhang, M.; Wang, W.; Zhuo, S.; Wu, Y.; Qiu, Q.; Yuan, Y.; Jiang, X. Ginsenoside Rh4 inhibits inflammation-related hepatocellular carcinoma progression by targeting HDAC4/IL-6/STAT3 signaling. *Mol. Genet. Genom.* **2023**, *298*, 1479–1492. [CrossRef] [PubMed]

75. Zhang, S.; Zhang, M.; Chen, J.; Zhao, J.; Su, J.; Zhang, X. Ginsenoside Compound K Regulates HIF-1α-Mediated Glycolysis Through Bclaf1 to Inhibit the Proliferation of Human Liver Cancer Cells. *Front. Pharmacol.* **2020**, *11*, 583334. [CrossRef]

76. Qu, L.; Liu, Y.; Deng, J.; Ma, X.; Fan, D. Ginsenoside Rk3 is a novel PI3K/AKT-targeting therapeutics agent that regulates autophagy and apoptosis in hepatocellular carcinoma. *J. Pharm. Anal.* **2023**, *13*, 463–482. [CrossRef] [PubMed]

77. Hu, Q.-R.; Huang, Q.-X.; Hong, H.; Pan, Y.; Luo, T.; Li, J.; Deng, Z.-Y.; Chen, F. Ginsenoside Rh2 and its octyl ester derivative inhibited invasion and metastasis of hepatocellular carcinoma via the c-Jun/COX2/PGE2 pathway. *Phytomedicine* **2023**, *121*, 155131. [CrossRef] [PubMed]

78. Wu, H.-C.; Hu, Q.-R.; Luo, T.; Wei, W.-C.; Wu, H.-J.; Li, J.; Zheng, L.-F.; Xu, Q.-Y.; Deng, Z.-Y.; Chen, F. The immunomodulatory effects of ginsenoside derivative Rh2-O on splenic lymphocytes in H22 tumor-bearing mice is partially mediated by TLR4. *Int. Immunopharmacol.* **2021**, *101*, 108316. [CrossRef] [PubMed]

79. Xiu, Z.; Zhu, Y.; Li, S.; Li, Y.; Yang, X.; Li, Y.; Song, G.; Jin, N.; Fang, J.; Han, J.; et al. Betulinic acid inhibits growth of hepatoma cells through activating the NCOA4-mediated ferritinophagy pathway. *J. Funct. Foods* **2023**, *102*, 105441. [CrossRef]

80. Wang, J.; Pu, J.; Zhang, Z.; Feng, Z.; Han, J.; Su, X.; Shi, L. Triterpenoids of Ganoderma lucidum inhibited S180 sarcoma and H22 hepatoma in mice by regulating gut microbiota. *Heliyon* **2023**, *9*, e16682. [CrossRef]

81. Zhang, J.; Ma, X.; Fan, D. Ginsenoside CK Inhibits Hypoxia-Induced Epithelial–Mesenchymal Transformation through the HIF-1α/NF-κB Feedback Pathway in Hepatocellular Carcinoma. *Foods* **2021**, *10*, 1195. [CrossRef] [PubMed]

82. You, M.; Fu, J.; Lv, X.; Wang, L.; Wang, H.; Li, R. Saikosaponin b2 inhibits tumor angiogenesis in liver cancer via down-regulation of VEGF/ERK/HIF-1α signaling. *Oncol. Rep.* **2023**, *50*, 8573. [CrossRef] [PubMed]

83. Lei, C.; Gao, Z.; Lv, X.; Zhu, Y.; Li, R.; Li, S. Saikosaponin-b2 Inhibits Primary Liver Cancer by Regulating the STK4/IRAK1/NF-κB Pathway. *Biomedicines* **2023**, *11*, 2859. [CrossRef] [PubMed]

84. Gong, Y.F.; Hou, S.; Xu, J.-C.; Chen, Y.; Zhu, L.-L.; Xu, Y.-Y.; Chen, Y.-Q.; Li, M.-M.; Li, L.-L.; Yang, J.-J.; et al. Amelioratory effects of astragaloside IV on hepatocarcinogenesis via Nrf2-mediated pSmad3C/3L transformation. *Phytomedicine* **2023**, *117*, 154903. [CrossRef]

85. Hu, X.; Jiao, F.; Zhang, L.; Jiang, Y. Dihydrotanshinone inhibits hepatocellular carcinoma by suppressing the JAK2/STAT3 pathway. *Front. Pharmacol.* **2021**, *12*, 654986. [CrossRef] [PubMed]

86. Liu, Z.; Wang, N.; Meng, Z.; Lu, S.; Peng, G. Pseudolaric acid B triggers cell apoptosis by activating AMPK/JNK/DRP1/mitochondrial fission pathway in hepatocellular carcinoma. *Toxicology* **2023**, *493*, 153556. [CrossRef] [PubMed]

87. Dai, X.; Sun, F.; Deng, K.; Lin, G.; Yin, W.; Chen, H.; Yang, D.; Liu, K.; Zhang, Y.; Huang, L. Mallotucin D, a Clerodane Diterpenoid from *Croton crassifolius*, Suppresses HepG2 Cell Growth via Inducing Autophagic Cell Death and Pyroptosis. *Int. J. Mol. Sci.* **2022**, *23*, 14217. [CrossRef]

88. Chang, W.-T.; Bow, Y.-D.; Fu, P.-J.; Li, C.-Y.; Wu, C.-Y.; Chang, Y.-H.; Teng, Y.-N.; Li, R.-N.; Lu, M.-C.; Liu, Y.-C.; et al. A Marine Terpenoid, Heteronemin, Induces Both the Apoptosis and Ferroptosis of Hepatocellular Carcinoma Cells and Involves the ROS and MAPK Pathways. *Oxidative Med. Cell. Longev.* **2021**, *2021*, 7689045. [CrossRef]

89. Liu, C.; Yang, S.; Wang, K.; Bao, X.; Liu, Y.; Zhou, S.; Liu, H.; Qiu, Y.; Wang, T.; Yu, H. Alkaloids from Traditional Chinese Medicine against hepatocellular carcinoma. *Biomed. Pharmacother.* **2019**, *120*, 109543. [CrossRef]

90. Wang, Y.; Zhang, S.; Liu, J.; Fang, B.; Yao, J.; Cheng, B. Matrine inhibits the invasive and migratory properties of human hepatocellular carcinoma by regulating epithelial-mesenchymal transition. *Mol. Med. Rep.* **2018**, *18*, 911–919. [CrossRef]

91. Li, P.; Liu, Y.; He, Q. Anisodamine Suppressed the Growth of Hepatocellular Carcinoma Cells, Induced Apoptosis and Regulated the Levels of Inflammatory Factors by Inhibiting NLRP3 Inflammasome Activation. *Drug Des. Dev. Ther.* **2020**, *14*, 1609–1620. [CrossRef] [PubMed]

92. Bao, X.; Liu, Y.; Huang, J.; Yin, S.; Sheng, H.; Han, X.; Chen, Q.; Wang, T.; Chen, S.; Qiu, Y.; et al. Stachydrine hydrochloride inhibits hepatocellular carcinoma progression via LIF/AMPK axis. *Phytomedicine* **2022**, *100*, 154066. [CrossRef] [PubMed]

93. Feng, F.; Pan, L.; Wu, J.; Li, L.; Xu, H.; Yang, L.; Xu, K.; Wang, C. Cepharanthine inhibits hepatocellular carcinoma cell growth and proliferation by regulating amino acid metabolism and suppresses tumorigenesis in vivo. *Int. J. Biol. Sci.* **2021**, *17*, 4340–4352. [CrossRef] [PubMed]

94. Hu, G.; Cao, C.; Deng, Z.; Li, J.; Zhou, X.; Huang, Z.; Cen, C. Effects of matrine in combination with cisplatin on liver cancer. *Oncol. Lett.* **2020**, *21*, 12327. [CrossRef] [PubMed]

95. Su, G.-F.; Huang, Z.-X.; Huang, D.-L.; Chen, P.-X.; Wang, Y.; Wang, Y.-F. Cepharanthine hydrochloride inhibits the Wnt/β-catenin/Hedgehog signaling axis in liver cancer. *Oncol. Rep.* **2022**, *47*, 8294. [CrossRef] [PubMed]

96. Yin, L.; Xia, Y.; Xu, P.; Zheng, W.; Gao, Y.; Xie, F.; Ji, Z. Veratramine suppresses human HepG2 liver cancer cell growth in vitro and in vivo by inducing autophagic cell death. *Oncol. Rep.* **2020**, *44*, 477–486. [CrossRef] [PubMed]

97. Hu, P.; Hu, L.; Chen, Y.; Wang, F.; Xiao, Y.; Tong, Z.; Li, H.; Xiang, M.; Tong, Q.; Zhang, Y. Chaetocochin J exhibits anti-hepatocellular carcinoma effect independent of hypoxia. *Bioorganic Chem.* **2023**, *139*, 106701. [CrossRef] [PubMed]

98. Zhao, Z.; Zhang, D.; Wu, F.; Tu, J.; Song, J.; Xu, M.; Ji, J. Sophoridine suppresses lenvatinib-resistant hepatocellular carcinoma growth by inhibiting RAS/MEK/ERK axis via decreasing VEGFR2 expression. *J. Cell. Mol. Med.* **2020**, *25*, 549–560. [CrossRef] [PubMed]

99. Zhang, J.; Chen, Y.; Lin, J.; Jia, R.; An, T.; Dong, T.; Zhang, Y.; Yang, X. Cyclovirobuxine D Exerts Anticancer Effects by Suppressing the EGFR-FAK-AKT/ERK1/2-Slug Signaling Pathway in Human Hepatocellular Carcinoma. *DNA Cell Biol.* **2020**, *39*, 355–367. [CrossRef]

100. Zheng, L.; Fang, S.; Chen, A.; Chen, W.; Qiao, E.; Chen, M.; Shu, G.; Zhang, D.; Kong, C.; Weng, Q.; et al. Piperlongumine synergistically enhances the antitumour activity of sorafenib by mediating ROS-AMPK activation and targeting CPSF7 in liver cancer. *Pharmacol. Res.* **2022**, *177*, 106140. [CrossRef]

101. Kong, W.-S.; Shen, F.-X.; Xie, R.-F.; Zhou, G.; Feng, Y.-M.; Zhou, X. Bufothionine Induces Autophagy in H22 Hepatoma-bearing Mice by Inhibiting JAK2/STAT3 Pathway, a Possible Anti-cancer Mechanism of Cinobufacini. *J. Ethnopharmacol.* **2021**, *270*, 113848. [CrossRef] [PubMed]

102. Jiang, Y.; Yang, Y.; Hu, Y.; Yang, R.; Huang, J.; Liu, Y.; Wu, Y.; Li, S.; Ma, C.; Humphries, F.; et al. Gasdermin D restricts anti-tumor immunity during PD-L1 checkpoint blockade. *Cell Rep.* **2022**, *41*, 111553. [CrossRef] [PubMed]

103. Liu, Y.; Cheng, D.-H.; Lai, K.-D.; Su, H.; Lu, G.-S.; Wang, L.; Lv, J.-H. Ventilagolin Suppresses Migration, Invasion and Epithelial-Mesenchymal Transition of Hepatocellular Carcinoma Cells by Downregulating Pim-1. *Drug Des. Dev. Ther.* **2021**, *15*, 4885–4899. [CrossRef] [PubMed]

104. Sufianova, G.; Gareev, I.; Beylerli, O.; Wu, J.; Shumadalova, A.; Sufianov, A.; Chen, X.; Zhao, S. Modern aspects of the use of natural polyphenols in tumor prevention and therapy. *Front. Cell Dev. Biol.* **2022**, *10*, 1011435. [CrossRef] [PubMed]

105. Huang, X.; Wang, Y.; Yang, W.; Dong, J.; Li, L. Regulation of dietary polyphenols on cancer cell pyroptosis and the tumor immune microenvironment. *Front. Nutr.* **2022**, *9*, 974896. [CrossRef] [PubMed]

106. Chang, Z.; Jian, P.; Zhang, Q.; Liang, W.; Zhou, K.; Hu, Q.; Liu, Y.; Liu, R.; Zhang, L. Tannins in Terminalia bellirica inhibit hepatocellular carcinoma growth by regulating EGFR-signaling and tumor immunity. *Food Funct.* **2021**, *12*, 3720–3739. [CrossRef] [PubMed]

107. Jiang, Y.; Nan, H.; Shi, N.; Hao, W.; Dong, J.; Chen, H. Chlorogenic acid inhibits proliferation in human hepatoma cells by suppressing noncanonical NF-κB signaling pathway and triggering mitochondrial apoptosis. *Mol. Biol. Rep.* **2021**, *48*, 2351–2364. [CrossRef]

108. Han, L.; Wang, Y.; Sun, S. Curcumin inhibits proliferation of hepatocellular carcinoma cells through down regulation of DJ-1. *Cancer Biomark.* **2020**, *29*, 1–8. [CrossRef]

109. Zhang, Y.; Li, X.; Li, X. Curcumae Ameliorates Diethylnitrosamine-Induced Hepatocellular Carcinoma via Alteration of Oxidative Stress, Inflammation and Gut Microbiota. *J. Inflamm. Res.* **2021**, *14*, 5551–5566. [CrossRef]

110. Dai, H.; Li, M.; Yang, W.; Sun, X.; Wang, P.; Wang, X.; Su, J.; Wang, X.; Hu, X.; Zhao, M. Resveratrol inhibits the malignant progression of hepatocellular carcinoma via MARCH1-induced regulation of PTEN/AKT signaling. *Aging* **2020**, *12*, 11717–11731. [CrossRef]

111. Liu, G.; Shi, A.; Wang, N.; Li, M.; He, X.; Yin, C.; Tu, Q.; Shen, X.; Tao, Y.; Wang, Q.; et al. Polyphenolic Proanthocyanidin-B2 suppresses proliferation of liver cancer cells and hepatocellular carcinogenesis through directly binding and inhibiting AKT activity. *Redox Biol.* **2020**, *37*, 101701. [CrossRef]

112. Na, W.; Ma, B.; Shi, S.; Chen, Y.; Zhang, H.; Zhan, Y.; An, H. Procyanidin B1, a novel and specific inhibitor of Kv10.1 channel, suppresses the evolution of hepatoma. *Biochem. Pharmacol.* **2020**, *178*, 114089. [CrossRef]

113. Gong, H.; He, Z.; Peng, A.; Zhang, X.; Cheng, B.; Sun, Y.; Zheng, L.; Huang, K. Effects of several quinones on insulin aggregation. *Sci. Rep.* **2014**, *4*, srep05648. [CrossRef]

114. Qin, B.; Zeng, Z.; Xu, J.; Shangwen, J.; Ye, Z.J.; Wang, S.; Wu, Y.; Peng, G.; Wang, Q.; Gu, W.; et al. Emodin inhibits invasion and migration of hepatocellular carcinoma cells via regulating autophagy-mediated degradation of snail and β-catenin. *BMC Cancer* **2022**, *22*, 671. [CrossRef]

115. Khan, H.; Jia, W.; Yu, Z.; Zaib, T.; Feng, J.; Jiang, Y.; Song, H.; Bai, Y.; Yang, B.; Feng, H. Emodin succinyl ester inhibits malignant proliferation and migration of hepatocellular carcinoma by suppressing the interaction of AR and EZH2. *Biomed. Pharmacother.* **2020**, *128*, 110244. [CrossRef]

116. Tadros, S.A.; Attia, Y.M.; Maurice, N.W.; Fahim, S.A.; Abdelwahed, F.M.; Ibrahim, S.; Badary, O.A. Thymoquinone Suppresses Angiogenesis in DEN-Induced Hepatocellular Carcinoma by Targeting miR-1-3p. *Int. J. Mol. Sci.* **2022**, *23*, 15904. [CrossRef]

117. Yao, L.; Yan, D.; Jiang, B.; Xue, Q.; Chen, X.; Huang, Q.; Qi, L.; Tang, D.; Chen, X.; Liu, J. Plumbagin is a novel GPX4 protein degrader that induces apoptosis in hepatocellular carcinoma cells. *Free Radic. Biol. Med.* **2023**, *203*, 1–10. [CrossRef]

118. Li, Y.; Tian, Y.; Zhong, W.; Wang, N.; Wang, Y.; Zhang, Y.; Zhang, Z.; Li, J.; Ma, F.; Zhao, Z.; et al. Artemisia argyi Essential Oil Inhibits Hepatocellular Carcinoma Metastasis via Suppression of DEPDC1 Dependent Wnt/β-catenin Signaling Pathway. *Front. Cell Dev. Biol.* **2021**, *9*, 664791. [CrossRef]

119. Costa, E.V.; de Souza, C.A.S.; Galvão, A.F.C.; Silva, V.R.; Santos, L.d.S.; Dias, R.B.; Rocha, C.A.G.; Soares, M.B.P.; da Silva, F.M.A.; Koolen, H.H.F.; et al. *Duguetia pycnastera* Sandwith (Annonaceae) Leaf Essential Oil Inhibits HepG2 Cell Growth In Vitro and In Vivo. *Molecules* **2022**, *27*, 5664. [CrossRef]

120. Lima, E.J.S.P.d.; Fontes, S.S.; Nogueira, M.L.; Silva, V.R.; Santos, L.d.S.; D'Elia, G.M.A.; Dias, R.B.; Sales, C.B.S.; Rocha, C.A.G.; Vannier-Santos, M.A.; et al. Essential oil from leaves of *Conobea scoparioides* (Cham. & Schltdl.) Benth. (Plantaginaceae) causes cell death in HepG2 cells and inhibits tumor development in a xenograft model. *Biomed. Pharmacother.* **2020**, *129*, 110402. [CrossRef]

121. de Oliveira, F.P.; da C Rodrigues, A.C.B.; de Lima, E.J.S.P.; Silva, V.R.; de S Santos, L.; da Anunciação, T.A.; Nogueira, M.L.; Soares, M.B.P.; Dias, R.B.; Gurgel Rocha, C.A.; et al. Essential Oil from Bark of *Aniba parviflora* (M$_{eisn}$.) M$_{ez}$ (Lauraceae) Reduces HepG2 Cell Proliferation and Inhibits Tumor Development in a Xenograft Model. *Chem. Biodivers.* **2021**, *18*, e2000938. [CrossRef]

122. Nogueira, M.L.; Lima, E.J.S.P.d.; Adrião, A.A.X.; Fontes, S.S.; Silva, V.R.; Santos, L.d.S.; Soares, M.B.P.; Dias, R.B.; Rocha, C.A.G.; Costa, E.V.; et al. *Cyperus articulatus* L. (Cyperaceae) Rhizome Essential Oil Causes Cell Cycle Arrest in the G2/M Phase and Cell Death in HepG2 Cells and Inhibits the Development of Tumors in a Xenograft Model. *Molecules* **2020**, *25*, 2687. [CrossRef]

123. Wang, Y.-F.; Zheng, Y.; Cha, Y.-Y.; Feng, Y.; Dai, S.-X.; Zhao, S.; Chen, H.; Xu, M. Essential oil of lemon myrtle (*Backhousia citriodora*) induces S-phase cell cycle arrest and apoptosis in HepG2 cells. *J. Ethnopharmacol.* **2023**, *312*, 116493. [CrossRef]

124. Huang, X.-F.; Sheu, G.-T.; Chang, K.-F.; Huang, Y.-C.; Hung, P.-H.; Tsai, N.-M. Pogostemon cablin Triggered ROS-Induced DNA Damage to Arrest Cell Cycle Progression and Induce Apoptosis on Human Hepatocellular Carcinoma In Vitro and In Vivo. *Molecules* **2020**, *25*, 5639. [CrossRef]

125. Wen, Y.; Zhou, X.; Lu, M.; He, M.; Tian, Y.; Liu, L.; Wang, M.; Tan, W.; Deng, Y.; Yang, X.; et al. Bclaf1 promotes angiogenesis by regulating HIF-1α transcription in hepatocellular carcinoma. *Oncogene* **2018**, *38*, 1845–1859. [CrossRef]

126. Cappuyns, S.; Llovet, J.M. Combination Therapies for Advanced Hepatocellular Carcinoma: Biomarkers and Unmet Needs. *Clin. Cancer Res.* **2022**, *28*, 3405–3407. [CrossRef]

127. Yao, H.; Liu, N.; Lin, M.C.; Zheng, J. Positive feedback loop between cancer stem cells and angiogenesis in hepatocellular carcinoma. *Cancer Lett.* **2016**, *379*, 213–219. [CrossRef]

128. Cheng, W.; Cheng, Z.; Weng, L.; Xing, D.; Zhang, M. Asparagus Polysaccharide inhibits the Hypoxia-induced migration, invasion and angiogenesis of Hepatocellular Carcinoma Cells partly through regulating HIF1α/VEGF expression via MAPK and PI3K signaling pathway. *J. Cancer* **2021**, *12*, 3920–3929. [CrossRef]

129. Yucui, L.; Hongxing, Z. Research Overview on the Mechanism of Traditional Chinese Medicine in Preventing and Treating Liver Cancer. *J. Pract. Tradit. Chin. Intern. Med.* **2023**, 1–8.

130. Su, W.; Wu, L.; Liang, Q.; Lin, X.; Xu, X.; Yu, S.; Lin, Y.; Zhou, J.; Fu, Y.; Gao, X.; et al. Extraction Optimization, Structural Characterization, and Anti-Hepatoma Activity of Acidic Polysaccharides From *Scutellaria barbata* D. Don. *Front. Pharmacol.* **2022**, *13*, 827782. [CrossRef]

131. Li, Q.-Z.; Chen, Y.-Y.; Liu, Q.-P.; Feng, Z.-H.; Zhang, L.; Zhang, H. Cucurbitacin B suppresses hepatocellular carcinoma progression through inducing deoxyribonucleic acid (DNA) damage-dependent cell cycle arrest. *Phytomedicine* **2024**, *126*, 155177. [CrossRef] [PubMed]

132. Elmore, S. Apoptosis: A review of programmed cell death. *Sage J.* **2007**, *35*, 495–516. [CrossRef] [PubMed]

133. Carneiro, B.A.; El-Deiry, W.S. Targeting apoptosis in cancer therapy. *Nat. Rev. Clin. Oncol.* **2020**, *17*, 395–417. [CrossRef] [PubMed]

134. Fernald, K.; Kurokawa, M. Evading apoptosis in cancer. *Trends Cell Biol.* **2013**, *23*, 620–633. [CrossRef]

135. Nie, C.; Wang, B.; Wang, B.; Lv, N.; Yu, R.; Zhang, E. Protopine triggers apoptosis via the intrinsic pathway and regulation of ROS/PI3K/Akt signalling pathway in liver carcinoma. *Cancer Cell Int.* **2021**, *21*, 396. [CrossRef] [PubMed]

136. Ichimiya, T.; Yamakawa, T.; Hirano, T.; Yokoyama, Y.; Hayashi, Y.; Hirayama, D.; Wagatsuma, K.; Itoi, T.; Nakase, H. Autophagy and autophagy-related diseases: A review. *Int. J. Mol. Sci.* **2020**, *21*, 8974. [CrossRef] [PubMed]

137. Li, Y.; Li, Y.; Zhang, J.; Ji, L.; Li, M.; Sun, X.; Feng, H.; Yu, Z.; Gao, Y. Current Perspective of Traditional Chinese Medicines and Active Ingredients in the Therapy of Hepatocellular Carcinoma. *J. Hepatocell. Carcinoma* **2022**, *9*, 41–56. [CrossRef] [PubMed]

138. Wang, Y.; Gao, W.; Shi, X.; Ding, J.; Liu, W.; He, H.; Wang, K.; Shao, F. Chemotherapy drugs induce pyroptosis through caspase-3 cleavage of a gasdermin. *Nature* **2017**, *547*, 99–103. [CrossRef]

139. Chen, X.; He, W.-t.; Hu, L.; Li, J.; Fang, Y.; Wang, X.; Xu, X.; Wang, Z.; Huang, K.; Han, J. Pyroptosis is driven by non-selective gasdermin-D pore and its morphology is different from MLKL channel-mediated necroptosis. *Cell Res.* **2016**, *26*, 1007–1020. [CrossRef]

140. Sun, Z.; Ma, C.; Zhan, X. Ajmalicine induces the pyroptosis of hepatoma cells to exert the antitumor effect. *J. Biochem. Mol. Toxicol.* **2023**, *38*, e23614. [CrossRef]

141. Wang, Y.; Zhao, M.; Zhao, L.; Geng, Y.; Li, G.; Chen, L.; Yu, J.; Yuan, H.; Zhang, H.; Yun, H. HBx-induced HSPA8 stimulates HBV replication and suppresses ferroptosis to support liver cancer progression. *Cancer Res.* **2023**, *83*, 1048–1061. [CrossRef] [PubMed]

142. Hu, W.; Zhou, C.; Jing, Q.; Li, Y.; Yang, J.; Yang, C.; Wang, L.; Hu, J.; Li, H.; Wang, H. FTH promotes the proliferation and renders the HCC cells specifically resist to ferroptosis by maintaining iron homeostasis. *Cancer Cell Int.* **2021**, *21*, 709. [CrossRef] [PubMed]

143. Warburg, O.; Wind, F.; Negelein, E. The metabolism of tumors in the body. *J. Gen. Physiol.* **1927**, *8*, 519. [CrossRef] [PubMed]

144. Cascone, T.; McKenzie, J.A.; Mbofung, R.M.; Punt, S.; Wang, Z.; Xu, C.; Williams, L.J.; Wang, Z.; Bristow, C.A.; Carugo, A. Increased tumor glycolysis characterizes immune resistance to adoptive T cell therapy. *Cell Metab.* **2018**, *27*, 977–987.e974. [CrossRef] [PubMed]

145. Greene, L.I.; Bruno, T.C.; Christenson, J.L.; D'Alessandro, A.; Culp-Hill, R.; Torkko, K.; Borges, V.F.; Slansky, J.E.; Richer, J.K. A role for tryptophan-2, 3-dioxygenase in CD8 T-cell suppression and evidence of tryptophan catabolism in breast cancer patient plasma. *Mol. Cancer Res.* **2019**, *17*, 131–139. [CrossRef] [PubMed]

146. Jiang, L.; Fang, X.; Wang, H.; Li, D.; Wang, X. Ovarian cancer-intrinsic fatty acid synthase prevents anti-tumor immunity by disrupting tumor-infiltrating dendritic cells. *Front. Immunol.* **2018**, *9*, 2927. [CrossRef] [PubMed]

147. Zhang, J.; Hong, Y.; Jiang, L.; Yi, X.; Chen, Y.; Liu, L.; Chen, Z.; Wu, Y.; Cai, Z. Global Metabolomic and Lipidomic Analysis Reveal the Synergistic Effect of Bufalin in Combination with Cinobufagin against HepG2 Cells. *J. Proteome Res.* **2020**, *19*, 873–883. [CrossRef] [PubMed]

148. Alam, Z.; Devalaraja, S.; Li, M.; To, T.K.J.; Folkert, I.W.; Mitchell-Velasquez, E.; Dang, M.T.; Young, P.; Wilbur, C.J.; Silverman, M.A.; et al. Counter Regulation of Spic by NF-κB and STAT Signaling Controls Inflammation and Iron Metabolism in Macrophages. *Cell Rep.* **2020**, *31*, 107825. [CrossRef]
149. Xie, X.; Shen, W.; Zhou, Y.; Ma, L.; Xu, D.; Ding, J.; He, L.; Shen, B.; Zhou, C. Characterization of a polysaccharide from Eupolyphaga sinensis walker and its effective antitumor activity via lymphocyte activation. *Int. J. Biol. Macromol.* **2020**, *162*, 31–42. [CrossRef]
150. Zhang, Q.; Huang, H.; Zheng, F.; Liu, H.; Qiu, F.; Chen, Y.; Liang, C.-L.; Dai, Z. Resveratrol exerts antitumor effects by downregulating CD8+CD122+ Tregs in murine hepatocellular carcinoma. *OncoImmunology* **2020**, *9*, 1829346. [CrossRef]
151. Anfray, C.; Ummarino, A.; Andón, F.T.; Allavena, P. Current Strategies to Target Tumor-Associated-Macrophages to Improve Anti-Tumor Immune Responses. *Cells* **2019**, *9*, 46. [CrossRef] [PubMed]
152. Degroote, H.; Van Dierendonck, A.; Geerts, A.; Van Vlierberghe, H.; Devisscher, L. Preclinical and Clinical Therapeutic Strategies Affecting Tumor-Associated Macrophages in Hepatocellular Carcinoma. *J. Immunol. Res.* **2018**, *2018*, 7819520. [CrossRef] [PubMed]
153. Wang, M.; Li, Y.; Li, S.; Wang, T.; Wang, M.; Wu, H.; Zhang, M.; Luo, S.; Zhao, C.; Li, Q.; et al. Cinobufacini injection delays hepatocellular carcinoma progression by regulating lipid metabolism via SREBP1 signaling pathway and affecting macrophage polarization. *J. Ethnopharmacol.* **2024**, *321*, 117472. [CrossRef]
154. Xu, Y.; Yu, H.; Wang, J.; Gao, W.; Guan, Y.; Yi, S.; Ren, G.; Bai, C.; Zhu, Y. Research progress on the mechanism of Traditional Chinese Medicine regulating intestinal functionin the treatment of hepatocellular carcinoma based on the gut liver axis. *Chin. J. Integr. Tradit. West. Med. Dig.* **2024**, *32*, 222–225. [CrossRef]
155. Liu, S.; Dong, W.; Yang, L.; Zhou, J.; Liu, D. Improvement and Mechanism of Ganoderma lucidum Polysaccharides and Its Flora Metabolites on Insulin Resistance in HepG2 Cells. *Sci. Technol. Food Ind.* **2023**, *44*, 314–321. [CrossRef]
156. Ren, Z.; Chen, X.; Hong, L.; Zhao, X.; Cui, G.; Li, A.; Liu, Y.; Zhou, L.; Sun, R.; Shen, S. Nanoparticle conjugation of ginsenoside Rg3 inhibits hepatocellular carcinoma development and metastasis. *Small* **2020**, *16*, 1905233. [CrossRef] [PubMed]
157. Ponziani, F.R.; Bhoori, S.; Castelli, C.; Putignani, L.; Rivoltini, L.; Del Chierico, F.; Sanguinetti, M.; Morelli, D.; Paroni Sterbini, F.; Petito, V. Hepatocellular carcinoma is associated with gut microbiota profile and inflammation in nonalcoholic fatty liver disease. *Hepatology* **2019**, *69*, 107–120. [CrossRef]
158. Xie, G.; Wang, X.; Liu, P.; Wei, R.; Chen, W.; Rajani, C.; Hernandez, B.Y.; Alegado, R.; Dong, B.; Li, D. Distinctly altered gut microbiota in the progression of liver disease. *Oncotarget* **2016**, *7*, 19355. [CrossRef] [PubMed]
159. Huo, R.; Yang, W.-J.; Liu, Y.; Liu, T.; Li, T.; Wang, C.-Y.; Pan, B.-S.; Wang, B.-L.; Guo, W. Stigmasterol: Remodeling Gut Microbiota and Suppressing Tumor Growth through Treg and CD8+ T Cells in Hepatocellular Carcinoma. *Phytomedicine* **2023**, *129*, 155225. [CrossRef] [PubMed]
160. Zhao, J.; He, R.; Zhong, H.; Liu, S.; Liu, X.; Hussain, M.; Sun, P. A cold-water extracted polysaccharide-protein complex from Grifola frondosa exhibited anti-tumor activity via TLR4-NF-κB signaling activation and gut microbiota modification in H22 tumor-bearing mice. *Int. J. Biol. Macromol.* **2023**, *239*, 124291. [CrossRef]
161. Dahmani, R.; Just, P.-A.; Perret, C. The Wnt/β-catenin pathway as a therapeutic target in human hepatocellular carcinoma. *Clin. Res. Hepatol. Gastroenterol.* **2011**, *35*, 709–713. [CrossRef] [PubMed]
162. Li, Y.; Wang, T.; Sun, Y.; Huang, T.; Li, C.; Fu, Y.; Li, Y.; Li, C. p53-Mediated PI3K/AKT/mTOR Pathway Played a Role in PtoxDpt-Induced EMT Inhibition in Liver Cancer Cell Lines. *Oxidative Med. Cell. Longev.* **2019**, *2019*, 2531493. [CrossRef]
163. Lei, R.-E.; Shi, C.; Zhang, P.-L.; Hu, B.-L.; Jiang, H.-X.; Qin, S.-Y. IL-9 promotes proliferation and metastasis of hepatocellular cancer cells by activating JAK2/STAT3 pathway. *Int. J. Clin. Exp. Pathol.* **2017**, *10*, 7940. [PubMed]
164. Cai, H.; Zhang, Y.; Meng, F.; Cui, C.; Li, H.; Sui, M.; Zhang, H.; Lu, S. Preoperative Serum IL6, IL8, and TNF-α May Predict the Recurrence of Hepatocellular Cancer. *Gastroenterol. Res. Pract.* **2019**, *2019*, 6160783. [CrossRef] [PubMed]
165. Zhao, S.; Xu, K.; Jiang, R.; Li, D.-Y.; Guo, X.-X.; Zhou, P.; Tang, J.-F.; Li, L.-S.; Zeng, D.; Hu, L.; et al. Evodiamine inhibits proliferation and promotes apoptosis of hepatocellular carcinoma cells via the Hippo-Yes-Associated Protein signaling pathway. *Life Sci.* **2020**, *251*, 117424. [CrossRef] [PubMed]
166. Li, X.; Liu, Q.; Xie, S.; Wu, X.; Fu, J. Mechanism of action of cordycepin in the treatment of hepatocellular carcinoma via regulation of the Hippo signaling pathway. *Food Sci. Hum. Wellness* **2024**, *13*, 1040–1054. [CrossRef]
167. Wu, D.-L.; Liao, Z.-D.; Chen, F.-F.; Zhang, W.; Ren, Y.-S.; Wang, C.-C.; Chen, X.-X.; Peng, D.-Y.; Kong, L.-Y. Benzophenones from Anemarrhena asphodeloides Bge. Exhibit Anticancer Activity in HepG2 Cells via the NF-κB Signaling Pathway. *Molecules* **2019**, *24*, 2246. [CrossRef] [PubMed]
168. Nüsslein-Volhard, C.; Wieschaus, E. Mutations affecting segment number and polarity in Drosophila. *Nature* **1980**, *287*, 795–801. [CrossRef]
169. Wu, J.; Tan, H.-Y.; Chan, Y.-T.; Lu, Y.; Feng, Z.; Yuan, H.; Zhang, C.; Feng, Y.; Wang, N. PARD3 drives tumorigenesis through activating Sonic Hedgehog signalling in tumour-initiating cells in liver cancer. *J. Exp. Clin. Cancer Res.* **2024**, *43*, 42. [CrossRef]
170. Ekiert, H.M.; Szopa, A. Biological Activities of Natural Products. *Molecules* **2020**, *25*, 5769. [CrossRef]
171. Wang, M.; Peng, P.; Chen, Z.; Deng, X. Nanoparticle delivery of active traditional Chinese medicine ingredients: A new strategy for the treatment of liver cancer. *Curr. Pharm. Biotechnol.* **2023**, *24*, 1630–1644. [CrossRef]
172. Zheng, Y.; Kong, F.; Liu, S.; Liu, X.; Pei, D.; Miao, X. Membrane protein-chimeric liposome-mediated delivery of triptolide for targeted hepatocellular carcinoma therapy. *Drug Deliv.* **2021**, *28*, 2033–2043. [CrossRef]

173. Shelash Al-Hawary, S.I.; Abdalkareem Jasim, S.; M Kadhim, M.; Jaafar Saadoon, S.; Ahmad, I.; Romero Parra, R.M.; Hasan Hammoodi, S.; Abulkassim, R.; M Hameed, N.; K Alkhafaje, W.; et al. Curcumin in the treatment of liver cancer: From mechanisms of action to nanoformulations. *Phytother. Res.* **2023**, *37*, 1624–1639. [CrossRef]
174. Chen, X.-P.; Li, Y.; Zhang, Y.; Li, G.-W. Formulation, Characterization And Evaluation Of Curcumin- Loaded PLGA- TPGS Nanoparticles For Liver Cancer Treatment. *Drug Des. Dev. Ther.* **2019**, *13*, 3569–3578. [CrossRef]
175. Zhu, M.; Shi, X.; Gong, Z.; Su, Q.; Yu, R.; Wang, B.; Yang, T.; Dai, B.; Zhan, Y.; Zhang, D.; et al. Cantharidin treatment inhibits hepatocellular carcinoma development by regulating the JAK2/STAT3 and PI3K/Akt pathways in an EphB4-dependent manner. *Pharmacol. Res.* **2020**, *158*, 104868. [CrossRef]
176. Yao, H.; Zhao, J.; Wang, Z.; Lv, J.; Du, G.; Jin, Y.; Zhang, Y.; Song, S.; Han, G. Enhanced anticancer efficacy of cantharidin by mPEG-PLGA micellar encapsulation: An effective strategy for application of a poisonous traditional Chinese medicine. *Colloids Surf. B Biointerfaces* **2020**, *196*, 111285. [CrossRef]
177. Wang, P.; Yang, J.; Zhu, Z.; Zhang, X. Marsdenia tenacissima: A Review of Traditional Uses, Phytochemistry and Pharmacology. *Am. J. Chin. Med.* **2018**, *46*, 1449–1480. [CrossRef]
178. Wang, X.; Yan, Y.; Chen, X.; Zeng, S.; Qian, L.; Ren, X.; Wei, J.; Yang, X.; Zhou, Y.; Gong, Z.; et al. The Antitumor Activities of *Marsdenia tenacissima*. *Front. Oncol.* **2018**, *8*, 473. [CrossRef]
179. Lin, S.; Sheng, Q.; Ma, X.; Li, S.; Xu, P.; Dai, C.; Wang, M.; Kang, H.; Dai, Z. *Marsdenia tenacissima* Extract Induces Autophagy and Apoptosis of Hepatocellular Cells via MIF/mToR Signaling. *Evid.-Based Complement. Altern. Med.* **2022**, *2022*, 7354700. [CrossRef]
180. Li, L.; Zhang, W.; Desikan Seshadri, V.D.; Cao, G. Synthesis and characterization of gold nanoparticles from *Marsdenia tenacissima* and its anticancer activity of liver cancer HepG2 cells. *Artif. Cells Nanomed. Biotechnol.* **2019**, *47*, 3029–3036. [CrossRef]
181. Huang, X.; Zhang, J.; Song, Y.; Zhang, T.; Wang, B. Combating liver cancer through GO-targeted biomaterials. *Biomed. Mater.* **2021**, *16*, 065003. [CrossRef] [PubMed]

Review

The Anticancer Effects and Therapeutic Potential of Kaempferol in Triple-Negative Breast Cancer

Sukhmandeep Kaur [1], Patricia Mendonca [2,*] and Karam F. A. Soliman [1,*]

[1] Division of Pharmaceutical Sciences, College of Pharmacy and Pharmaceutical Sciences, Institute of Public Health, Florida A&M University, Tallahassee, FL 32307, USA; sukhmandeep1.kaur@famu.edu

[2] Department of Biology, College of Science and Technology, Florida A&M University, Tallahassee, FL 32307, USA

* Correspondence: patricia.mendonca@famu.edu (P.M.); karam.soliman@famu.edu (K.F.A.S.); Tel.: +1-850-599-3306 (P.M. & K.F.A.S.); Fax: +1-850-599-3667 (P.M. & K.F.A.S.)

Abstract: Breast cancer is the second-leading cause of cancer death among women in the United States. Triple-negative breast cancer (TNBC), a subtype of breast cancer, is an aggressive phenotype that lacks estrogen (ER), progesterone (PR), and human epidermal growth (HER-2) receptors, which is challenging to treat with standardized hormonal therapy. Kaempferol is a natural flavonoid with antioxidant, anti-inflammatory, neuroprotective, and anticancer effects. Besides anti-tumorigenic, antiproliferative, and apoptotic effects, kaempferol protects non-cancerous cells. Kaempferol showed anti-breast cancer effects by inducing DNA damage and increasing caspase 3, caspase 9, and pAMT expression, modifying ROS production by Nrf2 modulation, inducing apoptosis by increasing cleaved PARP and Bax and downregulating Bcl-2 expression, inducing cell cycle arrest at the G2/M phase; inhibiting immune evasion by modulating the JAK-STAT3 pathway; and inhibiting the angiogenic and metastatic potential of tumors by downregulating MMP-3 and MMP-9 levels. Kaempferol holds promise for boosting the efficacy of anticancer agents, complementing their effects, or reversing developed chemoresistance. Exploring novel TNBC molecular targets with kaempferol could elucidate its mechanisms and identify strategies to overcome limitations for clinical application. This review summarizes the latest research on kaempferol's potential as an anti-TNBC agent, highlighting promising but underexplored molecular pathways and delivery challenges that warrant further investigation to achieve successful clinical translation.

Keywords: kaempferol; flavonoids; breast cancer; TNBC; apoptosis; metastasis inhibition; chemoresistance; oxidative stress

Citation: Kaur, S.; Mendonca, P.; Soliman, K.F.A. The Anticancer Effects and Therapeutic Potential of Kaempferol in Triple-Negative Breast Cancer. *Nutrients* **2024**, *16*, 2392. https://doi.org/10.3390/nu16152392

Academic Editors: Yi-Wen Liu and Ching-Hsein Chen

Received: 15 June 2024
Revised: 16 July 2024
Accepted: 21 July 2024
Published: 23 July 2024

1. Introduction

Breast cancer is one of the most common cancers affecting women around the globe. It is a highly heterogeneous disease with phenotypically and genetically diverse cell groups, making it challenging to treat [1,2]. There are more than 4 million cases in the United States as of January 2022, and about 13% of women in the United States are likely to have breast cancer in their lifetime [3,4]. The molecular subtypes of breast cancer are further classified based on the expression of specific genes and proteins within the tumor cells [5]. The key risk factors for the occurrence of this disease remain gender, ethnicity, genetic makeup, lifestyle, geographical factors, and radiation exposure. There are a range of treatment strategies available to treat breast cancer, including surgical resection, chemotherapeutic agents, radiation therapy, immunotherapy, and natural compounds [6]. However, with the constant evolution and development of more resistant molecular subtypes of breast cancer, typically triple-negative breast cancer (TNBC), it has been challenging for healthcare professionals to have positive treatment outcomes [7,8]. TNBC accounts for about 10 to 15% of all diagnosed breast cancer cases. TNBC patients are typically diagnosed in the later stages of the disease due to their aggressive nature. Even with initial responsiveness towards the treatment, TNBC cases show poor prognosis, a higher rate of recurrence, and shorter disease-free survival, especially in younger women diagnosed with TNBC [9]. The conventional therapies used in the treatment of such cancer types usually fail to eradicate the cancer tumors and, hence, achieve the desired outcome, causing the cells in the tumors to persist and ultimately leading to relapse [10].

The treatment of such aggressive and resistant breast cancer types has advanced with the use of numerous multi-modality treatment strategies, making the use of combinatorial therapies or adjuvant therapies an option to fight against the resistant subtypes of cancer and, more specifically, breast cancer. However, the uncertainties associated with these treatments concerning efficacy, adverse effects, resistance, and disease recurrence remain the biggest concern. These undesirable and unavoidable treatment-related outcomes significantly lower the quality of life in the short and long term. Therefore, developing new therapies and therapeutic strategies is one of the top priorities of the healthcare sector globally [6].

In this never-ending fight against chronic diseases like cancer, natural compounds have paved their way with a vast potential to be part of the standard treatments. Natural compounds derived from multiple sources have been reported to benefit the human body for thousands of years. Still, because of the lack of reported literature and clinical evidence, they were always sidelined in the modern world of healthcare. However, in the past few decades, with the availability of resources and published literature, natural compounds have gained enormous attention for their ability to target chronic diseases like cancer uniquely; unlike synthetic drugs, which usually act on one or limited mechanisms, they can simultaneously affect multiple pathways involved in cancer development and progression [11,12]. In an exploratory concept against resistance in cancer, combinatorial therapy is also a favorable way to overcome standard therapies' limitations. Natural compounds have been reported to target different pathways in cancer when combined with standard anticancer agents, providing synergistic effects, enhancing overall efficacy, and overcoming resistance [13].

Kaempferol, a natural flavonoid found commonly in various food items, namely tomato, tea, and broccoli, has been shown to have a variety of protective effects, including antioxidant, anti-inflammatory, cardioprotective, neuroprotective, anti-diabetic, and anti-cancer effects [14]. In the last few decades, kaempferol has become an important natural compound with potent anticancer activity, specifically against breast cancer. Kaempferol exhibits most of its anticancer activity by modulating multiple pathways, such as inducing apoptosis, inhibiting cell growth, and inhibiting cancer cell migration and invasion [15]. Besides its anticancer potential, kaempferol has also been studied for its synergistic potential when combined with other flavonoids or standard chemotherapeutic agents. It also protects against the adverse reactions caused by potential chemotherapeutic agents in various cancer types [16–19]. The potential of kaempferol as an anti-breast cancer agent has come a long way with an understanding of its action by modulating different molecular pathways. Still, its clinical translation as a therapy needs more robust validation to be considered part of standard adjuvant therapy [20].

In this review, we focus on the recently recognized work performed using kaempferol as an anticancer and chemo-preventive in human breast cancer using various in vitro and in vivo models, as well as the clinical studies that have used kaempferol as a part of their therapy. This review also aims to present the potential of kaempferol to target epigenetic modifications that render breast cancer resistant. We will further address the challenges related to its poor solubility, stability, and bioavailability and, subsequently, the efforts to address them.

2. Breast Cancer

The average risk of an American woman being diagnosed with breast cancer once in her lifetime is about 13 percent, and the incidence of new cases has been increasing every year by 0.6% recently. According to the American Cancer Society's statistics, there is an estimate of approximately 320,000 new cases of breast cancer to be diagnosed in the year 2024. Although the breast cancer-related death rate has declined by about 42% from 1989 to 2021, there are estimated to be about 42,000 deaths in women due to breast cancer in 2024 [21]. Concerning the racial and ethnic differences in the occurrence and progression of the disease, African American women are more likely to die from breast cancer at any age than Caucasians or women of any other ethnicity. The median age of African American women diagnosed with breast cancer is about 60 years, while it is 64 years for an average Caucasian woman. However, considering all these factors, about 4 million cancer survivors are living in the United States as of 2024 [4,21].

2.1. Breast Cancer Types

The most common classification of breast cancer divides it immunohistochemically into four subtypes: estrogen receptor (ER)-positive (+), ER-negative (−), human epidermal growth receptor (HER)+, and HER−, based on the hormonal subtype and their responsiveness towards the hormonal therapy. Another molecular subtype of breast cancer that does not express any of the aforementioned molecular receptors is TNBC [22]. TNBC has been further classified based on the genetic expression profiling of the tumor cells. Lehmann et al. (2011) explored the gene profiling of about 587 TNBC cases and performed gene cluster analysis of their genetic expressions. There are six different TNBC types with unique genetic expressions and ontologies. In this classification, there are two basal-like (BL) types, BL1 and BL2, immunomodulatory (IM) type, mesenchymal (M) type, mesenchymal stem cell-like (MSL), and luminal androgen receptor (LAR) type [23]. The authors further divided PAM50 subtyping into these six TNBC subtypes and reported their intrinsic subdivision and composition. They noted that except for MSL and LAR, all other types of TNBC are mostly composed of BL1 (99%), BL2 (95%), M (97%), and IM (84%). Whereas MSL is composed of BL (50%), normal-like (28%), and luminal B (14%), and LAR consists of HER2 (74%), and luminal B (14%). The subsequent evaluation of the subtypes for their differential prognosis revealed BL2 and M as the subtypes with the worst prognosis. LAR has the most prolonged progression and higher distant metastasis-free survival, with the maximum overall survival rate [24]. The absence of any responsive molecular targets and receptors makes TNBC relatively unresponsive to the targeted hormonal therapy and, hence, challenging to treat. It also contributes to the heterogeneous presentation of the disease. Burstein et al., 2015 also classified TNBC based on quantitative DNA expression and further classified it into four subtypes based on potential targets. Based on these targets, the classification is the LAR subtype, MSL subtype, basal-like immune-activated (BLIA) subtype (involved in signal transduction and transcription of STAT), and basal-like immunosuppressed subtype (BLIS) expressing the immunosuppressive molecule V-set domain containing T-cell activator inhibitor 1 (VTCN1) [25]. Similarly, TNBC was classified by different researchers, and each classification subgroup is characterized by unique gene expressions linked to the immune system. For example, Liu et al. used the combination of mRNA and long non-coding RNA (lncRNA) expression profiles to create the Fudan University classification (FUSCC) system, and this classification has four subtypes, namely: IM type, LAR type, M type, and BLIS type [26].

2.2. Treatment Approaches for TNBC

TNBC, being the aggressive and invasive breast cancer subtype with higher rates of recurrence and a poor prognosis, remains with adjuvant chemotherapy as the only treatment option. The standard treatment usually involves surgical resection, radiation therapy, and chemotherapy but is associated with some severe adverse effects. Apart from that, TNBC has been reported to have a higher tendency to resist the standard chemotherapeutic agents, leaving only limited therapeutic benefits from the available treatments. Also, these standard chemotherapeutic agents target only one target, such as a specific pathway, a protein, or some nucleic acids. Despite the availability of these drugs targeting various mechanisms, aggressive cancers like TNBC can still evolve and become resistant to them, which highlights the need for agents offering a more sustainable therapeutic approach [20].

2.3. Natural Compounds Used in Breast Cancer Therapy

Natural compounds have been reported and have gained immense attention in the last two decades due to their ability to target heterogeneous families of cancer cell targets and signaling pathways and simultaneously have minimal adverse effects. They are also reported to enhance the anticancer effects and lower the adverse effects of chemotherapy [27]. Many different classes of natural compounds have been evaluated for their anticancer potential. Natural products hold immense potential due to their unique mechanisms of action, but their inconsistent quality and difficulty in modification can present obstacles to treatments. The inherent structures of natural compounds do not obey the Lipinski rule of five and have a high molecular weight. Also, the notable increase in the average molecular weight of approved oral drugs over the past 20 years aligns with the growing interest in exploring natural compounds for their potential antitumor activity [28].

Nearly 50% of the antitumor drugs originate from natural compounds such as taxanes, vinblastine, vincristine, and podophyllotoxin analogs [29]. Also, beyond their anticancer properties, natural compounds can be used for chemoprevention since they can suppress cell growth, regulate cell division, and disrupt critical tumorigenic pathways like PI3K, MMP, MAPK/ERK, TLR, and AKT. They can activate DNA repair pathways (p21, p27, p53) and trigger the production of protective proteins (Bax, Bak, and Bid), ultimately leading to the synthesis of protective enzymes (caspases) and boosted antioxidant activity (GST, GSH, and GPx), shielding cells from damage. The chemo-preventive action of natural compounds usually comes from their potential to induce immunosurveillance (which eradicates the transforming cells), DNA repair mechanisms, apoptosis, antioxidant activity, inhibition of cell proliferation, tumor progression, and angiogenesis [29]. Natural compounds have also been found to overcome the chemoresistance developed towards standard chemotherapeutic agents by chemo-sensitizing the cells to augment the effect of chemotherapeutic agents. Natural compounds combined with chemotherapeutic agents have been shown to chemo-sensitize resistant cells and synergize the overall effect [20].

3. Kaempferol

Flavonoids are a diverse polyphenolic group of compounds found widely distributed in various natural substances that have recently gained attention for their potential health and dietary benefits [30]. Flavonoids have a low molecular weight and are formed as a product of the shikimic acid pathway, a metabolic pathway in plants' plastids. Flavonoids are based on carbon rings and the -OH group present in the structure and are divided into subgroups: flavanols, flavones, chalcones, flavanones, flavanonols, and isoflavones. All flavonoids from all the subclasses have a 15-carbon benzopyranone ring forming a C6-C3-C6 flavan nucleus [31–33] (Figure 1).

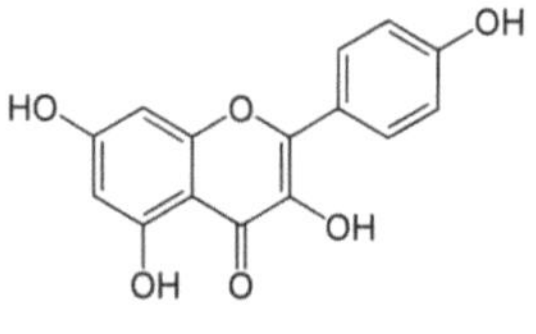

Figure 1. The chemical structure of kaempferol.

Kaempferol is one of the most common and widely studied flavonoids. It belongs to the flavanol subclass and is structurally characterized by a flavone backbone with 3 4′ 5 7-tetrahydroxy groups [34]. Kaempferol was initially extracted from the rhizomes of the plant *Kaempferia galanga*. The pure compound can also be found in various plant types within the Kingdom Plantae, including Pteridophyta, Coniferophyta, and Angiosperms, while the kaempferol glycosides can be found in multiple families. It is commonly found in widely consumed vegetables like broccoli, cabbage, onion, green peas, and spinach and in berries like strawberries, gooseberries, and blackberries (Figure 2). The most consumed green leafy vegetable, spinach, contains around 55 mg of kaempferol per 100 g, and cabbage contains about 47 mg per 100 g of the vegetable. In comparison, onion has about 4.5 mg per 100 g [35]. Among spices, capers have the highest quantity, about 104.29 mg per 100 mg, compared to cloves and cumin, which have about 23.8 mg and 38.6 mg per 100 g, respectively [35].

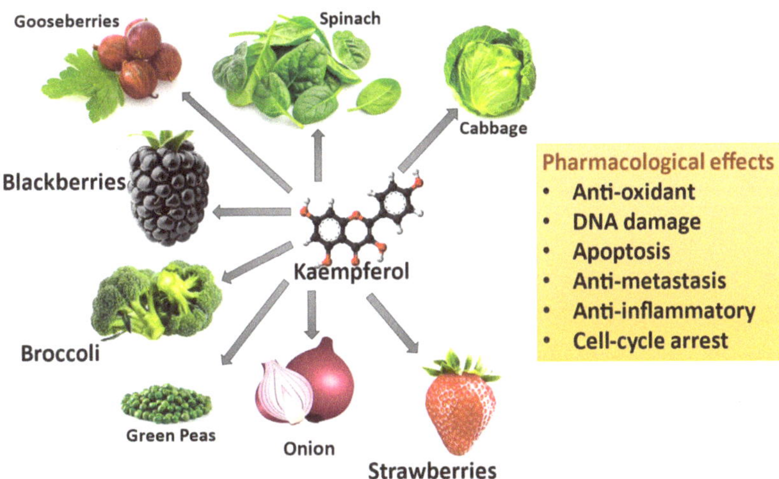

Figure 2. Sources and pharmacological effects of kaempferol.

3.1. Kaempferol Absorption and Metabolism

After being absorbed into the body, kaempferol undergoes extensive metabolism in the liver and converts to either methyl or sulfate salts as glucuronide-conjugated metabolites, which then circulate in the blood. Kaempferol is also metabolized by the intestinal microbiota into 4- 4-phloroglucinol, hydroxyphenyl acetic acid, and 4-methylphenol compounds, and then enters the systemic circulation. It has also been reported that kaempferol acts as a precursor of the CoQ ring in renal cells, thereby increasing concentration in the kidney cells and contributing to the production of ubiquinone. It is proposed to be responsible for its antioxidant properties. Although kaempferol has also been seen being excreted unchanged in the urine, the other metabolic products of kaempferol, both formed as free aglycone products or as O- or C-glycosides, are reported to be excreted in urine and feces [31,35,36].

3.2. Kaempferol Pharmacokinetics

Kaempferol has poor water solubility and has been studied extensively for its pharmacokinetic properties both in vitro and in vivo. It is mainly absorbed through the small intestine and has passive diffusion because of its lipophilic nature. Because of the multiple sugars in its structure, the membrane-bound beta-glucosidase in enterocytes removes the terminal saccharides and exposes the glucose, which is absorbed [32]. Kaempferol is metabolized in the liver and undergoes phase I (oxidation) and phase II (glucuronidation, sulfation) metabolism. The flora in the large intestine metabolizes the kaempferol glycosides into aglycones and then to some phenolic compounds, which are absorbed in the systemic circulation, distributed in various tissues, and excreted in feces and urine. It has been reported that kaempferol offers a vast therapeutic advantage, even with limited oral bioavailability [31].

3.3. Pharmacological and Toxicological Properties of Kaempferol

Kaempferol has been proven to have a variety of health benefits ranging from cardio-protective, neuroprotective, chemo-preventive, anticancer effects, and protective effects against various metabolic disorders, even when taken as a dietary substitute or as a part of adjuvant therapy [37,38]. It has been traditionally used as a dietary supplement for its antioxidant and anti-inflammatory properties [39].

Kaempferol has been explored for many years as a treatment for its anti-inflammatory properties. Chronic inflammation, which is the culprit of multiple diseases, namely cardio-vascular diseases, metabolic and autoimmune disorders, and even cancer, can be targeted by using natural products like kaempferol [40,41]. Kaempferol specifically targets oxidative stress and modulates pro-inflammatory enzyme activities, genetic expression of the genes involved in inflammation, inhibition of transcription factors involved in inflammatory pathways, and other mechanisms involved in inflammation [40].

Besides its active moiety, kaempferol can modulate physiological effects with its other bioactive metabolites. With its interaction with free radical generation in an oxidative stress environment, kaempferol tends to have pro-oxidant activity and produce genotoxic effects [40]. Various antioxidants and pro-oxidant enzymes regulate these pro-oxidant mechanisms. Kaempferol has also been reported to have an impact by reducing metal ion concentrations with its antioxidant activities and affecting iron concentrations and folic acid bioavailability. Despite the multiple studies addressing kaempferol's mutagenic and carcinogenic potential, there is limited in vivo evidence corroborating the effects [37,40,42].

Kaempferol is relatively non-toxic to healthy human cells, unlike the standard chemotherapeutic agents. It has been evaluated for its activity in healthy human breast epithelial cells (MCF-10A cells) and other class flavonoids. As indicated in a cytotoxicity study by Mohd. Afzal et al., 2013, kaempferol in combination with fisetin was safe for the cells and showed no significant toxicity [16]. They also analyzed the activity of kaempferol alone towards MCF-10A cells and found it to be safe with no cytotoxic potential towards healthy breast epithelium cells [16]. Interestingly, clinical studies performed on healthy individuals to analyze the nutritional benefits of consuming food items rich in kaempferol also signify the potential benefits of kaempferol against chronic inflammation, which has emerged as a culprit for multiple chronic diseases like cancer [32].

4. Anticancer Effects of Kaempferol in Human Cancers

The consumption of flavonoids like kaempferol in the form of fruits and vegetables is linked to a decreased incidence of various human cancers. Kaempferol and other flavonoids exert anticancer effects by modulating multiple physiological pathways. As part of various traditional medicinal systems and dietary habits across different communities, kaempferol emerged to have anti-tumorigenic and chemo-preventive effects in the 1970s and 1980s [43,44]. The subsequent development in understanding the mechanisms underlying the development and recurrence of cancers allowed the researchers to explore the potential benefits of using kaempferol against various human cancer types [45].

Initially, scavenging oxidative stress and modulating inflammatory pathways were primarily targeted for the anticancer activity of kaempferol [46–48]. Its antioxidant and anti-inflammatory properties were observed in hydrazine- and H_2O_2-induced colorectal and hepatoma cancer cell models [48,49]. Besides exerting anti-tumorigenic, antiproliferative, and apoptotic effects against cancer cells, kaempferol also showed protective activity towards healthy, non-cancerous cells. Early in evaluating kaempferol as an anticancer agent, Nguyen et al., 2003, showed inhibition in the growth of A549 lung cancer cells with kaempferol treatment in a dose-dependent manner. They also showed that kaempferol-induced apoptosis was validated by increased and decreased pro-apoptotic and anti-apoptotic protein expression, respectively. Further, they confirmed the activation of MEK-MAPK-induced apoptosis in A549 cells [50]. Subsequently, Zhang et al., 2008, reported similar results with kaempferol obtained from ginkgo biloba extract in pancreatic cancer cells [51,52]. Wang et al., 2021, in their study, showed ROS-dependent induction of apoptosis in in vitro and in vivo models of pancreatic cancer. They mainly showed that this induction of anticancer activity with ROS generation was via tissue transglutaminase (TGM2)-mediated Akt/mTOR signaling [53]. Subsequently, the potential anti-tumorigenic effects of kaempferol were further seen in different in vitro and in vivo models of human cancer, targeting other more cancer-specific molecular targets. Lee et al., 2016, in their in vitro study using kaempferol as a treatment against human pancreatic cancer cell lines Panc-1, Miapaca-2, and SNU-213, showed significant cytotoxic action in a dose-dependent manner. They also showed that the anti-migratory action of kaempferol was mediated by the inhibition of EGFR-related Src and ERK1/2/AKT pathways [54]. Similarly, several other studies showed the dose-dependent inhibition of cell viability and kaempferol-induced apoptosis via the P13/AKT hTERT pathway in various human cancer types [54–56]. Yao et al., 2016 showed that kaempferol induced G0/G1 cell cycle arrest. Their mechanistic studies showed a significant downregulation of EGFR signaling in an in vitro model of esophagus squamous cell carcinoma (ESCC). They also demonstrated decreased aerobic glycolysis, markedly seen in the tumor environment, with kaempferol treatment in the ESCC in vitro model. The in vivo xenograft model of ESCC further confirmed their observations [57]. Similarly, another study showed kaempferol-inducing apoptosis effects and cell cycle arrest in in vitro models of human renal carcinoma cells. They showed the EGFR/p38 signaling pathway and upregulation and downregulation of p21 and cyclin B1 expressions, respectively. Their cell growth inhibition and apoptosis were linked to the activation of PARP cleavage [58]. In an in vitro model of osteosarcoma cells (US-2 OS), apart from showing specific cytotoxic action against the OS cells and sparing the human fetal osteoblast progenitor cells (hFOB cells), the authors reported the induction of apoptosis with a significant increase in cytoplasmic Ca^{2+} and decrease in mitochondria membrane potential ($\Delta\psi$m) levels in US-2 OS cells, as demonstrated by the DiOC6-based flow-cytometric assay. They also showed the induction of various endoplasmic reticulum stress-related proteins and apoptotic proteins, including GRP78, GRP94, GADD153, ATF-6α, ATF-6β, caspase-4, caspase-12, calpain 1, caspase 3, and caspase 6 activity [57]. Yet, another group reported the induction of apoptosis and G2/M cell cycle arrest in human ovarian carcinoma A2780/CP70 cells with kaempferol treatment. They particularly observed that kaempferol-induced G2/M cell cycle arrest in their in vitro model of ovarian carcinoma was mediated by the Chk2/p21/Cdc2 and Chk2/Cdc25C/Cdc2 pathways [59]. Kaempferol could also induce anti-angiogenic potential, which is another major clinical challenge faced in cancer therapy. In 2009, a study by Luo and colleagues focused on analyzing the anti-angiogenic effect of kaempferol against ovarian cancer and reported that although kaempferol could not affect the viability of ovarian cancer cells significantly, it could inhibit angiogenesis and angiogenic proteins. The marked decrease in the mRNA and protein levels of Vascular Endothelial Growth Factor (VEGF), a potent anti-angiogenic factor caused by kaempferol, was also able to downregulate the expression of HIF-α (a regulator of VEGF) [60]. In a study by Chin et al., 2018, kaempferol decreased VEGF-stimulated human umbilical vein endothelial cell (HUVECs) viability, invasion, migration, and tube formation, and the angiogenic inhibition of kaempferol was related to the regulation of VEGF/VEGFR-2 and PI3K/AKT, MEK, and ERK pathways in VEGF-stimulated HUVECs [61]. Subsequently, kaempferol has also been shown to inhibit the migration and invasion of different human cancer cell types. Ju et al., 2021, in their study on human hepatocellular carcinoma cells (HCC; Huh-7 and SK-Hep-1 cells), reported that the treatment of kaempferol significantly reduced and suppressed the viability, migration, and invasion of the HCCs, and that this action of kaempferol was backed by decreased activity of MMP9, Cathepsin C, Cathepsin D, and phosphorylated AKT (pAKT) [62]. Amble literature indicates kaempferol's anti-cancer effect in many cancer cell types, making it an ideal candidate to be developed for its clinical use.

4.1. Kaempferol in Combinatorial Drug Therapy

Kaempferol has also been explored for its synergistic activity with other therapeutic agents, especially when given with chemotherapeutic agents. It has specifically been reported to increase chemo-sensitization in resistant cancer cells. The idea of combining kaempferol with standard chemotherapeutic agents came when some of the structurally related flavonoids were studied when combined with standard chemotherapeutic drugs in a study by Cipak et al., 2003 [63]. They reported that flavonoids differentially modulate the anticancer activity of doxorubicin by decreasing its cytotoxic activity and by shifting the cell cycle arrest from the G2/M phase to the S phase of the cell cycle [63]. It was then reported to downregulate the expression of cMyc, a regulator of cell proliferation and apoptosis but involved in drug resistance, in ovarian cancer cells [64]. Similarly, kaempferol, when used in combination with 5-fluorouracil, docetaxel, and cisplatin, showed additive effects towards pancreatic cancer, prostate cancer cells, and head and neck squamous cell carcinoma [52,65]. Additionally, the potential of kaempferol, used in combination with other natural drugs to target chronic diseases like cancer, was also explored, as it may affect multiple pathways involved in the disease process [66]. Kaempferol, in combination with another dietary flavonol, fisetin, showed synergistic cytotoxic activity and induction of apoptosis at lesser doses than when used individually [16].

4.2. Kaempferol Reversal of the Chemoresistance of Chemotherapeutic Agents

The standard anticancer therapies, including chemotherapy and radiotherapy, offer generous clinical outcomes, increasing overall survival rates in patients, but are associated with reduced efficacy and the development of resistance over time. The induction of drug-resistance proteins and epigenetic mechanisms are involved in the development of chemoresistance [14]. For many years, kaempferol has been evaluated by multiple research groups for its potential to reverse chemoresistance and synergize the anticancer activity of agents [67]. Using various mechanisms, including the inhibition of ROS generation and oxidative stress, kaempferol can reverse the chemoresistance of standard chemotherapeutic agents. In a study by Ichrak Riahi-Chebbi et al., 2019, apart from inhibiting and reversing the resistance, kaempferol also showed synergism with increased anticancer activity when given in combination with 5-fluorouracil (5-FU) in resistant human LS174 colon cancer cells [68]. Their findings demonstrated that kaempferol, alone or combined with 5-FU, induced apoptosis, cell cycle arrest, and suppressed reactive oxygen species (ROS) production. Additionally, kaempferol modulated the expression of key signaling pathways, including JAK/STAT3, MAPK, PI3K/AKT, and NF-κB, suggesting its potential role in reversing drug resistance [68]. Another study did so by suppressing glucose uptake and lactate production in human colorectal cancer cells. Mechanistically, it induced a pronounced upregulation of microRNA-326 (miR-326) and inhibited the expression of pyruvate kinase M2 isoform (PKM2) [69]. Similarly, kaempferol reversed the oxaliplatin resistance in human colon cancer cells by inhibiting the expression of A Jun and Fos heterodimers (AP-1), which are involved in growth factor-mediated cell cycle progression and cell proliferation [70]. Interestingly, Sourav Kumar Nandi et al., 2023, in their study, used the combination of kaempferol and verapamil to impede the chemoevasion in their ex vivo model of breast cancer, as demonstrated by the downregulation of resistance-related markers. In particular, the authors demonstrated that the combination of kaempferol and verapamil induced a significant overproduction of ROS and downregulation of chemoresistance and tumor acidosis markers. The treatment with kaempferol resulted in decreased expression of ATP1B1 (a regulator of cellular membrane potential) and disrupted lysosomal function [71]. Kaempferol's pleiotropic effects on reversing chemoresistance in human cancers, mediated through modulation of multiple molecular pathways, warrant further investigation as a potential therapeutic strategy to overcome resistance in TNBC.

5. Kaempferol Anticancer Mechanisms of Action in Breast Cancer

Kaempferol has been used to target multiple molecular pathways in human breast cancer. Eating foods rich in kaempferol and kaempferol-like flavanols is associated with a reduced risk of breast cancer [72,73]. It was in the 1990s that the varied effects of kaempferol and other flavonoids on human breast cancer were initially evaluated, followed by a determination and analysis of their potential. Kaempferol has been reported to target multiple pathways to produce the desired anticancer activity. Interestingly, being the most commonly used dietary phytoestrogen, kaempferol also possesses the ability to modulate estrogen-mediated activity, which acts as one of the principal mediators of hormone-dependent breast cancers. Similarly, kaempferol has been shown to have promising anticancer effects against TNBC by modulating more than one molecular pathway [74,75].

5.1. Effect of Kaempferol on DNA Synthesis Inhibition

As the development of breast cancer has also been historically linked to lifetime exposure to both endogenous and exogenous estrogen, kaempferol has been widely studied for its anticancer activity mediated by estrogen-dependent pathways and in cancer types independent of estrogen [76]. Wang et al., 1997, analyzed the effect of kaempferol on DNA synthesis in human estrogen-dependent (ER+) (MCF-7) and estrogen-independent (ER−) (MDA-MB-231) breast cancer cells. They showed the dose-dependent effects of kaempferol on DNA synthesis in breast cancer cells, in which kaempferol promoted DNA synthesis at very low concentrations. At the same time, it significantly inhibited DNA synthesis at a concentration of around 50 μM compared to the control cells [77]. Similarly, Zava et al., 1997 and 2023, replicated the action of kaempferol alongside other flavonoids in other ER+ (MCF-7 and T47D) and ER− (MDA-MB-468) breast cancer cells and reported that kaempferol shows estrogen agonistic activity. It has concentration-dependent DNA inhibition and cell growth inhibition [78].

Interestingly, kaempferol displayed notable estrogenic effects, enhancing the transcriptional activity of the estrogen receptor among other flavonoids when evaluated for estrogenic activity using an engineered yeast strain with human estrogen receptor integration. Although its potency was 4000–4,000,000 times lower than natural estrogen, its activity was comparable to certain established isoflavonoid estrogens, and its estrogenic potential was validated in estrogen-dependent MCF7 breast cancer cells [78]. Interestingly, kaempferol exhibited both estrogenic and antiestrogenic properties, showing a two-phase response on the estrogen receptor. In a study by Min oh et al., 2006, the authors reported that at lower concentrations (as low as 10–12 M), it acts as an estrogenic agent through the estrogen receptor-mediated pathway, and at higher concentrations (as high as 10^{-4} M), kaempferol exerts strong antiproliferative effects, independent of estrogen receptor activity. This also suggests that kaempferol can inhibit cancer cell proliferation via an estrogen-dependent pathway, indicating its potential to prevent malignant transformations driven by estrogen exposure and, hence, its ability to maintain a balanced estrogenic activity [74]. Similarly, Hung et al., 2004, showed with MCF-7 cells that kaempferol decreased cell viability in a dose- and time-dependent manner. It also notably decreased ER-alpha mRNA and protein levels, reducing estrogen-responsive gene expression. At the same time, they showed reduced expression of the progesterone receptor, cyclin D1, and insulin receptor substrate and that kaempferol induced ER-alpha protein aggregation and degradation through a distinct pathway compared to that of estradiol [79].

Interestingly, kaempferol has also been reported to have anticancer and antiproliferative properties in ER receptor-mediated triclosan-induced cell growth in MCF-7 cells. Kim et al., 2016 demonstrated the antiproliferative activity of kaempferol in triclosan-induced cell viability of MCF-7 cells and in a xenograft mouse model by using various combinations of an ER antagonist (ICI 182,780), an ERα agonist (Propyl pyrazole triol, PPT), and an ERβ agonist (diaryl propionitrile, DPN) [80]. They further analyzed the effect of kaempferol on triclosan or estradiol-induced activation of the IGF-1 signaling pathway and observed that the treatment of kaempferol in combination with triclosan, and estradiol significantly downregulated the expression of the proteins pAkt, pMEK1/2, pIRS-1, and pERK1/2 [80]. In another study, kaempferol showed inhibition of cancer cell migration in the triclosan and estradiol-induced cell growth, demonstrated using a wound healing assay and suppressed trans-well, and inhibited cell invasion when given in combination with the triclosan and estradiol in a cell invasion assay. This suppression demonstrates the inhibition of metastatic and epithelial–mesenchymal transition (EMT), which was further confirmed by the downregulation of the increased expression of EMT-related markers: N-cadherin, Snail, and Slug, and metastasis-related proteins: Cathepsin B, Cathepsin D, MMP-2, and MMP-9 in triclosan and estradiol-induced MCF-7 cells [81].

5.2. Effect of Kaempferol on ROS Production

Various anticancer treatments have been studied for their anti-oxidative properties to neutralize free radicals and the oxidative stress caused by irregular cellular function and metabolism. Mitochondrial dysfunction has emerged as one of the prominent culprits for the increased oxidative stress in the body and in and around the tumor sites, which results in cancer initiation and progression. The ROS encompasses a group of oxygen-containing molecules exhibiting high chemical reactivity. Prominent members of the ROS family include free radicals like hydroxyl and superoxide, along with hydrogen peroxide [82]. Cellular production of ROS occurs physiologically within the mitochondrial electron transport chain, peroxisomes, and phagosomes, contributing to energy generation and phagocytosis. In addition to this, the NADPH oxidase family in the plasma membrane generates ROS and participates in various intracellular signaling pathways, including those involving Ras, c-Jun N-terminal kinase (JNK), p38, mitogen-activated protein kinase (MAPK), and PI3K/AKT/mTOR. ROS also play a significant role in cell proliferation, along with processes like autophagy, apoptosis, and inflammation mediated by the NLRP3 inflammasome and the nuclear factor-κB (NF-κB) pathway [83]. The malignant transformation of the cells, the formation of tumors, and their progression and metastasis are part of a vicious cycle with oxidative stress. Tumors undergo vascularization via angiogenesis to meet the growing demands of oxygen and nutrients for sustained proliferation. ROS, mainly hydrogen peroxide (H_2O_2), plays a crucial role in this process. ROS promote vascularization by enhancing the activation of vascular endothelial growth factor (VEGF) and increasing the production of matrix proteins. Endothelial cells primarily generate ROS through NADPH oxidase, further contributing to the upregulation of hypoxia-inducible factor 1α (HIF-1α) expression [84].

Additionally, the generation of estrogen-dependent oxidative metabolism has also been reported to be involved in generating ROS implicated in the carcinogenic transformation and growth of cancer cells, which suggests the existence of additional mechanisms, independent of ER status, that mediate estrogen-induced cell signaling leading to malignant transformation and growth of mammary epithelial cells. Estrogen-induced mitochondrial ROS is a mechanism involved in mammary carcinogenesis [85]. In a study by Quentin Felty et al., 2005, the physiologically relevant estrogen concentrations corresponding to the menstrual peak demonstrated the rapid mitochondrial stimulation and generation of ROS in MCF-7 cells [86]. Interestingly, a convergence exists between mitogenic pathways sensitive to ROS levels and those regulated by carcinogenic estrogen concentrations. The inhibitors of mitochondrial ROS production abrogate the E2-induced expression of cell cycle genes harboring nuclear respiratory factor-1 (NRF1) binding sites, such as cyclin B1, PCNA, and PRC1, and these inhibitors suppress E2-induced NRF1 expression and, consequently, delay cellular proliferation [86].

Over the decades of cancer research, multiple therapeutic strategies modulate ROS levels and have demonstrated promising efficacy in both in vitro and in vivo cancer models. These strategies encompass approaches that either scavenge ROS or promote their generation within cancer cells, as shown in both in vitro and in vivo models of different human cancers [84]. Kaempferol's mechanism of action appears to be multifaceted, and multiple studies have demonstrated its ability to considerably impede the development of various inflammatory processes by suppressing ROS generation and exhibiting high anti-oxidative properties. It also inhibits the expression of pro-inflammatory cytokines IL-1β and TNF-α and disrupts the translocation of NF-κB into the nucleus, thereby hindering the production of inflammatory proteins [87]. Apart from the multiple molecular targets it acts on, the antioxidant activity of kaempferol is also attributed to the presence of hydroxyl groups within its molecular structure, particularly the one located at the C-3 position. Apart from the direct elimination of ROS, kaempferol also becomes involved in preserving endogenous antioxidant enzymes like glutathione peroxidase, superoxide dismutase, and catalase at normal physiological levels [87]. In another study by Jie Zeng et al., 2020, kaempferol inhibited breast tumor metastasis in both in vitro and in vivo models of 4T1 breast cancer cells by inhibiting the neutrophil extracellular trap. Neutrophils are the first cells to be recruited at the site of inflammation; the high neutrophil-to-lymphocyte ratio has been reported to be associated with higher metastasis and poor disease outcomes in breast cancer [88]. Moreover, kaempferol has also exhibited synergism when combined with other anticancer agents by inhibiting intracellular ROS generation as one of their mechanisms of action using various in vitro models of MDA-MB-231 cells. The combination of kaempferol and fisetin induced the activation of γ-H2AX, a histone variant, which, by causing DNA damage, leads to apoptosis in TNBC cells [16].

Interestingly, kaempferol also helps alleviate the serious adverse effects of standard chemotherapeutic agents. The standard chemotherapeutic drug doxorubicin is reported to have excessive ROS-mediated cardiotoxicity and endotheliotoxicity as its prominent adverse reactions. Wu et al., 2020, in their study, showed that kaempferol reversed the vascular doxorubicin-induced vascular toxicity by reducing oxidative stress, improving mitochondrial function by regulating the levels of 14-3-3 proteins, conserving regulatory molecules of eukaryotic cells and controlling the levels of various oxidative stress molecules [19]. Conversely, kaempferol has also been reported to induce apoptosis, with the production of ROS leading to apoptosis. In a study by Bong-Woo Kim et al., 2008, kaempferol induced apoptosis by modulating the ROS-mediated ERK/MEK1/ELK1 signaling pathway in breast cancer cells [89].

Many studies have investigated the anti-oxidative potential of kaempferol and its application as an anti-breast cancer agent. The multi-faceted mechanism of action associated with kaempferol allows it to effectively inhibit multiple ROS species, making it an ideal antioxidant agent.

5.3. Effect of Kaempferol on Nrf2 Activation

Nuclear factor-2 (Nrf2) is a protein regulator of cellular oxidative response in both normal and malignant cells. It protects the healthy cells from oxidative stress, thereby preventing their malignant transformation, and guards the malignant cells against radiation and chemotherapy, resulting in the development of chemoresistance. Nrf2 neutralizes ROS or repairs cellular damage caused by oxidative stress. It facilitates the rapid enzymatic detoxification and elimination of carcinogenic chemicals, but once the cancerous tumors have formed, Nrf2 becomes involved in the cancer progression and metastasis. This surprising phenomenon was later described as the "dark side" of Nrf2 [90]. In the absence of cellular stress, Nrf2 is localized within the cytoplasm. In this basal state, Nrf2 remains tethered to its negative regulatory protein, Kelch-like ECH-associated protein 1 (Keap1) [91]. On becoming exposed to oxidative stress, it undergoes nuclear translocation, where it heterodimerizes with avian musculoaponeurotic fibrosarcoma oncogene homolog (sMAF), forming a complex that subsequently binds to specific DNA sequences known as antioxidant-responsive elements (AREs), and with this interaction, Nrf2 regulates the transcription of genes involved in intracellular redox balance, metabolism, apoptosis, and autophagy. On the dark side, the chronic activation of Nrf2 leads to increased gene expression in drug metabolism, fostering resistance to chemotherapeutic agents and radiotherapy [92].

Furthermore, Nrf2 hyperactivation promotes cellular proliferation by inducing metabolic reprogramming towards anabolic pathways, modulating the pentose phosphate pathway (PPP, crucial for generating precursors for nucleic acid synthesis), and augmenting purine synthesis, which is essential for DNA replication and cell division. The dysregulation of the Nrf2-Keap1 signaling pathway has been observed in a broad spectrum of human malignancies, including breast cancer, and multiple downstream signaling pathways are involved in disease pathogenesis, including increased Nrf2 levels, decreased Keap1 levels, and blocked Nrf2 ubiquitination. Many molecules of natural origin have been used to modulate Nrf2 activity by disrupting the intermolecular disulfide bonds formed between two cysteine residues (Cys273 and Cys288) within the Keap1 protein, which weakens its ability to sequester Nrf2 in the cytoplasm, thereby facilitating Nrf2's nuclear accumulation and subsequent activation of antioxidant response pathways [92,93]. Kaempferol has been reported to activate Nrf2 and its downstream signaling pathways in multiple human cancer cell models, including MCF-7 breast cancer cells [15,53]. It modulates the Nrf2-ARE signaling pathway, thereby activating Nrf2 expression and the antioxidant response. However, in other models of human cancer, kaempferol has been seen to downregulate the expression of Nrf2, demonstrating kaempferol's action against the dark side of Nrf2. In a study by Foudzer et al., 2021, the authors showed that in NSCLC cells, kaempferol inhibited Nrf2 and induced ROS accumulation after 48 h of treatment, thereby making the NSCLC cells sensitive to apoptosis at physiological concentrations [94]. Nrf2 has attracted significant interest as a potential therapeutic target for several years, owing to its role in cancer progression. Notably, kaempferol exhibits a paradoxical effect on Nrf2 expression, demonstrating upregulation and inhibition in various human cancers [53,94]. This intriguing observation underscores the need for further exploration to elucidate the underlying mechanisms by which kaempferol modulates Nrf2 expression. Understanding these mechanisms is crucial for harnessing the full therapeutic potential of kaempferol in cancer treatment.

5.4. Effect of Kaempferol on Cell Cycle Arrest

Kaempferol causes cell growth inhibition by inducing cell cycle arrest in TNBC cells. It particularly induces cell cycle arrest at the G2/M phase of the cell cycle when treated for about 48 h in MDA-MB-468 and MDA-MB-231 cells [43,95]. It also produces the same effect on MDA-MB-453, androgen-responsive human breast carcinoma cells extensively used in TNBC research, particularly by downregulating the cyclin-dependent kinase 1 (CDK1) and the associated proteins cyclin A and cyclin B [96,97]. In another study, the chloroform extract of the *Butea monosperma* (Lam.) Taub bark, an Indian medicinal plant rich in kaempferol, showed arrest in the G1 phase of the cell cycle in MCF-7 cells in a concentration-dependent manner [98]. In Kim et al., 2016, the authors represented the efficacy of kaempferol in inducing cell cycle arrest by downregulating the expression of cyclin D1 and cyclin E and upregulating the expression of p21 in triclosan and estradiol-induced ER-mediated increased cell proliferation of MCF-7 cells and the in vivo mouse model [80].

5.5. Effect of Kaempferol on Apoptosis Induction

Induction of apoptosis in cancer cells is one of the principal mechanisms of kaempferol anticancer activity. Induction of apoptosis is also linked with cdc2 dephosphorylation and cyclin B1 downregulation, which alter cell cycle kinetics in breast cancer cells. Kaempferol was initially shown to induce apoptosis in MDA-MB-231 cells using the trypan blue staining method when treated in a different concentration-dependent manner, and it showed maximum effect at 48 h of incubation. The results were confirmed with kaempferol-induced chromatin condensation and formation of oligonucleotides due to nuclei fragmentation at a 50 μM concentration, which showed increased expression of cleaved PARP in kaempferol (50 μM) treated MDA-MB-231 cells at 24 h [99]. PARP is a protein that is involved in both DNA repair and apoptosis, is cleaved by caspases, and acts as an apoptotic marker.

Pro-apoptotic protein Bax and anti-apoptotic protein Bcl2 also serve as indicators of programmed cell death. Kaempferol extracted from various natural sources commonly used by humans and non-human primates has also been reported to show its antitumor activity via the induction of apoptosis demonstrated by activation of multiple caspases, increased expression of cleaved PARP, phosphorylated DNA damage associated protein ATM (pATM), and pro-apoptotic protein Bax, and decreased expression of anti-apoptotic protein Bcl-2 in MCF-7 and MDA-MB-231 cells [16,95,100–104].

Kaempferol apoptotic effect has been confirmed with the formation of condensed DNA, DAPI staining, and increased expression of phosphorylated H2A histone family member X [75]. Diverse subfamilies of MAPKs involved in apoptosis have been reported to be activated and responsible for the induction of apoptosis. Along with the activation of cleaved PARP, treatment of kaempferol in MDA-MB-231 and MCF-7 cells showed activation of extracellular signal-regulated kinase (ERK), evident with increased expression of ERK and phosphorylated ERK (pERK). The study showed the simultaneous activation of the upstream kinase and a substrate of ERK, as evidenced by the increased expressions of MEK1 and pMEK1 and ELK1 and pELK1. The inactivation of ERK using a MEK1 inhibitor (PD98059) and transfecting the MCF-7 cells with a kinase-inactive ERK mutant (ERK-DN (K52R)) showed marked suppression of apoptosis, as demonstrated by the decreased expression of cleaved PARP in the kaempferol treated cells. The induction of apoptosis via the modulation of ERK and its suppression by its activation was also reported to be more profoundly evident in 3D-cultured MCF-7 cells [89]. Kaempferol also downregulates polo-like kinase-1 (PLK-1) expression in MCF-7 cells, another mammalian protein kinase and a key regulator of mitosis. It has been reported to be involved in tumor induction and tumor progression and is overexpressed in a variety of tumors, including breast cancer [105]. Interestingly, kaempferol has also been reported to cause induction of apoptosis, as demonstrated by the increased expression of Bax and decreased expression of Cathepsin D (an essential lysosomal aspartic protease involved in breast cancer metastasis) in triclosan and estradiol-treated MCF-7 cells and a xenograft mouse model [80].

Alternatively, studies have also demonstrated the potential of kaempferol to reverse the ability of aggressive breast cancer tumors to escape apoptosis and possess stemness. The TCGA database analysis revealed a correlation between the levels of p53, a tumor suppressor protein, and caspase 3, where the two levels were higher in breast tumors than in normal cells [71,106]. Whereas, in a study by Nandi et al., 2022 on a cohort of 271 female breast cancer tissues, higher levels of p53 and lower levels of caspase 3 have been reported with advanced stages (stage II/IV) of the disease, while the vice versa is associated with lower stages (stage I/II). The high expression levels of p53 and downregulated caspase 3 expression have also been reported in multiple studies to be related to more advanced metastases with nodular-involved cancer stages. In usual conditions, the activation of caspase 3 leads to the induction of apoptosis and results in cell death. However, in p53-mutated cells, the activation of caspase 3 does not initiate apoptosis, which causes chemoresistance. Mutant p53 leading to its deactivation is also associated with increased stemness and expression of MDR1 and various anti-apoptotic proteins, including ALDH1, NANOG, NF-κB, CD 44, Ki-67, and Bcl2. The former cohort of 217 breast cancer patients showed more co-expression between p53 mutant and Ki-67 than other anti-apoptotic proteins. Interestingly, the treatment of ex vivo neo-adjuvant chemotherapy-treated primary human breast cancer tissues with kaempferol showed a notable downregulation of p53, ki-67, NANOG, NF-κB, CD 44, ALDH1, Bcl2, and upregulation in the expression of caspase 3 in comparison to the treatment with carboplatin, which indicates the ability of kaempferol to reverse the acquired chemo-tolerance in advanced stage breast cancer tumors [71].

Subsequently, kaempferol was further shown to inhibit breast cancer cell proliferation via diversified mechanisms. For instance, Brusselmans et al., 2005, showed that kaempferol inhibited fatty acid synthase, a prominent lipogenic enzyme found overexpressed in human cancers, correlating the effect of kaempferol with reduced cell growth and increased apoptosis [99].

5.6. Effect of Kaempferol on Cell Invasion and Metastasis Inhibition

Concerning inhibiting migration and invasion, kaempferol has also been described as having anti-metastatic activity [107]. Astin et al., 2014, in their study, reported, for the first time, kaempferol to be a novel inhibitor of Vascular Endothelial Growth Factor Receptor (VEGFR) kinases, which are involved in increased vascular permeability and angiogenesis, and they reported kaempferol anti-lymphangiogenic activity in their zebrafish model of lymphangiogenesis [108]. Kaempferol, especially in its lower doses, has shown promising effects in TNBC compared to other non-TNBC in vitro models. In a study by Shoushan Li et al., 2017, the authors showed the suppression of migration with low-dose kaempferol treatment in an in vitro model of MDA-MB-231 and MDA-MB-453 cells in comparison to MCF-7 and SK-BR-3 cells. They mainly reported the inhibition of activation of RhoA and Rac, small GTP-binding proteins primarily involved in microfilament rearrangement and cancer cell migration, with low doses of kaempferol treatment in TNBC cells [95]. In another study, kaempferol inhibited cell adhesion, cell motility, and cell migration in various in vitro assays using MDA-MB-231 cells. Kaempferol also significantly downregulated the expression of matrix metalloproteinases-2 (MMP-2) and MMP-9, molecules involved in ECM degradation, cell invasion, and cancer metastasis, compared to the control. The activity of MMPs is regulated by either the signal transducer and activator of transcription 3 (STAT3) or activator protein-1 (Ap-1) pathways. Kaempferol-treated MDA-MB-231 showed inhibition in the nuclear translocation of cJun and cFos to increased cytoplasmic levels, and both cJun and cFos are components of Ap-1. The further downstream pathway analysis showed downregulation of MMPs via AP-1 induction mediated activation of MAPK and the protein kinase Cδ (PKC-δ) signaling pathway in kaempferol-treated cells. Li et al., 2015, also reported the inhibition of lung metastasis with kaempferol when used in doses as high as 200 mg/kg compared to decarbazine at 100 mg/kg. They also reported a notable decrease in the expression of MMP-9 in kaempferol (200 mg/kg)-treated lung tissue nodules [109].

5.7. Kaempferol Epigenetic Modulation

The complex nature of cancer due to the underlying complex mechanisms makes it difficult to target and find an ultimate cure for the disease. Various pathways and mechanisms have been explored using a variety of interventions for years to fight multiple aspects of cancer. Cancer research to date has investigated diverse strategies ranging from cure (eradication, reduced aggressiveness), prevention (risk factor targeting, pre-cancer intervention), and management (reduced side effects, preventing recurrence) post-treatment [110].

Apart from genetics, epigenetic factors are broader in implicating cancer development and influencing its aggressiveness. Epigenetic changes are transmissible changes in chromatin and gene expression without causing any real change in the DNA sequence. However, they can still make characteristic changes in the cellular processes and the phenotype. Typical epigenetic modifications include chromatin remodeling, histone modifications, acetylation, methylation, phosphorylation, ADP-ribosylation, ubiquitination, and non-coding RNAs (ncRNAs), which can either result in the induction or suppression of gene transcription, specifically oncogene transcription, in cancer and therefore, play a crucial role in the induction and progression of cancer [111].

Due to its remarkable anticancer activity, kaempferol has been evaluated in multiple studies for its potential activity in modulating epigenetic regulation. In an in silico survey by Berger et al., kaempferol was reported to have HDAC inhibitory activity in human hepatoma cells (HepG2 and Hep3B). The in silico docking reported that kaempferol fits in the binding pocket of HDAC-2, 4, 7, or 8 and binds to the Zn ion in the enzyme's catalytic core. The authors further evaluated their results in in vitro models of human hepatoma cells [112]. Kim et al., 2020, explored kaempferol effects on gastric cancer cells, revealing its ability to induce autophagy and cell death through the IRE1-JNK-CHOP signaling pathway, potentially via epigenetic modulation involving G9a (a histone methyltransferase inhibitor) inhibition in human gastric cell lines (AGS, SNU-216, NCI-N87, SNU-638, and MKN-74) [89].

Moreover, kaempferol has also been evaluated for its potential to target histone modifications in aggressive TNBC. Their study indicated that, as per their network pharmacology analysis, the anticancer activity of kaempferol in their model of MDA-MB-468 cells, demonstrated by DNA damage, S-phase cell cycle arrest, and suppression of cancer stemness, might be related to the inhibition of sirtuins (SIRT4 and 6), nicotine adenine dinucleotide-dependent HDACs, involved in various cell signaling pathways. They confirmed their molecular dynamics by downregulating SIRT3 and SIRT6 in kaempferol-treated MDA-MB-468 cells [113].

The development of various synthetic modulators of epigenetic modifications across multiple cancer types indicates the dire need to further evaluate further the ability of kaempferol to target various epigenetic pathways in breast cancer.

5.8. Effect of Kaempferol on the Tumor Microenvironment and Immune Response

Immune cells have been researched as potential targets for many anticancer therapies. The involvement of immune cells and evasion of immunosurveillance by cancer cells is prominently involved in the development and progression of cancer, where immunosurveillance or immunoediting is a process where the immune cells perform their function and reject the malignant transformation of the cells, thereby inhibiting tumor growth. Prolonged immune responses and chronic inflammation in the body are responsible for cancer development by critically modulating the protective effects of immune cells [114]. In their tumor microenvironment (TME), the cancer cells interact with various other conditions of TME, such as oxidative stress, dysregulated immune function, inflammation, and an acidic environment, resulting in cancer progression [115]. Within the TME, the immune cells, helper CD4 + T cells (Th1), and cytotoxic CD8 cells by their cytokine IFN-γ, produce significant antitumor effects, and conversely, the myeloid-derived suppressive cells (MDSC) and tumor-associated macrophages (TAM) along with their associated cytokines IL-1beta, IL-6, and TNF-alpha become involved in tumor progression [114]. However, a subpopulation of T cells, the regulatory T cells (Tregs), having a surface expression of transcription factors Foxp3 and CD25, conversely inhibits the activation of immune cells against cancer. The physiological function of Tregs is to suppress the effector immune response against the self-antigens in the body, thereby preventing the autoimmunity and inflammation associated with it. In TME, Tregs inhibit the activation of CD4 and CD8 T cells, compromising their anticancer activity and promoting cancer progression [114].

5.8.1. Effect of Kaempferol on Tumor-Associated Macrophages (TAMs)

TME also consists of other immune cells, like macrophages and dendritic cells. TAMs are a population of the most abundant macrophages infiltrating tumor cells and have a prominent role in tumor progression. TAMs have two major populations of macrophages, M1 and M2, where M1 is called antitumor or good macrophages, while M2 is called tumor-promoting or bad macrophages [116]. TNF-α and high levels of iNOS characterize M1 macrophages, whereas M2 is associated with very high levels of cytokines, growth factors, and protease, along with high expression of Arginase 1 [117]. TAMs play a complex role in tumor development by secreting molecules and growth factors like platelet-derived growth factor (PDGF), VEGF, M-CSF, IL-10, and the chemokine C-X-C motif ligand (CXCL). The heterogeneity of the TAMs depends on the tumor type, polarization requirements in the TME, and variations in the TME. A single neoplasm can have M1 macrophage factors and other TAM phenotypes in different regions within the tumor. Within TME of solid tumors like TNBC, M1 possesses the ability to switch from the M1 subtype with antitumor activity to the M2 subtype, which suppresses the function of M1 and provides tissue repair function and limits the inherent ability of immune cells to recognize and kill the transforming cells. Additionally, another regulatory factor present in large amounts in a variety of T cells in TNBC is PD-1 and its receptor PD-L1, where PD-1 is a surface receptor on activated T cells and PD-L1 is its ligand expressed by various cells, including cancer cells. The binding of PD-1/PD-L1 limits the antitumor abilities of the immune cells, and their upregulation is associated with advanced stages of disease and lower survival rates in TNBC patients. TAMs with higher PD-1 expression have been reported to be related to immune suppression and reduced antitumor activity of the immune cells. The blockade of PD-1/PD-L1 binding using targeted therapy has been explored to revive the phagocytic activity of TAMs and thereby increase their antitumor action. TAMs have also been reported to be involved in enhancing the migration and invasion of tumor cells in TNBC. They have been shown to increase the secretion of serine proteases, MMPs, and cathepsin and are involved in tumor cell invasion of the tumor surroundings [117]. STAT3 is the prominent mediator of TAM expansion and polarization across the TME in breast cancer. Tumors release lactic acid, induce the expression of Arginase 1 (Arg1), CD206, and mannose receptor C-type 1, and activate the (ERK)/STAT3 signaling pathway, which further stimulates colony-stimulating factor 1 (CSF-1) and polarizes the TAMs. Apart from this, activating the STAT3 signaling pathway also triggers the release of IL-6, IL-4, and progranulin (a multifunctional growth factor), which are also involved in polarizing the TAMs [118]. Interestingly, breast cancer tumors have also been found to have unique TAMs in various breast cancer tissues. Breast cancer TAMs are also associated with CCL-2, CCl-5, and integrin-mediated disease progression to nearby organs such as the lungs [119]. Many compounds of natural origin have been described as inhibiting TAM infiltration by targeting CCL-2/CCR-2 signaling [120].

Many studies have explored a variety of natural compounds for their immunomodulatory properties by targeting infiltrating TAMs by modulating the (PI3K)/Akt/mTOR pathway. Alkaloids, flavonoids, and terpenoids have been used in multiple studies to target the STAT3/IL-6/Arg1 signaling pathway within TAMs, thereby inhibiting their expansion, infiltration, and immunosuppressive activities and exhibiting antitumor properties by blocking the production of pro-inflammatory molecules and signaling mediators within the TME [121]. Similarly, kaempferol has also shown anticancer activity in various cancer cell models by inhibiting STAT3 [122–124]. Likewise, in a study by Qian Yu et al., 2016, a kaempferol derivative called resokaempferol showed significant inhibition of lipopolysaccharide (LPS)-induced production of key inflammatory mediators including cyclooxygenase-2 (COX-2), prostaglandin E2 (PGE2), CCL2/MCP-1, nitric oxide (NO), iNOS, IL-1β, TNF-α, and IL-6 in primary murine macrophage culture. Resokaempferol also abrogated the activation of the JAK2/STAT3 signaling pathway in murine macrophages stimulated with exogenous interleukin-6 (IL-6) [125]. The immunomodulatory properties of kaempferol against TAMs by modulating multiple signaling pathways provide sufficient evidence for its ability to work against the different population macrophages in TNBC tumor TME. However, studies to analyze the exact mechanism adopted by kaempferol to inhibit the polarization of macrophages in TNBC TME need to be explored in future studies.

5.8.2. Effect of Kaempferol on the Expression of CCL2

Inflammatory mediators like chemokines and cytokines guide the immune cells to the tumor sites. CCL2 is a chemokine that binds to GPCRs and plays a crucial role in regulating the recruitment of macrophages during processes such as wound healing, inflammation, and infection [126]. The higher expression of the chemokine CCL2/Monocyte chemoattractant proteins (MCP-1) and its receptor CCR2 is related to the primary breast tumor cells undergoing malignant transformation, and it was found to be expressed by the multiple cells of TME. This chemokine is upregulated in pleural effusions, serum, and interstitial fluids in breast TME. It is associated with high-grade disease states and a poor prognosis [127]. Multiple cell types in the TME—endothelial cells, stromal cells, and tumor cells—tend to produce CCL2, which recruits monocytes and TAMs to the tumor sites. The chemokine, through its CCL2-CCR2 axis, polarizes the monocytes to TAMs, resulting in tumor cell survival, and the inhibition of CCL2-CCR2 signaling blocks the recruitment of inflammatory cells to the tumor site, reducing tumor progression and metastasis [128]. A clinical study performed by Xiangzhou Chen, 2020 revealed that CCL2 secreted by TAMs in the TNBC TME activates AKT/β-catenin signaling, which is prominently involved in cell growth, survival, proliferation, and stem cell properties [129]. In a study by Zvi G. Fridlender et al., 2010, the authors analyzed the ability of CCL2-CCR2 blockade to augment the therapeutic efficacy of immunotherapy by suppressing TME-associated local immunosuppression in addition to boosting the T-cell immune response, which is the general effect of standard immunotherapy [130]. CCL2 has also been reported to induce resistance against immunotherapy in aggressive cancers like TNBC. Junyoung Choi et al., 2020 demonstrated that activation of the PI3K/AKT signaling pathway with NF-κB activation induces CCL2 secretion and PD-L1 inhibitor resistance, which is a promising immunotherapy [131].

Interestingly, many compounds of natural origin have been seen to produce antiproliferative effects against genetically different TNBC cell types by inhibiting the release of CCL2 [132–134]. Interestingly, a computational molecular dynamic study of the analysis of the effect of flavonol-CCL2 interactions showed that flavanol compounds can attenuate the CCL2-mediated recruitment of leukocytes to the site of inflammation. In this study, kaempferol, among other flavonols (quercetin and myricetin) with multiple -OH groups in their structures, showed increased affinity towards the CCL-2 structure, demonstrating that kaempferol has the potential to reduce the CCL2-CCR2-cell surface glycosaminoglycans (GAGs)-mediated pathogenesis of inflammatory infirmities in cancer TME. More studies are still needed to explore the possibility of kaempferol modulating CCL2 in TNBC, which could reverse the drug resistance observed in the current treatments (Figure 3 and Table 1) [135].

Table 1. The anti-breast cancer properties of kaempferol in different signaling pathways.

Study	Pathway Targeted	Outcomes
DNA Synthesis Inhibition		
Zava et al., 2023 [78]	DNA synthesis inhibition and cell growth	Kaempferol demonstrated estrogen agonistic activity and showed cell growth and DNA inhibition (at 10 μM concentration, the 0.25 μg DNA in T47D cells).
Apoptosis		
Balabhadrapathruni et al., 2000 [43]	Cell proliferation, cell cycle, and apoptosis	Kaempferol inhibits cellular proliferation by targeting the G2/M cell cycle and apoptosis in MDA-MB-468 cells.
Brusselmans et al., 2005 [102]	Fatty acid synthetase pathway	Kaempferol inhibited the fatty acid synthetase enzyme, which was overexpressed in human breast cancers, reduced cell growth, and increased apoptosis.
Kim et al., 2008 [89]	Cell viability and apoptosis in 2D and 3D cultures	Kaempferol-induced apoptosis by modulating the ERK/MEK1/ELK1 signaling pathway.
Targeting the Estrogenic pathway		
Kim et al., 2016 [80]	E2-mediated breast cancer cell proliferation, cell cycle, and apoptosis	Kaempferol antagonized the Triclosan stimulated cell proliferation in MCF-7 cells, upregulated the expression of cathepsin, cyclin D1, and cyclin E, and downregulated the expression of Bax and p21.
Oh et al., 2006 [74]	Estrogen-dependent and estrogen-independent pathways in breast cancer cell proliferation and malignant cell transformation	Kaempferol inhibited cell proliferation via an estrogen-dependent pathway, preventing malignant transformation of the human breast cells.
Hung et al., 2004 [79]	Cell viability	Kaempferol reduced cell viability (IC$_{50}$: 35.0 mM and 70 mM for ER-positive and ER-negative breast cancer cells, respectively), decreased ER-alpha mRNA and protein expression, and decreased the progesterone receptor, cyclin D1, and insulin receptor expression.

Table 1. *Cont.*

Study	Pathway Targeted	Outcomes
Anti-Oxidative stress		
Zeng et al., 2020 [88]	Neutrophil extracellular traps, ROS	Kaempferol inhibited the formation of NETs, thereby reducing the formation of ROS and inhibiting metastasis in breast tumors.
Afzal et al., 2023 [16]	ROS, agonism with another potential anticancer agent: Fisetin, DNA damage, and apoptosis	Kaempferol synergized the anti-oxidative properties of fisetin and activated γ-H2AX, leading to DNA damage and apoptosis.
Wu et al., 2020 [19]	ROS, ROS-mediated cardiotoxicity, endotheliotoxicity	Kaempferol reversed the vascular toxicity and cardiotoxicity caused by doxorubicin's adverse effects by modulating the levels of the 14-3-3 s protein, which regulates ROS levels.
Inhibition of Metastasis and Invasion		
Li et al., 2017 [97]	Migration	Kaempferol inhibited the activation of RhoA, Rac, and GTP-binding proteins involved in microfilament arrangement, thereby inhibiting the cancer cell migration of MDA-MB-231 cells.
Li et al., 2015 [98]	TNBC cell adhesion, motility, and migration	Kaempferol (IC$_{50}$: 204.7 mol/L) decreased the expression of MMP-2 and MMP-9 in MDA-MB-231 cells.
Cell Cycle Arrest		
Choi et al., 2008 [100]	Cell cycle arrest	Kaempferol downregulated the levels of CCDK1, cyclin A, and cyclin B and induced cell cycle arrest at the G2/M phase of the cell cycle in MDA-MB-453 cells.
Varinder Kaur et al., 2018 [101]	Cell cycle arrest	Chloroform extract of *Butea monosperma* (Lam.) Taub bark rich in kaempferol produced cell cycle arrest at the G1 phase of the cell cycle in MCF-7 cells.
Zhu et al., 2019 [75]	Cell cycle arrest	Lower concentration of kaempferol is required to produce cell cycle arrest at G2/M phase in MDA-MB-231 (IC$_{50}$: 43 µmol/L) cells than in estrogen receptor-positive BT474 breast cancer cells (IC$_{50}$: 100 µmol/L).
Kim et al., 2016 [80]	Cell cycle arrest	Kaempferol downregulated the expression of cyclin D1 and cyclin E and upregulated the expression of p21 in triclosan-treated MCF-7 cell models.

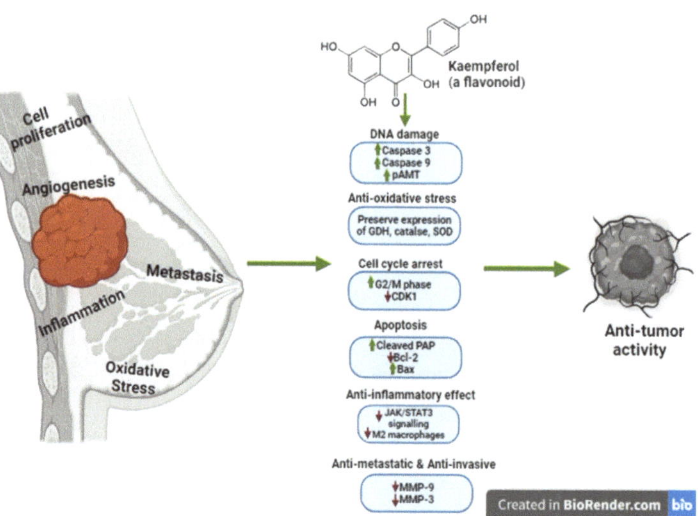

Figure 3. The effect of the flavonoid kaempferol on the development and progression of breast cancer. Green arrows indicate induction and red arrows indicate inhibition.

6. Addressing Kaempferol's Poor Solubility: Pharmaceutical Formulations of Kaempferol

Despite the promising anticancer activity against cancer, kaempferol still struggles with its biomedical applications due to its unstable chemical characteristics, poor water solubility and dissolution, and limited bioavailability. Quite a few efforts have been made to address these delivery-related issues of kaempferol with the development of various delivery systems, and surprisingly, they have been developed specifically for their delivery across multiple cancer types, including breast cancer. Drug delivery formulations like liposomes and nanoparticles have been extensively reported to overcome the challenges mentioned above of hydrophobic compounds like kaempferol, and they significantly improve the overall effectiveness of drugs, allowing for reduced dosages. Nanoformulations, particularly nanoparticles, have shown promising results for the biological delivery of various therapeutic agents by aiding in their bioavailability. Subsequently, the delivery of kaempferol in nanoparticles showed increased kaempferol dissolution with decreased particle size while retaining its therapeutic properties [136]. Luo et al., in 2012, developed five different types of kaempferol nanoparticle formulations and showed a significant reduction in the cell viability of A2780/CP70 and OVCAR-3 cancer cells with a lower concentration of kaempferol (25 µM) with one of their nonionic poly(ethylene oxide)-poly (propylene oxide)-poly(ethylene oxide) (PEO-PPO-PEO) nanoparticles [60]. In another study by Kazmi et al., 2021, the authors developed kaempferol-loaded nanoparticles (KFP-Np) with a quasi-emulsion solvent diffusion technique using two polymers (hydroxypropyl methylcellulose acetate succinate ((HPMC-AS) and Kollicoat MAE 30 DP) to analyze the potential of kaempferol to treat hepatocellular carcinoma (HCC) and its associated liver damage. In their in vivo model of Cadmium chloride ($CdCl_2$)-induced HCC, KFP-Np demonstrated a significant decrease in the levels of elevated liver enzymes, oxidative stress markers, and antioxidant enzymes (MDA, SOD, GST, and Catalase), and also significantly downregulated pro-inflammatory cytokine expression (IL-1β, IL-6, and TNF-α) and NF-κB in comparison to HPMC-AS and free kaempferol [137]. Later, Srinivas Raghvan et al., 2015, developed kaempferol gold nanoparticles (KAuNPs) using the reactive -OH group in the catechol ring of kaempferol and showed excellent biocompatibility with biological systems. They further observed a significant increase in the cytotoxic potential of KAuNPs compared to individual kaempferols in MCF-7 cells. Also, they demonstrated the induction of apoptosis and the anti-angiogenic potential of KAuNPs [138]. Similarly, Govindaraju et al., 2019, demonstrated the cytotoxic activity of kaempferol-conjugated gold nanoclusters (K-AuNCs) in A549 lung cancer cells and showed the induction of apoptosis and inhibition of cell migration in A549 cells [139].

Interestingly, the more advanced version of nano-formulations has also been developed to deliver kaempferol across the blood–brain barrier (BBB). Colombo et al., 2018, prepared kaempferol-loaded nanoemulsions and targeted the delivery of their payload, the kaempferol, to the brain intra-nasally to show its anticancer activity in their in vitro and in vivo models of glioma. The formulations were prepared with chitosan (for mucoadhesion and called mucoadhesive nanoemulsion (MNE) and without chitosan, called nonemulsion (NE)) to assess the impact on nasal targeting and antitumor activity against glioma cells. Their ex vivo diffusion studies revealed significantly higher kaempferol permeation across the mucosa with MNE compared to NE. The histopathological evaluation indicated the safety of both nanoemulsions for the nasal mucosa, with no compromise with the antioxidant activity of kaempferol. MNE significantly enhanced drug delivery in the in vivo model following intranasal administration, achieving 5 and 4.5-fold higher levels compared to free kaempferol and NE, respectively [140].

Furthermore, MNE exhibited superior antitumor activity against C6 glioma cells, inducing apoptosis more than free KPF or KPF-NE. The oral delivery of kaempferol using a carrier system to increase its bioavailability has also been reported in various studies with promising results. Qian Du et al., 2019, prepared different formulations of N-trimethyl chitosan (TMC) grafted with medium- and long-chain fatty acids and observed that TMC nanoparticles grafted with decylic acids (a medium-chain fatty acid) showed enhanced cellular uptake and intestinal absorption and showed great potential for the development of such formulations of hydrophobic compounds like kaempferol [141]. Furthermore, developing more advanced formulation types, like extracellular vesicles or exosomes, for delivering synthetic and hydrophobic natural compounds has also been explored in cancer and other chronic diseases. Exosomes as a delivery system mimic the cells of their origin and provide more targeted delivery of the therapeutic agent by offering better biocompatibility and low immunogenicity and toxicity [142]. Developing smart/engineered exosomes is another upcoming technology that can deliver the desired therapeutic agents very precisely at the targeted sites, eliminating the undesired toxicity. Great options must be explored to provide a compound like kaempferol, potentially targeting the deadliest cancer type (Table 2).

Table 2. Pharmaceutical formulations and delivery agents of kaempferol.

Study	Formulation	Results
Luo et al., 2012 [60]	Prepared PEO-PPO-PEO nanoparticles, poly(lactic acid-co-glycolic acid) (PLGA) nanoparticles, PLGA polyethyleneimine (PEI), poly(amidoamine) (PAMAM) dendrimer, and glycol chitosan nanoparticles	Out of all, PEO-PPO-PEO nanoparticles of kaempferol (25 µM concentration of kaempferol) showed significant cytotoxicity towards A2780/CP70 and OVCAR-3 cancer cells in comparison to kaempferol alone.
Raghvan et al., 2015 [138]	Developed KAuNPs.	KAuNPs showed tremendous biocompatibility, demonstrated increased cytotoxic potential towards MCF-7 cells, and exhibited induction of apoptosis and anti-angiogenic activity compared to the individual compound.
Govindaraju et al., 2019 [139]	Developed kaempferol-conjugated gold nanoclusters (K-AuNCs).	K-AuNCs showed preferential nuclear localization and enhanced cytotoxicity towards A549 lung carcinoma cells compared to normal human cells.
Colombo et al., 2018 [140]	Developed kaempferol-loaded nanoemulsions, MNE (made with chitosan for mucoadhesion to achieve nasal delivery) and NE (made without chitosan), for the targeted delivery of kaempferol to the glioma cells.	Both ex vivo and in vivo studies revealed significantly higher permeation across the nasal mucosa with MNE while maintaining its antioxidant activities. They also showed higher antitumor activity against C6 glioma cells than NE and kaempferol alone.
Kazmi et al., 2021 [137]	Developed KFP-Np with HPMC-AS and Kollicoat MAE 30 DP using a quasi-emulsion solvent diffusion technique.	In their in vivo model of $CdCl_2$-induced HCC, KFP-Np significantly improved liver function by lowering liver enzymes and oxidative stress markers, boosting physiological antioxidant enzymes, and reducing inflammatory markers.

7. Clinical Translation of Kaempferol as an Anticancer Agent

Despite much literature on the therapeutic efficacy of kaempferol against cancer and other chronic diseases, few studies address kaempferol's potential to be evaluated and used clinically. The widely reported efficacy of kaempferol in various in vitro and in vivo studies, mainly due to its anti-inflammatory, antioxidant properties, and anticancer activity due to induction of cell cycle arrest, apoptosis, and suppression of angiogenesis, demands the dire need to replicate the results in clinical studies. A toxicological, clinical study has addressed the toxicology profile of kaempferol in clinical settings [143]. Minoru Akiyama et al., 2023, in their study, investigated the safety of high-dose kaempferol aglycone (a kaempferol derivative) in healthy adults in a randomized, double-blind, placebo-controlled study in which the participants were given either a daily 50 mg kaempferol aglycone or a placebo for 4 weeks. The authors reported no significant differences between the kaempferol and placebo groups for measurements like body size and proportions, blood pressure, blood tests, urine tests, or any adverse events. They suggested that 50 mg of kaempferol aglycone daily for four weeks is safe for healthy adults [143].

Various researchers have investigated the anti-inflammatory effects of kaempferol by using cruciferous vegetables with high kaempferol content as an intervention. In a clinical trial, healthy adults were fed diets containing different amounts of kaempferol from cauliflower, broccoli, and radish and were evaluated for various markers at 14 days. In the study, those who consumed more kaempferol (broccoli) showed significantly lower levels of the inflammatory markers IL-6 and IL-8 [144]. A study examined male smokers who consumed a broccoli diet for 10 days. This group showed decreased inflammatory markers of TNF-α and IL-6 compared to the control group [32]. Kaempferol has also shown promising anticancer activity through its anti-inflammatory activity in various in vitro and in vivo models. While all this preclinical data supports kaempferol's role in human cancers, its clinical efficacy remains uncertain [32,35,145].

Although noticeable progress has been made to show the anticancer activity of kaempferol, resistance to standard chemotherapeutic drugs like 5-FU is still the major reason for treatment failure in human cancers like colorectal cancer. In this regard, studies have shown that combinatory therapy is a successful approach to inhibiting drug resistance. Riahi-Chebbi et al., 2019 examined the anticancer activity of kaempferol, alone or in combination with 5-FU, on human 5-FU-resistant LS174-R colorectal cells. The data indicated that kaempferol can reverse 5-FU resistance in LS174-R cells, promoting apoptosis and cell cycle arrest, hindering the production of ROS, and regulating the activation of several signaling pathways, indicating that kaempferol could be used as a possible chemotherapeutic agent to be used solely or in combination with 5-FU to reverse drug resistance [68]. In their review of kaempferol, Nejabati et al., 2022 emphasize the need for more clinical studies to unravel the therapeutic potentials of kaempferol in cancers because, although there are many preclinical studies proving kaempferol effects against different cancers, there are still numerous doubts regarding its therapeutic potency in cancer therapy [145].

To fully realize the therapeutic potential of kaempferol as an anticancer agent, further clinical trials are warranted for different human cancers, including breast cancer. Additionally, kaempferol-loaded delivery systems like nanoparticles also present a promising avenue for enhancing its bioavailability and therapeutic efficacy in cancer treatment, which can serve as another avenue to develop kaempferol as a full-blown anticancer agent.

8. Critical Areas for Future Research

8.1. Extrachromosomal Circular DNA (ecDNA) as a Target in Cancer Therapy

Extrachromosomal circular DNA (ecDNA) is a new target emerging in the scientific community as a potential genomic modifier resulting in aberrant oncogenic expressions. ecDNAs are a particular class of circular DNA that exists outside the chromosomes and is shed from the genomic DNA. It is almost 1 MB and can be non-coding and oncogene-expressing DNA [146]. They have been reported to be involved in increased copy numbers of the oncogenes, tumorigenesis, tumor heterogeneity, and drug resistance in the advanced stages of various cancer types [147]. Many proto-oncogenes in aggressive tumors have been reported to be linked to ecDNA. Additionally, it has been reported that the heterogeneity in resistant tumors is also related to ecDNA. With the evolution of recent technologies to evaluate novel mechanisms like ecDNA and its potential to modulate the occurrence and progression of cancer, it has become an important target to be studied extensively. The ecDNA presents itself as a novel target for cancer-specific therapies. It holds potential for either standalone therapeutic intervention or as a synergistic partner to augment the efficacy of established treatment modalities like radiotherapy, chemotherapy, and conventional anticancer drugs. The unique structural and functional characteristics of ecDNA in tumor development make it an attractive avenue for developing targeted drugs [148].

Furthermore, intratumorally and intertumoral variations in ecDNA levels could be exploited to design personalized treatment strategies for individual cancer patients [149,150]. A limited number of studies address the development of ecDNA as a target for various treatment modalities. Still, with the emerging information and data across multiple databases, ecDNA can be considered a novel therapeutic target for different treatment options. Similarly, using a natural compound like kaempferol, with immense potential and promising activity to modulate the activity of ecDNA in aggressive breast cancer stages, can be a therapeutic breakthrough in the treatment of breast cancer [151].

8.2. The Regulatory Role of MicroRNAs (miRNAs) in Breast Cancer

MicroRNAs (miRNAs) are small sets of non-coding RNA that regulate an array of molecular functions and are considered post-transcriptional gene regulators in normal cellular processes and carcinogenesis. They are widely found in the eukaryotic genome and are believed to constitute about 1 to 2% of the known eukaryotic genome. A single miRNA can modulate the gene expression of multiple genes and vice versa; more than one miRNA can coordinate multiple pathways to alter the gene expression of only one gene (oncogene in the case of carcinogenesis) and hence holds a considerable role in mediating normal eukaryotic cells and cancer cell functions. Breast cancer has a unique set of miRNAs regulating survival and cancer cell death [152]. Prominently, three miRNAs, miR-21, miR-210, and miR-221, have been seen to be considerably upregulated in TNBC and have also been associated with poor progression-free survival. At the same time, miR-10b, miR-145, miR-205, and miR-122a have been reported to be under-expressed in breast cancer cases [153].

Moreover, different miRNAs are linked to modulating different cellular functions; for example, the upregulation of miR-7-5p is related to the induction of apoptotic cell death and suppression of cell proliferation in breast cancers; over-expression of miR15a and miR-16 is linked to increased release of cytochrome c in the cytosol, activation of caspase-3 and caspase-6, and ultimately resulting in the induction of apoptosis; over-expression of miR-20a is linked to the decreased activity of the autophagy pathway; this is also related to the high copy number variations and mutations in the TNBC cells. Similarly, the down-regulation of miR-23a is associated with decreased invasiveness and migration of TNBC cells. Also, the levels of miRNA regulated the chemo-sensitization of the breast cancer cells towards standard chemotherapeutic agents like carboplatin, paclitaxel, etc. [152,154].

Interestingly, kaempferol has been seen regulating the levels of various miRNAs, thereby affecting the survival of cancer cells, in quite a few studies. Kaempferol was reported to inhibit the cell proliferation and migration of the HeG2 liver cancer cells by downregulating the expression of miR-21, which is upregulated in advanced and aggressive breast cancer stages. They showed that by reducing the expression of miR-21, kaempferol upregulated the expression of PTEN, thereby inhibiting the signaling of the PI3K/AKT/mTOR pathway in HeG2 cells. Janet Alejandra Gutierrez-Uribe et al., 2020 reported that kaempferol-3-O-glycoside isolated from black beans showed downregulated expression of miR-31 and miR-92a along with the KRAS oncogene and increased expression of tumor suppressor genes, APC and AMPK in RKO colon cancer cells [155]. Similarly, kaempferol has been seen to upregulate the expression of miR-181a and inhibit the MAPK/ERK and PI3K pathways, resulting in inhibiting cell proliferation and inducing autophagy in SNU-216 gastric cancer cells [124]. Kaempferol has also been seen as chemo-sensitizing the resistant HCT8-R colorectal cancer cells by upregulating the expression of miR-326 and inhibiting the expression of PKM2 in glycolysis [69].

Although the literature shows different studies on the kaempferol effect on miRNA, none of them show the anticancer effect of kaempferol by modulating the expression of miRNAs in TNBC. The potential of kaempferol in modifying the expression of various miRNAs in different human cancer types fuels the idea of utilizing kaempferol to target miRNA-based anticancer mechanisms.

9. Conclusions

Breast cancer is one of the leading causes of morbidity and mortality in adult females of different age groups. Its heterogeneous nature, among other aggressive types of human cancers, makes it particularly challenging to treat, particularly at later stages of the disease. TNBC is an even more aggressive subtype of breast cancer, which further limits the availability of treatments and clinical outcomes of standardized therapies due to its differential molecular nature. Advanced research in the last few years has led us to a better understanding of the immunological and molecular heterogeneity in TNBC. It has increased the dire need to develop and explore more targeted and effective treatments with limited compromise on patients' overall health. As a potent natural compound of enormous importance, kaempferol has been evaluated by different groups for different cancer types. Like all compounds of natural origin, kaempferol modulates multiple pathways to provide various properties, including anti-oxidative, anti-inflammatory, antiproliferative, and anti-tumorigenic properties. In human cancers, and specifically in human breast cancer, it works primarily by inhibiting DNA synthesis, modulating ROS production, inducing apoptosis, and inhibiting the angiogenic and metastatic potential of breast cancer tumors (Figure 4). Besides its potential antitumor properties, kaempferol has poor solubility and dissolution properties and limited bioavailability. This review highlights some prominent delivery systems and formulations developed to better deliver kaempferol in various cancerous and non-cancerous conditions, with increased bioavailability and targeted delivery. This review also demonstrates some of the critical mechanisms targeted with kaempferol in breast cancer, particularly in TNBC. It also emphasizes the potential of kaempferol to target various epigenetic mechanisms in different human cancers, and it fuels the idea of further exploring kaempferol in modifying epigenetics in TNBC. With the scientific advances and discovery of various novel therapeutic targets, this review highlights ways of exploring the activity of kaempferol in targeting some of the most novel targets. Overall, this article re-scrutinizes the available literature on the efficacy of kaempferol as an anti-breast cancer agent and stresses the need for further investigation to better understand and fully acknowledge the capacity of kaempferol as an antitumor agent.

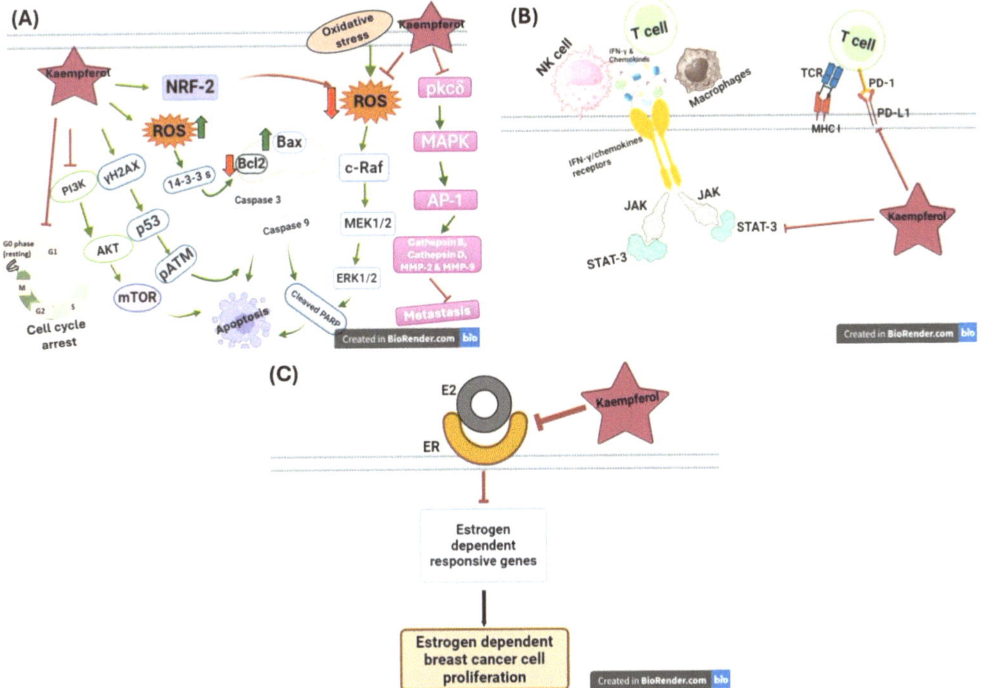

Figure 4. Summary of Pharmacological Effects of Kaempferol. (**A**) Oxidative stress modulation, apoptosis induction, and cell cycle arrest; (**B**) effect on inflammation and immune cells; (**C**) anti-estrogenic effect. Red arrows indicate inhibition, and green arrows indicate induction.

Author Contributions: Conceptualization: S.K., P.M., and K.F.A.S.; manuscript draft: S.K. and P.M.; figure creation and design: S.K.; review and editing: P.M. and K.F.A.S.; funding acquisition: K.F.A.S.; project administration: K.F.A.S.; resources: K.F.A.S.; supervision: K.F.A.S. All authors have read and agreed to the published version of the manuscript.

Funding: The research reported in this project was supported by the National Institute on Minority Health and Health Disparities of the National Institutes of Health under Award Number U54 MD007582.

Conflicts of Interest: The authors declare no conflicts of interest.

References

1. Al-Hajj, M.; Wicha, M.S.; Benito-Hernandez, A.; Morrison, S.J.; Clarke, M.F. Prospective identification of tumorigenic breast cancer cells. *Proc. Natl. Acad. Sci. USA* **2003**, *100*, 3983–3988. [CrossRef] [PubMed]
2. Dai, X.; Cheng, H.; Bai, Z.; Li, J. Breast Cancer Cell Line Classification and Its Relevance with Breast Tumor Subtyping. *J. Cancer* **2017**, *8*, 3131–3141. [CrossRef] [PubMed]
3. Breast Cancer Facts & Statistics 2023. Available online: https://www.breastcancer.org/facts-statistics (accessed on 13 June 2023).
4. Breast Cancer Facts & Figures | American Cancer Society. Available online: https://www.cancer.org/research/cancer-facts-statistics/breast-cancer-facts-figures.html (accessed on 13 June 2023).
5. Waks, A.G.; Winer, E.P. Breast Cancer Treatment: A Review. *JAMA* **2019**, *321*, 288–300. [CrossRef] [PubMed]
6. Moulder, S.; Hortobagyi, G. Advances in the Treatment of Breast Cancer. *Clin. Pharmacol. Ther.* **2008**, *83*, 26–36. [CrossRef] [PubMed]
7. Barzaman, K.; Karami, J.; Zarei, Z.; Hosseinzadeh, A.; Kazemi, M.H.; Moradi-Kalbolandi, S.; Safari, E.; Farahmand, L. Breast cancer: Biology, biomarkers, and treatments. *Int. Immunopharmacol.* **2020**, *84*, 106535. [CrossRef] [PubMed]
8. Fisusi, F.A.; Akala, E.O. Drug Combinations in Breast Cancer Therapy. *Pharm. Nanotechnol.* **2019**, *7*, 3–23. [CrossRef] [PubMed]
9. Mahmoud, R.; Ordóñez-Morán, P.; Allegrucci, C. Challenges for Triple Negative Breast Cancer Treatment: Defeating Heterogeneity and Cancer Stemness. *Cancers* **2022**, *14*, 4280. [CrossRef] [PubMed]
10. Chaudhary, L.N.; Wilkinson, K.H.; Kong, A. Triple-Negative Breast Cancer: Who Should Receive Neoadjuvant Chemotherapy? *Surg. Oncol. Clin. N. Am.* **2018**, *27*, 141–153. [CrossRef] [PubMed]
11. Lin, S.R.; Chang, C.H.; Hsu, C.F.; Tsai, M.J.; Cheng, H.; Leong, M.K.; Sung, P.J.; Chen, J.C.; Weng, C.F. Natural compounds as potential adjuvants to cancer therapy: Preclinical evidence. *Br. J. Pharmacol.* **2020**, *177*, 1409–1423. [CrossRef]
12. Nobili, S.; Lippi, D.; Witort, E.; Donnini, M.; Bausi, L.; Mini, E.; Capaccioli, S. Natural compounds for cancer treatment and prevention. *Pharmacol. Res.* **2009**, *59*, 365–378. [CrossRef]
13. Liang, Z.; Xie, H.; Shen, W.; Shao, L.; Zeng, L.; Huang, X. The Synergism of Natural Compounds and Conventional Therapeutics against Colorectal Cancer Progression and Metastasis. *Front. Biosci. Landmark Ed.* **2022**, *27*, 263. [CrossRef] [PubMed]
14. Imran, M.; Salehi, B.; Sharifi-Rad, J.; Gondal, T.A.; Saeed, F.; Imran, A.; Shahbaz, M.; Fokou, P.V.T.; Arshad, M.U.; Khan, H.; et al. Kaempferol: A Key Emphasis to Its Anticancer Potential. *Molecules* **2019**, *24*, 2277. [CrossRef] [PubMed]
15. Wang, X.; Yang, Y.; An, Y.; Fang, G. The mechanism of anticancer action and potential clinical use of kaempferol in the treatment of breast cancer. *Biomed. Pharmacother.* **2019**, *117*, 109086. [CrossRef] [PubMed]
16. Afzal, M.; Alarifi, A.; Karami, A.M.; Ayub, R.; Abduh, N.A.Y.; Saeed, W.S.; Muddassir, M. Antiproliferative Mechanisms of a Polyphenolic Combination of Kaempferol and Fisetin in Triple-Negative Breast Cancer Cells. *Int. J. Mol. Sci.* **2023**, *24*, 6393. [CrossRef] [PubMed]
17. Wu, Q.; Chen, J.; Zheng, X.; Song, J.; Yin, L.; Guo, H.; Chen, Q.; Liu, Y.; Ma, Q.; Zhang, H.; et al. Kaempferol attenuates doxorubicin-induced renal tubular injury by inhibiting ROS/ASK1-mediated activation of the MAPK signaling pathway. *Biomed. Pharmacother.* **2023**, *157*, 114087. [CrossRef] [PubMed]
18. Yang, G.; Xing, J.; Aikemu, B.; Sun, J.; Zheng, M. Kaempferol exhibits a synergistic effect with doxorubicin to inhibit proliferation, migration, and invasion of liver cancer. *Oncol. Rep.* **2021**, *45*, 32. [CrossRef] [PubMed]
19. Wu, W.; Yang, B.; Qiao, Y.; Zhou, Q.; He, H.; He, M. Kaempferol protects mitochondria and alleviates damages against endotheliotoxicity induced by doxorubicin. *Biomed. Pharmacother.* **2020**, *126*, 110040. [CrossRef]
20. Naeem, A.; Hu, P.; Yang, M.; Zhang, J.; Liu, Y.; Zhu, W.; Zheng, Q. Natural Products as Anticancer Agents: Current Status and Future Perspectives. *Molecules* **2022**, *27*, 8367. [CrossRef]
21. Siegel, R.L.; Giaquinto, A.N.; Jemal, A. Cancer statistics, 2024. *CA Cancer J. Clin.* **2024**, *74*, 12–49. [CrossRef]
22. Yin, L.; Duan, J.-J.; Bian, X.-W.; Yu, S.-C. Triple-negative breast cancer molecular subtyping and treatment progress. *Breast Cancer Res.* **2020**, *22*, 61. [CrossRef]
23. Lehmann, B.D.; Bauer, J.A.; Chen, X.; Sanders, M.E.; Chakravarthy, A.B.; Shyr, Y.; Pietenpol, J.A. Identification of human triple-negative breast cancer subtypes and preclinical models for selection of targeted therapies. *J. Clin. Investig.* **2011**, *121*, 2750–2767. [CrossRef] [PubMed]
24. Masuda, H.; Baggerly, K.A.; Wang, Y.; Zhang, Y.; Gonzalez-Angulo, A.M.; Meric-Bernstam, F.; Valero, V.; Lehmann, B.D.; Pietenpol, J.A.; Hortobagyi, G.N.; et al. Differential Response to Neoadjuvant Chemotherapy among 7 Triple-Negative Breast Cancer Molecular Subtypes. *Clin. Cancer Res.* **2013**, *19*, 5533–5540. [CrossRef] [PubMed]

25. Kudelova, E.; Smolar, M.; Holubekova, V.; Hornakova, A.; Dvorska, D.; Lucansky, V.; Koklesova, L.; Kudela, E.; Kubatka, P. Genetic Heterogeneity, Tumor Microenvironment and Immunotherapy in Triple-Negative Breast Cancer. *Int. J. Mol. Sci.* **2022**, *23*, 14937. [CrossRef] [PubMed]
26. Liu, X.; Li, Z.; Liu, S.; Sun, J.; Chen, Z.; Jiang, M.; Zhang, Q.; Wei, Y.; Wang, X.; Huang, Y.-Y.; et al. Potential therapeutic effects of dipyridamole in the severely ill patients with COVID-19. *Acta Pharm. Sin. B* **2020**, *10*, 1205–1215. [CrossRef] [PubMed]
27. Sauter, E.R. Cancer prevention and treatment using combination therapy with natural compounds. *Expert Rev. Clin. Pharmacol.* **2020**, *13*, 265–285. [CrossRef]
28. Huang, B.; Zhang, Y. Teaching an old dog new tricks: Drug discovery by repositioning natural products and their derivatives. *Drug Discov. Today* **2022**, *27*, 1936–1944. [CrossRef]
29. Egbuna, C.; Kumar, S.; Ifemeje, J.C.; Ezzat, S.M.; Kaliyaperumal, S. *Phytochemicals as Lead Compounds for New Drug Discovery*; Elsevier: London, UK, 2019.
30. Wang, J.; Fang, X.; Ge, L.; Cao, F.; Zhao, L.; Wang, Z.; Xiao, W. Antitumor, antioxidant and anti-inflammatory activities of kaempferol and its corresponding glycosides and the enzymatic preparation of kaempferol. *PLoS ONE* **2018**, *13*, e0197563. [CrossRef]
31. Periferakis, A.; Periferakis, K.; Badarau, I.A.; Petran, E.M.; Popa, D.C.; Caruntu, A.; Costache, R.S.; Scheau, C.; Caruntu, C.; Costache, D.O. Kaempferol: Antimicrobial Properties, Sources, Clinical, and Traditional Applications. *Int. J. Mol. Sci.* **2022**, *23*, 15054. [CrossRef]
32. Alam, W.; Khan, H.; Shah, M.A.; Cauli, O.; Saso, L. Kaempferol as a Dietary Anti-Inflammatory Agent: Current Therapeutic Standing. *Molecules* **2020**, *25*, 4073. [CrossRef]
33. Lee, H.S.; Cho, H.J.; Yu, R.; Lee, K.W.; Chun, H.S.; Park, J.H.Y. Mechanisms Underlying Apoptosis-Inducing Effects of Kaempferol in HT-29 Human Colon Cancer Cells. *Int. J. Mol. Sci.* **2014**, *15*, 2722–2737. [CrossRef]
34. Rho, H.S.; Ghimeray, A.K.; Yoo, D.S.; Ahn, S.M.; Kwon, S.S.; Lee, K.H.; Cho, D.H.; Cho, J.Y. Kaempferol and Kaempferol Rhamnosides with Depigmenting and Anti-Inflammatory Properties. *Molecules* **2011**, *16*, 3338–3344. [CrossRef] [PubMed]
35. Chen, J.; Zhong, H.; Huang, Z.; Chen, X.; You, J.; Zou, T. A Critical Review of Kaempferol in Intestinal Health and Diseases. *Antioxidants* **2023**, *12*, 1642. [CrossRef]
36. Fernández-Del-Río, L.; Soubeyrand, E.; Basset, G.J.; Clarke, C.F. Metabolism of the Flavonol Kaempferol in Kidney Cells Liberates the B-ring to Enter Coenzyme Q Biosynthesis. *Molecules* **2020**, *25*, 2955. [CrossRef] [PubMed]
37. Chen, A.Y.; Chen, Y.C. A review of the dietary flavonoid, kaempferol on human health and cancer chemoprevention. *Food Chem.* **2013**, *138*, 2099–2107. [CrossRef] [PubMed]
38. Calderon-Montaño, J.M.; Burgos-Morón, E.; Perez-Guerrero, C.; Lopez-Lazaro, M. A Review on the Dietary Flavonoid Kaempferol. *Mini-Reviews Med. Chem.* **2011**, *11*, 298–344. [CrossRef] [PubMed]
39. Ashrafizadeh, M.; Tavakol, S.; Ahmadi, Z.; Roomiani, S.; Mohammadinejad, R.; Samarghandian, S. Therapeutic effects of kaempferol affecting autophagy and endoplasmic reticulum stress. *Phytotherapy Res.* **2020**, *34*, 911–923. [CrossRef] [PubMed]
40. Devi, K.P.; Malar, D.S.; Nabavi, S.F.; Sureda, A.; Xiao, J.; Nabavi, S.M.; Daglia, M. Kaempferol and inflammation: From chemistry to medicine. *Pharmacol. Res.* **2015**, *99*, 1–10. [CrossRef]
41. Duan, L.; Rao, X.; Sigdel, K.R. Regulation of Inflammation in Autoimmune Disease. *J. Immunol. Res.* **2019**, *2019*, 7403796. [CrossRef]
42. Lemos, C.; Peters, G.J.; Jansen, G.; Martel, F.; Calhau, C. Modulation of folate uptake in cultured human colon adenocarcinoma Caco-2 cells by dietary compounds. *Eur. J. Nutr.* **2007**, *46*, 329–336. [CrossRef]
43. Balabhadrapathruni, S.; Thomas, T.J.; Yurkow, E.J.; Amenta, P.S.; Thomas, T. Effects of genistein and structurally related phytoestrogens on cell cycle kinetics and apoptosis in MDA-MB-468 human breast cancer cells. *Oncol. Rep.* **2000**, *7*, 3–12. [CrossRef]
44. Lee, K.-H.; Tagahara, K.; Suzuki, H.; Wu, R.-Y.; Haruna, M.; Hall, I.H.; Huang, H.-C.; Ito, K.; Iida, T.; Lai, J.-S. Antitumor Agents. 49. Tricin,Kaempferol-3-0-β-D-Glucopyranoside and (+)-Nortrachelogenin, Antileukemic Principles from Wikstroemia indica. *J. Nat. Prod.* **1981**, *44*, 530–535. [CrossRef]
45. Lin, J.K.; Chen, Y.C.; Huang, Y.T.; Lin-Shiau, S.Y. Suppression of protein kinase C and nuclear oncogene expression as possible molecular mechanisms of cancer chemoprevention by apigenin and curcumin. *J. Cell. Biochem. Suppl.* **1997**, *28–29*, 39–48. [CrossRef]
46. Stoner, G.D.; Chen, T.; Kresty, L.A.; Aziz, R.M.; Reinemann, T.; Nines, R. Protection Against Esophageal Cancer in Rodents with Lyophilized Berries: Potential Mechanisms. *Nutr. Cancer* **2006**, *54*, 33–46. [CrossRef] [PubMed]
47. Kowalski, A.; Brandis, D. Shock Resuscitation. In *StatPearls*; StatPearls Publishing: Treasure Island, FL, USA, 2023. Available online: http://www.ncbi.nlm.nih.gov/books/NBK534830/ (accessed on 11 December 2023).
48. Niering, P.; Michels, G.; Wätjen, W.; Ohler, S.; Steffan, B.; Chovolou, Y.; Kampkötter, A.; Proksch, P.; Kahl, R. Protective and detrimental effects of kaempferol in rat H4IIE cells: Implication of oxidative stress and apoptosis. *Toxicol. Appl. Pharmacol.* **2005**, *209*, 114–122. [CrossRef] [PubMed]
49. Nirmala, P.; Ramanathan, M. Effect of kaempferol on lipid peroxidation and antioxidant status in 1,2-dimethyl hydrazine induced colorectal carcinoma in rats. *Eur. J. Pharmacol.* **2011**, *654*, 75–79. [CrossRef] [PubMed]
50. Nguyen, N.H. A Protocol for Flavonols, Kaempferol and Quercetin, Staining in Plant Root Tips. *Bio-Protocol* **2020**, *10*, e3781. [CrossRef] [PubMed]

51. Zhang, Y.; Chen, A.Y.; Li, M.; Chen, C.; Yao, Q. Ginkgo biloba Extract Kaempferol Inhibits Cell Proliferation and Induces Apoptosis in Pancreatic Cancer Cells. *J. Surg. Res.* **2008**, *148*, 17–23. [CrossRef] [PubMed]
52. Kang, J.W.; Kim, J.H.; Song, K.; Kim, S.H.; Yoon, J.; Kim, K. Kaempferol and quercetin, components of *Ginkgo biloba* extract (EGb 761), induce caspase-3-dependent apoptosis in oral cavity cancer cells. *Phytotherapy Res.* **2010**, *24*, S77–S82. [CrossRef]
53. Wang, F.; Wang, L.; Qu, C.; Chen, L.; Geng, Y.; Cheng, C.; Yu, S.; Wang, D.; Yang, L.; Meng, Z.; et al. Kaempferol induces ROS-dependent apoptosis in pancreatic cancer cells via TGM2-mediated Akt/mTOR signaling. *BMC Cancer* **2021**, *21*, 396. [CrossRef]
54. Hung, T.-W.; Chen, P.-N.; Wu, H.-C.; Wu, S.-W.; Tsai, P.-Y.; Hsieh, Y.-S.; Chang, H.-R. Kaempferol Inhibits the Invasion and Migration of Renal Cancer Cells through the Downregulation of AKT and FAK Pathways. *Int. J. Med Sci.* **2017**, *14*, 984–993. [CrossRef]
55. Kashafi, E.; Moradzadeh, M.; Mohamadkhani, A.; Erfanian, S. Kaempferol increases apoptosis in human cervical cancer HeLa cells via PI3K/AKT and telomerase pathways. *Biomed. Pharmacother.* **2017**, *89*, 573–577. [CrossRef] [PubMed]
56. Choi, J.-B.; Kim, J.-H.; Lee, H.; Pak, J.-N.; Shim, B.S.; Kim, S.-H. Reactive Oxygen Species and p53 Mediated Activation of p38 and Caspases is Critically Involved in Kaempferol Induced Apoptosis in Colorectal Cancer Cells. *J. Agric. Food Chem.* **2018**, *66*, 9960–9967. [CrossRef] [PubMed]
57. Yao, S.; Wang, X.; Li, C.; Zhao, T.; Jin, H.; Fang, W. Kaempferol inhibits cell proliferation and glycolysis in esophagus squamous cell carcinoma via targeting EGFR signaling pathway. *Tumor Biol.* **2016**, *37*, 10247–10256. [CrossRef] [PubMed]
58. Song, W.; Dang, Q.; Xu, D.; Chen, Y.; Zhu, G.; Wu, K.; Zeng, J.; Long, Q.; Wang, X.; He, D.; et al. Kaempferol induces cell cycle arrest and apoptosis in renal cell carcinoma through EGFR/p38 signaling. *Oncol. Rep.* **2014**, *31*, 1350–1356. [CrossRef]
59. Gao, Y.; Yin, J.; Rankin, G.O.; Chen, Y.C. Kaempferol Induces G2/M Cell Cycle Arrest via Checkpoint Kinase 2 and Promotes Apoptosis via Death Receptors in Human Ovarian Carcinoma A2780/CP70 Cells. *Molecules* **2018**, *23*, 1095. [CrossRef] [PubMed]
60. Luo, H.; Jiang, B.; Li, B.; Li, Z.; Jiang, B.-H.; Chen, Y.C. Kaempferol nanoparticles achieve strong and selective inhibition of ovarian cancer cell viability. *Int. J. Nanomed.* **2012**, *7*, 3951–3959. [CrossRef]
61. Chin, H.; Horng, C.; Liu, Y.; Lu, C.; Su, C.; Chen, P.; Chiu, H.; Tsai, F.; Shieh, P.; Yang, J. Kaempferol inhibits angiogenic ability by targeting VEGF receptor-2 and downregulating the PI3K/AKT, MEK and ERK pathways in VEGF-stimulated human umbilical vein endothelial cells. *Oncol. Rep.* **2018**, *39*, 2351–2357. [CrossRef]
62. Ju, P.; Ho, Y.; Chen, P.; Lee, H.; Lai, S.; Yang, S.; Yeh, C. Kaempferol inhibits the cell migration of human hepatocellular carcinoma cells by suppressing MMP-9 and Akt signaling. *Environ. Toxicol.* **2021**, *36*, 1981–1989. [CrossRef] [PubMed]
63. Čipák, L.; Novotný, L.; Čipáková, I.; Rauko, P. Differential modulation of cisplatin and doxorubicin efficacies in leukemia cells by flavonoids. *Nutr. Res.* **2003**, *23*, 1045–1057. [CrossRef]
64. Luo, H.; Daddysman, M.K.; O Rankin, G.; Jiang, B.-H.; Chen, Y.C. Kaempferol enhances cisplatin's effect on ovarian cancer cells through promoting apoptosis caused by down regulation of cMyc. *Cancer Cell Int.* **2010**, *10*, 16. [CrossRef]
65. Zhou, Q.; Fang, G.; Pang, Y.; Wang, X. Combination of Kaempferol and Docetaxel Induces Autophagy in Prostate Cancer Cells In Vitro and In Vivo. *Int. J. Mol. Sci.* **2023**, *24*, 14519. [CrossRef] [PubMed]
66. Al-Nour, M.Y.; Ibrahim, M.M.; Elsaman, T. Ellagic Acid, Kaempferol, and Quercetin from Acacia nilotica: Promising Combined Drug with Multiple Mechanisms of Action. *Curr. Pharmacol. Rep.* **2019**, *5*, 255–280. [CrossRef] [PubMed]
67. To, K.K.; Cho, W.C. Flavonoids Overcome Drug Resistance to Cancer Chemotherapy by Epigenetically Modulating Multiple Mechanisms. *Curr. Cancer Drug Targets* **2021**, *21*, 289–305. [CrossRef]
68. Riahi-Chebbi, I.; Souid, S.; Othman, H.; Haoues, M.; Karoui, H.; Morel, A.; Srairi-Abid, N.; Essafi, M.; Essafi-Benkhadir, K. The Phenolic compound Kaempferol overcomes 5-fluorouracil resistance in human resistant LS174 colon cancer cells. *Sci. Rep.* **2019**, *9*, 195. [CrossRef] [PubMed]
69. Wu, H.; Du, J.; Li, C.; Li, H.; Guo, H.; Li, Z. Kaempferol Can Reverse the 5-Fu Resistance of Colorectal Cancer Cells by Inhibiting PKM2-Mediated Glycolysis. *Int. J. Mol. Sci.* **2022**, *23*, 3544. [CrossRef] [PubMed]
70. Park, J.; Lee, G.-E.; An, H.-J.; Lee, C.-J.; Cho, E.S.; Kang, H.C.; Lee, J.Y.; Lee, H.S.; Choi, J.-S.; Kim, D.J.; et al. Kaempferol sensitizes cell proliferation inhibition in oxaliplatin-resistant colon cancer cells. *Arch. Pharmacal Res.* **2021**, *44*, 1091–1108. [CrossRef] [PubMed]
71. Nandi, S.K.; Pradhan, A.; Das, B.; Das, B.; Basu, S.; Mallick, B.; Dutta, A.; Sarkar, D.K.; Mukhopadhyay, A.; Mukhopadhyay, S.; et al. Kaempferol attenuates viability of ex-vivo cultured post-NACT breast tumor explants through downregulation of p53 induced stemness, inflammation and apoptosis evasion pathways. *Pathol.-Res. Pr.* **2022**, *237*, 154029. [CrossRef] [PubMed]
72. Adebamowo, C.A.; Cho, E.; Sampson, L.; Katan, M.B.; Spiegelman, D.; Willett, W.C.; Holmes, M.D. Dietary flavonols and flavonol-rich foods intake and the risk of breast cancer. *Int. J. Cancer* **2005**, *114*, 628–633. [CrossRef]
73. Touillaud, M.S.; Pillow, P.C.; Jakovljevic, J.; Bondy, M.L.; Singletary, S.E.; Li, D.; Chang, S. Effect of Dietary Intake of Phytoestrogens on Estrogen Receptor Status in Premenopausal Women with Breast Cancer. *Nutr. Cancer* **2005**, *51*, 162–169. [CrossRef]
74. Oh, S.M.; Kim, Y.P.; Chung, K.H. Biphasic effects of kaempferol on the estrogenicity in human breast cancer cells. *Arch. Pharmacal Res.* **2006**, *29*, 354–362. [CrossRef]
75. Zhu, L.; Xue, L. Kaempferol Suppresses Proliferation and Induces Cell Cycle Arrest, Apoptosis, and DNA Damage in Breast Cancer Cells. *Oncol. Res. Featur. Preclin. Clin. Cancer Ther.* **2019**, *27*, 629–634. [CrossRef] [PubMed]

76. Yue, W.; Wang, J.; Li, Y.; Fan, P.; Liu, G.; Zhang, N.; Conaway, M.; Wang, H.; Korach, K.S.; Bocchinfuso, W.; et al. Effects of estrogen on breast cancer development: Role of estrogen receptor independent mechanisms. *Int. J. Cancer* **2010**, *127*, 1748–1757. [CrossRef] [PubMed]

77. Wang, C.; Kurzer, M.S. Phytoestrogen concentration determines effects on DNA synthesis in human breast cancer cells. *Nutr. Cancer* **1997**, *28*, 236–247. [CrossRef] [PubMed]

78. Zava, D.T.; Duwe, G. Estrogenic and antiproliferative properties of genistein and other flavonoids in human breast cancer cells in vitro. *Nutr. Cancer* **1997**, *27*, 31–40. [CrossRef] [PubMed]

79. Hung, H. Inhibition of estrogen receptor alpha expression and function in MCF-7 cells by kaempferol. *J. Cell. Physiol.* **2004**, *198*, 197–208. [CrossRef] [PubMed]

80. Kim, S.-H.; Hwang, K.-A.; Choi, K.-C. Treatment with kaempferol suppresses breast cancer cell growth caused by estrogen and triclosan in cellular and xenograft breast cancer models. *J. Nutr. Biochem.* **2016**, *28*, 70–82. [CrossRef] [PubMed]

81. Lee, G.-A.; Choi, K.-C.; Hwang, K.-A. Kaempferol, a phytoestrogen, suppressed triclosan-induced epithelial-mesenchymal transition and metastatic-related behaviors of MCF-7 breast cancer cells. *Environ. Toxicol. Pharmacol.* **2017**, *49*, 48–57. [CrossRef] [PubMed]

82. Sarmiento-Salinas, F.L.; Delgado-Magallón, A.; Montes-Alvarado, J.B.; Ramírez-Ramírez, D.; Flores-Alonso, J.C.; Cortés-Hernández, P.; Reyes-Leyva, J.; Herrera-Camacho, I.; Anaya-Ruiz, M.; Pelayo, R.; et al. Breast Cancer Subtypes Present a Differential Production of Reactive Oxygen Species (ROS) and Susceptibility to Antioxidant Treatment. *Front. Oncol.* **2019**, *9*, 480. [CrossRef] [PubMed]

83. Oshi, M.; Gandhi, S.; Yan, L.; Tokumaru, Y.; Wu, R.; Yamada, A.; Matsuyama, R.; Endo, I.; Takabe, K. Abundance of reactive oxygen species (ROS) is associated with tumor aggressiveness, immune response, and worse survival in breast cancer. *Breast Cancer Res. Treat.* **2022**, *194*, 231–241. [CrossRef]

84. Raza, M.H.; Siraj, S.; Arshad, A.; Waheed, U.; Aldakheel, F.; Alduraywish, S.; Arshad, M. ROS-modulated therapeutic approaches in cancer treatment. *J. Cancer Res. Clin. Oncol.* **2017**, *143*, 1789–1809. [CrossRef]

85. Okoh, V.; Deoraj, A.; Roy, D. Estrogen-induced reactive oxygen species-mediated signalings contribute to breast cancer. *Biochim. Biophys. Acta (BBA)—Rev. Cancer* **2011**, *1815*, 115–133. [CrossRef] [PubMed]

86. Felty, Q.; Xiong, W.-C.; Sun, D.; Sarkar, S.; Singh, K.P.; Parkash, J.; Roy, D. Estrogen-Induced Mitochondrial Reactive Oxygen Species as Signal-Transducing Messengers. *Biochemistry* **2005**, *44*, 6900–6909. [CrossRef]

87. Liao, W.; Chen, L.; Ma, X.; Jiao, R.; Li, X.; Wang, Y. Protective effects of kaempferol against reactive oxygen species-induced hemolysis and its antiproliferative activity on human cancer cells. *Eur. J. Med. Chem.* **2016**, *114*, 24–32. [CrossRef] [PubMed]

88. Zeng, J.; Xu, H.; Fan, P.-Z.; Xie, J.; He, J.; Yu, J.; Gu, X.; Zhang, C.-J. Kaempferol blocks neutrophil extracellular traps formation and reduces tumour metastasis by inhibiting ROS-PAD4 pathway. *J. Cell. Mol. Med.* **2020**, *24*, 7590–7599. [CrossRef]

89. Kim, B.-W.; Lee, E.-R.; Min, H.-M.; Jeong, H.-S.; Ahn, J.-Y.; Kim, J.-H.; Choi, H.-Y.; Choi, H.; Kim, E.Y.; Park, S.P.; et al. Sustained ERK activation is involved in the kaempferol-induced apoptosis of breast cancer cells and is more evident under 3-D culture condition. *Cancer Biol. Ther.* **2008**, *7*, 1080–1089. [CrossRef] [PubMed]

90. Menegon, S.; Columbano, A.; Giordano, S. The Dual Roles of NRF2 in Cancer. *Trends Mol. Med.* **2016**, *22*, 578–593. [CrossRef]

91. Sajadimajd, S.; Khazaei, M. Oxidative Stress and Cancer: The Role of Nrf2. *Curr. Cancer Drug Targets* **2018**, *18*, 538–557. [CrossRef]

92. Telkoparan-Akillilar, P.; Panieri, E.; Cevik, D.; Suzen, S.; Saso, L. Therapeutic Targeting of the NRF2 Signaling Pathway in Cancer. *Molecules* **2021**, *26*, 1417. [CrossRef] [PubMed]

93. De La Vega, M.R.; Chapman, E.; Zhang, D.D. NRF2 and the Hallmarks of Cancer. *Cancer Cell* **2018**, *34*, 21–43. [CrossRef]

94. Fouzder, C.; Mukhuty, A.; Kundu, R. Kaempferol inhibits Nrf2 signalling pathway via downregulation of Nrf2 mRNA and induces apoptosis in NSCLC cells. *Arch. Biochem. Biophys.* **2021**, *697*, 108700. [CrossRef]

95. Li, S.; Yan, T.; Deng, R.; Jiang, X.; Xiong, H.; Wang, Y.; Yu, Q.; Wang, X.; Chen, C.; Zhu, Y. Low dose of kaempferol suppresses the migration and invasion of triple-negative breast cancer cells by downregulating the activities of RhoA and Rac1. *OncoTargets Ther.* **2017**, *10*, 4809–4819. [CrossRef] [PubMed]

96. Hall, R.; Birrell, S.; Tilley, W.; Sutherland, R. MDA-MB-453, an androgen-responsive human breast carcinoma cell line with high level androgen receptor expression. *Eur. J. Cancer* **1994**, *30*, 484–490. [CrossRef] [PubMed]

97. Choi, E.J.; Ahn, W.S. Kaempferol induced the apoptosis via cell cycle arrest in human breast cancer MDA-MB-453 cells. *Nutr. Res. Pr.* **2008**, *2*, 322–325. [CrossRef] [PubMed]

98. Kaur, V.; Kumar, M.; Kumar, A.; Kaur, S. *Butea monosperma* (Lam.) Taub. Bark fractions protect against free radicals and induce apoptosis in MCF-7 breast cancer cells *via* cell-cycle arrest and ROS-mediated pathway. *Drug Chem. Toxicol.* **2020**, *43*, 398–408. [CrossRef]

99. Brusselmans, K.; Vrolix, R.; Verhoeven, G.; Swinnen, J.V. Induction of Cancer Cell Apoptosis by Flavonoids Is Associated with Their Ability to Inhibit Fatty Acid Synthase Activity. *J. Biol. Chem.* **2005**, *280*, 5636–5645. [CrossRef] [PubMed]

100. Diantini, A.; Subarnas, A.; Lestari, K.; Halimah, E.; Susilawati, Y.; Supriyatna; Julaeha, E.; Achmad, T.H.; Suradji, E.W.; Yamazaki, C.; et al. Kaempferol-3-O-rhamnoside isolated from the leaves of Schima wallichii Korth. inhibits MCF-7 breast cancer cell proliferation through activation of the caspase cascade pathway. *Oncol. Lett.* **2012**, *3*, 1069–1072. [CrossRef] [PubMed]

101. Radhika, M.; Ghoshal, N.; Chatterjee, A. Comparison of effectiveness in antitumor activity between flavonoids and polyphenols of the methanolic extract of roots of Potentilla fulgens in breast cancer cells. *J. Complement. Integr. Med.* **2012**, *9*, Article 24. [CrossRef] [PubMed]

102. Tor, Y.S.; Yazan, L.S.; Foo, J.B.; Wibowo, A.; Ismail, N.; Cheah, Y.K.; Abdullah, R.; Ismail, M.; Ismail, I.S.; Yeap, S.K. Induction of Apoptosis in MCF-7 Cells via Oxidative Stress Generation, Mitochondria-Dependent and Caspase-Independent Pathway by Ethyl Acetate Extract of Dillenia suffruticosa and Its Chemical Profile. *PLOS ONE* **2015**, *10*, e0127441. [CrossRef]

103. Shoja, M.; Reddy, N.D.; Nayak, P.G.; Srinivasan, K.; Rao, C.M. Glycosmis pentaphylla (Retz.) DC arrests cell cycle and induces apoptosis via caspase-3/7 activation in breast cancer cells. *J. Ethnopharmacol.* **2015**, *168*, 50–60. [CrossRef]

104. Yi, X.; Zuo, J.; Tan, C.; Xian, S.; Luo, C.; Chen, S. Kaempferol, a flavonoid compound from GYNURA MEDICA induced apoptosis and growth inhibition in mcf-7 breast cancer cell. *Afr. J. Tradit. Complement. Altern. Med. AJTCAM* **2016**, *13*, 210–215. [CrossRef]

105. Kang, G.-Y.; Lee, E.-R.; Kim, J.-H.; Jung, J.W.; Lim, J.; Kim, S.K.; Cho, S.-G.; Kim, K.P. Downregulation of PLK-1 expression in kaempferol-induced apoptosis of MCF-7 cells. *Eur. J. Pharmacol.* **2009**, *611*, 17–21. [CrossRef]

106. Zhang, Y.; Chen, F.; Chandrashekar, D.S.; Varambally, S.; Creighton, C.J. Proteogenomic characterization of 2002 human cancers reveals pan-cancer molecular subtypes and associated pathways. *Nat. Commun.* **2022**, *13*, 2669. [CrossRef] [PubMed]

107. Phromnoi, K.; Yodkeeree, S.; Anuchapreeda, S.; Limtrakul, P. Inhibition of MMP-3 activity and invasion of the MDA-MB-231 human invasive breast carcinoma cell line by bioflavonoids. *Acta Pharmacol. Sin.* **2009**, *30*, 1169–1176. [CrossRef] [PubMed]

108. Astin, J.W.; Jamieson, S.M.; Eng, T.C.; Flores, M.V.; Misa, J.P.; Chien, A.; Crosier, K.E.; Crosier, P.S. An In Vivo Antilymphatic Screen in Zebrafish Identifies Novel Inhibitors of Mammalian Lymphangiogenesis and Lymphatic-Mediated Metastasis. *Mol. Cancer Ther.* **2014**, *13*, 2450–2462. [CrossRef] [PubMed]

109. Li, C.; Zhao, Y.; Yang, D.; Yu, Y.; Guo, H.; Zhao, Z.; Zhang, B.; Yin, X. Inhibitory effects of kaempferol on the invasion of human breast carcinoma cells by downregulating the expression and activity of matrix metalloproteinase-9. *Biochem. Cell Biol.* **2015**, *93*, 16–27. [CrossRef] [PubMed]

110. Izzo, S.; Naponelli, V.; Bettuzzi, S. Flavonoids as Epigenetic Modulators for Prostate Cancer Prevention. *Nutrients* **2020**, *12*, 1010. [CrossRef] [PubMed]

111. Bouyahya, A.; Mechchate, H.; Oumeslakht, L.; Zeouk, I.; Aboulaghras, S.; Balahbib, A.; Zengin, G.; Kamal, M.A.; Gallo, M.; Montesano, D.; et al. The Role of Epigenetic Modifications in Human Cancers and the Use of Natural Compounds as Epidrugs: Mechanistic Pathways and Pharmacodynamic Actions. *Biomolecules* **2022**, *12*, 367. [CrossRef] [PubMed]

112. Berger, A.; Venturelli, S.; Kallnischkies, M.; Böcker, A.; Busch, C.; Weiland, T.; Noor, S.; Leischner, C.; Weiss, T.S.; Lauer, U.M.; et al. Kaempferol, a new nutrition-derived pan-inhibitor of human histone deacetylases. *J. Nutr. Biochem.* **2013**, *24*, 977–985. [CrossRef]

113. Sharma, A.; Sinha, S.; Keswani, H.; Shrivastava, N. Kaempferol and Apigenin suppresses the stemness properties of TNBC cells by modulating Sirtuins. *Mol. Divers.* **2022**, *26*, 3225–3240. [CrossRef]

114. Zamarron, B.F.; Chen, W. Dual Roles of Immune Cells and Their Factors in Cancer Development and Progression. *Int. J. Biol. Sci.* **2011**, *7*, 651–658. [CrossRef]

115. Tiwari, A.; Trivedi, R.; Lin, S.-Y. Tumor microenvironment: Barrier or opportunity towards effective cancer therapy. *J. Biomed. Sci.* **2022**, *29*, 83. [CrossRef] [PubMed]

116. Chen, Y.; Song, Y.; Du, W.; Gong, L.; Chang, H.; Zou, Z. Tumor-associated macrophages: An accomplice in solid tumor progression. *J. Biomed. Sci.* **2019**, *26*, 78. [CrossRef] [PubMed]

117. Qiu, X.; Zhao, T.; Luo, R.; Qiu, R.; Li, Z. Tumor-Associated Macrophages: Key Players in Triple-Negative Breast Cancer. *Front. Oncol.* **2022**, *12*, 772615. [CrossRef] [PubMed]

118. Mu, X.; Shi, W.; Xu, Y.; Xu, C.; Zhao, T.; Geng, B.; Yang, J.; Pan, J.; Hu, S.; Zhang, C.; et al. Tumor-derived lactate induces M2 macrophage polarization via the activation of the ERK/STAT3 signaling pathway in breast cancer. *Cell Cycle* **2018**, *17*, 428–438. [CrossRef]

119. Yoshimur, T.; Li, C.; Wang, Y.; Matsukawa, A. The Chemokine Monocyte Chemoattractant Protein 1/CCL2 Is a Promoter of Breast Cancer Metastasis | Cellular & Molecular Immunology. Available online: https://www.nature.com/articles/s41423-023-01013-0 (accessed on 10 June 2024).

120. Lafta, H.A.; Hussein, A.H.A.; Al-Shalah, S.A.J.; Alnassar, Y.S.; Mohammed, N.M.; Akram, S.M.; Qasim, M.T.; Najafi, M. Tumor-associated Macrophages (TAMs) in Cancer Resistance; Modulation by Natural Products—PubMed. Available online: https://pubmed.ncbi.nlm.nih.gov/36722486/ (accessed on 10 June 2024).

121. Malla, R.; Padmaraju, V.; Kundrapu, D.B. Tumor-associated macrophages: Potential target of natural compounds for management of breast cancer. *Life Sci.* **2022**, *301*, 120572. [CrossRef] [PubMed]

122. Sonoki, H.; Tanimae, A.; Endo, S.; Matsunaga, T.; Furuta, T.; Ichihara, K.; Ikari, A. Kaempherol and Luteolin Decrease Claudin-2 Expression Mediated by Inhibition of STAT3 in Lung Adenocarcinoma A549 Cells. *Nutrients* **2017**, *9*, 597. [CrossRef] [PubMed]

123. Wonganan, O.; He, Y.-J.; Shen, X.-F.; Wongkrajang, K.; Suksamrarn, A.; Zhang, G.-L.; Wang, F. 6-Hydroxy-3-O-methyl-kaempferol 6-O-glucopyranoside potentiates the anti-proliferative effect of interferon α/β by promoting activation of the JAK/STAT signaling by inhibiting SOCS3 in hepatocellular carcinoma cells. *Toxicol. Appl. Pharmacol.* **2017**, *336*, 31–39. [CrossRef] [PubMed]

124. Zhang, Z.; Guo, Y.; Chen, M.; Chen, F.; Liu, B.; Shen, C. Kaempferol potentiates the sensitivity of pancreatic cancer cells to erlotinib via inhibition of the PI3K/AKT signaling pathway and epidermal growth factor receptor. *Inflammopharmacology* **2021**, *29*, 1587–1601. [CrossRef]

125. Yu, Q.; Zeng, K.; Ma, X.; Song, F.; Jiang, Y.; Tu, P.; Wang, X. Resokaempferol-mediated anti-inflammatory effects on activated macrophages via the inhibition of JAK2/STAT3, NF-κB and JNK/p38 MAPK signaling pathways. *Int. Immunopharmacol.* **2016**, *38*, 104–114. [CrossRef]

126. Fang, W.B.; Yao, M.; Brummer, G.; Acevedo, D.; Alhakamy, N.; Berkland, C.; Cheng, N. Targeted gene silencing of CCL2 inhibits triple negative breast cancer progression by blocking cancer stem cell renewal and M2 macrophage recruitment. *Oncotarget* **2016**, *7*, 49349–49367. [CrossRef]

127. Mohamed, H.T.; El-Ghonaimy, E.A.; El-Shinawi, M.; Hosney, M.; Götte, M.; Woodward, W.A.; El-Mamlouk, T.; Mohamed, M.M. IL-8 and MCP-1/CCL2 regulate proteolytic activity in triple negative inflammatory breast cancer a mechanism that might be modulated by Src and Erk1/2. *Toxicol. Appl. Pharmacol.* **2020**, *401*, 115092. [CrossRef] [PubMed]

128. Hao, Q.; Vadgama, J.V.; Wang, P. CCL2/CCR2 signaling in cancer pathogenesis. *Cell Commun. Signal.* **2020**, *18*, 82. [CrossRef] [PubMed]

129. Chen, X.; Yang, M.; Yin, J.; Li, P.; Zeng, S.; Zheng, G.; He, Z.; Liu, H.; Wang, Q.; Zhang, F.; et al. Tumor-associated macrophages promote epithelial–mesenchymal transition and the cancer stem cell properties in triple-negative breast cancer through CCL2/AKT/β-catenin signaling. *Cell Commun. Signal.* **2022**, *20*, 92. [CrossRef] [PubMed]

130. Fridlender, Z.G.; Buchlis, G.; Kapoor, V.; Cheng, G.; Sun, J.; Singhal, S.; Crisanti, M.C.; Wang, L.-C.S.; Heitjan, D.; Snyder, L.A.; et al. CCL2 Blockade Augments Cancer Immunotherapy. *Cancer Res.* **2010**, *70*, 109–118. [CrossRef] [PubMed]

131. Choi, J.; Lee, H.J.; Yoon, S.; Ryu, H.M.; Lee, E.; Jo, Y. Blockade of CCL2 expression overcomes intrinsic PD-1/PD-L1 inhibitor-resistance in transglutaminase 2-induced PD-L1 positive triple negative breast cancer. *Am. J. Cancer Res.* **2020**, *10*, 2878–2894. [PubMed]

132. Mendonca, P.; Hilliard, A.; Soliman, K. The Inhibitory Effects of Ganoderma lucium on Cell Proliferation, Apoptosis, and TNF-α-Induced CCL2 Release in Genetically Different Triple-Negative Breast Cancer Cells. *FASEB J.* **2021**, *35*, S1. [CrossRef]

133. Kanga, K.J.; Mendonca, P.; Soliman, K.F.; Ferguson, D.T.; Darling-Reed, S.F. Effect of Diallyl Trisulfide on TNF-α-induced CCL2/MCP-1 Release in Genetically Different Triple-negative Breast Cancer Cells. *Anticancer. Res.* **2021**, *41*, 5919–5933. [CrossRef] [PubMed]

134. Messeha, S.S.; Zarmouh, N.O.; Mendonca, P.; Alwagdani, H.; Kolta, M.G.; Soliman, K.F.A. The inhibitory effects of plumbagin on the NF-κB pathway and CCL2 release in racially different triple-negative breast cancer cells. *PLOS ONE* **2018**, *13*, e0201116. [CrossRef] [PubMed]

135. Joshi, N.; Tripathi, D.K.; Nagar, N.; Poluri, K.M. Hydroxyl Groups on Annular Ring-B Dictate the Affinities of Flavonol–CCL2 Chemokine Binding Interactions. *ACS Omega* **2021**, *6*, 10306–10317. [CrossRef]

136. Tzeng, C.-W.; Yen, F.-L.; Wu, T.-H.; Ko, H.-H.; Lee, C.-W.; Tzeng, W.-S.; Lin, C.-C. Enhancement of Dissolution and Antioxidant Activity of Kaempferol Using a Nanoparticle Engineering Process. *J. Agric. Food Chem.* **2011**, *59*, 5073–5080. [CrossRef]

137. Kazmi, I.; Al-Abbasi, F.A.; Afzal, M.; Altayb, H.N.; Nadeem, M.S.; Gupta, G. Formulation and Evaluation of Kaempferol Loaded Nanoparticles against Experimentally Induced Hepatocellular Carcinoma: In Vitro and In Vivo Studies. *Pharmaceutics* **2021**, *13*, 2086. [CrossRef] [PubMed]

138. Raghavan, B.S.; Kondath, S.; Anantanarayanan, R.; Rajaram, R. Kaempferol mediated synthesis of gold nanoparticles and their cytotoxic effects on MCF-7 cancer cell line. *Process. Biochem.* **2015**, *50*, 1966–1976. [CrossRef]

139. Govindaraju, S.; Roshini, A.; Lee, M.-H.; Yun, K. Kaempferol conjugated gold nanoclusters enabled efficient for anticancer therapeutics to A549 lung cancer cells. *Int. J. Nanomed.* **2019**, *14*, 5147–5157. [CrossRef]

140. Colombo, M.; Melchiades, G.d.L.; Michels, L.R.; Figueiró, F.; Bassani, V.L.; Teixeira, H.F.; Koester, L.S. Solid Dispersion of Kaempferol: Formulation Development, Characterization, and Oral Bioavailability Assessment. *Aaps Pharmscitech* **2019**, *20*, 106. [CrossRef]

141. Du, Q.; Chen, J.; Yan, G.; Lyu, F.; Huang, J.; Ren, J.; Di, L. Comparison of different aliphatic acid grafted N-trimethyl chitosan surface-modified nanostructured lipid carriers for improved oral kaempferol delivery. *Int. J. Pharm.* **2019**, *568*, 118506. [CrossRef] [PubMed]

142. Kaur, S.; Nathani, A.; Singh, M.; Kaur, S.; Nathani, A.; Singh, M.; Kaur, S.; Nathani, A.; Singh, M. Exosomal delivery of cannabinoids against cancer. *Cancer Lett.* **2023**, *566*, 216243. [CrossRef] [PubMed]

143. Akiyama, M.; Mizokami, T.; Ito, H.; Ikeda, Y. A randomized, placebo-controlled trial evaluating the safety of excessive administration of kaempferol aglycone. *Food Sci. Nutr.* **2023**, *11*, 5427–5437. [CrossRef] [PubMed]

144. Navarro, S.L.; Schwarz, Y.; Song, X.; Wang, C.-Y.; Chen, C.; Trudo, S.P.; Kristal, A.R.; Kratz, M.; Eaton, D.L.; Lampe, J.W. Cruciferous Vegetables Have Variable Effects on Biomarkers of Systemic Inflammation in a Randomized Controlled Trial in Healthy Young Adults. *J. Nutr.* **2014**, *144*, 1850–1857. [CrossRef]

145. Nejabati, H.R.; Roshangar, L. Kaempferol: A potential agent in the prevention of colorectal cancer. *Physiol. Rep.* **2022**, *10*, e15488. [CrossRef] [PubMed]

146. Noer, J.B.; Hørsdal, O.K.; Xiang, X.; Luo, Y.; Regenberg, B. Extrachromosomal circular DNA in cancer: History, current knowledge, and methods. *Trends Genet.* **2022**, *38*, 766–781. [CrossRef]

147. Li, R.; Wang, Y.; Li, J.; Zhou, X. Extrachromosomal circular DNA (eccDNA): An emerging star in cancer. *Biomark. Res.* **2022**, *10*, 53. [CrossRef] [PubMed]

148. Zeng, T.; Huang, W.; Cui, L.; Zhu, P.; Lin, Q.; Zhang, W.; Li, J.; Deng, C.; Wu, Z.; Huang, Z.; et al. The landscape of extrachromosomal circular DNA (eccDNA) in the normal hematopoiesis and leukemia evolution. *Cell Death Discov.* **2022**, *8*, 1–10. [CrossRef]

149. Wu, S.; Tao, T.; Zhang, L.; Zhu, X.; Zhou, X. Extrachromosomal DNA (ecDNA): Unveiling its role in cancer progression and implications for early detection. *Heliyon* **2023**, *9*, e21327. [CrossRef] [PubMed]

150. Pan, X.; Veroniaina, H.; Su, N.; Sha, K.; Jiang, F.; Wu, Z.; Qi, X. Applications and developments of gene therapy drug delivery systems for genetic diseases. *Asian J. Pharm. Sci.* **2021**, *16*, 687–703. [CrossRef] [PubMed]
151. Wang, T.; Zhang, H.; Zhou, Y.; Shi, J. Extrachromosomal circular DNA: A new potential role in cancer progression. *J. Transl. Med.* **2021**, *19*, 257. [CrossRef] [PubMed]
152. Singh, R.; Mo, Y.-Y. Role of microRNAs in breast cancer. *Cancer Biol. Ther.* **2013**, *14*, 201–212. [CrossRef] [PubMed]
153. Muñoz, J.P.; Pérez-Moreno, P.; Pérez, Y.; Calaf, G.M. The Role of MicroRNAs in Breast Cancer and the Challenges of Their Clinical Application. *Diagnostics* **2023**, *13*, 3072. [CrossRef]
154. Ghafouri-Fard, S.; Sasi, A.K.; Abak, A.; Shoorei, H.; Khoshkar, A.; Taheri, M. Contribution of miRNAs in the Pathogenesis of Breast Cancer. *Front. Oncol.* **2021**, *11*, 768949. [CrossRef]
155. Gutierrez-Uribe, J.A.; Salinas-Santander, M.; Serna-Guerrero, D.; Serna-Saldivar, S.R.O.; Rivas-Estilla, A.M.; Rios-Ibarra, C.P. Inhibition of miR31 and miR92a as Oncological Biomarkers in RKO Colon Cancer Cells Treated with Kaempferol-3-*O*-Glycoside Isolated from Black Bean. *J. Med. Food* **2020**, *23*, 50–55. [CrossRef]

Review

Pharmacological Features and Therapeutic Implications of Plumbagin in Cancer and Metabolic Disorders: A Narrative Review

Bhoomika Sharma [1], Chitra Dhiman [1], Gulam Mustafa Hasan [2], Anas Shamsi [3,*] and Md. Imtiyaz Hassan [4,*]

[1] Department of Biosciences, Jamia Millia Islamia, Jamia Nagar, New Delhi 110025, India;
bhoomika124@gmail.com (B.S.); dchitra1103@gmail.com (C.D.)

[2] Department of Basic Medical Science, College of Medicine, Prince Sattam Bin Abdulaziz University,
Al-Kharj 11942, Saudi Arabia; mgulam@jmi.ac.in

[3] Centre of Medical and Bio-Allied Health Sciences Research, Ajman University,
Ajman P.O. Box 346, United Arab Emirates

[4] Centre for Interdisciplinary Research in Basic Sciences, Jamia Millia Islamia, Jamia Nagar,
New Delhi 110025, India

* Correspondence: anas.shamsi18@gmail.com (A.S.); mihassan@jmi.ac.in (M.I.H.)

Abstract: Plumbagin (PLB) is a naphthoquinone extracted from *Plumbago indica*. In recent times, there has been a growing body of evidence suggesting the potential importance of naphthoquinones, both natural and artificial, in the pharmacological world. Numerous studies have indicated that PLB plays a vital role in combating cancers and other disorders. There is substantial evidence indicating that PLB may have a significant role in the treatment of breast cancer, brain tumours, lung cancer, hepatocellular carcinoma, and other conditions. Moreover, its potent anti-oxidant and anti-inflammatory properties offer promising avenues for the treatment of neurodegenerative and cardiovascular diseases. A number of studies have identified various pathways that may be responsible for the therapeutic efficacy of PLB. These include cell cycle regulation, apoptotic pathways, ROS induction pathways, inflammatory pathways, and signal transduction pathways such as PI3K/AKT/mTOR, STAT3/PLK1/AKT, and others. This review aims to provide a comprehensive analysis of the diverse pharmacological roles of PLB, examining the mechanisms through which it operates and exploring its potential applications in various medical conditions. In addition, we have conducted a review of the various formulations that have been reported in the literature with the objective of enhancing the efficacy of the compound. However, the majority of the reviewed data are based on in vitro and in vivo studies. To gain a comprehensive understanding of the safety and efficacy of PLB in humans and to ascertain its potential integration into therapeutic regimens for cancer and chronic diseases, rigorous clinical trials are essential. Finally, by synthesizing current research and identifying gaps in knowledge, this review seeks to enhance our understanding of PLB and its therapeutic prospects, paving the way for future studies and clinical applications.

Keywords: plumbagin; natural products; anticancer; anti-inflammatory; apoptosis; antioxidant; cardiovascular disease; neurodegenerative diseases

Citation: Sharma, B.; Dhiman, C.; Hasan, G.M.; Shamsi, A.; Hassan, M.I. Pharmacological Features and Therapeutic Implications of Plumbagin in Cancer and Metabolic Disorders: A Narrative Review. *Nutrients* **2024**, *16*, 3033. https://doi.org/10.3390/nu16173033

Academic Editors: Debasis Bagchi, Yi-Wen Liu and Ching-Hsein Chen

Received: 16 August 2024
Revised: 30 August 2024
Accepted: 6 September 2024
Published: 8 September 2024

1. Introduction

Plants synthesize complex chemical compounds known as plant secondary metabolites in response to specific biotic and abiotic stresses [1]. In addition to protecting the plant, secondary metabolites are essential for a plant's physiological tasks, such as pollination, vascular development of lignified walls, etc. The ability of these metabolites to elicit a toxicological response in humans has made these compounds the centre of attention for the pharmaceutical industry [2].

Quinones are a type of secondary metabolite synthesized in plants and are named after their quinone form [3]. Based on their benzene rings, quinones are primarily classified into four different categories: phenanthrenequinone, anthraquinone, benzoquinone, and naphthoquinone. Recently, quinones have attracted considerable interest among pharmacological enthusiasts, largely due to their therapeutic potential against various diseases [4].

Naphthoquinones, naturally occurring in plants, have been used in human life since ancient times [5]. More and more studies elucidating the role of natural and artificial naphthoquinones have realized their pharmacological and biological significance. The most widely documented and stable isoform is known as 1,4-naphthoquinone. Various analogues have been discovered based on chemical modifications of 1,4-naphthoquinone, such as Juglone, Plumbagin, Shikonin, anthraquinone, etc. [6].

One of the analogues, Plumbagin (PLB), is isolated from the families *Ebenceae*, *Plumbaginaceae*, and *Droseraceae* [7]. PLB, 2-methyl 1-5-hydroxy-1,4-naphthoquinone, is a Vitamin K3 analogue [8]. It was initially isolated from *Plumbago indica* and is a principal component in the roots of Plumbago *zeylanica*, also designated as Chitrak in Ayurvedic medicine. Additionally, it is frequently observed in the carnivorous plant genera *Drosera* and *Nepenthes* and constitutes a constituent of the black walnut drupe [9].

Many studies have demonstrated numerous beneficial properties of PLB, including antioxidant, antidiabetic, antifungal, and, most notably, anticancer effects. Recent data have shown that PLB exerts an inhibitory effect on cancer progression in a range of cancer cell lines [10,11]. It has been shown that important pathways involved in cancer progression, such as the Akt/NF-kB signalling pathway and the MMP-9 pathway, are prime targets involved in the PLB-mediated cancer response.

In addition, activation of Reactive Oxygen Species (ROS), cell cycle arrest, and apoptotic signalling are crucial for PLB-induced anticancer response [12,13]. In Lewis Lung carcinoma (LLC) lung cancer cells, evidence suggests that PLB interactions with Thioredoxin Reductase (TrxR) prevent its downstream interactions with intracellular substrates and directly inhibit Glutathione Reductase (GR), thereby mediating the increase of intracellular ROS levels and ultimately leading to apoptosis [12].

Furthermore, the compound demonstrated considerable inhibitory effects via the IL-6/STAT3 signalling pathway, thereby impeding the proliferation and invasion of lung cancer cells [14]. PLB exerts its effect via various signalling mechanisms. For instance, PLB upregulates the expression levels of p21$^{CIP1/WAF1}$, causing the cell cycle to arrest in the G2/M phase, thereby inducing apoptosis [15]. The administration of PLB has been demonstrated to result in the downregulation of cytokine expression and the suppression of NF-κB-regulated genes in both MM-231 and MM-468 triple-negative breast cancer cell lines [16].

Besides cancer, it has been reported that PLB possesses anti-viral properties against RNA viruses such as Hepatitis C Virus (HCV), as well as SARS-CoV-2, due to its ability to generate ROS-mediated oxidative stress against their highly susceptible single-stranded RNA genome [17,18]. Researchers have demonstrated the anti-inflammatory role of PLB in neurodegenerative diseases such as Parkinson's disease, where PLB could effectively provide neuroprotection to PD mice models by inhibiting inflammation via the TLR/NF-κB pathway and reduced expression of IL-1β, TNF-α, and IL-6 mRNA levels [19]. Its role is also evident in cardiovascular diseases, as PLB showed a significant drop in the blood pressure of the rats studied. In addition, PLB has an antibacterial effect on the organism, effectively limiting its growth [20,21].

In an attempt to address the vast spectrum of therapeutic properties of PLB, this review article summarizes different pharmacological attributes, such as chemical properties and the biosynthesis of PLB. It also highlights the importance of PLB as an emerging therapeutic candidate, as documented in the literature for various diseases such as cancer, SARS-CoV-2, diabetes, cardiovascular diseases, etc., as their prevalence has significantly impacted the lives of millions and has resulted in the loss of lives (Figure 1).

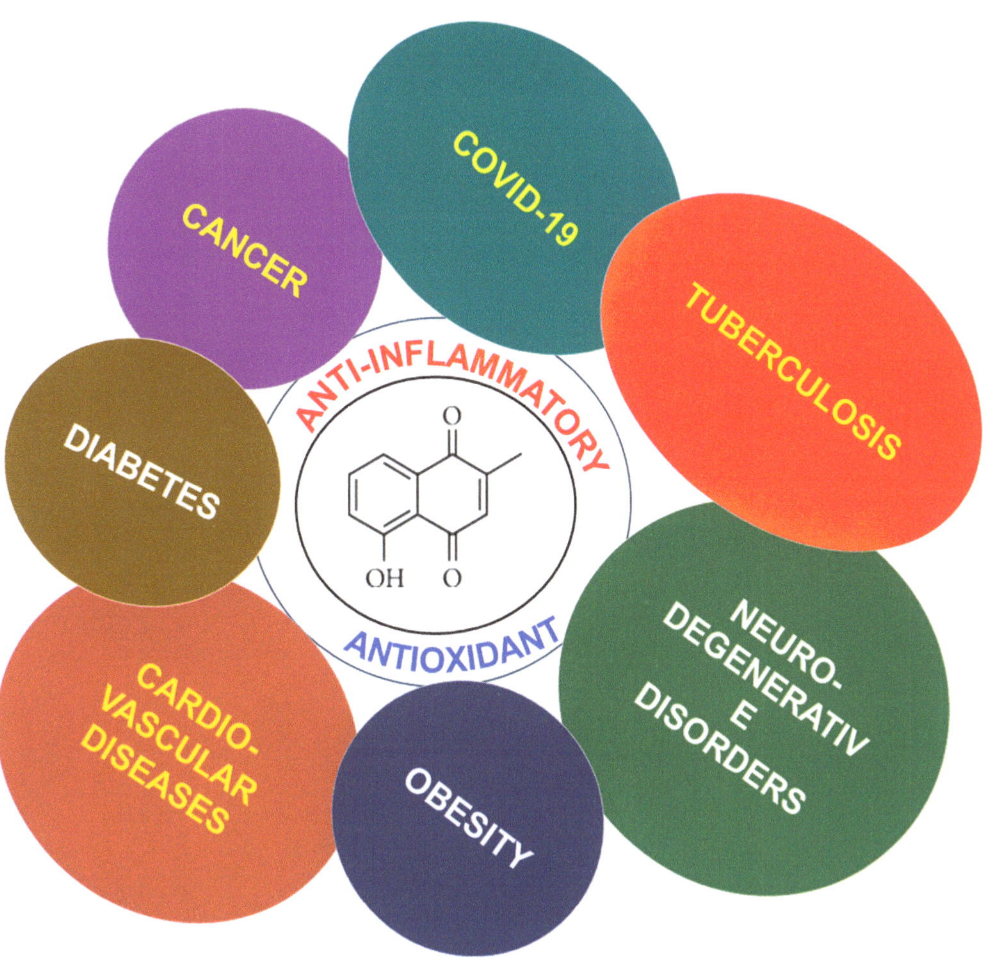

Figure 1. Structure of PLB with its role in preventing various diseases.

We further discuss the clinical formulations reported for PLB to increase its efficacy, which will further aid in its use as a potential drug and eventually support the healthcare system. In a nutshell, this review reflects on the potential therapeutic benefits and uses of PLB in the care and management of cancer, as well as other diseases.

2. Methods

For this review, a systematic literature search was conducted using several major databases, including PubMed, Google Scholar, and Science Direct. The search utilized specific keywords such as "Plumbagin", "cancer", "disorders", and "bioavailability" to identify relevant studies. As of 2024, a total of 337 articles were retrieved from these databases. Each article was carefully evaluated for relevance to the scope of the review. Following a thorough assessment, 200 articles were deemed pertinent and included in the review. This selection process ensured that the review encompasses a broad and relevant range of research on the therapeutic potential of PLB and its pharmacological properties.

3. Chemical Properties of PLB

PLB is a non-chiral, lipophilic compound with a molecular mass of 188.18 g/mol [22]. It exists in the form of a yellow crystalline substance. Experimentally, it is soluble to 100 mM in DMSO and to 50 mM in ethanol. Since it is sparingly soluble in water, its clinical translation requires formulations such as nanoemulsions [23,24]. The basic skeleton of the naphthoquinone is produced by a Polyketide Synthase (PKS) using six acetyl units [25] with 3-methyl-1,8-naphthalene-diol as an intermediary product. The PLB biosynthetic pathway also involves certain accessory enzymes, such as cyclase1 and AKR1 [26].

PLB is a part of the human exposome, having a role as a metabolite, an immunological adjuvant, an anticoagulant, and an antineoplastic agent. Predictions from the Swiss Absorption, Distribution, Metabolism, and Excretion tool (SwissADME) which suggested many drug-like features (http://www.swissadme.ch/ last accessed date: 8 August 2024). It has a positive LogP value of 1.72, which follows Lipinski's rule of 5 for drugs. PLB-labelled complex was found to have a short elimination half-life of about 2–3 h in blood. The excretion of PLB is primarily via the hepatobiliary and pulmonary routes, with limited pharmacokinetic properties resulting in inadequate systemic circulation. [27]. Nanoencapsulation of PLB overcomes the obstacles of low water, poor water-dissolving ability, and bioavailability, adding to pharmaceutical relevance with better therapeutic efficacy [28]. Important chemical and drug-like features of PLB are highlighted in Table 1.

Table 1. Chemical properties and drug-like features of Plumbagin.

Properties	Values
Molecular Weight	188.18
XLogP3	2.3
Log P	1.72
Hydrogen-Bond Donors	1
Hydrogen-Bond Acceptors	3
Rotatable Bonds	0
Topological Polar Surface Area	54.4 $Å^2$
Heavy Atoms	14
Formal Charge	0
Complexity	317
Isotope Atoms	0
λmax	
Melting point	77
Boiling point	383.927
Drug-like features	
Log Po/w	1.72
GI absorption	Yes, very high 96.258
BBB Permeability	Yes
Skin Permeability	No, −2.933
Caco2 permeability	Low, 1.192
Water Solubility	High, −2.655
Bioavailability score	0.55

4. Medicinal Properties of PLB

Ayurveda, a traditional system of medicine native to India, uses plant extracts, including PLB, to treat various diseases in the country [29]. However, it gained the attention of scientists worldwide only a few years ago, owing to the growing chemo-drug resistance in patients. Studies have shown that PLB possesses antifungal, antiviral, antibacterial, and antioxidant properties, as well as other biological properties such as anti-diabetic, analgesic, and anti-atherosclerosis [10]. It is widely established that inflammation represents a fundamental causal factor in a considerable number of pathological conditions, including cancer. PLB has been demonstrated to show strong anti-inflammatory properties, making it an interesting research candidate for treating cancer [30,31].

4.1. Role of PLB in Cancer Therapy

PLB has emerged as a promising compound in cancer therapy, due to its multifaceted therapeutic effects. The cytotoxicity of PLB against cancer cells has been observed in both in vitro and in vivo settings [28]. Anticancer activity of PLB has been reported against various cancer cell lines, including hepatoma, melanoma, leukaemia, breast carcinoma, prostate cancer, oral squamous-cell carcinoma, brain cancer, oesophageal cancer, lung cancer, kidney adenocarcinoma, cholangiocarcinoma, osteosarcoma, gastric cancer, and canine cancer [10]. The activity of PLB against cancer cells is summarized in Figure 2 and Table 2.

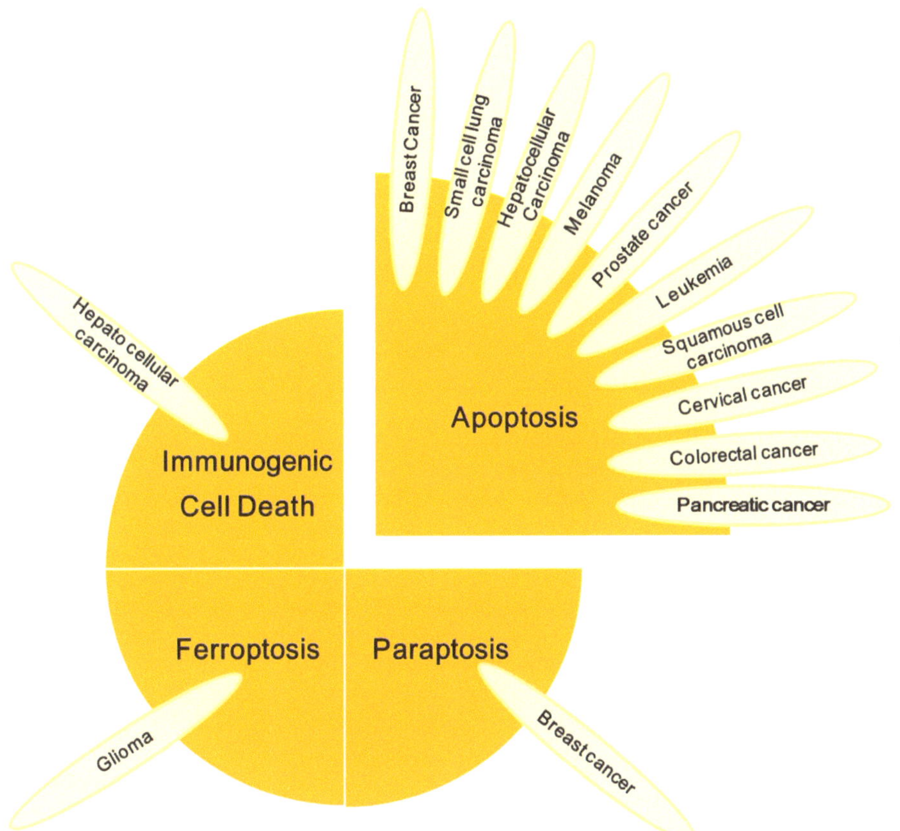

Figure 2. Modes of action of PLB in the treatment of various forms of Cancers. While apoptosis is the primary pathway involved in restricting cancer growth, other pathways, such as immunogenic cell death, ferroptosis, and proptosis, are also involved.

Table 2. The anticancer effects of plumbagin have been reported in different cell lines.

S. No.	Disease	Cell Line	IC$_{50}$ (nM)	Potential Mechanism(s)	Reference
1	Breast Cancer	MCF7	2840	Cytotoxic agent against human MCF7 cells expressing HER2 that showed reduced cell viability after 48 h via CellTiter-Glo assay	[32]
2	Breast cancer	MDA-MB-468	2500	Role as an antiproliferative agent against ER-negative human MDA-MB-468 cells after 48 h by MTT assay	[33]

Table 2. *Cont.*

S. No.	Disease	Cell Line	IC$_{50}$ (nM)	Potential Mechanism(s)	Reference
3	Breast Cancer	MDA-MB-231	3500	Reduced proliferation of ER-negative human MDA-MB-231 after 24 h analysed by MTT assay	[33]
4	Ductal breast carcinoma	BT-474	800	Cytotoxic for Her2-overexpressing human BT474 cells after 72 h analysed by MTT assay	[34]
5	Melanoma	SK-MEL-28	5000	Inhibited the growth of human SK-MEL-28 cells analysed by MTT assay	[35]
6	Colorectal cancer	HCT-116	9800	Cytotoxic for human HCT116 cells by inhibiting growth	[36]
7	Colon adenocarcinoma	SW480 SW-620	7300 7400	Cytotoxic role assessed by cell viability	[37]
8	Colorectal cancer	HT-29	4190	Cytotoxic against human HT-29 cells	[38]
9	Hepatocellular carcinoma	HepG2	9170	Cytotoxic against human HepG2 cells	[36]
10	Lung carcinoma	A549	3000	Inhibited growth of human A549 cells	[35]
11	Leukaemia	HL-60	1100	Cytotoxic against human HL60 cells; assessed reduced cell viability in cells	[39]
12	-	PBMC	2700	Cytotoxic against human PBMC cells	[39]
13	Cervical cancer	HeLa	10,200	Cytotoxicity	[40]

Numerous studies reported on the mechanism of the anticancer activity of PLB, which is attributed to its ability to exert an effect through different signal transduction pathways, including those involving PI3K/AKT/mTOR, Ras, Sirt1, AMPK, CDK1/CDC2, cyclin B1, cyclin D1, FOXM1, NF-κB, Nrf2/ARE, p53, p21 Waf1/Cip1, p27 Kip1, PI-5, STAT3/PLK1/AKT, and Wnt. Therefore, PLB may be a promising candidate for further investigation as a potential therapeutic agent for cancer treatment [22,41]. PLB has been shown to exert its effects primarily through the PI3/AKT/mTOR pathway in cancer cells, as illustrated in Figure 3. Self-regulated processes such as epithelial-to-mesenchymal transition (EMT) required for tissue repair may lead to conditions such as angiogenesis, fibrosis, loss of normal organ function, and even cancer in an uncontrolled state [42].

4.1.1. Breast Cancer

Breast cancer, a leading cause of cancer-related deaths among women worldwide [43], is driven by dysregulated pathways such as cell cycle, apoptosis, Wnt-signalling, DNA damage-repair pathways, inflammation, and hypoxia. Inflammation, in particular, is a known modulator of the progression and aggressive growth of breast cancer [44]. The NF-κB gene, a well-established regulator of the inflammatory response, plays a significant role in this context. Overstimulation of the NF-κB gene is a contributing factor to the aggressive nature of breast cancer [45]. One of the cytokines released in the inflammatory pathway of NF-κB is the C-C motif ligand, also known as CCL-2. CCL-2 is a chemokine that attracts immune cells to the site of inflammation and is purported to be implicated in most types of cancer, making it a crucial target for therapeutic intervention [46]. Endocrine therapy has shown promising avenues of treatment in early-stage breast cancer; however, advanced-stage breast cancer cells show poor response due to their increased chemoresistance [47]. Furthermore, the absence of the three essential endocrine receptors, progesterone (PR), oestrogen (ER), and human epidermal growth factor (Her2) receptors in triple-negative breast cancer (TNBC) cells significantly diminishes the efficacy of the treatment. Thus, studies for PLB for targeting inflammatory factors and cytokines have shown this secondary metabolite to be implicated in controlling breast cancer progression [47].

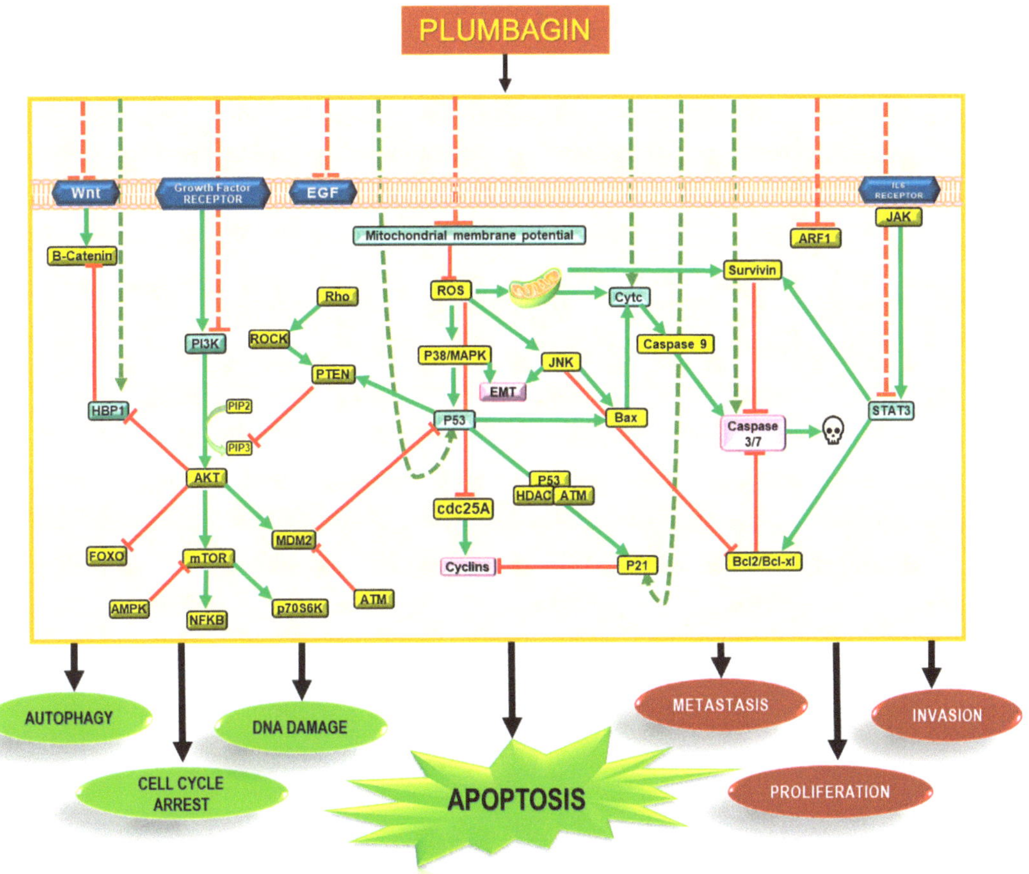

Figure 3. Molecular mechanisms of action of PLB. PLB downregulates several pathways in tumour cells to induce autophagy, cell cycle arrest, and DNA damage. These mechanisms inhibit metastasis and proliferation via Akt/PI3K, STAT3, NF-κB, and Wnt pathways. PLB introduces a cascade of signalling pathways that eventually stimulate apoptosis in cancerous cells.

PLB downregulates the expression of CCL2 cytokine and suppresses NF-κB-regulated genes in TNBC cell lines MM-231 and MM-468 [16]. The inhibitory effect of PLB on breast cancer progression in human endocrine-resistant cells (MCF-7/LCC2 and MCF-7/LCC9) is particularly pronounced, given that approximately half of breast cancer patients with advanced-stage ER-positive breast cancer develop endocrine resistance [48]. PLB targets the AKT signalling pathway and inhibits the progression by inhibiting AKT (pAKT) and pERK1/2 phosphorylation, regulated through Nuclear Receptor Coactivator 3 (NCOA3) in HER2-overexpressed endocrine-resistant cells. NCOA3 is a chromatin remodeler and a transcriptional factor involved in the progression of different cancers [49].

A further significant factor contributing to the growth of cancer is hypoxia. Due to the rapid development of cancer cells, the demand-supply equilibrium of oxygen is significantly disturbed, creating a hypoxic intra-tumour environment [50,51]. However, it is found that the cancer cells have adapted to maintain the equilibrium by increasing the expression of a transcriptional factor known as Hypoxia Inducible Factor (HIF-1α), which is responsible for preserving oxygen homeostasis in the body. But, in the case of cancer, it creates an environment conducive to the growth of tumour cells, resulting in rapid proliferation and, eventually, metastasis [51,52]. HIF-1α also upregulates important growth regulatory pathways such as AKT and ERK pathways. These further help tumour cells maintain growth, interfering with the cell's natural response of apoptosis, thus developing resistance against therapeutic agents. PLB downregulates the expression of HIF-1α under hypoxic conditions in MCF-7 cells via a pathway independent of PI3K/Akt/mTOR by abolishing HIF-1α at both the transcription and post-translational modification levels [53].

Among many cell signalling pathways reported through which PLB has been reported to exert its anticancer effect, one is mediated through p53, an anti-tumour molecule. A recent study has demonstrated a correlation between cell cycle regulators and PLB [15]. They reported that PLB upregulates p21$^{\text{CIP1/WAF1}}$, causing activation of apoptosis in a p53-dependent pathway, as well as the cell cycle, to arrest the G2/M phase via inhibition of cyclin B1 levels. PLB also down-regulates the anti-apoptotic genes pro-caspase 3 and Bcl-2 [54].

One of the most significant pathways in cancer treatment is the apoptosis of tumour cells. Nevertheless, cancer cells have evolved to successfully escape the event of apoptosis, leading to drug resistance [55]. Fortunately, PLB showed therapeutic potential for apoptosis-resistant cancers, i.e., PLB-induced paraptosis. This was evidenced by a trigger of extensive cytoplasmic vacuolation, which subsequently resulted in cell death. Notably, this process was neither apoptotic nor autophagic. Paraptosis is a caspase-dependent cell death pathway that involves cytoplasmic vacuolation caused by mitochondrial swelling and endoplasmic reticulum (ER) dilation [56]. Tumour cells actively synthesize new proteins to support their growth and proliferation. PLB has been demonstrated to induce cell death in such cells via a protein synthesis-dependent mechanism [55]. PLB reacts with the free thiol groups present on newly synthesized proteins, generating misfolded proteins and disrupting sulfhydryl homeostasis. This eventually generates a stress signal, resulting in ER-associated degradation (ERAD) and a subsequent loss of mitochondrial membrane potential (MMP), ultimately resulting in cell death [57,58].

Another noteworthy attribute of PLB is its ability to re-sensitize the chemo-resistant cancer cells, either in conjunction with other agents or alone [28]. It is established that a major limitation in using chemotherapy for cancer is the capacity of cells to develop resistance to these drugs. This acquired resistance is attributed to the MAPK/ERK pathway, which is a pro-survival pathway comprising a series of protein kinases, including BRAF and MEK, among others [59]. Once activated, these ERK protein kinases phosphorylate apoptotic regulators, leading to continued cell proliferation [60]. Paclitaxel is an anti-microtubule drug that inhibits tumour cell growth by causing tubulin dimerization and inhibiting microtubule depolymerization, thus preventing mitosis. It induces apoptosis independent of ERK signalling, but shows an affinity for GFs such as ERK and Ras, as they can co-localize with microtubules [61,62].

PLB reduces p-ERK levels and re-sensitizes paclitaxel-resistant breast cancer cells to paclitaxel-inducing cell death [63]. In a similar study, PLB successfully restored the sensitivity of breast cancer cells to the drug tamoxifen. Tamoxifen is an early-stage anti-cancer drug. It is a selective oestrogen receptor modulator [64]. Increased expression of an hsp70 class chaperone, GRP78, in tumour cells, reduces Bik's expression, a pro-apoptotic protein, eventually resulting in the development of drug resistance in cancer cells [65,66]. A total of 50% of advanced-stage cancer patients develop acquired resistance to the drug Tamoxifen [67]. PLB restores the sensitivity to Tamoxifen by inhibiting the anti-apoptotic activity of GRP78, thereby inducing Bik expression and rendering cancer cells susceptible to apoptosis. PLB also enhances sensitivity of endocrine-resistant breast cancer cells for Tamoxifen by suppressing the epithelial-to-mesenchymal transition (EMT) process, a crucial stage in the metastatic spread of cancer cells. Furthermore, it was reported that transcription factors are overexpressed at both transcriptional and protein levels, thereby triggering EMT. PLB restores the CDH1 expression levels by regulating the expression of snails, leading to reduced tumour progression in these patients [68–70]. A synergistic effect of PLB with paclitaxel and Tamoxifen was also observed [63].

All these studies suggest that PLB is a potent therapeutic agent for cancer treatment. However, studies have also shown that PLB treatment alone lacks cell specificity [71,72], and its weak lipophilic nature results in reduced intercellular uptake [73].

4.1.2. Lung Cancer

Lung cancer represents a significant global health burden, accounting for a considerable proportion of cancer-related mortalities [74]. Primarily caused by smoking and tobacco product usage, lung cancer can be categorized into two main histological types: small-cell lung carcinoma (SCLC) and non-small-cell lung carcinoma (NSCLC). Among these, NSCLC is the most frequently diagnosed form of lung cancer type and accounts for a significant proportion of lung cancer-related fatalities [75]. Targeted therapy is the most preferred therapy, with a good prognosis. However, with an increased risk of developing drug resistance, the therapy poses reduced therapeutic efficacy. Other treatments, such as chemotherapy, are associated with a risk of adverse reactions and harsh side effects [76]. With phytotherapeutic properties, PLB represents a promising candidate for treating patients affected with lung cancer. It works by targeting intrinsic mitochondrial apoptotic pathways and ROS generation [77].

A study on H195 and H160 cell populations shows the potent antitumor activity of PLB against NSCLC in a dosage-dependent manner. This is achieved by elevating ROS, Ca^{2+}, and CD8+T cells and downregulating ADP-ribosylation factor 1 (ARF1), which is a key target of PLB [78]. ARF1 belongs to a class of Ras-related GTP-binding proteins. Its overexpression is associated with a poor cancer prognosis [79]. Another experiment demonstrated that the elimination of ARF1 disrupts lipid metabolism, resulting in the aggregation of lipid droplets and the stimulation of an immune response. It was also observed to cause mitochondrial disruption, leading to an elevated ER stress response, inducing CD8+T cells in the circulation, a possible mechanism used by PLB to restrict cancer cell proliferation [78]. Mitochondrial apoptotic pathway-mediated cell death is important in eradicating cancer cells. A pivotal compound in this process is Caspase 9.

PLB treatment is found to upregulate Bax protein, facilitating the release of Caspase 9 via Cytochrome *c* (Cyt-*c*). Once released in the cytosol, it induces the expression of Apoptosis protease activating factor 1 (APAF1), which subsequently interacts with the apoptosis-associated factor 15 (Apaf-1) to create the apoptosome. The apoptosome activates caspase 9, which activates Caspase 3 downstream. Thus, PLB upregulates the expression levels of cytochrome c and induces caspase-9 and caspase-3 activity in NSCLC, ultimately resulting in the apoptosis of A549 cells [77,80]. In normal cells, PLB prevents radiation-induced apoptosis by inhibiting caspase-3 activity [81].

Matrix Metalloproteases (MMPs) promote tumour growth via matrix barrier degradation, and facilitate angiogenesis. PLB reduces MMP expression via increased ROS production. It has been demonstrated that PLB treatment results in the increased expression of antioxidant genes, specifically GSTP1 and SOD2, due to its intrinsic oxidative stress-inducing properties. The same study on A549 and NCI-H522 found that PLB treatment induces apoptosis at lower concentrations (0–6 μM). Change in morphologies of cancerous cells, including shrinkage of the cells and plate detachment, was depicted in tumour spheroid models. Following PLB treatment, A549, and NCI-H522 cells also demonstrated reduced single-cell colony-forming ability, along with reduced migration. The study showed that PLB restricts cell proliferation through cell cycle arrest, like other cancers. PLB inhibited A549 cells in the G2/M and S phases, whereas NCI-H522 cells were limited to the G2/M phase [82]. It was also noted that in A549 cells, PLB restricts the cell cycle by regulating the expression of cdc2 and cyclin B1, both promoting cell cycle progression through the G2/M phase.

PLB induces autophagy in lung cancer cell lines A549 and H23 via inhibition of the PI3K/Akt/mTOR pathway. PLB demonstrated pro-apoptotic and pro-autophagic properties in a dose-dependent manner. The study also shows that PLB could perform apoptosis via p53-dependent and p-53-independent pathways, as H23 cells are p53 mutant cell lines. A concentration-dependent increase in the expression of PUMA, a pro-apoptotic protein, was observed in A549 cells. However, it was not replicated in the H23 cell line. Elevated expression levels of PUMA inhibit anti-apoptotic proteins like Bcl2 and Bcl-xl, thereby promoting Bax- and Cas9-mediated apoptosis [83]. Even though PLB initiates programmed cell death in cancer cells via both apoptosis and autophagy, working in a coordinated manner, the latter plays a negligible role in determining the effectiveness of PLB [82]. In the A549 lung cancer cell line, PLB also initiates paraptosis via proteasome inhibition and disruption of sulfhydryl homeostasis [57].

PLB inhibits the Rho-associated kinase (ROCK) pathway via the FAK/AKT pathway, thereby suppressing lung metastasis in A549 cells treated with osteopontin (OPN) [13]. OPN is a tumour-microenvironment (TME) component that is actively secreted by tumour cells and plays a pivotal role in the progression of cancer cells by upregulating MMP2/9, VEGF, and other key factors. OPN actively phosphorylates FAK and AKT and induces the ROCK pathway [84]. PLB interacts with TrxR and GR in LLC lung cancer cells via intracellular ROS [14]. Furthermore, PLB considerably inhibits the proliferation and invasive potential of L9981 and NL9980 cells via the IL-6/STAT3 signalling pathway [14].

Despite being a common treatment for cancer, the collateral damage associated with radiation therapy requires attention to finding safer and less cytotoxic alternatives like phytochemicals. PLB, when given in combination with radiotherapy, resulted in a less cytotoxic response in normal cells [85]. Thus, it showed that PLB provides radioprotection to non-cancerous cells; however, it did not render any such protection to tumour cells, therefore increasing the efficacy of radiation therapy.

4.1.3. Hepatocellular Carcinoma

Hepatocellular carcinoma represents the most prevalent form of liver cancer, and is the fourth most commonly diagnosed neoplasm worldwide [86]. PLB inhibits hepatocellular cancer progression [87]. PLB induces ATM-p53 pathway-mediated G2/M cell cycle arrest and downregulation of 25C (cdc25C), resulting from ROS-mediated oxidative stress [88]. PLB inhibits the growth of HCC cells by downregulating Glutathione peroxidase 4 (GPX4) via its ubiquitination and resultant proteasomal degradation. GPX4 protects the cells from lipid peroxidation by reducing lipid peroxides, thus maintaining cell homeostasis [89]. It negatively regulates oxidative stress-induced apoptosis. Among different antioxidant enzymes studied, such as SOD1, CAT, and TXN, PLB showed selectivity toward GPX4. GPX4 has been identified as a novel substrate of USP31, a deubiquitinase (DUB) that plays a crucial role in regulating cell proliferation. Inhibition of USP31 via PLB makes GPX4 available for protein degradation, followed by tumour cell apoptosis. Contrary to other tumours, where USP31 is shown to act as a tumour suppressor, its inhibition by PLB in HCC showed reduced cancer progression [90].

PLB inhibits cell proliferation by preventing angiogenesis in HCC cell line SMMC-7721 in a dosage- and time-dependent manner. Stromal cell-derived factor (SDF1), a chemokine, promotes angiogenesis via VEGF/IL-8 mediated NF-kb activation. In this study, PLB abolishes SDF-1-induced tumour progression by inhibiting the expression of CXCR4 and CXCR7, transmembrane receptors of SDF-1 [91]. PLB also upregulates the caspase-3 protein level in the same cell line and causes cell apoptosis via signal-mediated EMT [92]. PLB inhibits proliferation by inhibiting the SIVA/mTOR signalling pathway in HCC cell lines HepG2 and LM3, resulting in apoptosis [87]. SIVA is an apoptosis regulator protein known to promote tumorigenesis via the mTOR pathway. The signalling pathway is crucial in cell proliferation and cycle progression. Therefore, PLB's ability to restrict tumour progression through multiple signalling pathways makes it a key therapeutic candidate for treating HCC. However, attempts have been made to enhance the activity of PLB in cases of immune resistance, including PLB-mediated immunogenic cell death (ICD). Nano co-delivery of PLB, along with Dihydrotanshinone (DIH), a natural compound from the roots of *Saliva miltiorrhiza*, helps overcome immunotherapy resistance caused by TME in HCC. PLB is an ICD inducer and Dihydrotanshinone ICD enhancer [93].

4.1.4. Melanoma

Melanoma is known as the most dangerous form of skin cancer, exhibiting high drug resistance and poor prognosis [94,95]. PLB exhibited a dose-dependent reduction in cell viability among melanoma cell lines [96]. In SK-MEL-28 cells, PLB induces metabolic alterations by inducing Mitochondrial Oxidative Phosphorylation (OXPHOS). In PLB-induced A375 cells, mitochondrial OXPHOS and ATP production levels were decreased, whereas Mitochondrial Membrane Potential (MMP) was elevated. PLB augmented the generation of ROS and increased proton leak in both cell lines [97]. In B16F10 melanoma cancer cells, PLB downregulates genes associated with the MAPK pathway, MMPs, and cell adhesion, whereas it elevates the expression levels of apoptotic, ROS response, and tumour suppressor genes, suggesting an anti-invasive and anti-metastatic role of PLB [94]. Furthermore, a combination of PLB and the antineoplastic drug celecoxib has been demonstrated to exert a synergistic inhibitory effect. This was evidenced by a reduction in the proliferation of melanoma cells and the suppression of vascular development in tumours, mediated by COX-2 and STAT3 inhibition. This subsequently led to a decrease in the levels of the key cyclins that are essential for melanoma cell survival [95].

4.1.5. Prostate Cancer

Prostate cancer represents the second most prevalent form of cancer diagnosed in males worldwide [98]. These cancers are most commonly adenocarcinomas. Currently, both non-surgical and surgical treatments exist for treatment. Like other cancers, prostate cancer has also responded to PLB treatment. PLB administration in mice harbouring PTEN-P2 tumours in the prostate induced changes in gene expression in a dihydrotestosterone (DHT)-dependent manner, suggesting that PLB activity could be mediated by androgen receptors (ARs). PLB also reduced the growth of cancer cells and induced apoptosis in DU145 and PC-3 cells [99]. The elevation of intracellular ROS by PLB results in the induction of a lethal endoplasmic reticulum stress response in Prostate Cancer (PCa) cells.

Furthermore, PLB also demonstrated the capacity to inhibit the growth of PCa xenografts without displaying any discernible toxic effects. The administration of PLB to mice with human PCa xenografts also induced ER stress activation [100]. Like melanoma cancer cells, PLB was also found to work in synergy with other drugs for PCa. PLB was shown to significantly augment the effect of androgen deprivation therapy (ADT) drugs currently in pharmaceutical use, with only a few side effects in mice. However, a similar effect was not seen in drugs that bind to AR [101].

4.1.6. Squamous-Cell Carcinoma

PLB exerts a suppressive effect on oral squamous-cell carcinoma (SCC) cells. This suppression is achieved through the induction of apoptosis, a process triggered by endoplasmic reticulum (ER) stress, ROS production, and mitochondrial dysfunction. In both drug-resistant (CAL27/RE) and drug-sensitive (CAL27) SCC cells, PLB stimulates apoptosis and autophagy by activating the JNK and AKT/mTOR pathways through ROS-mediated mechanisms. Notably, the effect on drug-resistant cells is more pronounced [102–104].

PLB also demonstrated a protective role by inducing autophagy and inhibiting EMT via the ROS/p38MAPK pathway in tongue squamous-cell carcinoma (TSCC) cells. In vitro studies have shown that PLB administration results in a reduction in cell viability and colony formation, accompanied by an increase in cell apoptosis. This is linked to the downregulation of GLUT1, a glucose transporter, which is mediated by the inhibition of the PI3K/Akt pathway. Studies have demonstrated that PLB can suppress tumour growth, which is correlated with the downregulation of GLUT1, in comparison to the control group [105].

Furthermore, PLB exhibits a synergistic effect when combined with existing cancer drugs. For instance, the combination of PLB and cisplatin effectively inhibits the growth of TSCC. Additionally, the combination of cisplatin, 5-fluorouracil (PF), and PLB has been reported to induce cell cycle arrest in the S phase, enhance cisplatin-induced cytotoxicity, autophagy, and apoptosis in both CAL27 and CAL27/CDDP (cisplatin-resistant) cells, and to result in higher apoptosis rates compared to PF treatment alone.

PLB has notably increased the sensitivity of TSCC cells to PF. This enhancement is achieved by inducing cell cycle arrest in the S phase and apoptosis through the PI3K/AKT/mTOR/p70S6K pathway [106]. Additionally, PLB has been demonstrated to enhance the radiosensitivity of TSCC cells. When combined with irradiation (IR), PLB leads to cell cycle arrest in the G2/M phase and, eventually, apoptosis, accompanied by the downregulation of ATM and NF-κB.

In SCC25 cells, PLB induces the G2/M-phase cell cycle arrest, inhibits EMT and stemness, and promotes extrinsic apoptosis. Notably, PLB suppresses the translocation of Nrf2 from the cytosol to the nucleus, thereby inhibiting the expression levels of downstream targets [107]. PLB has been reported to promote autophagy and cellular apoptosis in TSCC cells. This process involves the p38 MAPK and PI3K/Akt/mTOR pathways, with additional contributions from the ROS- and GSK3β-mediated pathways.

4.1.7. Colorectal Cancer

Colorectal cancer (CRC) is the third most common cancer, with a five-year survival rate of less than 20% [108]. PLB has shown its anticancer properties in CRC by targeting signalling pathways affected in cancer. Alteration in miR-22-3p levels by PLB induces apoptosis and inhibits Wnt signalling and colony formation in CRC cells. MiR-22-3p plays a significant role in apoptosis, autophagy, etc., and is down-regulated in several cancers. PLB also downregulates Wnt signalling in HCT116p53+/+ and HCT116p53−/− CRC cells in a p53-independent manner and upregulates HBP1, a negative modulator of Wnt signalling in these cells [37,109–111]. CRC samples have shown an upregulation in expression levels of neoplastic MAPK1 and PARP1 mRNAs and downregulation in EP300 mRNA levels, whereas PLB-treated CRC cells showed results otherwise, with suppressed cell proliferation [112].

PLB activates the AMPK/ASK1/TRAF2 association, leading to the activation of the pro-apoptotic c-Jun N-terminal kinases (JNK)-p53 signalling axis. Following PLB treatment, activated AMPK directly leads to phosphorylation of Raptor, inhibiting the activation of mTOR complex 1 (mTORC1) and Bcl-2 expression in colon cancer cells. Furthermore, the exogenous addition of short-chain ceramide (C6) has been proven to enhance PLB-induced AMPK activation, thereby facilitating cell apoptosis and growth inhibition. The PLB-induced apoptosis of colonic cancer cells is dependent on the TNF-α-mediated pathway, with the effect being contingent on the expression of COX-2 [113].

4.1.8. Pancreatic Cancer

Pancreatic cancer (PC) is a carcinoma of the pancreatic duct cells. Despite being relatively uncommon, it has a very high mortality rate [114]. PLB has shown potential inhibitory effects on PANC-1 pancreatic cancer cells. Pre-treatment with PLB significantly prevented EMT, EGF-induced survival, matrix protein hyaluron production, clonogenesis, migration, and MMP-2 gene expression and secretion in PANC-1 cells [115]. PLB significantly inhibited cell growth and induced ROS-mediated apoptosis through an intrinsic pathway, leading to the Bax and Bcl-2 being upregulated and downregulated, respectively. Additionally, PLB demonstrated an anti-migratory effect against PANC-1 cells, suggesting its potential as an antimetastatic agent [116]. Four main pharmacological targets regulated by PLB include TP53, MAPK1, BCL2, and IL6. The potential biological mechanisms involved include the advanced glycation end-product (AGE)-receptor for the advanced glycation end-product (RAGE) signalling pathway, the PI3K/Akt signalling pathway, and the hypoxia-inducible factor 1 (HIF-1) signalling pathway. These pathways play a pivotal role in regulating the survival of a cell, apoptosis, and metabolism [117]. In another study, HT in combination with PLB induced more PANC02 cell deaths than HT or PLB alone. The combination therapy demonstrated efficacy in inhibiting the accumulation of MDSCs, while simultaneously promoting the infiltration of CD4+ and CD8+ T cells within the tumour microenvironment [118].

4.1.9. Cervical Cancer

Cervical cancer is one of the primary causes of cancer deaths in females across the globe, and the most frequently occurring cancer in India [119]. PLB administration demonstrated a reduction in the viability of CaSki cells in a dose-dependent manner and suppressed the migratory potential of these cells. This was attributed to reduced expression of MMP-2 and an upregulation of TIMP2. It also led to downregulation in the expression levels of E2F1 and an upregulation in the levels of p21. PLB downregulated UHRF1 expression, enhanced apoptosis, and reduced the metastasizing potential of CaSki cells. PLB also caused strong G2/M-phase arrest in CaSki and SiHa cells and S-G2/M-phase cell cycle arrest in HeLa cells [120,121].

PLB treatment also decreases mitochondrial membrane potential, eventually decreasing the ATP in cervical cancer cells [57,120]. PLB has been demonstrated to attenuate the HA-CD44 pathway in HeLa cells, thereby exerting an inhibitory effect on tumour growth. [122]. PLB treatment causes DNA cleavage in SiHa cancer cells. The translocation of LC-3B protein from cytoplasm to autophagosome and morphological analysis showed that PLB treatment causes autophagic cell death [123]. PLB also induced paraptosis via inhibition of the proteasome and the disruption of sulfhydryl homeostasis in cervical cancer cells (HeLa) [57]. In SiHa and HeLa cells, PLB strongly induced ROS-mediated apoptosis. At non-cytotoxic doses, PLB possesses an anti-metastatic effect, accompanied by inhibition of EMT [121]. PLB interacts synergistically with cisplatin, reducing its IC_{50} value with improved effectivity in CaSki cells [120].

4.1.10. Ovarian Cancer

PLB treatment inhibited the growth of cancer cells and induced apoptosis in these cells. Furthermore, a reduction in the mRNA levels of intracellular OCT4 and PCNA was observed in PLB-treated cancer cells, accompanied by an increase in KLF4 mRNA activation [124]. Network pharmacology analysis revealed that PLB's anti-UEC/COVID-19 effects are achieved via a multifaceted approach, encompassing anti-proliferation, cytotoxicity, apoptosis induction, anti-inflammatory activity, and modulation of key molecular pathways associated with anti-inflammatory activity and immunomodulation. Molecular docking studies further supported these findings by identifying potential anti-UCEC/COVID-19 pharmacological biotargets of PLB, such as mitogen-activated protein kinase 3 (MAPK3), urokinase-type plasminogen activator (PLAU), and tumour necrosis factor (TNF) [125].

4.1.11. Leukaemia

The term leukaemia encompasses a diverse range of cancers, with the specific type depending on the type of blood cell affected and the rate of growth. A newly developed PLB dimer was observed to inhibit the growth of NALM-6 cells (human B cell precursor leukaemia). PLB augmented TRAIL-induced apoptosis in Kasumi-1 cells, concomitant with mitochondrial damage, up-regulation of death receptors (DRs), caspase activation, and decreased expression of cFLIP. The combination of PLB and rsTRAIL in both in vivo and in vitro models has been demonstrated to induce apoptosis of leukemic Kasumi-1 cells. Furthermore, each agent has been shown to elicit this effect independently. ROS Scavenger NAC could partially abolish the impact of PLB on the expression of Bax, DR5, and cFLIP. Conversely, glutathione (GSH) depletion by PLB was observed to increase the production of ROS [126]. PLB exerted a selective inhibitory effect on TrxR, resulting in the production of ROS in human promyelocytic leukaemia HL-60 cells. This, in turn, gives rise to an elevation of the GSSG/GSH ratio and a reduction in the cellular thiol pool [127]. The inhibition of c-Myb by PLB has been observed to suppress Myb target genes and induce differentiation of myeloid leukaemia cell line HL60 [128].

PLB demonstrated enhanced efficacy in the inhibition of Chronic lymphocytic leukaemia (CLL) cell lines in a dose-dependent manner. PLB also promoted the accumulation of MEC-1 cells in the S phase of the cell cycle and blocked the cell cycle transition of HG3 cells in G0/G1 to the S phase. PLB induced the apoptosis of CLL cells by increasing Bax protein expression and reducing Bcl-2 expression [129]. PLB induced a caspase-dependent apoptotic mechanism in the T-cell acute lymphoblastic leukaemia (T-ALL) cell line, MOLT-4, without any significant cytotoxicity observed in normal peripheral blood mononuclear cells (PBMCs). Furthermore, PLB has been shown to inhibit LPS-induced phosphorylation of p65 and the transcription of NF-κB target genes [130].

4.1.12. Brain Cancer

PLB has demonstrated inhibitory effects on glioma growth in both in vitro and in vivo models. These effects are ascribed to the drug's ability to target NQO1/GPX4-mediated ferroptosis. Specifically, PLB downregulates xCT and GPX4 while enhancing NQO1 activity, leading to NQO1-mediated cell death. Additionally, PLB promotes the degradation of GPX4 via the lysosomal pathway, resulting in GPX4-dependent cell death [131]. In nude mice with glioma cell xenografts, PLB treatment significantly reduced the tumour volume, by 54.48%. Furthermore, PLB downregulated FOXM1, along with its downstream targets (cyclin D1 and Cdc25B), while concurrently increasing the expression of p21 and p27. PLB-induced DNA damage, apoptosis, and cell cycle arrest, subsequently suppressing the ability of brain tumour cells to form a colony. This treatment also upregulated PTEN and TNFRSF1A, while downregulating E2F1 and reducing survivin, MDM2, cyclin B1, and BCL2 levels. Additionally, PLB increased caspase-3/7 activity and inhibited telomerase activity in brain tumour cells [132].

Studies revealed that PLB increases the transcriptional activity and nuclear localization of Nrf2, resulting in an increase in Nrf2/ARE-dependent genes in human neuroblastoma cells. PLB exposure has been shown to protect neuroblastoma cells and primary cortical neurons against oxidative and metabolic insults [133]. PLB administration significantly reduces brain damage and alleviates associated neurological deficits in a mouse model of focal ischemic stroke. PLB also directly interacts with Nox-4, inhibiting its activity in a time- and dosage-dependent manner in LN229 and HEK293 cell lines [134].

4.1.13. Oesophageal Cancer

PLB treatment suppressed the oesophageal squamous-cell carcinoma (ESCC) cell viability and proliferation. The PLB treatment increased the proportion of cells present in the G0/G1 phase, accompanied by a corresponding decrease in the number of cells present in the S phase. PLB also targeted STAT3 signalling and inhibited ESCC cell growth [135]. PLB has been demonstrated to inhibit growth and induce apoptosis in human ESCC cells through the regulation of STAT3-PLK1-AKT signalling [136].

4.2. Neurodegenerative Diseases

Neurodegenerative diseases are characterized by loss of neuronal functioning, and they represent a substantial contributor to mortality and disability [137,138]. Protein kinases are considered substantial regulators for important signalling pathways, and their aberrant expression is associated with various types of cancers, metabolic disorders, and neurodegenerative disorders [139–142]. Thus, they represent optimal candidates for the development of novel therapeutic strategies. One such kinase, known as Microtubule affinity-regulating kinase (MARK4), is an isoform of the MARK family [143,144]. It regulates the phosphorylation of tau protein at specific serine residues in the microtubule-binding domain. MARK 4 also plays a role in modulating different biological functions, such as glucose homeostasis, and diseases, such as breast cancer and neurodegeneration. Computational and spectroscopic methods to analyse the binding mechanism between PLB and MARK4 showed strong binding affinity at 10^6, as estimated. A simulation study using All-atom molecular dynamics also showed a strong binding affinity, with few conformational changes [145].

4.2.1. Parkinson's Disease

Parkinson's disease (PD) represents the second most prevalent neurological disorder, following AD. However, only 2% of individuals aged 65 and above are afflicted with this disease [146]. Two main pathogenic aspects of this disease include the gradual deterioration of dopaminergic neurons in the substantia nigra pars compacta (SNpc) and the formation of Lewy bodies, which are primarily composed of comprise α-synuclein [147,148]. However, it is known that major pathological conditions associated with neurological diseases like depression, AD, and PD are caused by neuroinflammation [149]. Microglia cells control NF-κB activity through toll-like receptors (TLRs), which are reported to have a significant role in mediating neuroinflammation. Nuclear factor erythroid 2-related factor 2 (Nrf 2) is a transcription factor that inhibits NF-κB activity by stabilizing NF-κB inhibitor (NF-κBi) [150,151].

In SD rats, it is shown that PLB treatment reduces NF-κB levels and downregulates the expression of inflammatory proteins such as COX-2. It was also shown to increase the expression levels of Nrf2. The research also demonstrated that in the case of subacute and chronic PD mice models, PLB effectively inhibits inflammatory pathway activation via the TLR/NF-κB pathway and decreases the expression of TNF-α, IL-6, and IL-1βmRNA, thus providing neuroprotection. Furthermore, in LPS-stimulated RAW 264.7 cells, PLB has also been demonstrated to elicit an anti-inflammatory effect by downregulating pro-inflammatory mediators via the inhibition of NF-κB and MAPK signalling [19].

In addition to neuroinflammation, autophagy is a crucial target for PD treatment. The accumulation of damaged organelles and abnormal proteins is a characteristic of PD, thus pointing to a possible dysregulation of autophagy. Autophagy is primarily modulated by the mammalian target of rapamycin (mTOR) in macrophages, and has become the most significant signalling system. PLB exerts its inhibitory effect on the Akt signalling pathway by blocking Akt activation, which in turn reduces the phosphorylation of downstream targets of mTOR.

Furthermore, it has been demonstrated that this agent reduces the level of p-mTOR protein in the substantia nigra of PD mouse models and facilitates the removal of autophagy substrate p62. Consequently, it can be concluded that this agent acts as a neuroprotective agent. Nevertheless, to substantiate PLB's anti-PD efficacy more rigorously, additional animal and cellular PD models could be employed for future investigations [152]. PLB caused significant downregulation in the levels of various cytokines, including IL-1α, IL-12 p40/p70, IL-1β, G-CSF, MCP-1 IL-6, IL-6, MCP-5, and TNF-α [152,153].

4.2.2. Alzheimer's Disease

Alzheimer's disease (AD) is a common neurological disorder that progresses to dementia in later stages [154]. Pathologically, AD is characterized by the accumulation of amyloidal protein deposits when β-secretase and γ-secretase act on amyloid precursor protein, which leads to the formation of β-amyloid peptide (Aβ) plaques [155,156]. AD pathogenesis is characterized by oxidative stress and subsequent neuronal damage. Usually, the antioxidant system can effectively manage oxidative stress in the cells. However, a significant increase in oxidative stress exhausts the cellular antioxidant defence mechanisms, leading to increased inflammation and subsequent neurotoxicity. Treatment with PLB in AD individuals showed that it restored the cellular antioxidant system and successfully reduced overall oxidative stress. PLB also enhances the expression of Nrf2, along with its downstream targets, HO-1 and NQO-1, in AD, to treat neuronal inflammation [157].

4.2.3. Amyloid Aggregation

Amyloid aggregation of soluble proteins is one of the key factors that can cause the onset of various neuronal disorders [158]. The formation of amyloid fibrils from different proteins, including insulin and bovine serum albumin (BSA), can result in multiple medical complications. For instance, there is a documented risk of insulin aggregation into cytotoxic amyloid fibrils, especially in diabetes patients. Repeated use of insulin injections and insulin inhalers can lead to insulin aggregates in these patients. In vitro studies have also revealed the inverse association between $A\beta_{25-35}$ positivity and serum albumin cross-linkage [159]. PLB was found to effectively inhibit insulin and serum albumin aggregation under in vitro conditions. In addition to preventing aggregate formation, it effectively promoted the disassembly of existing protein amyloids. PLB also disrupts and inhibits Human islet amyloid polypeptide (amylin) assembly into amyloid fibrils. Docking studies performed between insulin and PLB revealed strong interactions. Therefore, PLB-based formulations can be designed for such proteins, thereby preventing potential amyloid-related health risks [160,161]. Significant release of β-secretase enzyme leads to overproduction of amyloid-β. Docking studies have indicated the potential for PLB to exert an inhibitory effect on the enzyme β-secretase [162].

4.2.4. Depression

Depression is a psychiatric disorder caused due to increased oxidative damage and monoamine neurotransmitter imbalance. The monoamine hypothesis posits that the depletion of monoamines, including serotonin, norepinephrine, and dopamine, in specific regions of the brain, including the limbic system, frontal cortex, and hippocampus, can contribute to the development of depression [163]. Monoamine oxidase (MAO) is an important enzyme involved in the metabolism of monoamines. PLB has demonstrated antidepressant-like activity in both unstressed and unstressed mice. These effects may be attributed to its ability to inhibit brain MAO-A activity and enhance the antioxidant status, as well as its capacity to elevate corticosterone levels in response to stress, reversibly [164].

4.3. Cardiovascular Diseases

4.3.1. Hypertension

Hypertension is regarded as a potential risk factor and a major contributor to cardiovascular diseases, including heart failure, myocardial infarction, stroke, etc. Elevated cardiac output and increased vascular resistance are some characteristics of hypertension [165]. Disturbances in the regulatory mechanisms and normal compensatory mechanisms of blood pressure, including an imbalance in the retention of sodium during the excretion process, result in endothelial dysfunction and vasoconstriction, leading to hypertension. PLB is known to possess cardioprotective properties. A study by [166] investigated the role of PLB in lowering blood pressure using an invasive blood pressure (IBP) apparatus. Other parameters, viz., vascular tension and cardiac depressant effects, were monitored by performing experiments on isolated rats. PLB demonstrated the ability to reduce blood pressure, which may be attributed to a reduction in vascular resistance through the action of calcium antagonists, interference with calcium efflux, and the exertion of depressive effects on the force and rate of cardiac contraction.

4.3.2. Coronary Heart Diseases

Coronary heart diseases (CHDs) contribute significantly to morbidity and mortality rates globally, and thus result in socioeconomic burden [167]. Patients suffering from Acute Myocardial infarction (MI) (a subtype of CHD) may also face severe complications such as angina pectoris and, in the worst cases, heart failure in later stages of life. Significantly reduced body weight, as well as heart weight, is observed in such patients [168]. Doxorubicin is a potent drug administered during the treatment of various cancers. However, it is highly cardiotoxic and stimulates inflammatory responses in cardiac tissues [169]. Experiments to determine the effect of PLB in doxorubicin-affected animals showed that elevated cardiac markers associated with damage of the heart and inflammatory markers were significantly reduced following PLB treatment. The PLB treatment also alleviated the overexpression of proinflammatory proteins such as NF-κB, TNF-α, and IL-1β and apoptotic proteins in the heart tissues of doxorubicin-treated rats. PLB supplementation resulted in a considerable gain in body weight and heart weight of the animals investigated. The histopathological analysis revealed that the PLB exhibited a notable protective effect on cardiac tissues against damage induced by doxorubicin [170].

4.3.3. Myocardial Ischemia-Reperfusion

The role of PLB as an anti-inflammatory agent in cardiovascular diseases is further supported by its role in Myocardial Ischemia-Reperfusion (MI/R) injury, which leads to prolonged oxidative stress, loss of membrane phospholipids and endothelial dysfunction [171,172]. PLB reduced oxidative stress by decreasing lipid peroxide levels and ROS levels in rats with MI/R injury. PLB modulated the redox imbalance induced by MI/R injury by modulating the expression of transcription factors NF-κB and Nrf-2, eventually reducing the expression levels of their downstream targets. Furthermore, the expression levels of pro-inflammatory cytokines were markedly decreased by PLB treatment [173].

4.3.4. Pulmonary Arterial Hypertension

Chronic inflammation leads to the development of another significant cardiovascular disease known as Pulmonary Arterial Hypertension (PAH). The disease is clinically characterized by the aggressive proliferation and resistance to cell death of the pulmonary artery smooth muscle cells (PASMCs) [174,175]. The PAH-associated phenotype is primarily attributed to the expression of STAT3 and NFAT. STAT3 activation positively regulates Pim1, the NFAT activator, leading to increased expression levels of NFAT [176]. As in cancer, PLB can also successfully inhibit the progression of PAH by targeting STAT3 both in vivo and in vitro. Therefore, the STAT3/NFAT axis can be used as a therapeutic target by PLB in human PAH-PASMCs and experimental PAH rat models [177].

4.4. COVID-19

COVID-19, caused by the novel coronavirus SARS-CoV-2, has emerged as a major global health crisis, since its discovery in late 2019. SARS-CoV-2 relies on host signalling mechanisms to propagate through RNA replication and transcription [178]. Upon entering the host cell, viruses shift the oxidant–antioxidant equilibrium towards the antioxidant state induced by Nrf-2 expression to prevent the host cell from ROS cytotoxicity and to promote mitogenic activity. In addition to this, during viral infection, several host genes associated with redox mechanisms are modulated to enable viral propagation and pathogenesis [179,180]. PLB is a known ROS inducer that possesses the ability to alter redox potential. It has known antimicrobial, anti-inflammatory, antiviral, and anticancer properties. It is, therefore, conceivable that PLB could be employed in the future to combat a pandemic, thereby preventing a significant loss of life from a potential future outbreak of a disease such as that caused by the SARS-CoV-2 virus [17]. PLB was identified as a potent main protease (Mpro) inhibitor from a natural product library. Development of an optimized FRET-based HTS assay for the discovery of Mpro inhibitors could utilize PLB as the promising lead compound to generate an antiviral agent with better potential for targeting SARS-CoV-2 Mpro [181].

4.5. Tuberculosis

Tuberculosis is caused by *Mycobacterium tuberculosis*, which claims nearly 2 million lives yearly, worldwide. India is among the 22 countries with an exceptionally high tuberculosis burden. It accounts for about 40% of the total number of patients afflicted with TB worldwide [182]. The alarming increase in the rate of MDR-TB cases and increased chances of HIV co-infection in such patients underscores the urgent necessity to enhance the pharmacological regimen employed for the treatment of tuberculosis. A study investigating the therapeutic potential of quinoids against tuberculosis infection showed that PLB exhibited marked inhibitory effects on the growth of *Mycobacterium tuberculosis*. Consequently, it may be regarded as a promising candidate for the development of anti-TB drugs for the management of MDR and XDR tuberculosis [20,21].

However, its mode of action still needs to be elucidated. It has recently been discovered that PLB mediates its anti-TB response by targeting the enzyme thymidylate synthase (ThyX), which is responsible for synthesizing dTMP from dUMP. ThyX is found selectively in pathogenic bacteria like *Mtb*, and is an essential enzyme for their survival. PLB inhibits the activity of ThyX and causes cell death by disrupting the intracellular [dTTP]/[dATP] ratio [183].

PLB effectively alters the intracellular redox potential in *Mtb*, causing oxidative stress [184]. Earlier studies have shown iron to be crucial for the survival of mycobacterial species. Peroxidase activity has been demonstrated to be highly sensitive to iron limitation, with a significant decrease observed in the presence of low iron concentrations [185]. A reduction in peroxidase activity results in the inability of isoniazid, an anti-Mtb drug, to undergo the necessary activation process [186]. In a docking study, a novel PLB-Isoniazid Analogue (PLIHZ) was observed to exhibit MIC values of 0.5 and 2.0 µg/mL under high and low iron conditions. These findings suggest that combining PLB with INH may prove an advantageous strategy for overcoming resistance. The cyclodextrin conjugate β-cyclodextrin inclusion complex (PLIHZCD) offers improved aqueous solubility and thermal stability, which are advantages in the treatment protocol [187].

4.6. Diabetes

Diabetes mellitus is an endocrine disorder that is characterized by hyperglycaemia, which is caused by a deficiency in insulin secretion, insulin resistance, or a combination of both. Insulin stimulates the uptake of glucose from the circulation into muscle and fat tissues via the stimulation of glucose transporter subtype 4 (GLUT 4), a member of the glucose transporter family predominantly expressed in the skeletal muscle, heart, and adipose tissues. In diabetes, as a consequence of the lack of insulin, GLUT4 is not translocated from the internal membrane to the plasma membrane, rendering GLUT4 ineffective. Thus, decreased glucose uptake by these cells results in elevated glucose levels in the blood. PLB enhanced the protein and mRNA levels of GLUT4 in diabetic rats.

PLB also enhanced the translocation of GLUT4 and thus maintained glucose homeostasis. It also significantly reduced blood glucose levels and altered all other biochemical parameters to near-normal in STZ-induced diabetic rats. Further, PLB significantly increased hexokinase activity and decreased the activities of glucose-6-phosphatase and fructose-1,6-bisphosphatase in treated diabetic rats. The findings suggest that PLB may warrant further investigation as a potential therapeutic agent for the management of diabetes [188].

Increased glucose concentration, i.e., hyperglycaemia, leads to decreased collagen production, reduced chemotaxis, and adverse effects on wound healing [189]. Treatment with PLB in diabetic rats showed elevated levels of serum insulin, collagen deposition, and antioxidant status. PLB-treated mice showed increased protein content compared to the control, resulting in increased collagen content. High lipid levels in the blood further characterize diabetes, due to reduced levels of lipoprotein lipase in the blood. Diabetes causes the levels of total cholesterol, triglycerides, and LDL to increase and lowers the levels of HDL. Treatment with PLB significantly decreased lipid peroxides and lipid levels and increased HDL levels in the blood. Nrf2 plays an important role in wound healing. In hyperglycaemic conditions, dysfunction of keap1 leads to a decline in the levels of nrf2. This leads to impaired redox homeostasis and delayed wound healing. In diabetic mice, PLB is found to upregulate nrf2 expression levels and decrease keap1 mRNA levels. Thus, PLB administration could serve as a potent antidiabetic agent [190].

4.7. Other Diseases

In addition, PLB also plays a vital role as a therapeutic agent in diseases like malaria and obesity. A study on chloroquine-resistant and sensitive P. falciparum demonstrated that PLB administered at 25 mg/kg body weight for 4 days exhibits safe but low-intensity antimalarial efficacy. Thus, chemical formulations derived from the parent compound could help improve its bioavailability. In a study to investigate the role of PLB in obesity and NAFLD in mice, it was found that PLB treatment for 8 weeks improved insulin resistance and dyslipidaemia, and the mice showed a significant reduction in their body weight and obesity. This could be due to its anti-inflammatory and antioxidant properties and ability to suppress de novo lipogenesis and promote fat oxidation.

5. Anti-Oxidant Activity and Anti-Inflammatory Activity of H_2O_2

ROS, at optimum levels, is known to support cell proliferation. Still, at increased levels, it increases DNA damage and mutagenesis in cancer cells, a possible mechanism PLB uses to exert its effect. PLB is a highly potent ROS inducer that shifts the host redox potential [17]. PLB was observed to induce DNA damage and apoptosis in cells of diverse mutational backgrounds, with comparable potency [191].

Oxidative stress is significantly associated with almost all inflammatory conditions, certain cancers, aging, neurological disorders, etc. Oxidative stress is defined as an imbalance between the production of ROS and their elimination by protective mechanisms. Oxidative stress can lead to the activation of various transcription factors, causing differential expression of some genes involved in inflammatory pathways, eventually leading to chronic inflammation. The inflammation resulting from oxidative stress can lead to many chronic diseases [192]. ROS at optimum levels is known to support cell proliferation, but at increased levels it increases DNA damage and mutagenesis in cancer cells, a possible mechanism used by PLB to exert its effect.

PLB is a potent ROS inducer that acts by shifting the host redox potential. It induces DNA damage and apoptosis in cells with diverse mutational backgrounds. Despite its role as a ROS inducer, PLB also exhibits antioxidant, anti-inflammatory, and anticancer properties [17]. In PC12 cells, PLB increased cell viability against H_2O_2-induced cell death by reducing oxidative stress and activating p-Nrf-2 levels. Additionally, PLB demonstrated anti-inflammatory effects by suppressing and activating NF-κB p65, downregulating the expression of COX-2 [191].

PLB also modulated inflammatory cytokine expression in response to H_2O_2-induced neurotoxic effects. It exhibited hepatoprotective activity by dampening HMGB1 expression. In a study on murine schistosomiasis, PLB treatment significantly reduced cytokine levels, restored hepatic enzyme activity, and increased antioxidant levels [193]. PLB led to a reduction in MDA levels while increasing SOD and GSH-PX levels. It also downregulated NOX4 mRNA, procollagen I mRNA, and the protein expression of NOX4 and p-IκB, as well as decreased NF-κB transcriptional activity in liver fibrosis rats [194].

PLB lowered the expression levels of pro-inflammatory markers, indicating its potential as a therapeutic agent for neurodegenerative diseases. In H_2O_2-induced chondrocytes, PLB significantly reduced oxidative stress, modulated redox and inflammation regulation transcription factors, and enhanced antioxidant defences [195]. It also demonstrated anti-inflammatory effects by downregulating COX-2, iNOS, and pro-inflammatory cytokines [153].

These findings suggest a protective role for PLB against H_2O_2-induced oxidative stress and inflammation by modulating redox signalling transcription factors. In NPCs, PLB increased viability and reduced ROS production, lipid peroxidation, and pro-inflammatory cytokine levels while elevating GSH levels and enhancing antioxidant enzyme activities. It inhibited caspase-9 and caspase-3 activity, downregulated NF-κB, and upregulated Nrf-2 expression [196], and, in addition, the neuroprotective effects of PLB in NPCs by mitigating H_2O_2-induced oxidative stress, inflammation, and apoptosis through the regulation of NF-κB and Nrf-2 expression. PLB also modulates redox imbalance induced by I/R injury by regulating these transcription factors and their downstream targets. Despite its chemotoxic mechanism, PLB's activation of Nrf2 enables it to act as a chemopreventive agent.

6. Formulations and Binding Partners

Pharmaceutical formulation is a multi-step process. In this process, the active drug is mixed with all other components according to particle size, polymorphism, pH, and solubility, resulting in the final beneficial medicinal product. Novel formulations of PLB include liposomes, liposomes, microspheres, nanoparticles, micelles, metal nanoparticles, crystal modification, etc. Table 3 provides the list of formulations of PLB developed for therapeutic purposes.

Table 3. Plumbagin formulations.

S. No.	Category	Name of Formulation	Effect on the Therapeutic Role of PLB	Disease	Reference
1	Micelle	PLB-loaded micelles (M-PLB) PCL-PEG-PCL	There was an eightfold increase in anti-plasmodial activity.	Malaria	[197]
		Tween® 80	Sustained release of PLB, enhanced antitumor activity.	Breast Cancer	[198]
		PTM & PTEM	Increased bioavailability and circulation, no blood toxicity.	Breast Cancer	[199]
2	Liposomes	Lipo-PTEN-Plum nanoliposomes	Restoration of PTEN, G2/M cell cycle arrest, and cell death via inhibition of PI3K/AKT pathway.	Hepatocellular carcinoma	[200]
		PLB-loaded long circulating pegylated liposomes	No tissue toxicity.	Cancer	[201]
		Transferrin-bearing liposomes	Increased uptake, improved antiproliferative and apoptotic activity.	Cancer	[202]
		PLB and genistein	Inhibits tumour growth by ~80%.	Prostate cancer	[203]
		CelePlum-777	Stable release, decreased levels of key cyclins.	Melanoma	[95]
		Glycerosome	Deeper skin-layer penetration, higher drug accumulation.	Skin cancer	[204]
3	Nano-emulsion	Self-emulsifying drug-delivery system.	Higher bioavailability,	Anti-inflammatory	[205]
		Capryol 90-based and Oleic-acid-based nanoemulsion.	high drug-loading capacity with enhanced cytotoxicity.	Prostate cancer	[23]
4	Nanoparticles	BSA@PLB-NPs	Cytotoxicity against cancer cells.	Breast cancer	
		PLB-AgNPs	Enhanced internalization, antimitotic and antiproliferative.	Breast Cancer	[206]
		Plumbagin Entrapped in Transferrin-Conjugated, Lipid–Polymer Hybrid Nanoparticles	Disappearance, along with regression, of tumour in mice.	Melanoma	[80]
5	Niosome	P-Ns-Opt	Controlled release inhibits oxygen radicals, α-amylase, and α-glucosidase enzymes.	Diabetes.	[207]

Table 3. *Cont.*

S. No.	Category	Name of Formulation	Effect on the Therapeutic Role of PLB	Disease	Reference
6	Microspheres	Chitosan microspheres	Increase in elimination half-life of PLB.	Melanoma	[208]
7	Metal complex	Cu-PLB	Increased cell specificity and cytotoxicity, induction of ROS, and DNA damage.	Breast Cancer	[209]
		Cu1-Cu4	Mitochondria dysfunction, and apoptosis, cell cycle arrest at S phase.	Cervical carcinoma	[210]

6.1. Micelles

PLB release studies demonstrated a sustained-release pattern in PLB-loaded micelles (M-PLB). These micelles exhibited a higher drug release rate in acidic conditions compared to neutral conditions. In vivo, the M-PLB against *P. berghei* showed an eight-fold increase in anti-plasmodial activity compared to free PLB at the tested dosage level on the seventh day. These findings suggest that PCL-PEG-PCL micelles are promising carriers for PLB in malaria-targeting applications [197]. Tween® 80 micelles also demonstrated sustained release of PLB. These micelles caused a two-fold enhancement in in vitro antitumor activity of PLB towards MCF-7 cells. The micelles were safe for intravenous injection, as PLB remained stable at high pH, and their size and encapsulation efficiency were retained upon dilution [198].

PLB into TPGS micelles without folic acid conjugates and PLB into TPGS micelles with folic acid conjugates increased PLB bioavailability 3.8- and 4.8-fold, respectively. These micelles exhibited extended circulation time, slower plasma clearance, and no evidence of blood and tissue toxicity as compared to those of free PLB. Micelles also demonstrated higher in vitro anticancer activity in folate-overexpressing human breast cancer MCF-7 cells [199].

6.2. Liposomes and Nano-Liposomes

PLB has shown good antitumor efficacy as a long-lasting circulating liposome with no evidence of normal tissue toxicity [201]. Co-delivery of PLB with PTEN plasmids in nanoliposomes (Lipo-PTEN-Plum) causes G2/M cell-cycle-arrest DNA damage and inhibits PI3K/AKT pathway, leading to apoptosis in hepatic cancer cells [200]. Another novel nano-liposomal formulation, containing PLB and the genistein drug, synergistically inhibits xenograft prostate tumour growth by ~80% without any significant toxicity, as well as decreasing the number of Glut-1 transporters for retarding tumour growth [203].

PLB entrapped in transferrin-bearing liposomes resulted in increased PLB uptake by cancer cells. This enhanced uptake led to improved antiproliferative and apoptotic activity in B16-F10, A431, and T98G cell lines, compared to the drug solution. In vivo, intravenous injection of PLB-encapsulated transferrin-loaded liposomes resulted in tumour suppression in 10% of B16-F10 tumours and tumour regression in a further 10%, without any evidence of toxicity [202].

A nanoliposomal-based formulation of PLB with celecoxib, called CelePlum-777, was found to be stable and released these drugs in an optimal ratio for maximum synergistic killing of melanoma cells over normal cells. This formulation inhibited xenograft melanoma tumour growth by up to 72%, without any evident toxicity. The drug reduced levels of key cyclins involved in cancer cell proliferation and survival, a phenomenon that was not observed with the individual agents [201]. PLB-loaded glycerosome gel-treated rat skin showed significantly higher drug accumulation in the dermis, higher cytotoxicity, and higher antioxidant activity compared to conventional liposome gel and PLB suspension [204].

6.3. Nanoemulsions

A self-nanoemulsifying drug delivery technique improved the solubility and oral bioavailability of PLB. A study confirmed enhanced activity compared to pure PLB. A pharmacokinetic study in rats showed that a solid self-nanoemulsifying drug delivery system had more than 4-fold higher bioavailability than that of PLB alone. An ex vivo permeation study revealed that the self-nanoemulsifying drug delivery system had almost twice the intestinal permeability as that of pure PLB [205]. Novel nanoemulsion formulation of PLB based on Capryol 90 and oleic acid has shown high drug loading capacity and enhanced cytotoxicity against prostate cancer cells, PTEN-P2, compared to free PLB [23,24].

6.4. Nanoparticles

PLB-loaded Bovine Serum Albumin nanoparticles enhanced the bioavailability and decreased the toxicity of the hydrophobic drug PLB. In in silico studies, stable binding interactions between PLB and BSA are observed. BSA@PLB-NPs showed potential cytotoxicity against breast cancer cells [211]. Enhanced internalization of PLB into the HeLa cells was observed in PLB-AgNPs. PLB inhibited the proliferation of cells in a concentration-dependent manner. PLB was also observed to inhibit the clonogenic survival of cells following drug exposure, and to induce apoptosis. Furthermore, the antiproliferative, antimitotic, and apoptotic activities were enhanced when cells were treated with PLB AgNPs [195]. The intravenous administration of transferrin-loaded lipid–polymer hybrid nanoparticles loaded with PLB resulted in the disappearance of 40% of B16-F10 tumours and the regression of 10% of tumours. The mice exhibited no adverse reactions to the treatment. [212].

6.5. Niosomes

PLB-loaded niosomes (P-Ns-Opt) reveal a controlled release system and potential antidiabetic activity by inhibiting oxygen radicals, α-amylase, and α-glucosidase enzymes [207].

6.6. Microspheres

The results of the pharmacokinetic studies demonstrated a 22.2-fold increase in the elimination half-life (t(1/2)) of PLB from chitosan microspheres, in comparison to free PLB. The administration of PLB microspheres was observed to result in significant inhibition of tumour growth and a reduction in systemic toxicity [208].

6.7. Metal Complexes

Studies show that the Cu-PLB complex has increased cell specificity, better pharmacokinetic profile, and increased cytotoxicity. The copper complex of PLB (Cu-PLB) shows antiproliferative activity in human breast cancer cells (MCF-7) with an IC-50 of 2.3 ± 0.1, stronger than PLB alone (8.2 ± 0.2), as well as Cisplatin, a widely used anticancer drug. Cu-PLB inhibits the proliferation of HeLa, MCF-7, and murine melanoma (B16F10) cells with half-maximal inhibitory concentrations (IC50), lower than those achieved with PLB alone. Cu-PLB caused microtubule disassembly, induction of ROS, and DNA damage. In breast cancer, Cu-PLB induces apoptosis through MAPK-mediated inhibition of anti-apoptotic protein Mcl-1, thereby inhibiting cancer progression [209,213]. Four copper(II)-plumbagin and -bipyridine complexes (Cu1-Cu4) showed enhanced anti-tumour activity by accumulation in mitochondria, causing their dysfunction, activating caspase-9/3, and inducing apoptosis of cancer cells [210].

7. Limitations and Challenges

Like many phytochemicals, PLB exhibits both pro-oxidant and antioxidant properties, necessitating careful consideration of its effects on cells and tissues, particularly with respect to dosage and in vitro factors such as pH, media composition, oxidative stress, and oxygen amounts. These factors can reveal unexpected biological effects by uncovering hidden mechanisms of action [31]. Therefore, before any clinical application, it is crucial to identify bioavailable and safe doses that minimize risks for patients. Further chemical modifications of PLB may enable targeted specificity for particular cell types, or enhance its bioavailability for in vivo studies.

However, PLB faces several challenges that hinder its clinical translation, including poor water solubility (79 µg/mL), high lipophilicity (log P 3.04), unstable nature (spontaneous sublimation), and low oral bioavailability (less than 40%). Additionally, its therapeutic concentration in tumours is difficult to achieve, due to its low specificity and rapid elimination [202]. PLB accelerated xanthine oxidation in mouse liver S9 (MLS9), human liver S9 (HLS9), and XO monoenzyme system. PLB was shown to be well-bound to XO. In addition, in vivo studies have demonstrated that PLB significantly increased serum uric acid levels and enriched serum XO activity in mice [214]. Despite being a potential bioactive molecule, PLB still finds limited use in medicine. While many formulations have provided a solution to overcome the limitations, further research will help put PLB on the market as an efficient drug.

8. Conclusions and Prospects

PLB, a member of the naphthoquinones, has been broadly utilized as conventional medicine, owing to its various health benefits. Through multiple studies, it has been demonstrated as a potential therapeutic drug for diseases including cancer, neurodegenerative disorders, genetic disorders, and lifestyle diseases. PLB has been reported to play a significant role in preventing a broad range of malignancies, including cancers with high mortality rates, such as breast cancer, lung cancer, etc. PLB has shown promising results via its anti-proliferative, anti-angiogenesis, anti-metastatic activities, and induction of apoptosis in cancer cells.

Several studies have demonstrated that when PLB is used in combination with other drugs, it displays enhanced antitumor effects, reinforcing its potential in cancer therapy. The therapeutic roles of PLB are not limited to cancer. The versatility of its medicinal properties lies in its antioxidant and anti-inflammatory nature. PLB has been found to play curative and restorative roles in neurodegenerative diseases, genetic diseases, lifestyle diseases, and viral and bacterial diseases.

Despite its benefits and cytoprotective activity in various diseases, PLB remains in the cradle of drug research. Its poor pharmacokinetics limit human use. Low bioavailability, lack of specificity, and some reports on toxicity in mice have restricted the promotion of PLB as an industrially important drug. Various strategies have been implemented to overcome these limitations, such as nanoparticles, emulsions, metal complexes, etc. These formulations have ensured more efficient and safer delivery of PLB with enhanced therapeutic power.

PLB represents a novel anti-tumour drug with the potential to yield promising results in both pre-clinical and clinical trials. Therefore, future research should concentrate on laboratory studies, with the aim of expanding in-depth research on the molecular mechanism of PLB and developing a strategy for creating potent formulations that minimize toxicity and maximize efficacy.

Author Contributions: Conceptualization, B.S. and M.I.H.; methodology, B.S. and C.D.; validation, A.S. and M.I.H.; formal analysis, G.M.H.; investigation, B.S. and C.D.; resources, M.I.H.; data curation, G.M.H. and A.S.; writing—original draft preparation, B.S. and C.D.; writing—review and editing, G.M.H., A.S. and M.I.H.; visualization, G.M.H.; supervision, M.I.H.; project administration, M.I.H.; funding acquisition, C.D. and G.M.H. All authors have read and agreed to the published version of the manuscript.

Funding: This work is supported by the Central Council for Research in Unani Medicine (CCRUM), Ministry of AYUSH, Government of India (Grant No. 3-69/2020-CCRUM/Tech). Ajman University funded the APC.

Institutional Review Board Statement: Not applicable.

Informed Consent Statement: Not applicable.

Data Availability Statement: All data generated or analysed during this study are included in this manuscript.

Conflicts of Interest: The authors declare no conflicts of interest.

References

1. Atanasov, A.G.; Zotchev, S.B.; Dirsch, V.M.; Supuran, C.T. Natural products in drug discovery: Advances and opportunities. *Nat. Rev. Drug Discov.* **2021**, *20*, 200–216. [CrossRef]
2. Guerriero, G.; Berni, R.; Muñoz-Sanchez, J.A.; Apone, F.; Abdel-Salam, E.M.; Qahtan, A.A.; Alatar, A.A.; Cantini, C.; Cai, G.; Hausman, J.F.; et al. Production of Plant Secondary Metabolites: Examples, Tips and Suggestions for Biotechnologists. *Genes* **2018**, *9*, 309. [CrossRef]
3. Zhang, L.; Zhang, G.; Xu, S.; Song, Y. Recent advances of quinones as a privileged structure in drug discovery. *Eur. J. Med. Chem.* **2021**, *223*, 113632. [CrossRef]
4. Lu, J.-J.; Bao, J.-L.; Wu, G.-S.; Xu, W.-S.; Huang, M.-Q.; Chen, X.-P.; Wang, Y.-T. Quinones Derived from Plant Secondary Metabolites as Anti-cancer Agents. *Anti-Cancer Agents Med. Chem.* **2013**, *13*, 456–463.
5. Dar, U.A.; Kamal, M.; Shah, S.A. 1, 4-Naphthoquinone: A Privileged Structural Framework in Drug Discovery. *Chem. Biol. Potent Nat. Prod. Synth. Compd.* **2021**, 133–153. [CrossRef]
6. Qiu, H.Y.; Wang, P.F.; Lin, H.Y.; Tang, C.Y.; Zhu, H.L.; Yang, Y.H. Naphthoquinones: A continuing source for the discovery of therapeutic antineoplastic agents. *Chem. Biol. Drug Des.* **2018**, *91*, 681–690. [CrossRef]
7. van der Vijver, L.M. Distribution of plumbagin in the mplumbaginaceae. *Phytochemistry* **1972**, *11*, 3247–3248. [CrossRef]
8. Gupta, A.C.; Mohanty, S.; Saxena, A.; Maurya, A.K.; Bawankule, D.U. Plumbagin, a vitamin K3 analogue, ameliorates malaria pathogenesis by inhibiting oxidative stress and inflammation. *Inflammopharmacology* **2018**, *26*, 983–991. [CrossRef] [PubMed]
9. Rajalakshmi, S.; Vyawahare, N.; Pawar, A.; Mahaparale, P.; Chellampillai, B. Current development in novel drug delivery systems of bioactive molecule plumbagin. *Artif. Cells Nanomed. Biotechnol.* **2018**, *46*, 209–218. [CrossRef]
10. Yin, Z.; Zhang, J.; Chen, L.; Guo, Q.; Yang, B.; Zhang, W.; Kang, W. Anticancer Effects and Mechanisms of Action of Plumbagin: Review of Research Advances. *Biomed. Res. Int.* **2020**, *2020*, 6940953. [CrossRef]
11. Roy, A. Plumbagin: A Potential Anti-cancer Compound. *Mini-Rev. Med. Chem.* **2021**, *21*, 731–737. [CrossRef] [PubMed]
12. Hwang, G.H.; Ryu, J.M.; Jeon, Y.J.; Choi, J.; Han, H.J.; Lee, Y.M.; Lee, S.; Bae, J.S.; Jung, J.W.; Chang, W.; et al. The role of thioredoxin reductase and glutathione reductase in plumbagin-induced, reactive oxygen species-mediated apoptosis in cancer cell lines. *Eur. J. Pharmacol.* **2015**, *765*, 384–393. [CrossRef]
13. Kang, C.G.; Im, E.; Lee, H.J.; Lee, E.O. Plumbagin reduces osteopontin-induced invasion through inhibiting the Rho-associated kinase signaling pathway in A549 cells and suppresses osteopontin-induced lung metastasis in BalB/c mice. *Bioorg Med. Chem. Lett.* **2017**, *27*, 1914–1918. [CrossRef]
14. Yu, T.; Xu, Y.Y.; Zhang, Y.Y.; Li, K.Y.; Shao, Y.; Liu, G. Plumbagin suppresses the human large cell lung cancer cell lines by inhibiting IL-6/STAT3 signaling in vitro. *Int. Immunopharmacol.* **2018**, *55*, 290–296. [CrossRef] [PubMed]
15. De, U.; Son, J.Y.; Jeon, Y.; Ha, S.Y.; Park, Y.J.; Yoon, S.; Ha, K.T.; Choi, W.S.; Lee, B.M.; Kim, I.S.; et al. Plumbagin from a tropical pitcher plant (Nepenthes alata Blanco) induces apoptotic cell death via a p53-dependent pathway in MCF-7 human breast cancer cells. *Food Chem. Toxicol.* **2019**, *123*, 492–500. [CrossRef] [PubMed]
16. Messeha, S.S.; Zarmouh, N.O.; Mendonca, P.; Alwagdani, H.; Kolta, M.G.; Soliman, K.F.A. The inhibitory effects of plumbagin on the NF-κB pathway and CCL2 release in racially different triple-negative breast cancer cells. *PLoS ONE* **2018**, *13*, e0201116. [CrossRef] [PubMed]
17. Nadhan, R.; Patra, D.; Krishnan, N.; Rajan, A.; Gopala, S.; Ravi, D.; Srinivas, P. Perspectives on mechanistic implications of ROS inducers for targeting viral infections. *Eur. J. Pharmacol.* **2021**, *890*, 173621. [CrossRef]
18. Hassan, M.I.; Mohamed, A.F.; Taher, F.A.; Kamel, M.R. Antimicrobial Activities of Chitosan Nanoparticles Prepared-from Lucila Cuprina Maggots (Diptera: Calliphoridae). *J. Egypt Soc. Parasitol.* **2016**, *46*, 563–570.
19. Wang, T.; Wu, F.; Jin, Z.; Zhai, Z.; Wang, Y.; Tu, B.; Yan, W.; Tang, T. Plumbagin inhibits LPS-induced inflammation through the inactivation of the nuclear factor-kappa B and mitogen activated protein kinase signaling pathways in RAW 264.7 cells. *Food Chem. Toxicol.* **2014**, *64*, 177–183. [CrossRef]
20. Sharma, D.; Yadav, J.P. An Overview of Phytotherapeutic Approaches for the Treatment of Tuberculosis. *Mini Rev. Med. Chem.* **2017**, *17*, 167–183. [CrossRef]
21. Dey, D.; Ray, R.; Hazra, B. Antitubercular and antibacterial activity of quinonoid natural products against multi-drug resistant clinical isolates. *Phytother. Res.* **2014**, *28*, 1014–1021. [CrossRef] [PubMed]
22. Hafeez, B.B.; Zhong, W.; Fischer, J.W.; Mustafa, A.; Shi, X.; Meske, L.; Hong, H.; Cai, W.; Havighurst, T.; Kim, K.; et al. Plumbagin, a medicinal plant (Plumbago zeylanica)-derived 1,4-naphthoquinone, inhibits growth and metastasis of human prostate cancer PC-3M-luciferase cells in an orthotopic xenograft mouse model. *Mol. Oncol.* **2013**, *7*, 428–439. [CrossRef] [PubMed]
23. Chrastina, A.; Welsh, J.; Borgström, P.; Baron, V.T. Propylene Glycol Caprylate-Based Nanoemulsion Formulation of Plumbagin: Development and Characterization of Anticancer Activity. *Biomed. Res. Int.* **2022**, *2022*, 3549061. [CrossRef]

24. Chrastina, A.; Baron, V.T.; Abedinpour, P.; Rondeau, G.; Welsh, J.; Borgström, P. Plumbagin-Loaded Nanoemulsion Drug Delivery Formulation and Evaluation of Antiproliferative Effect on Prostate Cancer Cells. *Biomed. Res. Int.* **2018**, *2018*, 9035452. [CrossRef] [PubMed]

25. Springob, K.; Samappito, S.; Jindaprasert, A.; Schmidt, J.; Page, J.E.; De-Eknamkul, W.; Kutchan, T.M. A polyketide synthase of Plumbago indica that catalyzes the formation of hexaketide pyrones. *Febs J.* **2007**, *274*, 406–417. [CrossRef]

26. Vasav, A.P.; Meshram, B.G.; Pable, A.A.; Barvkar, V.T. Artificial microRNA mediated silencing of cyclase and aldo–keto reductase genes reveal their involvement in the plumbagin biosynthetic pathway. *J. Plant Res.* **2023**, *136*, 47–62. [CrossRef]

27. Sumsakul, W.; Plengsuriyakarn, T.; Na-Bangchang, K. Pharmacokinetics, toxicity, and cytochrome P450 modulatory activity of plumbagin. *BMC Pharmacol. Toxicol.* **2016**, *17*, 50. [CrossRef]

28. Tripathi, S.K.; Panda, M.; Biswal, B.K. Emerging role of plumbagin: Cytotoxic potential and pharmaceutical relevance towards cancer therapy. *Food Chem. Toxicol.* **2019**, *125*, 566–582. [CrossRef]

29. Singh, A.; Agrawal, S.; Patwardhan, K.; Gehlot, S. Overlooked contributions of Ayurveda literature to the history of physiology of digestion and metabolism. *Hist. Philos. Life Sci.* **2023**, *45*, 13. [CrossRef]

30. Bakery, H.H.; Allam, G.A.; Abuelsaad, A.S.A.; Abdel-Latif, M.; Elkenawy, A.E.; Khalil, R.G. Anti-inflammatory, antioxidant, anti-fibrotic and schistosomicidal properties of plumbagin in murine schistosomiasis. *Parasite Immunol.* **2022**, *44*, 6. [CrossRef]

31. Petrocelli, G.; Marrazzo, P.; Bonsi, L.; Facchin, F.; Alviano, F.; Canaider, S. Plumbagin, a Natural Compound with Several Biological Effects and Anti-Inflammatory Properties. *Life* **2023**, *13*, 1303. [CrossRef] [PubMed]

32. Sridhar, J.; Sfondouris, M.E.; Bratton, M.R.; Nguyen, T.-L.K.; Townley, I.; Klein Stevens, C.L.; Jones, F.E. Identification of quinones as HER2 inhibitors for the treatment of trastuzumab resistant breast cancer. *Bioorganic Med. Chem. Lett.* **2014**, *24*, 126–131. [CrossRef] [PubMed]

33. Dandawate, P.; Khan, E.; Padhye, S.; Gaba, H.; Sinha, S.; Deshpande, J.; Venkateswara Swamy, K.; Khetmalas, M.; Ahmad, A.; Sarkar, F.H. Synthesis, characterization, molecular docking and cytotoxic activity of novel plumbagin hydrazones against breast cancer cells. *Bioorganic Med. Chem. Lett.* **2012**, *22*, 3104–3108. [CrossRef]

34. Kawiak, A.; Zawacka-Pankau, J.; Lojkowska, E. Plumbagin Induces Apoptosis in Her2-Overexpressing Breast Cancer Cells through the Mitochondrial-Mediated Pathway. *J. Nat. Prod.* **2012**, *75*, 747–751. [CrossRef]

35. Fiorito, S.; Genovese, S.; Taddeo, V.A.; Mathieu, V.; Kiss, R.; Epifano, F. Novel juglone and plumbagin 5-O derivatives and their in vitro growth inhibitory activity against apoptosis-resistant cancer cells. *Bioorganic Med. Chem. Lett.* **2016**, *26*, 334–337. [CrossRef] [PubMed]

36. Bao, N.; Ou, J.; Li, N.; Zou, P.; Sun, J.; Chen, L. Novel anticancer hybrids from diazen-1-ium-1,2-diolate nitric oxide donor and ROS inducer plumbagin: Design, synthesis and biological evaluations. *Eur. J. Med. Chem.* **2018**, *154*, 1–8. [CrossRef]

37. Raghu, D.; Karunagaran, D. Plumbagin Downregulates Wnt Signaling Independent of p53 in Human Colorectal Cancer Cells. *J. Nat. Prod.* **2014**, *77*, 1130–1134. [CrossRef]

38. Bhasin, D.; Chettiar, S.N.; Etter, J.P.; Mok, M.; Li, P.-K. Anticancer activity and SAR studies of substituted 1,4-naphthoquinones. *Bioorganic Med. Chem.* **2013**, *21*, 4662–4669. [CrossRef]

39. Schepetkin, I.A.; Karpenko, A.S.; Khlebnikov, A.I.; Shibinska, M.O.; Levandovskiy, I.A.; Kirpotina, L.N.; Danilenko, N.V.; Quinn, M.T. Synthesis, anticancer activity, and molecular modeling of 1,4-naphthoquinones that inhibit MKK7 and Cdc25. *Eur. J. Med. Chem.* **2019**, *183*, 111719. [CrossRef]

40. Shen, C.-C.; Afraj, S.N.; Hung, C.-C.; Barve, B.D.; Kuo, L.-M.Y.; Lin, Z.-H.; Ho, H.-O.; Kuo, Y.-H. Synthesis, biological evaluation, and correlation of cytotoxicity versus redox potential of 1,4-naphthoquinone derivatives. *Bioorganic Med. Chem. Lett.* **2021**, *41*, 127976. [CrossRef]

41. Panichayupakaranant, P.; Ahmad, M.I. Plumbagin and its role in chronic diseases. In *Drug Discovery from Mother Nature*; Springer: Cham, Switzerland, 2016; pp. 229–246.

42. Avila-Carrasco, L.; Majano, P.; Sánchez-Toméro, J.A.; Selgas, R.; López-Cabrera, M.; Aguilera, A.; González Mateo, G. Natural Plants Compounds as Modulators of Epithelial-to-Mesenchymal Transition. *Front. Pharmacol.* **2019**, *10*, 715. [CrossRef] [PubMed]

43. Wilkinson, L.; Gathani, T. Understanding breast cancer as a global health concern. *Br. J. Radiol.* **2022**, *95*, 20211033. [CrossRef]

44. Kuper, H.; Adami, H.O.; Trichopoulos, D. Infections as a major preventable cause of human cancer. *J. Intern. Med.* **2000**, *248*, 171–183. [CrossRef] [PubMed]

45. Zubair, A.; Frieri, M. Role of nuclear factor-κB in breast and colorectal cancer. *Curr. Allergy Asthma Rep.* **2013**, *13*, 44–49. [CrossRef] [PubMed]

46. Matsushima, K.; Larsen, C.G.; DuBois, G.C.; Oppenheim, J.J. Purification and characterization of a novel monocyte chemotactic and activating factor produced by a human myelomonocytic cell line. *J. Exp. Med.* **1989**, *169*, 1485–1490. [CrossRef]

47. Liu, P.; Kumar, I.S.; Brown, S.; Kannappan, V.; Tawari, P.E.; Tang, J.Z.; Jiang, W.; Armesilla, A.L.; Darling, J.L.; Wang, W. Disulfiram targets cancer stem-like cells and reverses resistance and cross-resistance in acquired paclitaxel-resistant triple-negative breast cancer cells. *Br. J. Cancer* **2013**, *109*, 1876–1885. [CrossRef]

48. Sakunrangsit, N.; Ketchart, W. Plumbagin inhibits cancer stem-like cells, angiogenesis and suppresses cell proliferation and invasion by targeting Wnt/β-catenin pathway in endocrine resistant breast cancer. *Pharmacol. Res.* **2019**, *150*, 3. [CrossRef]

49. Li, Y.; Li, L.; Chen, M.; Yu, X.; Gu, Z.; Qiu, H.; Qin, G.; Long, Q.; Fu, X.; Liu, T.; et al. MAD2L2 inhibits colorectal cancer growth by promoting NCOA3 ubiquitination and degradation. *Mol. Oncol.* **2018**, *12*, 391–405. [CrossRef]

50. Koh, M.Y.; Powis, G. Passing the baton: The HIF switch. *Trends Biochem. Sci.* **2012**, *37*, 364–372. [CrossRef]

51. Muz, B.; de la Puente, P.; Azab, F.; Azab, A.K. The role of hypoxia in cancer progression, angiogenesis, metastasis, and resistance to therapy. *Hypoxia* **2015**, *3*, 83–92. [CrossRef]
52. Jing, X.; Yang, F.; Shao, C.; Wei, K.; Xie, M.; Shen, H.; Shu, Y. Role of hypoxia in cancer therapy by regulating the tumor microenvironment. *Mol. Cancer* **2019**, *18*, 157. [CrossRef]
53. Jampasri, S.; Reabroi, S.; Tungmunnithum, D.; Parichatikanond, W.; Pinthong, D. Plumbagin Suppresses Breast Cancer Progression by Downregulating HIF-1α Expression via a PI3K/Akt/mTOR Independent Pathway under Hypoxic Condition. *Molecules* **2022**, *27*, 5716. [CrossRef] [PubMed]
54. Sakunrangsit, N.; Ketchart, W. Plumbagin inhibited AKT signaling pathway in HER-2 overexpressed-endocrine resistant breast cancer cells. *Eur. J. Pharmacol.* **2020**, *868*, 19. [CrossRef] [PubMed]
55. Hanahan, D.; Weinberg, R.A. Hallmarks of cancer: The next generation. *Cell* **2011**, *144*, 646–674. [CrossRef]
56. Sperandio, S.; de Belle, I.; Bredesen, D.E. An alternative, nonapoptotic form of programmed cell death. *Proc. Natl. Acad. Sci. USA* **2000**, *97*, 14376–14381. [CrossRef] [PubMed]
57. Binoy, A.; Nedungadi, D.; Katiyar, N.; Bose, C.; Shankarappa, S.A.; Nair, B.G.; Mishra, N. Plumbagin induces paraptosis in cancer cells by disrupting the sulfhydryl homeostasis and proteasomal function. *Chem. Biol. Interact.* **2019**, *310*, 2. [CrossRef]
58. Klotz, L.O.; Hou, X.; Jacob, C. 1,4-naphthoquinones: From oxidative damage to cellular and inter-cellular signaling. *Molecules* **2014**, *19*, 14902–14918. [CrossRef]
59. Yoon, S.; Seger, R. The extracellular signal-regulated kinase: Multiple substrates regulate diverse cellular functions. *Growth Factors* **2006**, *24*, 21–44. [CrossRef]
60. Balmanno, K.; Cook, S.J. Tumour cell survival signalling by the ERK1/2 pathway. *Cell Death Differ.* **2009**, *16*, 368–377. [CrossRef]
61. Perez, E.A. Doxorubicin and paclitaxel in the treatment of advanced breast cancer: Efficacy and cardiac considerations. *Cancer Investig.* **2001**, *19*, 155–164. [CrossRef]
62. Okano, J.; Rustgi, A.K. Paclitaxel induces prolonged activation of the Ras/MEK/ERK pathway independently of activating the programmed cell death machinery. *J. Biol. Chem.* **2001**, *276*, 19555–19564. [CrossRef] [PubMed]
63. Kawiak, A.; Domachowska, A.; Lojkowska, E. Plumbagin Increases Paclitaxel-Induced Cell Death and Overcomes Paclitaxel Resistance in Breast Cancer Cells through ERK-Mediated Apoptosis Induction. *J. Nat. Prod.* **2019**, *82*, 878–885. [CrossRef] [PubMed]
64. Lumachi, F.; Luisetto, G.; Basso, S.M.; Basso, U.; Brunello, A.; Camozzi, V. Endocrine therapy of breast cancer. *Curr. Med. Chem.* **2011**, *18*, 513–522. [CrossRef]
65. Zhou, H.; Zhang, Y.; Fu, Y.; Chan, L.; Lee, A.S. Novel mechanism of anti-apoptotic function of 78-kDa glucose-regulated protein (GRP78): Endocrine resistance factor in breast cancer, through release of B-cell lymphoma 2 (BCL-2) from BCL-2-interacting killer (BIK). *J. Biol. Chem.* **2011**, *286*, 25687–25696. [CrossRef] [PubMed]
66. Fu, Y.; Li, J.; Lee, A.S. GRP78/BiP inhibits endoplasmic reticulum BIK and protects human breast cancer cells against estrogen starvation-induced apoptosis. *Cancer Res.* **2007**, *67*, 3734–3740. [CrossRef] [PubMed]
67. Musgrove, E.A.; Sutherland, R.L. Biological determinants of endocrine resistance in breast cancer. *Nat. Rev. Cancer* **2009**, *9*, 631–643. [CrossRef]
68. Drake, J.M.; Strohbehn, G.; Bair, T.B.; Moreland, J.G.; Henry, M.D. ZEB1 enhances transendothelial migration and represses the epithelial phenotype of prostate cancer cells. *Mol. Biol. Cell* **2009**, *20*, 2207–2217. [CrossRef]
69. Manu, K.A.; Shanmugam, M.K.; Rajendran, P.; Li, F.; Ramachandran, L.; Hay, H.S.; Kannaiyan, R.; Swamy, S.N.; Vali, S.; Kapoor, S.; et al. Plumbagin inhibits invasion and migration of breast and gastric cancer cells by downregulating the expression of chemokine receptor CXCR4. *Mol. Cancer* **2011**, *10*, 1476–4598. [CrossRef]
70. Sakunrangsit, N.; Kalpongnukul, N.; Pisitkun, T.; Ketchart, W. Plumbagin Enhances Tamoxifen Sensitivity and Inhibits Tumor Invasion in Endocrine Resistant Breast Cancer through EMT Regulation. *Phytother. Res.* **2016**, *30*, 1968–1977. [CrossRef]
71. Ramirez, O.; Motta-Mena, L.B.; Cordova, A.; Estrada, A.; Li, Q.; Martinez, L.; Garza, K.M. A small library of synthetic di-substituted 1,4-naphthoquinones induces ROS-mediated cell death in murine fibroblasts. *PLoS ONE* **2014**, *9*, e106828. [CrossRef]
72. Sagar, S.; Esau, L.; Moosa, B.; Khashab, N.M.; Bajic, V.B.; Kaur, M. Cytotoxicity and apoptosis induced by a plumbagin derivative in estrogen positive MCF-7 breast cancer cells. *Anticancer. Agents Med. Chem.* **2014**, *14*, 170–180. [CrossRef] [PubMed]
73. Alanazi, M.Q.; Al-Jeraisy, M.; Salam, M. Severity scores and their associated factors among orally poisoned toddlers: A cross sectional single poison center study. *BMC Pharmacol. Toxicol.* **2016**, *17*, 1. [CrossRef]
74. Zhang, J.; Li, J.; Xiong, S.; He, Q.; Bergin, R.J.; Emery, J.D.; IJzerman, M.J.; Fitzmaurice, C.; Wang, X.; Li, C. Global burden of lung cancer: Implications from current evidence. *Ann. Cancer Epidemiol.* **2021**, *5*. [CrossRef]
75. Lemjabbar-Alaoui, H.; Hassan, O.U.; Yang, Y.W.; Buchanan, P. Lung cancer: Biology and treatment options. *Biochim. Biophys. Acta* **2015**, *2*, 189–210. [CrossRef]
76. Heavey, S.; Godwin, P.; Baird, A.M.; Barr, M.P.; Umezawa, K.; Cuffe, S.; Finn, S.P.; O'Byrne, K.J.; Gately, K. Strategic targeting of the PI3K-NFκB axis in cisplatin-resistant NSCLC. *Cancer Biol. Ther.* **2014**, *15*, 1367–1377. [CrossRef]
77. Tripathi, S.K.; Rengasamy, K.R.R.; Biswal, B.K. Plumbagin engenders apoptosis in lung cancer cells via caspase-9 activation and targeting mitochondrial-mediated ROS induction. *Arch. Pharm. Res.* **2020**, *43*, 242–256. [CrossRef] [PubMed]
78. Jiang, Z.B.; Xu, C.; Wang, W.; Zhang, Y.Z.; Huang, J.M.; Xie, Y.J.; Wang, Q.Q.; Fan, X.X.; Yao, X.J.; Xie, C.; et al. Plumbagin suppresses non-small cell lung cancer progression through downregulating ARF1 and by elevating CD8(+) T cells. *Pharmacol. Res.* **2021**, *169*, 6. [CrossRef] [PubMed]

79. Casalou, C.; Ferreira, A.; Barral, D.C. The Role of ARF Family Proteins and Their Regulators and Effectors in Cancer Progression: A Therapeutic Perspective. *Front. Cell Dev. Biol.* **2020**, *8*, 217. [CrossRef] [PubMed]
80. Itharat, A.; Rattarom, R.; Hansakul, P.; Sakpakdeejaroen, I.; Ooraikul, B.; Davies, N.M. The effects of Benjakul extract and its isolated compounds on cell cycle arrest and apoptosis in human non-small cell lung cancer cell line NCI-H226. *Res. Pharm. Sci.* **2021**, *16*, 129–140. [CrossRef]
81. Checker, R.; Pal, D.; Patwardhan, R.S.; Basu, B.; Sharma, D.; Sandur, S.K. Modulation of Caspase-3 activity using a redox active vitamin K3 analogue, plumbagin, as a novel strategy for radioprotection. *Free Radic. Biol. Med.* **2019**, *143*, 560–572. [CrossRef]
82. Li, Y.C.; He, S.M.; He, Z.X.; Li, M.; Yang, Y.; Pang, J.X.; Zhang, X.; Chow, K.; Zhou, Q.; Duan, W.; et al. Plumbagin induces apoptotic and autophagic cell death through inhibition of the PI3K/Akt/mTOR pathway in human non-small cell lung cancer cells. *Cancer Lett.* **2014**, *344*, 239–259. [CrossRef]
83. Nakano, K.; Vousden, K.H. PUMA, a novel proapoptotic gene, is induced by p53. *Mol. Cell* **2001**, *7*, 683–694. [CrossRef] [PubMed]
84. Ahmed, M.; Behera, R.; Chakraborty, G.; Jain, S.; Kumar, V.; Sharma, P.; Bulbule, A.; Kale, S.; Kumar, S.; Mishra, R.; et al. Osteopontin: A potentially important therapeutic target in cancer. *Expert. Opin. Ther. Targets* **2011**, *15*, 1113–1126. [CrossRef]
85. Koukourakis, M.I. Radiation damage and radioprotectants: New concepts in the era of molecular medicine. *Br. J. Radiol.* **2012**, *85*, 313–330. [CrossRef] [PubMed]
86. El-Serag, H.B. Hepatocellular carcinoma. *N. Engl. J. Med.* **2011**, *365*, 1118–1127. [CrossRef] [PubMed]
87. Li, T.; Lv, M.; Chen, X.; Yu, Y.; Zang, G.; Tang, Z. Plumbagin inhibits proliferation and induces apoptosis of hepatocellular carcinoma by downregulating the expression of SIVA. *Drug Des. Devel Ther.* **2019**, *13*, 1289–1300. [CrossRef]
88. Liu, H.; Zhang, W.; Jin, L.; Liu, S.; Liang, L.; Wei, Y. Plumbagin Exhibits Genotoxicity and Induces G2/M Cell Cycle Arrest via ROS-Mediated Oxidative Stress and Activation of ATM-p53 Signaling Pathway in Hepatocellular Cells. *Int. J. Mol. Sci.* **2023**, *24*, 6279. [CrossRef]
89. Imai, H.; Nakagawa, Y. Biological significance of phospholipid hydroperoxide glutathione peroxidase (PHGPx, GPx4) in mammalian cells. *Free Radic. Biol. Med.* **2003**, *34*, 145–169. [CrossRef]
90. Yao, L.; Yan, D.; Jiang, B.; Xue, Q.; Chen, X.; Huang, Q.; Qi, L.; Tang, D.; Liu, J. Plumbagin is a novel GPX4 protein degrader that induces apoptosis in hepatocellular carcinoma cells. *Free Radic. Biol. Med.* **2023**, *203*, 1–10. [CrossRef]
91. Zhong, J.; Li, J.; Wei, J.; Huang, D.; Huo, L.; Zhao, C.; Lin, Y.; Chen, W.; Wei, Y. Plumbagin Restrains Hepatocellular Carcinoma Angiogenesis by Stromal Cell-Derived Factor (SDF-1)/CXCR4-CXCR7 Axis. *Med. Sci. Monit.* **2019**, *25*, 6110–6119. [CrossRef]
92. Wei, Y.; Lv, B.; Xie, J.; Zhang, Y.; Lin, Y.; Wang, S.; Zhong, J.; Chen, Y.; Peng, Y.; Ma, J. Plumbagin promotes human hepatoma SMMC-7721 cell apoptosis via caspase-3/vimentin signal-mediated EMT. *Drug Des. Devel Ther.* **2019**, *13*, 2343–2355. [CrossRef] [PubMed]
93. Xue, B.; DasGupta, D.; Alam, M.; Khan, M.S.; Wang, S.; Shamsi, A.; Islam, A.; Hassan, M.I. Investigating binding mechanism of thymoquinone to human transferrin, targeting Alzheimer's disease therapy. *J. Cell Biochem.* **2022**, *123*, 1381–1393. [CrossRef] [PubMed]
94. Alem, F.Z.; Bejaoui, M.; Villareal, M.O.; Rhourri-Frih, B.; Isoda, H. Elucidation of the effect of plumbagin on the metastatic potential of B16F10 murine melanoma cells via MAPK signalling pathway. *Exp. Dermatol.* **2020**, *29*, 427–435. [CrossRef] [PubMed]
95. Gowda, R.; Sharma, A.; Robertson, G.P. Synergistic inhibitory effects of Celecoxib and Plumbagin on melanoma tumor growth. *Cancer Lett.* **2017**, *385*, 243–250. [CrossRef]
96. Anuf, A.R.; Ramachandran, R.; Krishnasamy, R.; Gandhi, P.S.; Periyasamy, S. Antiproliferative effects of Plumbago rosea and its purified constituent plumbagin on SK-MEL 28 melanoma cell lines. *Pharmacogn. Res.* **2014**, *6*, 312–319.
97. Zhang, H.; Zhang, A.; Gupte, A.A.; Hamilton, D.J. Plumbagin Elicits Cell-Specific Cytotoxic Effects and Metabolic Responses in Melanoma Cells. *Pharmaceutics* **2021**, *13*, 706. [CrossRef]
98. Ahmadieh, N.; Zeidan, T.; Chaaya, C.; Cain, D.; Aoude, M.; Abouchahla, A.; Kourie, H.R.; Nemer, E. Biomarkers in Prostate Cancer: A Review. *Gulf J. Oncol.* **2024**, *1*, 81–93.
99. Rondeau, G.; Abedinpour, P.; Chrastina, A.; Pelayo, J.; Borgstrom, P.; Welsh, J. Differential gene expression induced by anti-cancer agent plumbagin is mediated by androgen receptor in prostate cancer cells. *Sci. Rep.* **2018**, *8*, 2694. [CrossRef]
100. Huang, H.; Xie, H.; Pan, Y.; Zheng, K.; Xia, Y.; Chen, W. Plumbagin Triggers ER Stress-Mediated Apoptosis in Prostate Cancer Cells via Induction of ROS. *Cell Physiol. Biochem.* **2018**, *45*, 267–280. [CrossRef]
101. Abedinpour, P.; Baron, V.T.; Chrastina, A.; Rondeau, G.; Pelayo, J.; Welsh, J.; Borgström, P. Plumbagin improves the efficacy of androgen deprivation therapy in prostate cancer: A pre-clinical study. *Prostate* **2017**, *77*, 1550–1562. [CrossRef]
102. Xue, D.; Zhou, X.; Qiu, J. Cytotoxicity mechanisms of plumbagin in drug-resistant tongue squamous cell carcinoma. *J. Pharm. Pharmacol.* **2021**, *73*, 98–109. [CrossRef] [PubMed]
103. Li, B.; Pan, S.T.; Qiu, J.X. [Effect of plumbagin on epithelial-mesenchymal transition and underlying mechanisms in human tongue squamous cell carcinoma cells]. *Zhonghua Kou Qiang Yi Xue Za Zhi* **2017**, *52*, 421–426. [PubMed]
104. Chen, P.H.; Lu, H.K.; Renn, T.Y.; Chang, T.M.; Lee, C.J.; Tsao, Y.T.; Chuang, P.K.; Liu, J.F. Plumbagin Induces Reactive Oxygen Species and Endoplasmic Reticulum Stress-related Cell Apoptosis in Human Oral Squamous Cell Carcinoma. *Anticancer. Res.* **2024**, *44*, 1173–1182. [CrossRef] [PubMed]
105. Na, S.; Zhang, J.; Zhou, X.; Tang, A.; Huang, D.; Xu, Q.; Xue, D.; Qiu, J. Plumbagin-mediating GLUT1 suppresses the growth of human tongue squamous cell carcinoma. *Oral. Dis.* **2018**, *24*, 920–929. [CrossRef]

106. Pan, S.T.; Huang, G.; Wang, Q.; Qiu, J.X. Plumbagin Enhances the Radiosensitivity of Tongue Squamous Cell Carcinoma Cells via Downregulating ATM. *J. Oncol.* **2021**, *2021*, 8239984. [CrossRef]

107. Pan, S.T.; Qin, Y.; Zhou, Z.W.; He, Z.X.; Zhang, X.; Yang, T.; Yang, Y.X.; Wang, D.; Zhou, S.F.; Qiu, J.X. Plumbagin suppresses epithelial to mesenchymal transition and stemness via inhibiting Nrf2-mediated signaling pathway in human tongue squamous cell carcinoma cells. *Drug Des. Devel Ther.* **2015**, *9*, 5511–5551.

108. Alzahrani, S.M.; Al Doghaither, H.A.; Al-Ghafari, A.B. General insight into cancer: An overview of colorectal cancer. *Mol. Clin. Oncol.* **2021**, *15*, 271. [CrossRef]

109. Yadav, P.; Sharma, P.; Chetlangia, N.; Mayalagu, P.; Karunagaran, D. Upregulation of miR-22-3p contributes to plumbagin-mediated inhibition of Wnt signaling in human colorectal cancer cells. *Chem. Biol. Interact.* **2022**, *368*, 20. [CrossRef]

110. Yang, X.; Su, W.; Li, Y.; Zhou, Z.; Zhou, Y.; Shan, H.; Han, X.; Zhang, M.; Zhang, Q.; Bai, Y.; et al. MiR-22-3p suppresses cell growth via MET/STAT3 signaling in lung cancer. *Am. J. Transl. Res.* **2021**, *13*, 1221–1232.

111. Shao, H.; Shen, P.; Chen, J. Expression Profile Analysis and Image Observation of miRNA in Serum of Patients with Obstructive Sleep Apnea-Hypopnea Syndrome. *Contrast Media Mol. Imaging* **2021**, *2021*, 9731502. [CrossRef]

112. Liang, Y.; Zhou, R.; Liang, X.; Kong, X.; Yang, B. Pharmacological targets and molecular mechanisms of plumbagin to treat colorectal cancer: A systematic pharmacology study. *Eur. J. Pharmacol.* **2020**, *881*, 4. [CrossRef] [PubMed]

113. Subramaniya, B.R.; Srinivasan, G.; Sadullah, S.S.; Davis, N.; Subhadara, L.B.; Halagowder, D.; Sivasitambaram, N.D. Apoptosis inducing effect of plumbagin on colonic cancer cells depends on expression of COX-2. *PLoS ONE* **2011**, *6*, 0018695. [CrossRef] [PubMed]

114. Thapa, P. Epidemiology of Pancreatic and Periampullary Cancer. *Indian. J. Surg.* **2015**, *77*, 358–361. [CrossRef]

115. Periyasamy, L.; Murugantham, B.; Muthusami, S. Plumbagin binds to epidermal growth factor receptor and mitigate the effects of epidermal growth factor micro-environment in PANC-1 cells. *Med. Oncol.* **2023**, *40*, 184. [CrossRef]

116. Pan, Q.; Zhou, R.; Su, M.; Li, R. The Effects of Plumbagin on Pancreatic Cancer: A Mechanistic Network Pharmacology Approach. *Med. Sci. Monit.* **2019**, *25*, 4648–4654. [CrossRef]

117. Palanisamy, R.; Indrajith Kahingalage, N.; Archibald, D.; Casari, I.; Falasca, M. Synergistic Anticancer Activity of Plumbagin and Xanthohumol Combination on Pancreatic Cancer Models. *Int. J. Mol. Sci.* **2024**, *25*, 2340. [CrossRef]

118. Wang, B.; Yang, L.; Liu, T.; Xun, J.; Zhuo, Y.; Zhang, L.; Zhang, Q.; Wang, X. Hydroxytyrosol Inhibits MDSCs and Promotes M1 Macrophages in Mice With Orthotopic Pancreatic Tumor. *Front. Pharmacol.* **2021**, *12*, 759172. [CrossRef]

119. Kaarthigeyan, K. Cervical cancer in India and HPV vaccination. *Indian J. Med. Paediatr. Oncol.* **2012**, *33*, 7–12. [CrossRef] [PubMed]

120. Sidhu, H.; Capalash, N. Plumbagin downregulates UHRF1, p-Akt, MMP-2 and suppresses survival, growth and migration of cervical cancer CaSki cells. *Toxicol. In Vitro* **2023**, *86*, 4. [CrossRef] [PubMed]

121. Jaiswal, A.; Sabarwal, A.; Narayan Mishra, J.P.; Singh, R.P. Plumbagin induces ROS-mediated apoptosis and cell cycle arrest and inhibits EMT in human cervical carcinoma cells. *RSC Adv.* **2018**, *8*, 32022–32037. [CrossRef]

122. Roy, R.; Mandal, S.; Chakrabarti, J.; Saha, P.; Panda, C.K. Downregulation of Hyaluronic acid-CD44 signaling pathway in cervical cancer cell by natural polyphenols Plumbagin, Pongapin and Karanjin. *Mol. Cell Biochem.* **2021**, *476*, 3701–3709. [CrossRef]

123. Periasamy, V.S.; Athinarayanan, J.; Ramankutty, G.; Akbarsha, M.A.; Alshatwi, A.A. Plumbagin triggers redox-mediated autophagy through the LC3B protein in human papillomavirus-positive cervical cancer cells. *Arch. Med. Sci.* **2020**, *18*, 171–182. [CrossRef] [PubMed]

124. Liang, K.; Pan, X.; Chen, Y.; Huang, S. Anti-ovarian cancer actions and pharmacological targets of plumbagin. *Naunyn Schmiedebergs Arch. Pharmacol.* **2023**, *396*, 1205–1210. [CrossRef]

125. Li, Y.; Yu, S.; Liang, X.; Su, M.; Li, R. Medical Significance of Uterine Corpus Endometrial Carcinoma Patients Infected With SARS-CoV-2 and Pharmacological Characteristics of Plumbagin. *Front. Endocrinol.* **2021**, *12*, 714909. [CrossRef] [PubMed]

126. Kong, X.; Luo, J.; Xu, T.; Zhou, Y.; Pan, Z.; Xie, Y.; Zhao, L.; Lu, Y.; Han, X.; Li, Z.; et al. Plumbagin enhances TRAIL-induced apoptosis of human leukemic Kasumi-1 cells through upregulation of TRAIL death receptor expression, activation of caspase-8 and inhibition of cFLIP. *Oncol. Rep.* **2017**, *37*, 3423–3432. [CrossRef] [PubMed]

127. Zhang, J.; Peng, S.; Li, X.; Liu, R.; Han, X.; Fang, J. Targeting thioredoxin reductase by plumbagin contributes to inducing apoptosis of HL-60 cells. *Arch. Biochem. Biophys.* **2017**, *619*, 16–26. [CrossRef]

128. Uttarkar, S.; Piontek, T.; Dukare, S.; Schomburg, C.; Schlenke, P.; Berdel, W.E.; Müller-Tidow, C.; Schmidt, T.J.; Klempnauer, K.H. Small-Molecule Disruption of the Myb/p300 Cooperation Targets Acute Myeloid Leukemia Cells. *Mol. Cancer Ther.* **2016**, *15*, 2905–2915. [CrossRef]

129. Fu, C.; Gong, Y.; Shi, X.; Sun, Z.; Niu, M.; Sang, W.; Xu, L.; Zhu, F.; Wang, Y.; Xu, K. Plumbagin reduces chronic lymphocytic leukemia cell survival by downregulation of Bcl-2 but upregulation of the Bax protein level. *Oncol. Rep.* **2016**, *36*, 1605–1611. [CrossRef]

130. Bae, K.J.; Lee, Y.; Kim, S.A.; Kim, J. Plumbagin exerts an immunosuppressive effect on human T-cell acute lymphoblastic leukemia MOLT-4 cells. *Biochem. Biophys. Res. Commun.* **2016**, *473*, 272–277. [CrossRef]

131. Niu, M.; Cai, W.; Liu, H.; Chong, Y.; Hu, W.; Gao, S.; Shi, Q.; Zhou, X.; Liu, X.; Yu, R. Plumbagin inhibits growth of gliomas in vivo via suppression of FOXM1 expression. *J. Pharmacol. Sci.* **2015**, *128*, 131–136. [CrossRef]

132. Khaw, A.K.; Sameni, S.; Venkatesan, S.; Kalthur, G.; Hande, M.P. Plumbagin alters telomere dynamics, induces DNA damage and cell death in human brain tumour cells. *Mutat. Res. Genet. Toxicol. Environ. Mutagen.* **2015**, *793*, 86–95. [CrossRef] [PubMed]

133. Son, T.G.; Camandola, S.; Arumugam, T.V.; Cutler, R.G.; Telljohann, R.S.; Mughal, M.R.; Moore, T.A.; Luo, W.; Yu, Q.S.; Johnson, D.A.; et al. Plumbagin, a novel Nrf2/ARE activator, protects against cerebral ischemia. *J. Neurochem.* **2010**, *112*, 1316–1326. [CrossRef] [PubMed]
134. Ding, Y.; Chen, Z.J.; Liu, S.; Che, D.; Vetter, M.; Chang, C.H. Inhibition of Nox-4 activity by plumbagin, a plant-derived bioactive naphthoquinone. *J. Pharm. Pharmacol.* **2005**, *57*, 111–116. [CrossRef]
135. Cao, Y.; Yin, X.; Jia, Y.; Liu, B.; Wu, S.; Shang, M. Plumbagin, a natural naphthoquinone, inhibits the growth of esophageal squamous cell carcinoma cells through inactivation of STAT3. *Int. J. Mol. Med.* **2018**, *42*, 1569–1576. [CrossRef]
136. Cao, Y.Y.; Yu, J.; Liu, T.T.; Yang, K.X.; Yang, L.Y.; Chen, Q.; Shi, F.; Hao, J.J.; Cai, Y.; Wang, M.R.; et al. Plumbagin inhibits the proliferation and survival of esophageal cancer cells by blocking STAT3-PLK1-AKT signaling. *Cell Death Dis.* **2018**, *9*, 17. [CrossRef]
137. Kumar, V.; Sami, N.; Kashav, T.; Islam, A.; Ahmad, F.; Hassan, M.I. Protein aggregation and neurodegenerative diseases: From theory to therapy. *Eur. J. Med. Chem.* **2016**, *124*, 1105–1120. [CrossRef]
138. Sami, N.; Rahman, S.; Kumar, V.; Zaidi, S.; Islam, A.; Ali, S.; Ahmad, F.; Hassan, M.I. Protein aggregation, misfolding and consequential human neurodegenerative diseases. *Int. J. Neurosci.* **2017**, *127*, 1047–1057. [CrossRef]
139. Basheer, N.; Smolek, T.; Hassan, I.; Liu, F.; Iqbal, K.; Zilka, N.; Novak, P. Does modulation of tau hyperphosphorylation represent a reasonable therapeutic strategy for Alzheimer's disease? From preclinical studies to the clinical trials. *Mol. Psychiatry* **2023**, *2*, 2197–2214. [CrossRef] [PubMed]
140. Fatima, U.; Roy, S.; Ahmad, S.; Ali, S.; Elkady, W.M.; Khan, I.; Alsaffar, R.M.; Adnan, M.; Islam, A.; Hassan, M.I. Pharmacological attributes of Bacopa monnieri extract: Current updates and clinical manifestation. *Front. Nutr.* **2022**, *9*, 972379. [CrossRef]
141. Khan, P.; Rahman, S.; Queen, A.; Manzoor, S.; Naz, F.; Hasan, G.M.; Luqman, S.; Kim, J.; Islam, A.; Ahmad, F.; et al. Elucidation of Dietary Polyphenolics as Potential Inhibitor of Microtubule Affinity Regulating Kinase 4: In silico and In vitro Studies. *Sci. Rep.* **2017**, *7*, 9470. [CrossRef]
142. Kumar, A.; Nisha, C.M.; Silakari, C.; Sharma, I.; Anusha, K.; Gupta, N.; Nair, P.; Tripathi, T. Current and novel therapeutic molecules and targets in Alzheimer's disease. *J. Formos. Med. Assoc.* **2016**, *115*, 3–10. [CrossRef] [PubMed]
143. Naz, F.; Anjum, F.; Islam, A.; Ahmad, F.; Hassan, M.I. Microtubule affinity-regulating kinase 4: Structure, function, and regulation. *Cell Biochem. Biophys.* **2013**, *67*, 485–499. [CrossRef] [PubMed]
144. Naz, F.; Shahbaaz, M.; Khan, S.; Bisetty, K.; Islam, A.; Ahmad, F.; Hassan, M.I. PKR-inhibitor binds efficiently with human microtubule affinity-regulating kinase 4. *J. Mol. Graph. Model.* **2015**, *62*, 245–252. [CrossRef] [PubMed]
145. Adnan, M.; DasGupta, D.; Anwar, S.; Patel, M.; Jamal Siddiqui, A.; Bardakci, F.; Snoussi, M.; Imtaiyaz Hassan, M. Investigating role of plumbagin in preventing neurodegenerative diseases via inhibiting microtubule affinity regulating kinase 4. *J. Mol. Liq.* **2023**, *384*, 122267. [CrossRef]
146. Xu, X.; Su, Y.; Zou, Z.; Zhou, Y.; Yan, J. Correlation between C9ORF72 mutation and neurodegenerative diseases: A comprehensive review of the literature. *Int. J. Med. Sci.* **2021**, *18*, 378–386. [CrossRef]
147. Chia, S.J.; Tan, E.K.; Chao, Y.X. Historical Perspective: Models of Parkinson's Disease. *Int. J. Mol. Sci.* **2020**, *21*, 2464. [CrossRef]
148. Flagmeier, P.; Meisl, G.; Vendruscolo, M.; Knowles, T.P.; Dobson, C.M.; Buell, A.K.; Galvagnion, C. Mutations associated with familial Parkinson's disease alter the initiation and amplification steps of α-synuclein aggregation. *Proc. Natl. Acad. Sci. USA* **2016**, *113*, 10328–10333. [CrossRef]
149. Mukhara, D.; Oh, U.; Neigh, G.N. Neuroinflammation. *Handb. Clin. Neurol.* **2020**, *175*, 235–259.
150. Qian, Y.F.; Xiong, Q.; Yang, S.P.; Xie, J. Formula optimization for melanosis-inhibitors of Pacific white shrimp (Litopenaeus vannamei) by response surface methodology. *Food Sci. Biotechnol.* **2019**, *28*, 1687–1692. [CrossRef]
151. Azam, S.; Jakaria, M.; Kim, I.S.; Kim, J.; Haque, M.E.; Choi, D.K. Regulation of Toll-Like Receptor (TLR) Signaling Pathway by Polyphenols in the Treatment of Age-Linked Neurodegenerative Diseases: Focus on TLR4 Signaling. *Front. Immunol.* **2019**, *10*, 1000. [CrossRef]
152. Su, Y.; Li, M.; Wang, Q.; Xu, X.; Qin, P.; Huang, H.; Zhang, Y.; Zhou, Y.; Yan, J. Inhibition of the TLR/NF-κB Signaling Pathway and Improvement of Autophagy Mediates Neuroprotective Effects of Plumbagin in Parkinson's Disease. *Oxid. Med. Cell Longev.* **2022**, *2022*, 1837278. [CrossRef] [PubMed]
153. Messeha, S.S.; Zarmouh, N.O.; Mendonca, P.; Kolta, M.G.; Soliman, K.F.A. The attenuating effects of plumbagin on pro-inflammatory cytokine expression in LPS-activated BV-2 microglial cells. *J. Neuroimmunol.* **2017**, *313*, 129–137. [CrossRef] [PubMed]
154. Khan, T.; Waseem, R.; Shahid, M.; Ansari, J.; Ahanger, I.A.; Hassan, I.; Islam, A. Recent advancement in therapeutic strategies for Alzheimer's disease: Insights from clinical trials. *Ageing Res. Rev.* **2023**, *92*, 102113. [CrossRef]
155. Wolfe, M.S. Therapeutic strategies for Alzheimer's disease. *Nat. Rev. Drug Discov.* **2002**, *1*, 859–866. [CrossRef]
156. Mattson, M.P. Pathways towards and away from Alzheimer's disease. *Nature* **2004**, *430*, 631–639. [CrossRef] [PubMed]
157. Wang, Y.; Miao, Y.; Mir, A.Z.; Cheng, L.; Wang, L.; Zhao, L.; Cui, Q.; Zhao, W.; Wang, H. Inhibition of beta-amyloid-induced neurotoxicity by pinocembrin through Nrf2/HO-1 pathway in SH-SY5Y cells. *J. Neurol. Sci.* **2016**, *368*, 223–230. [CrossRef]
158. Greenwald, J.; Riek, R. Biology of amyloid: Structure, function, and regulation. *Structure* **2010**, *18*, 1244–1260. [CrossRef]
159. Kim, J.W.; Byun, M.S.; Lee, J.H.; Yi, D.; Jeon, S.Y.; Sohn, B.K.; Lee, J.Y.; Shin, S.A.; Kim, Y.K.; Kang, K.M.; et al. Serum albumin and beta-amyloid deposition in the human brain. *Neurology* **2020**, *95*, e815–e826. [CrossRef]

160. Nabi, F.; Ahmad, O.; Khan, A.; Hassan, M.N.; Hisamuddin, M.; Malik, S.; Chaari, A.; Khan, R.H. Natural compound plumbagin based inhibition of hIAPP revealed by Markov state models based on MD data along with experimental validations. *Proteins* **2024**, *18*, 26682. [CrossRef]

161. Anand, B.G.; Prajapati, K.P.; Purohit, S.; Ansari, M.; Panigrahi, A.; Kaushik, B.; Behera, R.K.; Kar, K. Evidence of Anti-amyloid Characteristics of Plumbagin via Inhibition of Protein Aggregation and Disassembly of Protein Fibrils. *Biomacromolecules* **2021**, *22*, 3692–3703. [CrossRef]

162. Nakhate, K.T.; Bharne, A.P.; Verma, V.S.; Aru, D.N.; Kokare, D.M. Plumbagin ameliorates memory dysfunction in streptozotocin induced Alzheimer's disease via activation of Nrf2/ARE pathway and inhibition of β-secretase. *Biomed. Pharmacother.* **2018**, *101*, 379–390. [CrossRef]

163. Tanabe, A.; Nomura, S. [Pathophysiology of depression]. *Nihon Rinsho* **2007**, *65*, 1585–1590.

164. Dhingra, D.; Bansal, S. Antidepressant-like activity of plumbagin in unstressed and stressed mice. *Pharmacol. Rep.* **2015**, *67*, 1024–1032. [CrossRef]

165. Oparil, S.; Zaman, M.A.; Calhoun, D.A. Pathogenesis of Hypertension. *Ann. Intern. Med.* **2003**, *139*, 761–776. [CrossRef]

166. Ahmad, M.; Ahmad, T.; Irfan, H.M.; Noor, N. Blood pressure-lowering and cardiovascular effects of plumbagin in rats: An insight into the underlying mechanisms. *Curr. Res. Pharmacol. Drug Discov.* **2022**, *3*, 100139. [CrossRef] [PubMed]

167. Kainuma, S.; Miyagawa, S.; Fukushima, S.; Tsuchimochi, H.; Sonobe, T.; Fujii, Y.; Pearson, J.T.; Saito, A.; Harada, A.; Toda, K.; et al. Influence of coronary architecture on the variability in myocardial infarction induced by coronary ligation in rats. *PLoS ONE* **2017**, *12*, e0183323. [CrossRef]

168. Sahu, B.D.; Kumar, J.M.; Kuncha, M.; Borkar, R.M.; Srinivas, R.; Sistla, R. Baicalein alleviates doxorubicin-induced cardiotoxicity via suppression of myocardial oxidative stress and apoptosis in mice. *Life Sci.* **2016**, *144*, 8–18. [CrossRef] [PubMed]

169. Nitiss, K.C.; Nitiss, J.L. Twisting and ironing: Doxorubicin cardiotoxicity by mitochondrial DNA damage. *Clin. Cancer Res.* **2014**, *20*, 4737–4739. [CrossRef] [PubMed]

170. Li, Z.; Chinnathambi, A.; Ali Alharbi, S.; Yin, F. Plumbagin protects the myocardial damage by modulating the cardiac biomarkers, antioxidants, and apoptosis signaling in the doxorubicin-induced cardiotoxicity in rats. *Environ. Toxicol.* **2020**, *35*, 1374–1385. [CrossRef]

171. Maxwell, S.R.; Lip, G.Y. Reperfusion injury: A review of the pathophysiology, clinical manifestations and therapeutic options. *Int. J. Cardiol.* **1997**, *58*, 95–117. [CrossRef]

172. Ceconi, C.; Cargnoni, A.; Pasini, E.; Condorelli, E.; Curello, S.; Ferrari, R. Evaluation of phospholipid peroxidation as malondialdehyde during myocardial ischemia and reperfusion injury. *Am. J. Physiol.* **1991**, *260*, H1057–H1061. [CrossRef]

173. Wang, S.X.; Wang, J.; Shao, J.B.; Tang, W.N.; Zhong, J.Q. Plumbagin Mediates Cardioprotection Against Myocardial Ischemia/Reperfusion Injury Through Nrf-2 Signaling. *Med. Sci. Monit.* **2016**, *22*, 1250–1257. [CrossRef] [PubMed]

174. Krick, S.; Platoshyn, O.; Sweeney, M.; McDaniel, S.S.; Zhang, S.; Rubin, L.J.; Yuan, J.X.J. Nitric oxide induces apoptosis by activating K+ channels in pulmonary vascular smooth muscle cells. *Am. J. Physiol. -Heart Circ. Physiol.* **2002**, *282*, H184–H193. [CrossRef] [PubMed]

175. Bonnet, S.; Rochefort, G.; Sutendra, G.; Archer, S.L.; Haromy, A.; Webster, L.; Hashimoto, K.; Bonnet, S.N.; Michelakis, E.D. The nuclear factor of activated T cells in pulmonary arterial hypertension can be therapeutically targeted. *Proc. Natl. Acad. Sci. USA* **2007**, *104*, 11418–11423. [CrossRef] [PubMed]

176. Paulin, R.; Courboulin, A.; Meloche, J.; Mainguy, V.; Dumas de la Roque, E.; Saksouk, N.; Côté, J.; Provencher, S.; Sussman, M.A.; Bonnet, S. Signal Transducers and Activators of Transcription-3/Pim1 Axis Plays a Critical Role in the Pathogenesis of Human Pulmonary Arterial Hypertension. *Circulation* **2011**, *123*, 1205–1215. [CrossRef]

177. Courboulin, A.; Barrier, M.; Perreault, T.; Bonnet, P.; Tremblay, V.L.; Paulin, R.; Tremblay, E.; Lambert, C.; Jacob, M.H.; Bonnet, S.N.; et al. Plumbagin reverses proliferation and resistance to apoptosis in experimental PAH. *Eur. Respir. J.* **2012**, *40*, 618–629. [CrossRef]

178. Naqvi, A.A.T.; Fatima, K.; Mohammad, T.; Fatima, U.; Singh, I.K.; Singh, A.; Atif, S.M.; Hariprasad, G.; Hasan, G.M.; Hassan, M.I. Insights into SARS-CoV-2 genome, structure, evolution, pathogenesis and therapies: Structural genomics approach. *Biochim. Biophys. Acta Mol. Basis Dis.* **2020**, *1866*, 165878. [CrossRef]

179. Lee, C. Therapeutic Modulation of Virus-Induced Oxidative Stress via the Nrf2-Dependent Antioxidative Pathway. *Oxidative Med. Cell. Longev.* **2018**, *2018*, 6208067. [CrossRef]

180. Li, Z.; Wu, J.; DeLeo, C.J. RNA damage and surveillance under oxidative stress. *IUBMB Life* **2006**, *58*, 581–588. [CrossRef]

181. Yan, G.; Li, D.; Qi, H.; Fu, Z.; Liu, X.; Zhang, J.; Chen, Y. [Discovery of SARS-CoV-2 main protease inhibitors using an optimized FRET-based high-throughput screening assay]. *Sheng Wu Gong Cheng Xue Bao* **2022**, *38*, 2236–2249.

182. Floyd, K.; Glaziou, P.; Zumla, A.; Raviglione, M. The global tuberculosis epidemic and progress in care, prevention, and research: An overview in year 3 of the End TB era. *Lancet Respir. Med.* **2018**, *6*, 299–314. [CrossRef] [PubMed]

183. Sarkar, A.; Ghosh, S.; Shaw, R.; Patra, M.M.; Calcuttawala, F.; Mukherjee, N.; Das Gupta, S.K. Mycobacterium tuberculosis thymidylate synthase (ThyX) is a target for plumbagin, a natural product with antimycobacterial activity. *PLoS ONE* **2020**, *15*, e0228657. [CrossRef] [PubMed]

184. Arumugam, P.; Shankaran, D.; Bothra, A.; Gandotra, S.; Rao, V. The MmpS6-MmpL6 Operon Is an Oxidative Stress Response System Providing Selective Advantage to Mycobacterium tuberculosis in Stress. *J. Infect. Dis.* **2019**, *219*, 459–469. [CrossRef] [PubMed]

185. Yeruva, V.C.; Sundaram, C.A.; Sritharan, M. Effect of iron concentration on the expression and activity of catalase-peroxidases in mycobacteria. *Indian. J. Biochem. Biophys.* **2005**, *42*, 28–33.

186. Sritharan, M.; Yeruva, V.C.; Sivasailappan, S.C.; Duggirala, S. Iron enhances the susceptibility of pathogenic mycobacteria to isoniazid, an antitubercular drug. *World J. Microbiol. Biotechnol.* **2006**, *22*, 1357–1364. [CrossRef]

187. Dandawate, P.; Vemuri, K.; Venkateswara Swamy, K.; Khan, E.M.; Sritharan, M.; Padhye, S. Synthesis, characterization, molecular docking and anti-tubercular activity of Plumbagin-Isoniazid Analog and its β-cyclodextrin conjugate. *Bioorg Med. Chem. Lett.* **2014**, *24*, 5070–5075. [CrossRef]

188. Sunil, C.; Duraipandiyan, V.; Agastian, P.; Ignacimuthu, S. Antidiabetic effect of plumbagin isolated from Plumbago zeylanica L. root and its effect on GLUT4 translocation in streptozotocin-induced diabetic rats. *Food Chem. Toxicol.* **2012**, *50*, 4356–4363. [CrossRef]

189. Blakytny, R.; Jude, E. The molecular biology of chronic wounds and delayed healing in diabetes. *Diabet. Med.* **2006**, *23*, 594–608. [CrossRef]

190. Shao, Y.; Dang, M.; Lin, Y.; Xue, F. Evaluation of wound healing activity of plumbagin in diabetic rats. *Life Sci.* **2019**, *231*, 3. [CrossRef]

191. Kapur, A.; Beres, T.; Rathi, K.; Nayak, A.P.; Czarnecki, A.; Felder, M.; Gillette, A.; Ericksen, S.S.; Sampene, E.; Skala, M.C.; et al. Oxidative stress via inhibition of the mitochondrial electron transport and Nrf-2-mediated anti-oxidative response regulate the cytotoxic activity of plumbagin. *Sci. Rep.* **2018**, *8*, 1073. [CrossRef]

192. Hussain, T.; Tan, B.; Yin, Y.; Blachier, F.; Tossou, M.C.; Rahu, N. Oxidative Stress and Inflammation: What Polyphenols Can Do for Us? *Oxid. Med. Cell Longev.* **2016**, *2016*, 7432797. [CrossRef]

193. Kuan-Hong, W.; Bai-Zhou, L. Plumbagin protects against hydrogen peroxide-induced neurotoxicity by modulating NF-κB and Nrf-2. *Arch. Med. Sci.* **2018**, *14*, 1112–1118. [CrossRef] [PubMed]

194. Zaki, A.M.; El-Tanbouly, D.M.; Abdelsalam, R.M.; Zaki, H.F. Plumbagin ameliorates hepatic ischemia-reperfusion injury in rats: Role of high mobility group box 1 in inflammation, oxidative stress and apoptosis. *Biomed. Pharmacother.* **2018**, *106*, 785–793. [CrossRef] [PubMed]

195. Chen, Y.; Zhao, C.; Liu, X.; Wu, G.; Zhong, J.; Zhao, T.; Li, J.; Lin, Y.; Zhou, Y.; Wei, Y. Plumbagin ameliorates liver fibrosis via a ROS-mediated NF-κB signaling pathway in vitro and in vivo. *Biomed. Pharmacother.* **2019**, *116*, 30. [CrossRef]

196. Chu, H.; Yu, H.; Ren, D.; Zhu, K.; Huang, H. Plumbagin exerts protective effects in nucleus pulposus cells by attenuating hydrogen peroxide-induced oxidative stress, inflammation and apoptosis through NF-κB and Nrf-2. *Int. J. Mol. Med.* **2016**, *37*, 1669–1676. [CrossRef] [PubMed]

197. Rashidzadeh, H.; Zamani, P.; Amiri, M.; Hassanzadeh, S.M.; Ramazani, A. Nanoincorporation of Plumbagin in Micelles Increase Its in Vivo Anti-Plasmodial Properties. *Iran. J. Parasitol.* **2022**, *17*, 202–213. [CrossRef]

198. Bothiraja, C.; Kapare, H.S.; Pawar, A.P.; Shaikh, K.S. Development of plumbagin-loaded phospholipid-Tween® 80 mixed micelles: Formulation, optimization, effect on breast cancer cells and human blood/serum compatibility testing. *Ther. Deliv.* **2013**, *4*, 1247–1259. [CrossRef]

199. Pawar, A.; Patel, R.; Arulmozhi, S.; Bothiraja, C. d-α-Tocopheryl polyethylene glycol 1000 succinate conjugated folic acid nanomicelles: Towards enhanced bioavailability, stability, safety, prolonged drug release and synergized anticancer effect of plumbagin. *RSC Adv.* **2016**, *6*, 78106–78121. [CrossRef]

200. Bhagat, S.; Singh, S. Use of antioxidant nanoliposomes for co-delivery of PTEN plasmids and plumbagin to induce apoptosis in hepatic cancer cells. *Biomed. Mater.* **2024**, *19*, 025026. [CrossRef]

201. Kumar, M.R.; Aithal, B.K.; Udupa, N.; Reddy, M.S.; Raakesh, V.; Murthy, R.S.; Raju, D.P.; Rao, B.S. Formulation of plumbagin loaded long circulating pegylated liposomes: In vivo evaluation in C57BL/6J mice bearing B16F1 melanoma. *Drug Deliv.* **2011**, *18*, 511–522. [CrossRef]

202. Sakpakdeejaroen, I.; Somani, S.; Laskar, P.; Mullin, M.; Dufès, C. Transferrin-bearing liposomes entrapping plumbagin for targeted cancer therapy. *J. Interdiscip. Nanomed.* **2019**, *4*, 54–71. [CrossRef] [PubMed]

203. Song, Y.Y.; Yuan, Y.; Shi, X.; Che, Y.Y. Improved drug delivery and anti-tumor efficacy of combinatorial liposomal formulation of genistein and plumbagin by targeting Glut1 and Akt3 proteins in mice bearing prostate tumor. *Colloids Surf. B Biointerfaces* **2020**, *190*, 12. [CrossRef] [PubMed]

204. Md, S.; Alhakamy, N.A.; Aldawsari, H.M.; Husain, M.; Khan, N.; Alfaleh, M.A.; Asfour, H.Z.; Riadi, Y.; Bilgrami, A.L.; Akhter, M.H. Plumbagin-Loaded Glycerosome Gel as Topical Delivery System for Skin Cancer Therapy. *Polymers* **2021**, *13*, 923. [CrossRef] [PubMed]

205. Kamble, P.R.; Shaikh, K.S. Optimization and Evaluation of Self-nanoemulsifying Drug Delivery System for Enhanced Bioavailability of Plumbagin. *Planta Med.* **2022**, *88*, 79–90. [CrossRef]

206. Appadurai, P.; Rathinasamy, K. Plumbagin-silver nanoparticle formulations enhance the cellular uptake of plumbagin and its antiproliferative activities. *IET Nanobiotechnol.* **2015**, *9*, 264–272. [CrossRef]

207. Tyagi, R.; Waheed, A.; Kumar, N.; Ahad, A.; Bin Jardan, Y.A.; Mujeeb, M.; Kumar, A.; Naved, T.; Madan, S. Formulation and Evaluation of Plumbagin-Loaded Niosomes for an Antidiabetic Study: Optimization and In Vitro Evaluation. *Pharmaceuticals* **2023**, *16*, 1169. [CrossRef] [PubMed]

208. Mandala Rayabandla, S.K.; Aithal, K.; Anandam, A.; Shavi, G.; Nayanabhirama, U.; Arumugam, K.; Musmade, P.; Bhat, K.; Bola Sadashiva, S.R. Preparation, in vitro characterization, pharmacokinetic, and pharmacodynamic evaluation of chitosan-based plumbagin microspheres in mice bearing B16F1 melanoma. *Drug Deliv.* **2010**, *17*, 103–113. [CrossRef]
209. Mukherjee, S.; Sawant, A.V.; Prassanawar, S.S.; Panda, D. Copper-Plumbagin Complex Produces Potent Anticancer Effects by Depolymerizing Microtubules and Inducing Reactive Oxygen Species and DNA Damage. *ACS Omega* **2023**, *8*, 3221–3235. [CrossRef]
210. Zhang, H.Q.; Lu, X.; Wu, J.L.; Ou, M.Q.; Chen, N.F.; Liang, H.; Chen, Z.F. Discovery of mitochondrion-targeting copper(II)-plumbagin and -bipyridine complexes as chemodynamic therapy agents with enhanced antitumor activity. *Dalton Trans.* **2024**, *53*, 3244–3253. [CrossRef]
211. Solanki, R.; Saini, M.; Mochi, J.; Pappachan, A.; Patel, S. Synthesis, characterization, in-silico and in-vitro anticancer studies of Plumbagin encapsulated albumin nanoparticles for breast cancer treatment. *J. Drug Deliv. Sci. Technol.* **2023**, *84*, 104501. [CrossRef]
212. Sakpakdeejaroen, I.; Somani, S.; Laskar, P.; Mullin, M.; Dufès, C. Regression of Melanoma Following Intravenous Injection of Plumbagin Entrapped in Transferrin-Conjugated, Lipid-Polymer Hybrid Nanoparticles. *Int. J. Nanomed.* **2021**, *16*, 2615–2631. [CrossRef] [PubMed]
213. Kawiak, A.; Domachowska, A.; Krolicka, A.; Smolarska, M.; Lojkowska, E. 3-Chloroplumbagin Induces Cell Death in Breast Cancer Cells Through MAPK-Mediated Mcl-1 Inhibition. *Front. Pharmacol.* **2019**, *10*, 784. [CrossRef] [PubMed]
214. Yue, L.; Jiang, N.; Wu, A.; Qiu, W.; Shen, X.; Qin, D.; Li, H.; Lin, J.; Liang, S.; Wu, J. Plumbagin can potently enhance the activity of xanthine oxidase: In vitro, in vivo and in silico studies. *BMC Pharmacol. Toxicol.* **2021**, *22*, 45. [CrossRef] [PubMed]

Article

Hibiscus Anthocyanins Extracts Induce Apoptosis by Activating AMP-Activated Protein Kinase in Human Colorectal Cancer Cells

Ming-Chang Tsai [1,2], Ching-Chun Chen [3], Tsui-Hwa Tseng [4], Yun-Ching Chang [5], Yi-Jie Lin [5], I-Ning Tsai [3], Chi-Chih Wang [1,2,*] and Chau-Jong Wang [5,6,*]

1   Division of Gastroenterology and Hepatology, Department of Internal Medicine, Chung Shan Medical University Hospital, Taichung 402, Taiwan; tsaimc1110@gmail.com
2   School of Medical, Chung Shan Medical University, Taichung 402, Taiwan
3   Institute of Medicine, Chung Shan Medical University, Taichung 402, Taiwan; sandy985619@gmail.com (C.-C.C.); 0312elise@gmail.com (I.-N.T.)
4   Department of Medical Applied Chemistry, Chung Shan Medical University, Taichung 402, Taiwan; tht@csmu.edu.tw
5   Department of Health Industry Technology Management, Chung Shan Medical University, Taichung 402, Taiwan; changyc@csmu.edu.tw (Y.-C.C.); exist7730@gmail.com (Y.-J.L.)
6   Department of Medical Research, Chung Shan Medical University Hospital, Taichung 402, Taiwan
*   Correspondence: bananaudwang@gmail.com (C.-C.W.); wcj@csmu.edu.tw (C.-J.W.); Tel.: +886-4-247-30022 (ext. 34704) (C.-C.W. & C.-J.W.)

Abstract: Apoptosis, a programmed cell death process preventing cancer development, can be evaded by cancer cells. AMP-activated protein kinase (AMPK) regulates energy levels and is a key research topic in cancer prevention and treatment. Some bioactive components of *Hibiscus sabdariffa* L. (HAs), including anthocyanins, have potential anticancer properties. Our study investigated the in vitro cytotoxic potential and mode of action of HAs extracts containing anthocyanins in colorectal cancer cells. The results showed that Hibiscus anthocyanin-rich extracts induced apoptosis in human colorectal cancer cells through the activation of multiple signaling pathways of AMPK. We observed the dose–response and time-dependent induction of apoptosis with HAs. Subsequently, the activation of Fas-mediated proteins triggered apoptotic pathways associated with Fas-mediated apoptosis-related proteins, including caspase-8/tBid. This caused the release of cytochrome C from the mitochondria, resulting in caspase-3 cleavage and apoptosis activation in intestinal cancer cells. These data elucidate the relationship between Has' regulation of apoptosis-related proteins in colorectal cancer cells and apoptotic pathways.

Keywords: Hibiscus anthocyanidin extract; apoptosis; AMPK; Fas/Fas L; colorectal cancer

Citation: Tsai, M.-C.; Chen, C.-C.; Tseng, T.-H.; Chang, Y.-C.; Lin, Y.-J.; Tsai, I.-N.; Wang, C.-C.; Wang, C.-J. Hibiscus Anthocyanins Extracts Induce Apoptosis by Activating AMP-Activated Protein Kinase in Human Colorectal Cancer Cells. *Nutrients* **2023**, *15*, 3972. https://doi.org/10.3390/nu15183972

Academic Editor: Maria Annunziata Carluccio

Received: 21 August 2023
Revised: 6 September 2023
Accepted: 12 September 2023
Published: 14 September 2023

1. Introduction

Apoptosis is a programmed cell death process regulated through caspase activation to maintain normal cell populations. When cellular abnormalities arise, damaged cells undergo apoptosis, which prevents cancer development. However, cancer cells can avoid apoptosis and continuously divide. Most anticancer therapies trigger apoptosis induction to eliminate malignant cells. Apoptosis can be activated through the death receptor signal pathway or intrinsic apoptosis pathway wherein changes in the integrity of the mitochondrial membrane are regulated by Bcl-2 family proteins [1]. A crosstalk of these two major apoptotic pathways has also been identified [2,3]. Researchers have expressed interest in natural products that can modulate various cancer cell death signaling pathways and have investigated their modes of action against cancers.

AMP-activated protein kinase (AMPK) is a crucial mediator in the maintenance of cellular energy homeostasis. AMPK, which regulates metabolic energy, is a common research focus for the treatment of metabolic syndromes such as type-2 diabetes. Metformin is an AMPK activator that reduces the incidence of cancer [4,5]. The deregulation of cellular energetics is a hallmark of cancer. Activated AMPK may influence many effector proteins involved in the regulatory processes and pathogenesis of cancers. AMPK activation may suppress metabolic tumor growth through the regulation of energy levels, enforcement of metabolic checkpoints, and inhibition of cell growth. Thus, AMPKs have been widely investigated in studies on cancer prevention and treatment [6].

Colorectal cancer (CRC) is the most prevalent form of cancer in Taiwan. The incidence and mortality rates of CRC have increased, particularly among young adults [7,8], although treatment quality has improved. CRC therapy includes radiotherapy, chemotherapy, and surgery. After these treatments, delaying the onset and progression of CRC is the primary focus of doctors. CRC development is influenced by genetic, environmental, and nutritional factors, with nutritional factors playing both protective and causal roles. Certain health foods and beverages may contain components exhibiting antiproliferative effects on colorectal cancer cells [9–13]. *Hibiscus sabdariffa* L. (HAs), also known as Roselle, belongs to the Malvaceae family. Widely distributed and cultivated in tropical and subtropical regions, the flowering plant is used in cosmetics, food, and medicine. Its main bioactive components are anthocyanins, phenolic acids, polysaccharides, and flavonoids [14,15]. Many in vitro and in vivo studies have demonstrated the beneficial pharmacological properties and functions of anthocyanin extracts, including hepatoprotection, antioxidant activity, anti-atherosclerotic effects, and anti-inflammation and anticancer properties. Several previous research studies have explored the relationship between anthocyanins and cancer, specifically their role in inducing apoptosis in cancer cells. Some key findings include anticancer properties of berry anthocyanins, apoptotic effects of grape anthocyanins, the connection between blueberry anthocyanins and prostate cancer, etc. [16–18]. In addition, our previous studies showed that Hibiscus anthocyanins may protect against cancer by inhibiting the growth, proliferation, and migration of cancer cells and by reducing inflammation [19–22]. However, the extract's relationship with CRC remains unclear. This study analyzed the cytotoxic potential and mode of action of the anthocyanins from HAs in vitro.

2. Materials and Methods

2.1. Reagents and Chemicals

Dulbecco's Modified Eagle's Medium (DMEM) and fetal bovine serum (FBS) were purchased from Gibco Ltd. (Grand Island, NY, USA). Antibodies for the testing of p-AMPK (#2535), AMPK (#2532), and p-Akt (#9271) were obtained from Cell Signaling Technology (Danvers, MA, USA). Other antibodies, including Bcl-2 family AIF (sc-13116), Bad (sc-8044), Bax (sc-526), Bcl-xl (sc-8392), Bid (sc-6538), Cytochrome C (sc-13560) and PARP (sc-377015), caspases-3 (sc-7148), caspases-8 (sc-7890), caspases-9 (sc-73548), and Fas/FasL (sc-7886, sc-6237), were acquired from Santa Cruz (Santa Cruz, CA, USA). Antibiotic-antimycotic (100X) and trypsin-EDTA were purchased from Gibco Ltd. (Grand Island, NY, USA). 4,6-diamidino-2-phenylindole (DAPI), Tris base, propidium iodide (PI), and other materials were acquired from Sigma Chemical (St. Louis, MO, USA).

2.2. Preparation of HAs

First, extraction was performed on Hibiscus flowers [22]; specifically, 20 g of dried Hibiscus flower was combined with methanol and 1% HCl for 1 day at 4 °C. The extract was filtered and concentrated; subsequently, the collected precipitate was loaded into an Amberlife Diaion HP-20 resin column and allowed to settle for 24 h. Thereafter, we cleaned the column with distilled water containing 0.1% HCl solution and used methanol to elute. The filtrate was collected and subsequently lyophilized to obtain approximately 2 g of HAs, which was stored at −20 °C before use. 100 mg/mL HAs stock solution in ddH_2O water was prepared and stored at 4 °C, protected from light.

2.3. HPLC Analysis

Total HAs were extracted with the Fuleki and Francis method [23]. In particular, a 100 µL aliquot of HAs (10 mg/mL) was diluted with 3 mL of pH 1.0 and pH 4.5 buffers. The optical density (OD) of the sample was measured at 535 nm, using distilled water as a reference. The change in OD was calculated by subtracting the total OD at pH 4.5 from the total OD at pH 1.0. Both values were calculated with OD readings and standard dilution and calculation factors. For HAs standardization, cyanidin and delphinidin parameters were determined with high-performance liquid chromatography (HPLC) using the symmetry shield RP18 column (3.5 µm, 4.6 × 150 mm) and UV–VIS detector (monitored at 530 nm). The mobile phase comprised H_2O, 10% formic acid, and methanol (65/35, *v/v*.). We used 1 mL acidic methanol ($HCl:CH_3OH$ = 1:1, *v/v*.) to dissolve 1 mg sample and boiled it at 95 °C for 30 min. Subsequently, 10 µL of the solution was injected into a chromatography column, and the flow rate was maintained at 1 mL/min. The result was appraised with cyanidin and delphinidin.

2.4. Cell Culture

The human colon cancer cell line LoVo was obtained from the American Tissue Culture Collection. The LoVo cell line was cultured with DMEM complemented with 10% FBS, 2 mmol/L glutamine, 100 µg/mL antibiotic-antimycotic (100X), and 1 mmol/L HEPES buffer in an incubator with 5% CO_2 at 37 °C.

2.5. MTT Assay

For the MTT assay, we first used 3-[4,5-dimethylthiazol-2-yl]-2,5-diphenyltetrazolium bromide (MTT), a yellow water-soluble solid that can be metabolized by dehydrogenase in mitochondria in cells and cut off at the tetrazolium ring. The purple insoluble precipitate formazan (3-[4,5-dimethylthiazol-2-yl]-2,5-diphenyl-formazan) accumulated in the cells and subsequently dissolved in an organic solution. Its absorbance level was measured to be 563 nm (OD). LoVo cells were seeded in a culture dish overnight, and we treated them with drugs (0, 1, 2, 3 mg/mL) for a specified duration of 0, 12, 24, 36, and 48 h. Subsequently, the medium was aspirated, and another medium containing MTT was added for a two-hour incubation period. After removing the medium containing MTT, we dissolved the purple crystals with DMSO, and the optical density was measured to be 563 nm.

2.6. DAPI Staining for Cell Apoptosis

DAPI is a blue fluorescent dye that can penetrate cell membranes. DAPI fluorescent staining was used to observe chromatin changes during cell apoptosis. Human LoVo cancer cells (106 cells/mL) were treated with 1, 2, and 3 mg/mL HAs for 24 and 48 h. Post treatment, the cell monolayers were rinsed in PBS and fixed with 4% paraformaldehyde for 30 min at room temperature. After fixing, we prepared 0.2% Triton X-100 in PBS to treat the cells and permeabilized them three times. Subsequently, they were incubated with DAPI for 30 min and subjected to three additional PBS washes. The apoptotic nuclei (intensely stained with a fragmented nucleus and condensed chromatin) were examined and photographed under 200× magnification using a fluorescent microscope with a 340/380 nm excitation filter.

2.7. Flow Cytometry for Sub-G1 Phase

The DNA content of HAs-treated cells (3 mg/mL, 0–48 h) was determined with a Becton Dickinson flow cytometer with PI staining. The DNA content distribution was expressed as sub-G1, G0/G1, S, and G2/M phases. The percentage of hypodiploid cells (sub-G1) over the total cells was calculated and expressed in terms of the percent of apoptosis.

2.8. Measurement of Mitochondrial Membrane Potential

Mitochondrial membrane potential was assessed with JC-1 ($C_{25}H_{27}C_{14}IN_4$). LoVo Cells (1×10^6) were treated with HAs (1, 2, and 3 mg/mL) for 24 h, then subsequently harvested and washed with cold PBS twice. The cells were incubated with the fresh culture medium containing 2.5 μg/mL of JC-1 dye for 30 min at 37 °C. Cells were collected using centrifugation at 2000× *g* for 5 min and subsequently washed once with DMEM to enable observation. Red and green fluorescence emissions were photographed under fluorescence microscopy. Additionally, the cell was analyzed with flow cytometry using an excitation wavelength of 488 nm and an emission wavelength of 530 nm (green fluorescence). An increase in green fluorescent (FI) intensity represented mitochondrial swelling and loss of mitochondrial membrane potential. Quantitative analysis of the mitochondrial membrane potential was performed with a flow cytometer (FACS).

2.9. Western Blot Assay

After HAs treatment, total cell lysates were split into equal proteins in polyacrylamide gel (8–12%) and transferred to PVDF transfer membranes. Next, the blot membrane was washed with PBS three times after steeping in blocking buffer (PBS with 5% nonfat milk) for 1–3 h and incubated in primary antibody buffer at 4 °C overnight. Subsequently, the blot was washed with PBS three times and incubated in a secondary antibody buffer at 4 °C for 1 h. Lastly, the ECL detection system revealed the antigen-antibody complex. The relative band density of the image was quantitated with a densitometer.

2.10. Statistical Analysis

Data are reported as the means ± standard deviation of three independent experiments, and the groups were compared with one-way analysis of variance. Differences were considered significant at $p < 0.05$.

3. Results

3.1. Component Analysis

First, HPLC was used to detect the major components of hibiscus anthocyanins. The results of Figure 1 indicate that the peak of HAs most closely matched the retention time of standard cyanidin and delphinidin. The contents of cyanidin and delphinidin in HAs were 27% and 69%, respectively. These findings verified that cyanidin and delphinidin are the two primary components of HAs.

3.2. HAs Inducing Cytotoxicity Apoptosis

We investigated HAs' modulation of the cell viability of LoVo colorectal cancer cells. HAs with various concentrations (0, 1, 2, and 3 mg/mL) were administered over 0, 12, 24, and 48 h. HAs with the highest concentration (3 mg/mL) reached the IC_{50} after treatment for 24 h (Figure 2).

LoVo cells were cultured at 37 °C with various concentrations (0, 1, 2, and 3 mg/mL) of HAs. After 24 h and 48 h of action, the cell type and survival status were observed with an optical microscope and DAPI staining, respectively. Figure 3A indicates a negative correlation between HAs concentration and cell survival rate. However, higher HAs concentrations also increased cell shrinkage and apoptotic body generation. DAPI staining results indicated apoptosis through dense staining of the nucleus. Pro-apoptotic body presence was observed with flow cytometry to verify HAs causing apoptosis in LoVo cells. The cells were treated with the highest concentration (3 mg/mL) of HAs and LoVo cells were incubated at 37 °C. Figure 3B demonstrates that at various points in time (0, 12, 24, 36, and 48 h), the cells cultured for 48 h produced a substantial amount of pro-apoptotic bodies. DNA content during the sub-G1 phase of cells increased from 0.66% to 41.7%, representing an approximate 40% increase. These results indicate a time-dependent phenomenon (Figure 3C).

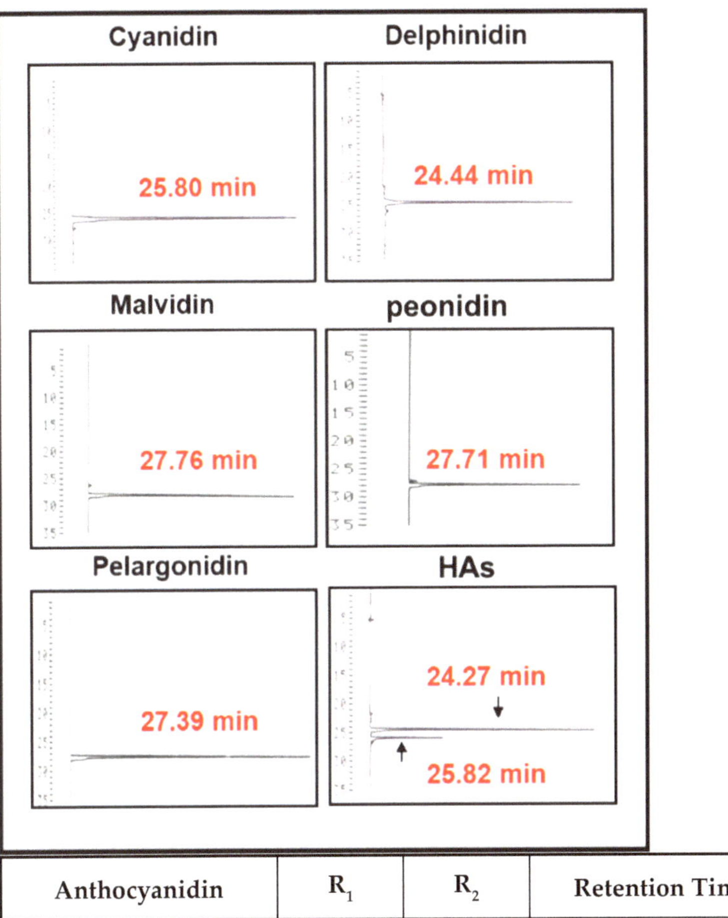

Anthocyanidin	R$_1$	R$_2$	Retention Time	% of HAs
Cyanidin(Cy)	OH	H	25.80 min	27%
Delphinidin(Dp)	OH	OH	24.44 min	69%
Malvidin(Mv)	OCH$_3$	OCH$_3$	27.76 min	
Pelargonidin(Pg)	H	H	27.39 min	4%
Peonidin(Pn)	OCH$_3$	H	27.71 min	
R$_3$, R$_4$ = H or sugar moiety				

Figure 1. The HPLC chromatogram of HAs. Arrow indicates peaks of delphinidin and cyanidin.

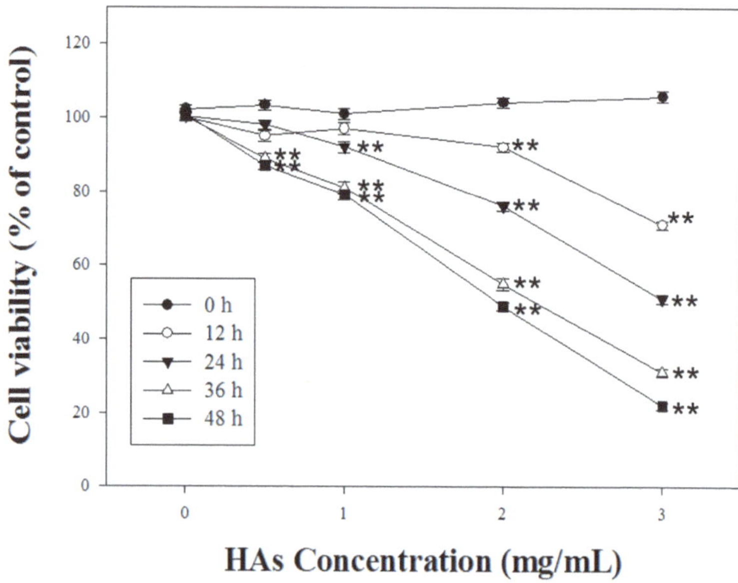

Figure 2. The cytotoxic effect of HAs on LoVo cells. The cells were treated with 0, 1, 2, and 3 mg/mL HAs for 0, 12, 24, 36, and 48 h and cell viability was measured with MTT assay. Data are shown as mean ± SD. Results were analyzed with Student's t-test. ** $p < 0.005$.

(A)

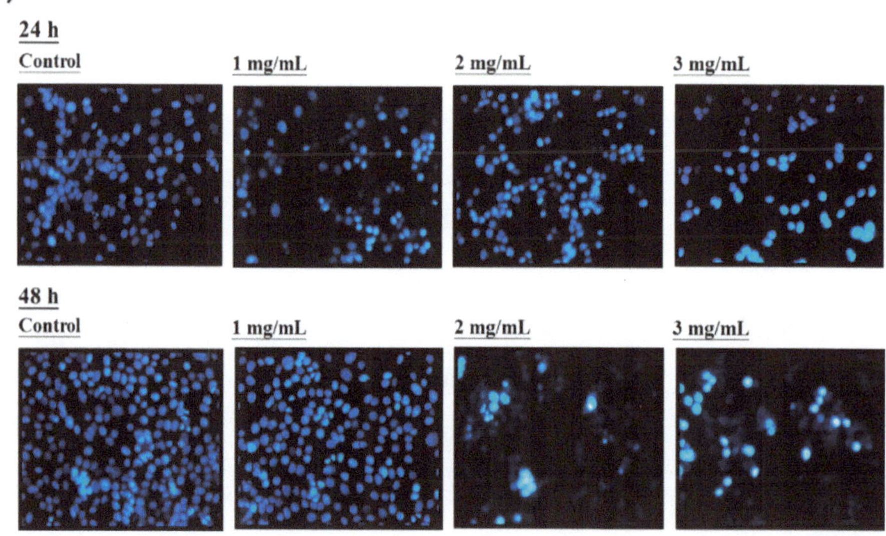

Figure 3. *Cont.*

(B)

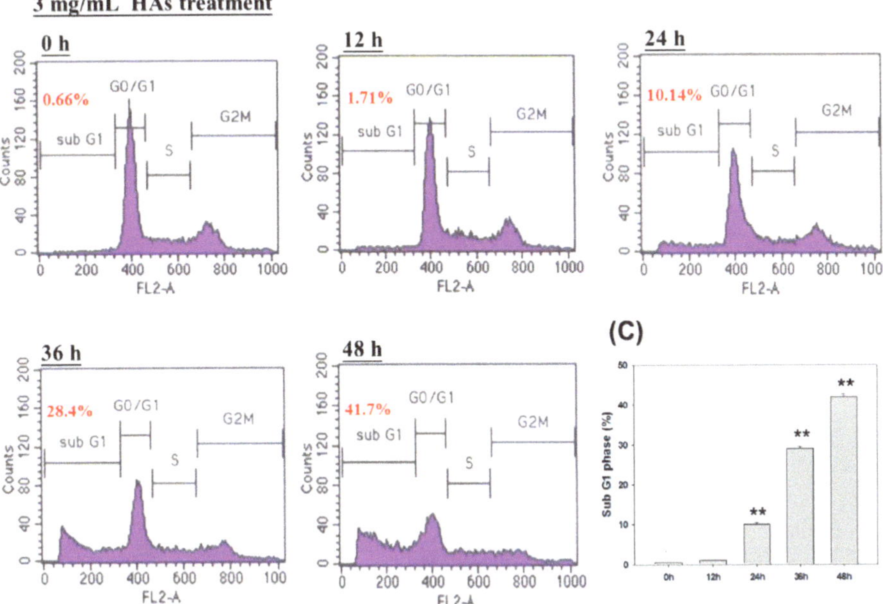

Figure 3. Apoptosis induction in HAs-treated LoVo cells. (**A**) LoVo cells treated with HAs under various concentrations (0, 1, 2, and 3 mg/mL) for 24 or 48 h. Apoptosis cells assayed by DAPI staining and observed with fluorescence microscopy (magnification 200×). (**B**) LoVo cells treated with HAs of 3 mg/mL for 0, 12, 24, 36, and 48 h. Cell apoptosis detected through PI staining and analyzed with flow cytometry. (**C**) DNA content in sub-G1 cell phase. Data shown as mean ± SD. Results analyzed with Student t test. ** $p < 0.005$.

3.3. Effects of HAs on the Mitochondrial Membrane Potential

LoVo colorectal cancer cells were treated with HAs at different concentrations (0, 1, 2, and 3 mg/mL). When JC-1 staining was conducted, LoVo cells not treated with HAs exhibited red fluorescence (Figure 4A). When Has concentration increased, the red fluorescence gradually transitioned to green fluorescence. Furthermore, the dye intensity of JC-1 was analyzed with flow cytometry to assess mitochondrial membrane potentials. The green absorption intensity of JC-1 dye was compared with that in the control group. Green fluorescence intensity increased from 0.04% to 77.56%. Treatment of LoVo cells with HAs resulted in the damage and disintegration of mitochondria in a dose-dependent manner. These results suggest that HAs treatment may induce a cell death program in colorectal cancer cells through the intrinsic apoptotic pathway.

3.4. The Expression of HAs on the Apoptosis-Associated Proteins of LoVo Cells

To investigate the mechanism underlying HAs-induced apoptosis in LoVo colorectal cancer cells, the Western immunoblotting method was employed. First, LoVo cells were treated with the highest concentration of HAs (3 mg/mL) at 37 °C for 0, 12, 24, and 36 h to assess the expression of apoptosis-related proteins. The results indicated that HAs elevated the expression of tBid, Bax, and Bad, triggering a cascade of mitochondrial death pathways. This process mediated an upregulation in the expression of the downstream protein apoptosis-inducing factor (AIF), culminating in cell apoptosis (Figure 5A). Bcl-2 family members that inhibit apoptosis, such as Bcl-xl, significantly decreased. Treatment with HAs induced a time-dependent increase in the level of cytochrome C and increased the expression of Fas/Fas L, which both activate the extrinsic apoptosis pathway (Figure 5B). Additionally, HAs triggered the activation of multiple members of the caspase family, namely caspase 3, 8, 9, and PARP involved in the DNA repair function (Figure 5B). Lastly, AMP-activated protein kinase (AMPK) includes the highly conserved serine/threonine protein kinase and may regulate cancer cell metabolic energy. Figure 5C indicates that increasing HAs concentration promoted p-AMPK expression and inhibited p-Akt expression. Therefore, HAs may regulate proteins associated with apoptotic pathways to induce apoptosis.

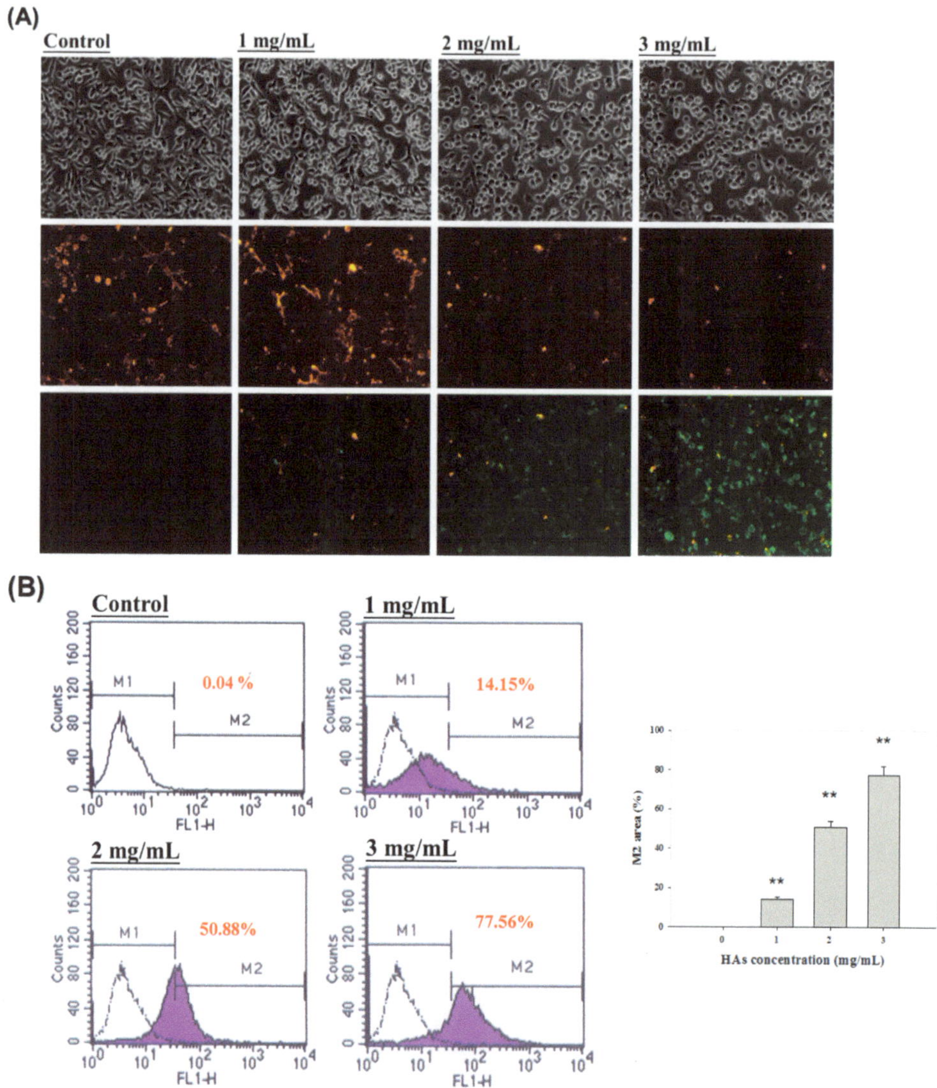

Figure 4. Mitochondria membrane potential ($\Delta\Psi$m) assessment of LoVo cells treated with HAs. (**A**) LoVo cells treated with HAs (1, 2, and 3 mg/mL) for 24 h; cells stained with JC-1. Monomeric form JC-1 with green fluorescence indicated the dissipation of MMP. (**B**) LoVo cells detected with flow cytometry. M2 area percentage represents the potential of mitochondrial membrane depolarization. Values are the mean SD ($n = 3$). ** $p < 0.005$, compared with the control group (0 mg/mL) of the HAs concentration, respectively.

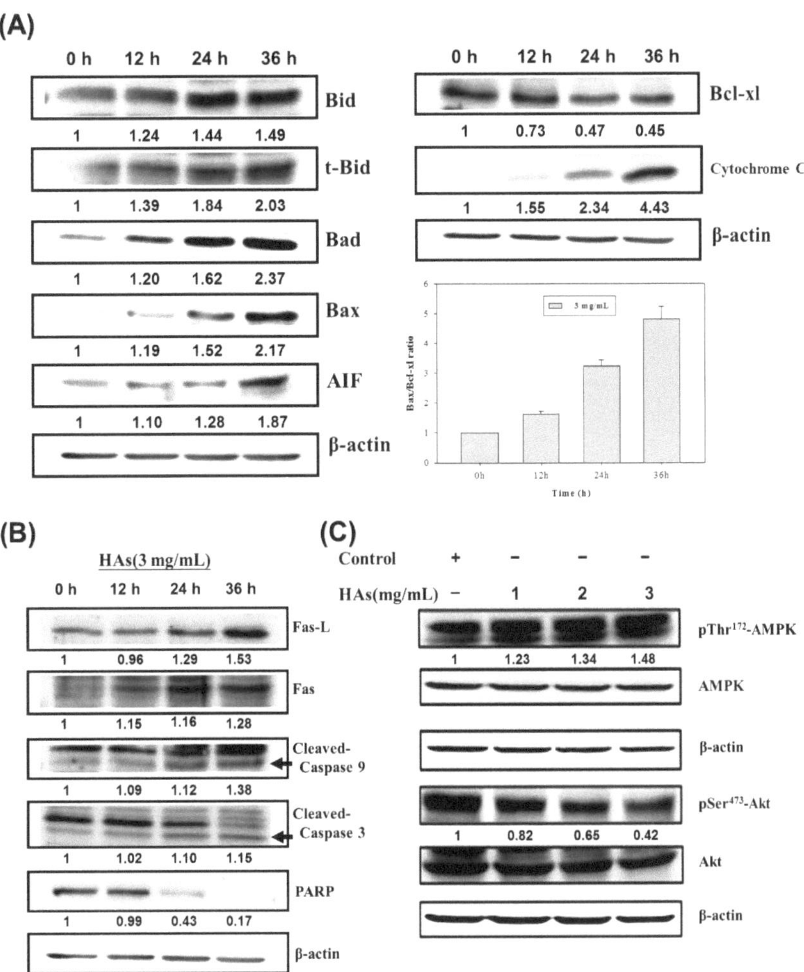

Figure 5. Effects of HAs on apoptosis-associated proteins in LoVo cells. (**A**) LoVo cells treated with HAs (3 mg/mL) for 24–48 h. Cells subjected to Western blotting to analyze signaling pathways related to mitochondria membrane potential. (**B**) Signaling pathways related to apoptosis receptors. (**C**) LoVo cells treated with 1, 2, and 3 mg/mL HAs for 24 h. Cell expression and phosphorylation of AMPK and Akt proteins were analyzed using Western blotting.

To elucidate why HAs triggered the apoptosis of LoVo colon cells, Compound C (an AMPK inhibitor) was added to the solution. The LoVo cells were treated with the highest concentration of HAs (3 mg/mL) at 37 °C for 24 h. Compound C (10 μM) was added one h before HAs to inhibit p-AMPK and verify the presumed relationships of other proteins. As illustrated in Figure 6A, the simultaneous addition of Compound C and HAs was associated with a lower p-AMPK expression compared with the use of HAs alone. Additionally, the p-AMPK expression significantly decreased when Compound C was used independently. These results demonstrate that HAs activated AMPK. Figure 6B indicates that p-Akt expression levels increased after p-AMPK inhibition, providing evidence that p-AMPK acts as an upstream regulator of p-Akt/Akt. Additionally, Figure 6B indicates that Compound C affected the expression of FasL and downstream proteins, including caspase, PARP, and AIF. This suggests that HAs modulated the exogenous and endogenous apoptosis pathways in LoVo cancer cells through the p-AMPK/FasL pathway to impede cell growth.

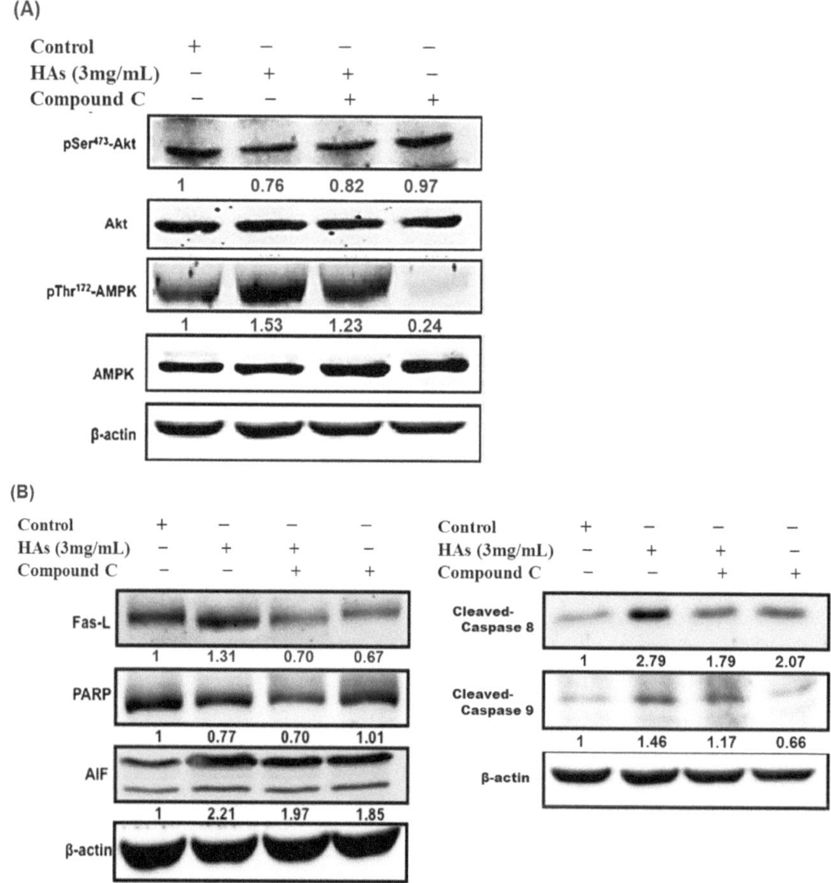

Figure 6. HAs-regulated expression of apoptotic proteins through the activation of the AMPK signaling pathway in LoVo cells. (**A**) LoVo cells treated with Compound C (10 μM) for 1 h and subsequently treated with HAs (3 mg/mL) for 24 h. Treated cells were subjected to Western blotting to analyze the phosphorylation of AMPK and Akt. (**B**) AMPK activating apoptotic proteins.

4. Discussion

Hibiscus extracts that are rich in anthocyanins exhibit a wide range of biological activities, including anti-apoptosis, anti-angiogenesis [14], antioxidation [24–26], anticancer, and anti-metastasis [21] in human cancer cells and animal models. Apoptosis is induced in cancer cells through the activation of extrinsic and intrinsic signaling pathways. AMPK and Akt proteins are pivotal in the apoptotic pathway, underscoring their relevance in cancer treatment [27,28]. Akt is a regulator of various biological functions closely related to cell growth, proliferation, and apoptosis. Our findings indicate that HAs upregulated the protein expression of AMPK and inhibited the phosphorylation of Akt protein, thereby inducing apoptosis in cells. AMPK is composed of trimeric protein polymers and is central to protein synthesis, cell growth, and apoptosis [29]. The experimental results indicated that AMPK protein expression increased the protein content of the death receptor Fas/Fas ligand on the cell membrane (Figure 5). The combination of Fas and Fas L caused FADD (Fas-associated protein with death domain) recruitment and death-inducing signaling complex (DISC) formation. This facilitated the binding of procaspase-8 to DISC, thereby activating caspase-8 and its downstream proteins and promoting a cascade of apoptosis reactions in cells (Figure 3).

Apoptosis is closely related to mitochondrial membrane integrity [30]. The cytochrome C and AIF were estimated using the lipophilic cationic probe JC-1. The results demonstrated the breakdown of the mitochondrial membrane potential (MMP), which occurs through the apoptotic pathway. The Bcl-2 family exerts its influence on the intrinsic apoptotic pathway by regulating the interplay between proapoptotic proteins and antiapoptotic proteins, thereby affecting the cell's commitment to the apoptotic pathway [31,32]. The BH3-only proapoptotic Bcl-2 members, namely Bax, Bad, and tBid, are antagonized by anti-apoptotic family members, including Bcl-xl [33]. Bcl-xl promotes cell survival through the maintenance of the integrity of the outer mitochondrial membrane and the prevention of the release of cytochrome c from mitochondria [34]. Caspases are a family of cysteine proteases that act as mediators of apoptosis and promote apoptotic morphology through the cleaving of various cellular substrates [35]. The intrinsic pathway of apoptosis involves the activation of caspase-9, which cleaves and activates caspase-3 [36]. Our experiments found that HAs enhanced the expression of pro-apoptotic members and decreased the expression of anti-apoptotic members to cause disruption of the MMP (Figures 4 and 5).

Cancer cells often have reduced mitochondrial oxidative phosphorylation, increased glycolysis, and altered metabolism of energy. AMPK activation can promote mitochondrial biogenesis, inhibit glycolysis, and restore energy balance in cancer cells, thereby regulating cell survival signaling pathways such as the Akt signaling pathway. Previous studies have indicated that natural compounds, including curcumin, gallocatechin, and paclitaxel, have been proven to be effective in treating breast cancer by inducing apoptosis [37]. Quercetin has also been shown to induce apoptosis in bladder cancer cells [38,39], while Baicalein induces apoptosis in human lung carcinoma A549 cells [40], all through the AMPK pathway. Therefore, we used Compound C to inhibit AMPK, which revealed that HAs induce apoptosis via the AMPK pathway and the apoptosis signaling pathway, which is triggered by Fas/Fas L. Figure 6 indicates that adding AMPK inhibitors increased the expression of p-Akt/Akt and significantly downregulated apoptosis-related proteins such as Fas L, PARP, AIF, and caspase-9, suggesting that HAs induced apoptosis of colorectal cancer LoVo cells through exogenous and intrinsic pathways. The inhibition of AMPK by Compound C highlights the inhibitor's critical role in mediating the pro-apoptotic effects of HAs. These findings reveal the underlying mechanism through which HAs affect cancer cell death and provide a potential avenue for therapy aimed at promoting cancer cell apoptosis through the manipulation of the AMPK pathway. Delphinidin, a compound found in anthocyanins, has declared protection against various chronic conditions, consisting of cancer, cardiovascular disease, and diabetes, by activating AMPK in vitro and in vivo studies. Delphinidin also regulates downstream targets in multiple cancer cell lines, including breast, colon, prostate, and liver cancer, thereby impeding cancer cell proliferation, inducing apoptosis, and suppressing tumor growth [41]. Previous studies have indicated that anthocyanins such as delphinidin may bind to enzymes and regulate their activity. One notable example is the binding of delphinidin to angiotensin-converting enzyme (ACE), enabling it to regulate ACE activity [15].

Further investigations are required to ascertain whether our experimental findings stem from the binding of delphinidin to AMPK kinase. Findings in the literature indicate that HAs are rich in polyphenols, flavonoids, vitamin C, and other active compounds that impart significant antioxidant, anti-inflammatory, and anticancer properties. However, despite the multifunctional and pharmacological characteristics of HAs, the anti-colorectal cancer effects are still unclear. Our study provides evidence that HAs inhibit the AMPK and Akt signaling pathways to promote apoptosis in colorectal cancer cells. Several mechanisms contribute to these effects, including the activation of the Fas protein, subsequent activation of Fas-mediated apoptosis-related protein pathways (such as the caspase-8/tBid pathway), modulation of the mitochondria membrane potential, and subsequent release of cytochrome

C from mitochondria. These effects suggest that the Hibiscus anthocyanidin extract can prevent and inhibit colorectal cancer by activating AMPK, inhibiting Akt, and increasing Fas/Fas L to produce intrinsic and extrinsic apoptosis. (Figure 7).

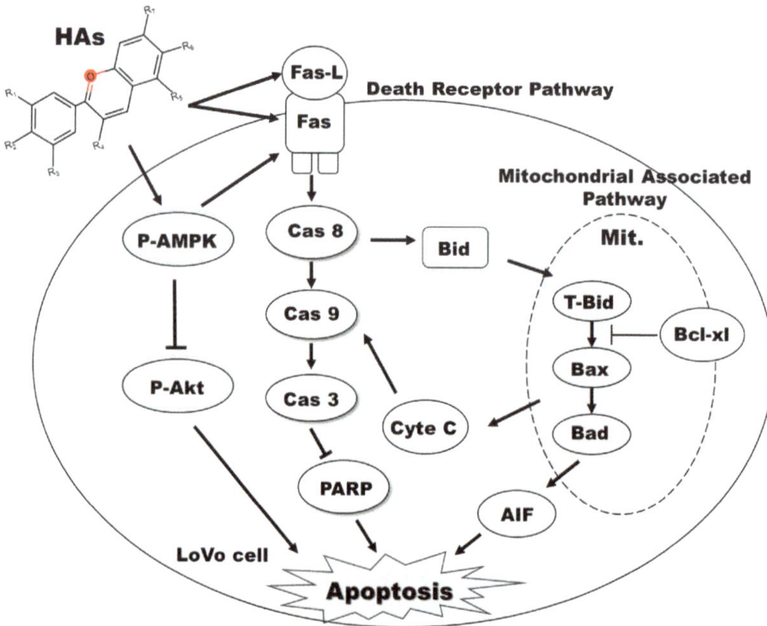

Figure 7. HAs induced LoVo cell apoptosis and inhibited the phosphorylation of Akt. AMPK activation increased expressions of Fas/Fas L through intrinsic and extrinsic apoptosis pathways.

Author Contributions: M.-C.T. and C.-C.C. carried out the experiments and summarized the results; T.-H.T. and Y.-C.C. assisted the process of the experiments; Y.-J.L. and I.-N.T. participated in the interpretation of the data; C.-C.W. and C.-J.W. provided supervision and wrote the manuscript. All authors have read and agreed to the published version of the manuscript.

Funding: This study was supported by grants from the Chung Shan Medical University Hospital (CSH-2023-C-007) and the Ministry of Science and Technology Grant (MOST 111-2320-B-040-025). Digital image analysis was performed in the Instrument Center of Chung Shan Medical University, which was supported by the Ministry of Science and Technology, the Ministry of Education, and Chung Shan Medical University Hospital, Taiwan.

Institutional Review Board Statement: Not applicable.

Informed Consent Statement: Not applicable.

Data Availability Statement: Not applicable.

Conflicts of Interest: The authors declare no conflict of interest.

References

1. Roy, S.; Nicholson, D.W. Cross-talk in cell death signaling. *J. Exp. Med.* **2000**, *192*, F21–F26. [CrossRef] [PubMed]
2. Hongmei, Z. Extrinsic and intrinsic apoptosis signal pathway review. In *Apoptosis and Medicine*; InTechOpen: London, UK, 2012.
3. Jin, Z.; El-Deiry, W.S. Overview of cell death signaling pathways. *Cancer Biol. Ther.* **2005**, *4*, 147–171. [CrossRef] [PubMed]
4. Li, W.; Saud, S.M.; Young, M.R.; Chen, G.; Hua, B. Targeting AMPK for cancer prevention and treatment. *Oncotarget* **2015**, *6*, 7365. [CrossRef] [PubMed]
5. Kim, H.-J.; Kim, S.-K.; Kim, B.-S.; Lee, S.-H.; Park, Y.-S.; Park, B.-K.; Kim, S.-J.; Kim, J.; Choi, C.; Kim, J.-S. Apoptotic effect of quercetin on HT-29 colon cancer cells via the AMPK signaling pathway. *J. Agric. Food Chem.* **2010**, *58*, 8643–8650. [CrossRef] [PubMed]

6. Hwang, J.-T.; Ha, J.; Park, I.-J.; Lee, S.-K.; Baik, H.W.; Kim, Y.M.; Park, O.J. Apoptotic effect of EGCG in HT-29 colon cancer cells via AMPK signal pathway. *Cancer Lett.* **2007**, *247*, 115–121. [CrossRef] [PubMed]

7. Rawla, P.; Sunkara, T.; Barsouk, A. Epidemiology of colorectal cancer: Incidence, mortality, survival, and risk factors. *Gastroenterol. Rev./Przegląd Gastroenterol.* **2019**, *14*, 89–103. [CrossRef]

8. Bhandari, A.; Woodhouse, M.; Gupta, S. Colorectal cancer is a leading cause of cancer incidence and mortality among adults younger than 50 years in the USA: A SEER-based analysis with comparison to other young-onset cancers. *J. Investig. Med.* **2017**, *65*, 311–315. [CrossRef]

9. Liang, Z.E.; Yi, Y.J.; Guo, Y.T.; Wang, R.C.; Hu, Q.L.; Xiong, X.Y. Inhibition of migration and induction of apoptosis in LoVo human colon cancer cells by polysaccharides from Ganoderma lucidum. *Mol. Med. Rep.* **2015**, *12*, 7629–7636. [CrossRef]

10. Hsu, H.H.; Chen, M.C.; Day, C.H.; Lin, Y.M.; Li, S.Y.; Tu, C.C.; Padma, V.V.; Shih, H.N.; Kuo, W.W.; Huang, C.Y. Thymoquinone suppresses migration of LoVo human colon cancer cells by reducing prostaglandin E2 induced COX-2 activation. *World J. Gastroenterol.* **2017**, *23*, 1171–1179. [CrossRef]

11. Lim, H.M.; Lee, J.; Nam, M.J.; Park, S.H. Acetylshikonin Induces Apoptosis in Human Colorectal Cancer HCT-15 and LoVo Cells via Nuclear Translocation of FOXO3 and ROS Level Elevation. *Oxid. Med. Cell Longev.* **2021**, *2021*, 6647107. [CrossRef]

12. D'Onofrio, N.; Cacciola, N.A.; Martino, E.; Borrelli, F.; Fiorino, F.; Lombardi, A.; Neglia, G.; Balestrieri, M.L.; Campanile, G. ROS-Mediated Apoptotic Cell Death of Human Colon Cancer LoVo Cells by Milk delta-Valerobetaine. *Sci. Rep.* **2020**, *10*, 8978. [CrossRef]

13. Atashrazm, F.; Lowenthal, R.M.; Woods, G.M.; Holloway, A.F.; Dickinson, J.L. Fucoidan and cancer: A multifunctional molecule with anti-tumor potential. *Mar. Drugs* **2015**, *13*, 2327–2346. [CrossRef]

14. Joshua, M.; Okere, C.; Yahaya, M.; Precious, O.; Dluya, T.; Um, J.-Y.; Neksumi, M.; Boyd, J.; Vincent-Tyndall, J.; Choo, D.-W. Disruption of angiogenesis by anthocyanin-rich extracts of *Hibiscus sabdariffa*. *Int. J. Sci. Eng. Res.* **2017**, *8*, 299. [CrossRef]

15. Da-Costa-Rocha, I.; Bonnlaender, B.; Sievers, H.; Pischel, I.; Heinrich, M. *Hibiscus sabdariffa* L.–A phytochemical and pharmacological review. *Food Chem.* **2014**, *165*, 424–443. [CrossRef]

16. Wang, L.-S.; Stoner, G.D. Anthocyanins and their role in cancer prevention. *Cancer Lett.* **2008**, *269*, 281–290. [CrossRef] [PubMed]

17. Zhou, K.; Raffoul, J.J. Potential anticancer properties of grape antioxidants. *J. Oncol.* **2012**, *2012*, 803294. [CrossRef] [PubMed]

18. Kaume, L.; Howard, L.R.; Devareddy, L. The blackberry fruit: A review on its composition and chemistry, metabolism and bioavailability, and health benefits. *J. Agric. Food Chem.* **2012**, *60*, 5716–5727. [CrossRef] [PubMed]

19. Huang, C.-C.; Hung, C.-H.; Chen, C.-C.; Kao, S.-H.; Wang, C.-J. *Hibiscus sabdariffa* polyphenol-enriched extract inhibits colon carcinoma metastasis associating with FAK and CD44/c-MET signaling. *J. Funct. Foods* **2018**, *48*, 542–550. [CrossRef]

20. Tsai, T.C.; Huang, H.P.; Chang, K.T.; Wang, C.J.; Chang, Y.C. Anthocyanins from roselle extract arrest cell cycle G2/M phase transition via ATM/Chk pathway in p53-deficient leukemia HL-60 cells. *Environ. Toxicol.* **2017**, *32*, 1290–1304. [CrossRef]

21. Su, C.-C.; Wang, C.-J.; Huang, K.-H.; Lee, Y.-J.; Chan, W.-M.; Chang, Y.-C. Anthocyanins from *Hibiscus sabdariffa* calyx attenuate in vitro and in vivo melanoma cancer metastasis. *J. Funct. Foods* **2018**, *48*, 614–631. [CrossRef]

22. Chang, Y.-C.; Huang, H.-P.; Hsu, J.-D.; Yang, S.-F.; Wang, C.-J. Hibiscus anthocyanins rich extract-induced apoptotic cell death in human promyelocytic leukemia cells. *Toxicol. Appl. Pharmacol.* **2005**, *205*, 201–212. [CrossRef] [PubMed]

23. Fuleki, T.F.F.J. Quantitative method for anthocyanins. extraction and determination of total anthcyanins in cranberries. *J. Food Sci.* **1968**, *33*, 266–274. [CrossRef]

24. Kao, E.S.; Hsu, J.D.; Wang, C.J.; Yang, S.H.; Cheng, S.Y.; Lee, H.J. Polyphenols extracted from *Hibiscus sabdariffa* L. inhibited lipopolysaccharide-induced inflammation by improving antioxidative conditions and regulating cyclooxygenase-2 expression. *Biosci. Biotechnol. Biochem.* **2009**, *73*, 385–390. [CrossRef] [PubMed]

25. Lin, H.H.; Chen, J.H.; Wang, C.J. Chemopreventive properties and molecular mechanisms of the bioactive compounds in *Hibiscus sabdariffa* Linne. *Curr. Med. Chem.* **2011**, *18*, 1245–1254. [CrossRef]

26. Liu, L.C.; Wang, C.J.; Lee, C.C.; Su, S.C.; Chen, H.L.; Hsu, J.D.; Lee, H.J. Aqueous extract of *Hibiscus sabdariffa* L. decelerates acetaminophen-induced acute liver damage by reducing cell death and oxidative stress in mouse experimental models. *J. Sci. Food Agric.* **2010**, *90*, 329–337. [CrossRef]

27. Tanaka, M.; Sato, A.; Kishimoto, Y.; Mabashi-Asazuma, H.; Kondo, K.; Iida, K. Gallic acid inhibits lipid accumulation via AMPK pathway and suppresses apoptosis and macrophage-mediated inflammation in hepatocytes. *Nutrients* **2020**, *12*, 1479. [CrossRef]

28. Zhao, Y.; Hu, X.; Liu, Y.; Dong, S.; Wen, Z.; He, W.; Zhang, S.; Huang, Q.; Shi, M. ROS signaling under metabolic stress: Cross-talk between AMPK and AKT pathway. *Mol. Cancer* **2017**, *16*, 79. [CrossRef]

29. Villanueva-Paz, M.; Cotán, D.; Garrido-Maraver, J.; Oropesa-Ávila, M.; de la Mata, M.; Delgado-Pavón, A.; de Lavera, I.; Alcocer-Gómez, E.; Álvarez-Córdoba, M.; Sánchez-Alcázar, J.A. AMPK regulation of cell growth, apoptosis, autophagy, and bioenergetics. *AMP-Act. Protein Kinase* **2016**, *107*, 45–71.

30. Green, D.R.; Reed, J.C. Mitochondria and apoptosis. *Science* **1998**, *281*, 1309–1312. [CrossRef]

31. Warren, C.F.; Wong-Brown, M.W.; Bowden, N.A. BCL-2 family isoforms in apoptosis and cancer. *Cell Death Dis.* **2019**, *10*, 177. [CrossRef]

32. Singh, R.; Letai, A.; Sarosiek, K. Regulation of apoptosis in health and disease: The balancing act of BCL-2 family proteins. *Nat. Rev. Mol. Cell Biol.* **2019**, *20*, 175–193. [CrossRef]

33. Ke, F.S.; Holloway, S.; Uren, R.T.; Wong, A.W.; Little, M.H.; Kluck, R.M.; Voss, A.K.; Strasser, A. The BCL-2 family member BID plays a role during embryonic development in addition to its BH3-only protein function by acting in parallel to BAX, BAK and BOK. *EMBO J.* **2022**, *41*, e110300. [CrossRef] [PubMed]

34. Kalpage, H.A.; Wan, J.; Morse, P.T.; Zurek, M.P.; Turner, A.A.; Khobeir, A.; Yazdi, N.; Hakim, L.; Liu, J.; Vaishnav, A. Cytochrome c phosphorylation: Control of mitochondrial electron transport chain flux and apoptosis. *Int. J. Biochem. Cell Biol.* **2020**, *121*, 105704. [CrossRef] [PubMed]

35. Boice, A.; Bouchier-Hayes, L. Targeting apoptotic caspases in cancer. *Biochim. Biophys. Acta (BBA)-Mol. Cell Res.* **2020**, *1867*, 118688. [CrossRef]

36. Asadi, M.; Taghizadeh, S.; Kaviani, E.; Vakili, O.; Taheri-Anganeh, M.; Tahamtan, M.; Savardashtaki, A. Caspase-3: Structure, function, and biotechnological aspects. *Biotechnol. Appl. Biochem.* **2022**, *69*, 1633–1645. [CrossRef]

37. Peng, B.; Zhang, S.Y.; Chan, K.I.; Zhong, Z.F.; Wang, Y.T. Novel Anti-Cancer Products Targeting AMPK: Natural Herbal Medicine against Breast Cancer. *Molecules* **2023**, *28*, 740. [CrossRef] [PubMed]

38. Wu, P.; Liu, S.; Su, J.; Chen, J.; Li, L.; Zhang, R.; Chen, T. Apoptosis triggered by isoquercitrin in bladder cancer cells by activating the AMPK-activated protein kinase pathway. *Food Funct.* **2017**, *8*, 3707–3722. [CrossRef]

39. Su, Q.; Peng, M.; Zhang, Y.; Xu, W.; Darko, K.O.; Tao, T.; Huang, Y.; Tao, X.; Yang, X. Quercetin induces bladder cancer cells apoptosis by activation of AMPK signaling pathway. *Am. J. Cancer Res.* **2016**, *6*, 498–508.

40. Kim, H.J.; Park, C.; Han, M.H.; Hong, S.H.; Kim, G.Y.; Hong, S.H.; Kim, N.D.; Choi, Y.H. Baicalein Induces Caspase-dependent Apoptosis Associated with the Generation of ROS and the Activation of AMPK in Human Lung Carcinoma A549 Cells. *Drug Dev. Res.* **2016**, *77*, 73–86. [CrossRef]

41. Wu, C.-H.; Huang, C.-C.; Hung, C.-H.; Yao, F.-Y.; Wang, C.-J.; Chang, Y.-C. Delphinidin-rich extracts of *Hibiscus sabdariffa* L. trigger mitochondria-derived autophagy and necrosis through reactive oxygen species in human breast cancer cells. *J. Funct. Foods* **2016**, *25*, 279–290. [CrossRef]

 *nutrients*

MDPI

Article

Molecular Evidence of Breast Cancer Cell Proliferation Inhibition by a Combination of Selected Qatari Medicinal Plants Crude Extracts

Nouralhuda Alateyah [1,†], **Mohammed Alsafran** [1,2,†], **Kamal Usman** [2] and **Allal Ouhtit** [1,*]

1 Biological Sciences Program, Department of Biological & Environmental Sciences, College of Arts and Science, Qatar University, Doha P.O. Box 2713, Qatar; na1601400@student.qu.edu.qa (N.A.); m.alsafran@qu.edu.qa (M.A.)

2 Agricultural Research Station (ARS), Office of VP for Research & Graduate Studies, Qatar University, Doha P.O. Box 2713, Qatar; kusman@qu.edu.qa

* Correspondence: aouhtit@qu.edu.qa

† These authors contributed equally to this work.

Abstract: Breast cancer (BC) is the most common malignancy, and conventional medicine has failed to establish efficient treatment modalities. Conventional medicine failed due to lack of knowledge of the mechanisms that underpin the onset and metastasis of tumors, as well as resistance to treatment regimen. However, Complementary and Alternative medicine (CAM) modalities are currently drawing the attention of both the public and health professionals. Our study examined the effect of a super-combination (SC) of crude extracts, which were isolated from three selected Qatari medicinal plants, on the proliferation, motility and death of BC cells. Our results revealed that SC attenuated cell growth and caused the cell death of MDA-MB-231 cancer cells when compared to human normal neonatal fibroblast cells. On the other hand, functional assays showed that SC reduced BC cell migration and invasion, respectively. SC-inhibited cell cycle and SC-regulated apoptosis was most likely mediated by p53/p21 pathway and p53-regulated Bax/BCL-2/Caspace-3 pathway. Our ongoing experiments aim to validate these in vitro findings in vivo using a BC-Xenograft mouse model. These findings support our hypothesis that SC inhibited BC cell proliferation and induced apoptosis. These findings lay the foundation for further experiments, aiming to validate SC as an effective chemoprevention and/or chemotherapeutic strategy that can ultimately pave the way towards translational research/clinical trials for the eradication of BC.

Keywords: breast tumors; cell growth and cell survival; cell death; cell motility and invasion

Citation: Alateyah, N.; Alsafran, M.; Usman, K.; Ouhtit, A. Molecular Evidence of Breast Cancer Cell Proliferation Inhibition by a Combination of Selected Qatari Medicinal Plants Crude Extracts. *Nutrients* **2023**, *15*, 4276. https://doi.org/10.3390/nu15194276

Academic Editors: Ching-Hsein Chen, Yi-Wen Liu and Edgard Delvin

Received: 6 September 2023
Revised: 29 September 2023
Accepted: 3 October 2023
Published: 7 October 2023

1. Introduction

Breast cancer (BC), a serious global health issue, is the most prominent type of tumors in women [1,2]. Malignant breast tumors can metastasize locally or to distant organs through the invasion of cells from the primary tumor to distant sites [3]. There has been major progress in technology that has led to breakthroughs in cancer research, paving the way towards designing specific-targeted therapies against cancer. However, conventional medicine (CM) has failed to establish the ultimate cure for cancer. This failure is due to the poor understanding of the specific molecular mechanisms associated with tumor development within various groups of patients, drug resistance, and the failure of clinical trials and the current therapies in curing cancer. Hence, the field of Complementary and Alternative medicine (CAM) is drawing more attention, especially in countries where CAM has been practiced for a long time [4,5]. CAM treatment modalities use crude extracts isolated from various medicinal plants. Each of these crude extracts encompass a variety of compounds or phytochemicals known for their anti-cancer properties [6,7].

Various natural compounds isolated from medicinal plants, commonly found in Qatar and the Gulf region, have been extensively explored for their anti-cancer properties in inhibiting cell proliferation and inducing apoptosis of different cancer cell lines. These plants include *Artemisia herba-alba* [8], Aloe vera [9], *Ferula asafetida* [10], Frankincense (*Boswellia sacra*) [11], Sidr (*Ziziphus spina-christi*) [12,13], Dates (*Phoenix dactylifera*) [14].

The induction of apoptosis (programmed cell death) by medicinal plant extracts involves complex mechanisms. These mechanisms can vary depending on the specific plant and its chemical composition. The common mechanisms that are often associated with the induction of apoptosis by plant extracts include the disruption of the mitochondrial membrane potential and favoring the pro-apoptotic proteins over the anti-apoptotic proteins to trigger apoptosis [15]. Plants extracts can also inhibit the cell proliferation of cancer cells by inducing cell cycle arrest and apoptosis [16]. On the other hand, phytochemical compounds isolated from plant extract can also inhibit the PI3K/Akt or the NF-κB (Phosphatidylinositol 3-kinase/Alpha serine/therionine-protein Kinase, Nuclear Factor-Kappa B) survival signaling pathways, which are commonly dysregulated in cancer cells to promote cell survival [17–19].

For the last two decades, our laboratory has focused on identifying an effective combination of phytochemicals/bioactives that can inhibit tumor cell proliferation and induce cancer cell death. The combination of bioactives can target, simultaneously, different signaling pathways that control cell proliferation and apoptosis [20]. Thus, we examined the effect of a super-combination (SC) of phyto-bioactives present in crude extracts, which were isolated from three selected Qatari medicinal plants (*Haplophyllum tuberculatum* (A), *Erodium glaucophyllum* (B), and *Heliotropium bacciferum* (C)) on the cell growth, cell survival, cell motility and invasion of the BC cell line MDA-MB-231. The plants used in the present study were selected based on our preliminary work, as well as their anti-cancer, anti-inflammatory properties, their content of antioxidants and well-known active compounds [21–26]. Moreover, the expression levels of proteins associated with cell cycle control and apoptosis were determined to identify the potential molecular mechanisms underlying SC-induced cell death.

2. Materials and Methods

2.1. Plants Collection

Fresh wild ABC plant species growing in Qatar's desert sandy soil were collected during April and June 2016. These plants were identified by the second author (AlSafran M.) according to the standard sampling guidelines for local Qatari Medicinal plants [27]. The samples were collected in a polyethylene bag, immediately transported to the laboratory, and kept at −80 °C. Later, the samples were ground into a fine powder using a coffee grinder and sieved through a 24-mesh sieve for homogeneity. Powdered materials were then kept at −20 °C and protected from light until use.

2.2. Plant Crude Extraction

Crude extracts were isolated from the plants in methanol, as previously described with slight modifications [28]. Firstly, stored plant samples were dried, then 2 g of dried material was ground into a powder and transferred into a 50 mL polyethylene centrifuge tube. Powdered material was then mixed in 30 mL methanol for 24 h at ambient temperature in water bath, followed by ultrasonic frequency (50–60 Hz) bath (FS100B, Decon Laboratories Ltd., Hove, UK) for 60 min at 25 °C and. Before sonication, the tubes were vortexed for 10–15 s to enhance the yield of the extracts. The extract was centrifuged (Centurion Scientific C2) for 10 min at 2000 rpm, and the supernatant was transferred to an empty tube for storage at 4 °C. Plant residues were re-extracted with 30 mL of methanol, using the same procedure after 48 h period. A rotary evaporator (Laborota 4000, Heidolph, Germany) was used to evaporate combined supernatants to dryness under vacuum at 40 °C. Afterwards, the extracted product was then dissolved in 4 mL methanol, transferred to an amber glass vial, and kept at 4 °C in the refrigerator until use.

2.3. Determination of Total Phenolic Contents

The total phenolic contents were measured by following Folin–Ciocalteau (FC) method [29], using plant (50 μL) extracts mixed with 2.5 mL of 1:10 Folin–Ciocalteau reagent. Briefly, upon adding saturated Na_2CO_3 (75 g/L) solution and incubation at 30 °C for 1.5 h with intermittent shaking, the resulting blue-colored solution was measured at 765 nm using a spectrophotometer (Lambda 25, Perkin Elmer, Waltham, MA, USA). The measurements were carried out using a calibration curve of gallic acid mixed in methanol, and total phenolic content was expressed as Gallic Acid Equivalent (GAE) in mg/g of dry weight (DW) of the sample.

2.4. Determination of Total Flavonoid Contents

The aluminum chloride method was used to determine the total flavonoid content, with a few modifications from Chia-Chi et al. (2002) [30]. Each plant extract (0.5 mL) was mixed separately with 95% ethanol (1.5 mL), 10% aluminum chloride (0.1 mL), 1 M potassium acetate (0.1 mL), and distilled water (2.8 mL). The calibration curve was established by preparing Quercetin (QE) solution into 80% ethanol at a concentration of 25–100 μg/mL. The absorbance of the reaction mixture was measured at 415 nm after incubation for 30 min at room temperature, using a spectrophotometer (Lambda 25, Perkin Elmer), and total flavonoid content was expressed as mg QE/g DW.

2.5. Cell Lines and Cell Culture

The highly metastatic aggressive triple negative BC cell MDA-MB-231 was obtained from the American Type Culture Collection (ATCC, Manassas, Virginia). The primary dermal normal human neonatal fibroblast (HDFn), a cell line isolated from neonatal dermis were obtained from Weill Cornell Medicine-Qatar [20,31,32]. The cells were supplemented with DMEM media containing 10% fetal bovine serum (FBS), 1% penicillin–streptomycin, and 5% L-glutamine and then cultured and incubated at 37 °C in a humidified incubator adjusted to 5% CO_2. All experiments were carried out until the cells reached 60 to 70% confluency.

2.6. Alamar-Blue Cell Proliferation Assay

Approximately 10,000 cells in each cell line were seeded in 96-wells culture plates in triplicates using DMEM medium to encompass all the ingredients, as described above. The cells were treated either with individual compounds, the SC combination, or with DMSO as a control. Based on preliminary optimization experiments, MDA-MB-231 and HDFn cells were treated with 0.798 mg/mL SC combination (the SC contained A = 0.4 mg/mL, B = 0.26 mg/mL, and C = 0.138 mg/mL). The SC killed more than 50% of the MDA-MB-231 cells compared to normal cells. This concentration of the SC was then selected for all the remaining experiments.

Following SC treatment for 24 and 48 h and PBS washing, the cells were incubated in 10% Alamar-Blue reagent (Invitrogen, Thermo Fisher Scientific, Waltham, MA, USA) for 4 h, as we have previously described [8]. Fluorescence was measured at 560 nm and 590 nm TECAN infinite M200 plate reader (Männedorf, Switzerland). The rate of cell proliferation/cell viability was determined based on the fluorescence of SC-treated cells compared to the control cells.

2.7. Morphological Study

Upon treatment with the sub-optimal dose of the SC, both MDA-MB-231 and human fibroblast cells were monitored alive under the microscope (OPTIKA Microscopes, Ponteranica, Italy) at 24 and 48 h post treatment [31,32].

2.8. Wound Healing Assay

To determine the effect of the SC treatment on BC cell migration, wound healing assay was applied to MDA-MB-231 cells, as we have previously reported [31]. Briefly, the cells were incubated in 6-well plates, and prior to the treatment, a straight scratch was made in each well using a sterile pipette tip. Following removal of floating cells and debris in a wash with PBS, the cells were treated with 0.798 mg/mL of SC and monitored for 48 h. During the treatment, the cells were observed and photographed at different time points (0, 24, and 48 h) to determine the wound closure of the scratch. Images were examined using the ImageJ application, and the data were analyzed and plotted in graphs.

2.9. Invasion Assay

The effect of the SC treatment on BC cell invasion was examined by Boyden cell invasion assay using Matrigel-coated chambers, as we have previously reported [32]. Briefly, treated cells were seeded onto the upper chambers of Matrigel plates, while Dulbeccos Modified Eagle Medium (DMEM) containing 10% FBS was added to the bottom chamber to serve as chemoattractant. Following cell incubation for a period of 48 h, cells that did not invade through the Matrigel were gently washed with phosphate buffered saline (PBS) and removed from the upper chambers using a sterile cotton swab. In contrast, cells that invaded to the lower chambers were fixed with methanol and formaldehyde for 10 min and then stained in 0.5% crystal violet. Once the crystal violet stain was removed with PBS, the cells were observed under the microscope, and photos were taken for analysis of cell invasion. The percentage of the inhibition of cell invasion was determined for the triplicates using Image J software version 1.53t (NIH, LOCI, University of Wisconsin, Madison, WI, USA).

2.10. Western Blot Analysis

The molecular mechanisms by which the SC inhibited cell growth and induced cell death were investigated by analyzing the expression levels of proteins associated with cell cycle (p53 and p21) as well as apoptosis (p53, BCL-2, Bax, Casp3 and Casp7), respectively, using Western blot analysis, as we have reported [4,8,18]. Pursuant to this goal, protein lysates were isolated from the 48 h SC-treated MDA-MB-231 cells (0.798 mg/mL), and 30 µg of total cell lysates were separated and transferred to PVDF membranes. Immunoblotting was carried out by probing the membranes overnight with the primary antibodies, including anti-rabbit p53 (Cell Signaling Technology, CST #2527S) anti-rabbit Bax (Cell Signaling Technology, CST #5023S), anti-mouse BcCL-2 (Cell Signaling Technology, CST #15071S), anti-rabbit Casp3 (Cell Signaling Technology, CST #9662), anti-mouse Casp7 (CST# AF823), and anti-rabbit Casp7 (CST# AF823). The anti-rabbit β-actin (Cell Signaling Technology, CST #4970S) served as a loading control. Following the incubation of the membranes with the secondary antibody for 2 h, immunoreactivity was visualized using ECL substrate (Pierce Biotechnology, Rockford, IL, USA). Images obtained with iBright developer (Thermo Fisher Scientific, Waltham, MA, USA), were analyzed using Image J software for the relative amounts of protein expression.

2.11. Statistical Analysis

Data obtained from the comparison between treated and untreated cells in the triplicated experiments were analyzed using Statistical Package of the Social Sciences (SPSS) and presented as mean $\pm$ Standard Error of the Mean (S.E.M). Means were compared using *t*-test, and differences were considered significant when $p < 0.05$.

3. Results

3.1. Total Phenolic Contents

Plant phenolic compounds have a variety of physiological activities [33], and many of these compounds are used in pharmaceuticals. Phenolic compounds exhibit health-improving properties, including reducing inflammation, free radicals, and cancer risk [34]. Figure 1a shows the total phenolic contents (TPC) measured using the FC method for the extracts with GAE as a standard sample. TPC values were calculated based on the calibration curve $y = 0.0042x - 0.0005$ $R^2 = 0.9993$. It is calculated from the absorbance "x" and the concentration of the solution "y" (µg/mL) expressed as mg GAE/g DW. The TPC values were 10.55 mg GAE/g DW, 75.76 mg GAE/g DW, 13.97 mg GAE/g DW for *Haplophyllum tuberculatum*, *Erodium glaucophyllum* and *Heliotropium bacciferum*, respectively. Previous studies have reported the TPC in *E. glaucophyllum* [35], *H. bacciferum* [36], and *H. tuberculatum* [37].

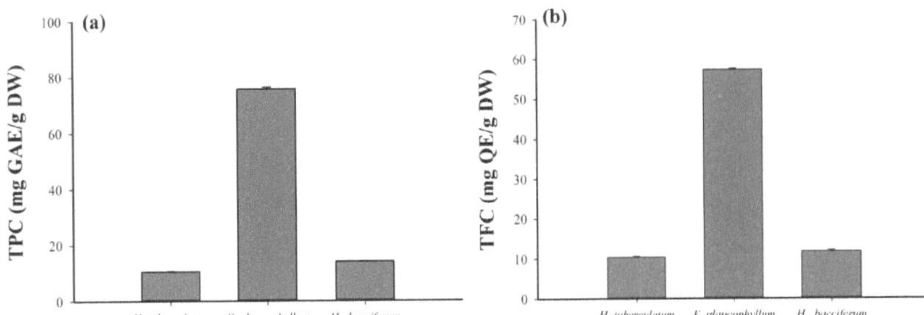

Figure 1. Total phenolic and flavonoid contents in the plants were used in the study. (**a**) Total phenolic contents (as mg gallic acid equivalents (GAE)/g of dry weight) and (**b**) total flavonoid content (as mg quercetin equivalents (QE)/g of dry weight) methanol extracts of selected plants.

3.2. Total Flavonoid Contents (TFC)

Flavonoids, which comprise the most abundant phenolic compound group in nature, are potent antioxidant, antimicrobial, antiulcer, anti-diabetic, hepato-protective, and anticarcinogenic agents [36]. A colorimetric assay was performed to evaluate the total flavonoid content (TFC) with QE as a standard sample. TFC values were calculated based on the calibration curve y = 0.001x + 0.0012 R^2 = 0.9994. It is calculated from the absorbance x and the solution concentration y (µg/mL) expressed as mg QE/g DW (Figure 1b). The TFC values were 10.42 mg QE/g DW, 57.28 mg QE/g DW, 11.75 mg QE/g DW) for *H. tuberculatum*, *E. glaucophyllum* and *H. bacciferum*, respectively.

3.3. Establishment of the Optimal Dose of the Super-Combination

To determine the optimal dose of the super-combination (SC) that will be used in all experiments, (i) we determined the optimal dose for each individual crude extract isolated from each of the three selected Qatari medicinal plants (*H. tuberculatum* (A), *E. glaucophyllum* (B), and *H. bacciferum* (C)) by examining the effect of a range of ascending doses on the proliferation of MDA-MB-231 BC cell lines, and (ii) we examined the effect of different combinations (SC1 to SC4) on the growth of both MDA-MB-231 and the normal human dermal neonatal fibroblasts (HDF), which were used as control normal cells, while DMSO was used as vehicle control. Both cells were treated for 48 h with varying concentrations of SC (A + B + C), as shown in Table 1. Our results revealed that while the combinations tested all significantly inhibited MDA-MB-231 cell proliferation (Figure 2), SC1 showed the highest effect among all the SCs (Figure 2A).

Table 1. The different concentrations of the super-combination (SC) examined on the proliferation of both MDA-MB-231 and the normal human dermal neonatal fibroblasts.

SC 1	A = 10%, B = 5%, C = 3%
SC 2	A = 10%, B = 10%, C = 3%
SC 3	A = 8%, B = 10%, C = 3%
SC 4	A = 5%, B = 10%, C = 4%

3.4. Effect of the Super-Combination on the Cell Morphology

Next, we examined and compared the morphology of MDA-MB-231 cells and the human neonatal fibroblasts in the absence or presence of SC1 optimal dose. In the absence of SC1 treatment, MDA-MB-231 cells exhibited smooth epithelial with prominent nuclei. Compared to control cells, SC1-treated MDA-MB-231 lost cell–cell contact and detached from the tissue culture dish, suggesting SC1-induced cell death. More interestingly, however, the neonatal fibroblast control cells showed normal cell–cell contact and were healthy (Figure 2C).

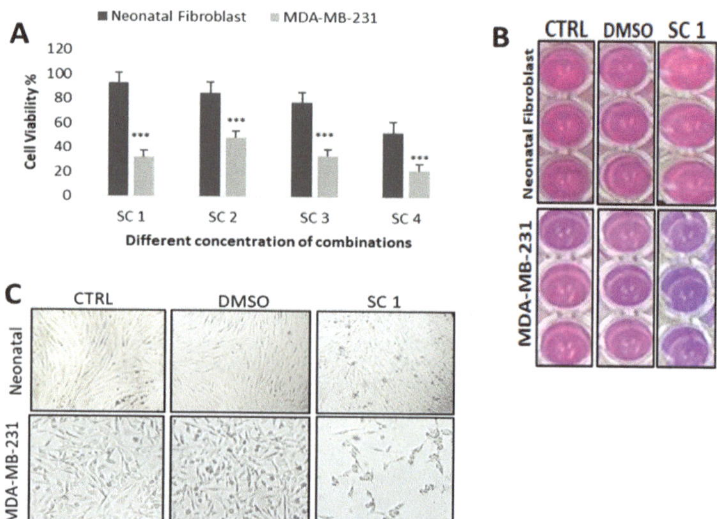

Figure 2. Effect of the optimal SC on breast cancer cell proliferation. (**A**,**C**) Effect of the SC on cell viability and morphology of MDA-231 cancer cells. (**B**) SC-treated cells on Alamar-blue colors (Pink: living cells; blue: dead cells). The results are statistically significant as *** $p < 0.0001$.

3.5. Effect of the Super-Combination on the Cell Migration and Invasion

Next, we explored the effect of SC1 on BC cell migration and invasion using wound-healing and invasion Boyden chamber assays, respectively. Our results showed that SC1 significantly reduced BC cell migration (Figure 3) and invasion (Figure 4) by ~73% (Figure 3) and 90% (Figure 4), respectively, when compared to the control ($p < 0.05$).

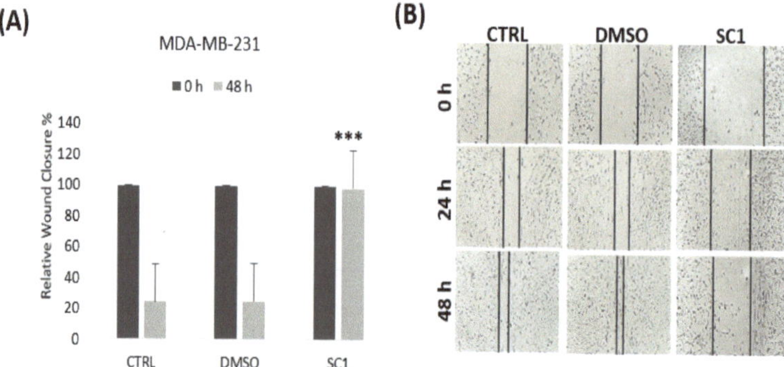

Figure 3. Effect of SC on MDA-MB-231 cell migration using scratch wound-healing assay. (**A**) SC1 inhibited MDA-MB-231 cell migration. (**B**) Semi-quantitative analysis of the relative wound closure (%) showing an increase in the wound closure by the SC, indicating an inhibition of MDA-231 cell migration. Representative images are mean values of the percentage of wound closure ± SEM ($n = 3$): *** $p < 0.001$.

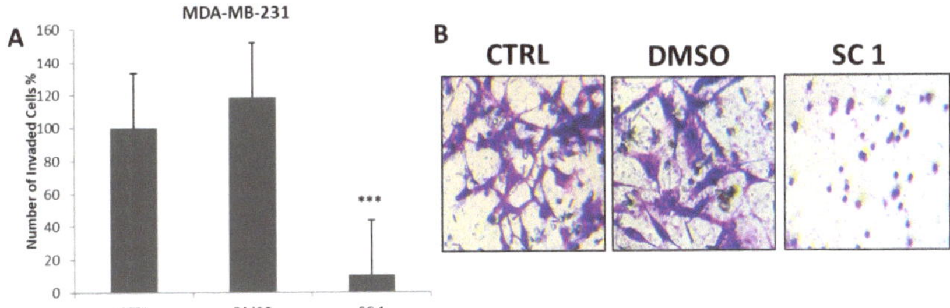

Figure 4. Effect of SC on MDA-MB-231 cell invasion using Boyden chamber assay. (**A**) Semi-quantitative analysis showing a decrease in invaded cells by the SC, indicating an inhibition of MDA-231 cell invasion. (**B**) Representative images of the invasion assay in MDA-MB-231 BC cells. SC 1 inhibited cell invasion of MDA-MB-231 by ~90% in comparison to control cells (*** $p < 0.0001$).

3.6. Molecular Mechanisms Mediating Super-Combination (SC)-Induced Apoptosis

Last but not least, we investigated the molecular mechanisms that underpin SC1-induced apoptosis. Therefore, after 48 h of SC treatment, protein lysates were collected from MDA-MB-231-treated cells as well as from their matched control DMSO untreated cells. The protein lysates were examined for the expression levels of the key genes associated with both cell cycle (p53 and p21) and apoptosis (p53, Bax, BCL-2, Casp3 and Casp7) using Western blot analysis (Figure 5). Interestingly, the expression of the mutant p53 was significantly attenuated by SC1 treatment in MDA-MB-231 as compared to the control cells (Figure 5A). Curiously, while MDA-MB-231 cells do express high levels of the stable mutant p53, the expression of p21 was remarkably inhibited, perhaps due to the inhibition of p53 (Figure 5A). However, SC1 significantly increased Bax and decreased BCL-2, indicating that, most likely, SC1 induced apoptosis via the intrinsic mitochondrial pathway by inducing Bax and inhibiting BCL-2 expression (Figure 5B). More interestingly, SC1 increased Casp7 but decreased Casp3, indicating that the apoptosis of MDA-MB-231 cells was executed via the induction of the effector Casp7, resulting in PARP-1 cleavage.

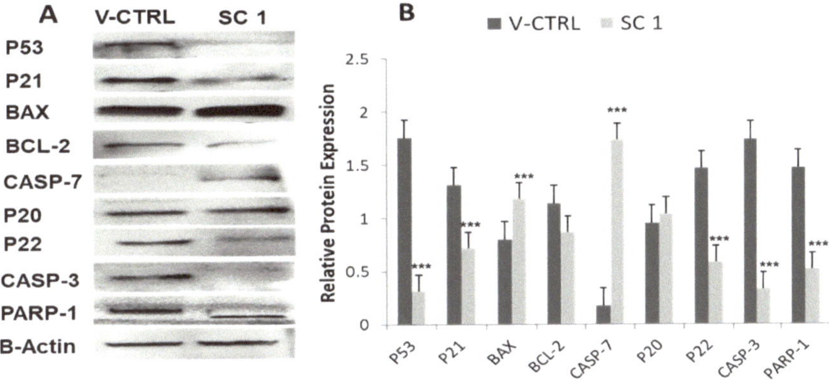

Figure 5. Expression levels of proteins associated with SC1-induced apoptosis. (**A**) Western blot analysis of SC1-inhibited mutant p53 expression, p21, BCL-2, and Casp3, while upregulating Bax and Casp7 compared to their vehicle control (V-CTRL). (**B**) Semi-quantitative analysis showing the relative expression level of each protein. β-actin served as control for the amount of proteins used in the assay. Density was analyzed using ImageJ. Mean Values ± SEM ($n = 3$) were compared using *t*-Test: *** $p < 0.001$.

4. Discussion

The phenolic compounds in plants have a variety of physiological activities, and many of these compounds are used in pharmaceuticals. Phenolic compounds are believed to possess health-improving properties, such as reducing inflammation, reducing free radicals, and reducing cancer risk [20]. The TPC contents has already been reported in *E. glaucophyllum* [38], *H. bacciferum* [11], *H. tuberculatum* [36]. Several differences exist between the current study and earlier studies, including the methods of preparation, the extraction solvent, and the equivalent compounds. It is therefore not possible to compare TPC contents between the present study and earlier studies. Flavonoids comprise the most abundant phenolic compound group in nature; they have been shown to have diverse biological activities, including antioxidant, antimicrobial, antiulcer, anti-diabetic, hepatoprotective, and anticarcinogenic properties [35].

Here, we explored the effects of the combination SC 1(SC1 = A + B + C) of crude extracts (Figure 2), isolated from three selected Qatari medicinal plants (*H. tuberculatum* (A), *E. glaucophyllum* (B), and *H. bacciferum* (C)), on cell proliferation, morphological changes, and the cell motility of the highly metastatic triple-negative BC cell line MDA-MB-231. The primary dermal normal human neonatal fibroblasts (HDFn) were used as control normal cells. Furthermore, we investigated the role of molecular mechanisms regulating the intrinsic pathway of apoptosis in SC-treated cells, as well as the expression of the key players mediating cell cycle and cell proliferation (Figure 3). Our findings revealed the SC-induced apoptosis and reduced cell migration and invasion of MDA-MB-231 when compared to normal control cells. More importantly, SC treatment resulted in apoptosis, which was most likely mediated by the mitochondrial Bax/BCL-2/Casp7 signaling pathway.

As indicated earlier, many studies have explored the medicinal potential of the plant crude extracts reported in this work or their active compounds, including their prospects as a source of cancer therapeutic agents. Although few of such studies investigated the anti-cancer potency of the individual crude extracts or their compounds, no study has investigated the combination of the crude extracts of the plant species used in our present investigation.

Several studies have isolated bioactive compounds and known active ingredients from plant species and characterized their anti-cancer properties. For instance, an analysis of crude extracts from *E. glaucophyllum* identified a number of known antioxidants bioactive, such as Geraniin [39–41], Gallocatechin [25], Quercetin and others [42]. An analysis of crude extracts from *H. bacciferum* showed the presence of a well-known carotenoid metabolite isololiolide [11,43]. More interestingly, the crude extracts from *H. tuberculatum* showed the presence of various alkaloids (e.g., haplotubinone, haplotubine, diphyllin, etc.) [44], polyphenols (e.g., lignans, arabelline, majidine, dictamine, and a qudsine derivative) [45,46], and flavonoids (e.g., resveratrol, kaempferol, myricetin, rutin, quercetin, etc.) [47].

It is evident from the literature described above that all the three medicinal plants (A, B and C) used in the present investigation are rich in various compounds (flavonoids, alkaloids, and polyphenols) that have already been characterized, individually, for their efficacy as anti-cancer agents. In fact, in our previous study, we combined in vitro cytotoxicity as well as microarray analysis, and identified, for the first time, a super-combination of six well-characterized bioactive compounds (resveratrol + indol-3-Carbinol + C-phycocyanin + isoflavone (genistein) + curcumin + quercetin), and showed that this super-cocktail induced apoptosis and inhibited the cell growth and motility of MDA-MB-231 BC cells [8]. More interestingly, our study revealed a myriad of major signaling pathways targeted by each of these individual compounds of the cocktail, in a combination, to exert synergistic anti-cancer activity [8]. Similarly, in the present study, it is obvious that mixing crude extracts from the three medicinal plants into an SC means a mix of various flavonoids, alkaloids, and polyphenols, as described above [11,43–47].

Several studies have already demonstrated the role of these compounds, individually, in inducing apoptosis and further inhibiting cancer cell migration and invasion (Figure 4). For instance, in addition to our previous studies on resveratrol and quercetin [8], geraniin, the main active ingredient isolated from *E. glaucophyllum*, was shown to inhibit human osteosarcoma cancer cell migration and invasion via the PI3K/Akt and ERK1/2 signaling pathways [39]. Furthermore, Geraniin decreased BCL-2 expression and increased Bax expression, leading to mitochondrial cytochrome c release and the subsequent activation of caspase-9 and caspase-3 cascades [39]. In vivo, geraniin resulted in tumor growth inhibition in A549 xenografts [39]. Isololiolide, which is abundant in *H. bacciferum*, induced significant apoptosis in the hepatocellular carcinoma HepG2 cells, when compared to non-malignant MRC-5 and HFF-1 human fibroblasts [43]. Isololiolide-induced apoptosis was associated with increased PARP cleavage and p53 expression and decreased procaspase-3 and BCL-2 levels [43]. In another study, lololide and isololiolide, isolated from *H. bacciferum*, were also found to inhibit the proliferation of colon cancer cells HCT116 and DLD1 [11]. Interestingly, Yang et al. reported that loliolide was a potent suppressor of the Epithelial-Mesenchymal Transition (EMT) process against colon and breast cancer cells [48]. Loliolide-suppressed EMT was associated with the inhibition of the chemokines CXCR4 and CXCR7, the suppression of expression of mesenchymal markers, and the induction of epithelial markers [48]. In addition, loliolide inhibited cancer cell invasion [48]. Based on the findings described above and the literature, the alkaloids and polyphenol compounds present in our crude extracts have been well characterized for their anti-cancer properties [49–51].

We also investigated the molecular mechanisms underlying SC-induced apoptosis and cell proliferation. Our results revealed that the SC attenuated cell growth and triggered apoptosis of the human TNBC cells most likely through the Bax/BCL-2 mitochondrial intrinsic pathway. As expected, and as shown in Figure 4, the SC led to the cleavage and activation of both Casp-3 and Casp-7, which in turn cleaved PARP-1, indicating that the SC induced the mitochondrial apoptosis of MDA-MB-231 BC cell lines. Interestingly, the SC inhibited the expression levels of the mutant p53 (Figure 4), while it induced Bax and inhibited BCL-2 (Figure 5). It must be noted that the stable mutant p53 is abundant in MDA-MB-231 cells and can suppress apoptosis via a dominant negative effect [52]. Also, the dysregulation of p53 function increases the Bax/BCL-2 ratio, rendering the cells vulnerable to apoptosis [53]. More importantly, regardless of the tumor suppression function of p53 [54], specific mutations confer a 'gain-of-function' or 'dominant negative' function to the new mutant p53 protein, which subsequently promote tumorigenesis and tumor progression [55–57]. Moreover, the abundance of mutant p53 protein in MDA-MB-231 [51] equip these cells with signals that promote cell survival by suppressing the pro-apoptotic effect that might originate from other p53 family members [52,53]. In summary, all our finding put together with the data described above from the literature support our hypothesis that the SC crude extract exhibited anti-cancer properties. In addition to its anti-migration/anti-invasive effects, it exhibited anti-proliferative effect, observed in MDA-MB-231, where it targeted and attenuated the expression of the mutant p53, subsequently resulting in apoptosis.

5. Conclusions

The findings support our hypothesis that SC inhibited cell proliferation, cell migration, cell invasion, and induced apoptosis. SC-promoted apoptosis appears to be mediated via the attenuation of the mutant p53 protein as well as upregulated Bax and downregulated BCL-2. This suggests that the super-combination triggered BC cell death via the p53/Bax/BCL-2/Casp7 pathway. Our ongoing experiments aim to validate these in vitro findings in vivo using a BC-Xenograft mouse model. These findings support our hypothesis that the SC of crude extracts exhibited promising anti-cancer properties and lays the foundation for further experiments to validate SC as an effective chemoprevention and/or chemotherapeutic strategy that can ultimately pave the way towards translational research/clinical trials for the eradication of BC.

Author Contributions: Conceptualization, A.O. and M.A.; methodology, N.A. and K.U.; software, N.A.; validation, N.A. and A.O.; formal analysis, A.O., M.A., N.A. and K.U.; investigation, N.A. and A.O.; resources, A.O. and M.A.; data curation, A.O. and N.A.; writing—original draft preparation, N.A.; writing—review and editing, A.O. and M.A.; visualization, N.A. and K.U.; supervision, A.O. and M.A.; project administration, A.O. and M.A.; funding acquisition, A.O. and M.A. All authors have read and agreed to the published version of the manuscript.

Funding: This research was made possible by support from the Qatar University's-Japan Research Collaboration Grant (M-QJRC-2020-10) and the Qatar National Research Fund (UREP21-080-1-015 and UREP29-186-3-059).

Institutional Review Board Statement: Not applicable.

Data Availability Statement: All the data of the research is shown in the article. No other data is available.

Conflicts of Interest: The authors declare no conflict of interest. The funders had no role in the design of the study; in the collection, analyses, or interpretation of the data; in the writing of the manuscript, or in the decision to publish the results.

References

1. Ferlay, J.; Colombet, M.; Soerjomataram, I.; Mathers, C.; Parkin, D.M.; Piñeros, M.; Znaor, A.; Bray, F. Estimating the Global Cancer Incidence and Mortality in 2018: GLOBOCAN Sources and Methods. *Int. J. Cancer* **2019**, *144*, 1941–1953. [CrossRef]
2. Narayan, A.K.; Al-Naemi, H.; Aly, A.; Kharita, M.H.; Khera, R.D.; Hajaj, M.; Rehani, M.M. Breast Cancer Detection in Qatar: Evaluation of Mammography Image Quality Using A Standardized Assessment Tool. *Eur. J. Breast Health* **2020**, *16*, 124–128. [CrossRef]
3. McSherry, E.A.; Donatello, S.; Hopkins, A.M.; McDonnell, S. Molecular Basis of Invasion in Breast Cancer. *Cell. Mol. Life Sci.* **2007**, *4*, 3201–3218. [CrossRef]
4. Ouhtit, A.; Ismail, M.F.; Othman, A.; Fernando, A.; Abdraboh, M.E.; El-Kott, A.F.; Azab, Y.A.; Abdeen, S.H.; Gaur, R.L.; Gupta, I.; et al. Chemoprevention of Rat Mammary Carcinogenesis by Spirulina. *Am. J. Pathol.* **2014**, *184*, 296–303. [CrossRef]
5. Longley, D.B.; Johnston, P.G. Molecular Mechanisms of Drug Resistance. *J. Pathol.* **2005**, *205*, 275–292. [CrossRef]
6. Kawasaki, B.T.; Hurt, E.M.; Mistree, T.; Farrar, W.L. Targeting Cancer Stem Cells with Phytochemicals. *Mol. Interv.* **2008**, *8*, 174–184. [CrossRef]
7. Moiseeva, E.P.; Manson, M.M. Dietary Chemopreventive Phytochemicals: Too Little or Too Much? *Cancer Prev. Res.* **2009**, *2*, 611–616. [CrossRef]
8. Khlifi, D.; Sghaier, R.M.; Amouri, S.; Laouini, D.; Hamdi, M.; Bouajila, J. Composition and anti-oxidant, anti-cancer and anti-inflammatory activities of *Artemisia herba-alba*, *Ruta chalpensis* L. and *Peganum harmala* L. *Food Chem. Toxicol.* **2013**, *55*, 202–208. [CrossRef]
9. Farshori, N.N.; Siddiqui, M.A.; Al-Oqail, M.M.; Al-Sheddi, E.S.; Al-Massarani, S.M.; Saquib, Q.; Ahmad, J.; Al-Khedhairy, A.A. Aloe vera-induced apoptotic cell death through ROS generation, cell cycle arrest, and DNA damage in human breast cancer cells. *Biologia* **2022**, *77*, 2751–2761. [CrossRef]
10. Alafnan, A.; Alamri, A.; Alanazi, J.; Hussain, T. Farnesiferol C Exerts Antiproliferative Effects on Hepatocellular Carcinoma HepG2 Cells by Instigating ROS-Dependent Apoptotic Pathway. *Pharmaceuticals* **2022**, *15*, 1070. [CrossRef]
11. Azzazy, H.M.E.; Abdelnaser, A.; Al Mulla, H.; Sawy, A.M.; Shamma, S.N.; Elhusseiny, M.; Alwahibi, S.; Mahdy, N.K.; Fahmy, S.A. Essential Oils Extracted from *Boswellia sacra* Oleo Gum Resin Loaded into PLGA-PCL Nanoparticles: Enhanced Cytotoxic and Apoptotic Effects against Breast Cancer Cells. *ACS Omega* **2022**, *8*, 1017–1025. [CrossRef]
12. Qanash, H.; Bazaid, A.S.; Binsaleh, N.K.; Patel, M.; Althomali, O.W.; Sheeha, B.B. In Vitro Antiproliferative Apoptosis Induction and Cell Cycle Arrest Potential of Saudi Sidr Honey against Colorectal Cancer. *Nutrients.* **2023**, *15*, 3448. [CrossRef]
13. Hosmani, J.V.; Al Shahrani, A.; Hosmani, J.; AlShahrani, I.; Togoo, R.A.; Ain, T.S.; Syed, S.; Mannakandath, M.L.; Addas, M.K.A. Cytotoxic and antitumor properties of Ziziphus spina-christi found in Al Bahah Region of Hejaz area: An in vitro study on oral cancer cell lines. *Pharmacogn. Mag.* **2021**, *17*, 793–801. [CrossRef]
14. Habib, H.M.; El-Fakharany, E.M.; El-Gendi, H.; El-Ziney, M.G.; El-Yazbi, A.F.; Ibrahim, W.H. Palm Fruit (*Phoenix dactylifera* L.) Pollen Extract Inhibits Cancer Cell and Enzyme Activities and DNA and Protein Damage. *Nutrients* **2023**, *15*, 2614. [CrossRef]
15. Fouzat, A.; Hussein, O.J.; Gupta, I.; Al-Farsi, H.F.; Khalil, A.; Al Moustafa, A.E. Elaeagnus, angustifolia Plant Extract Induces Apoptosis via P53 and Signal Transducer and Activator of Transcription 3 Signaling Pathways in Triple-Negative Breast Cancer Cells. *Front. Nutr.* **2022**, *9*, 871667. [CrossRef]
16. Roudsari, M.T.; Bahrami, A.R.; Dehghani, H.; Iranshahi, M.; Matin, M.M.; Mahmoudi, M. Bracken-fern extracts induce cell cycle arrest and apoptosis in certain cancer cell lines. *Asian Pac. J. Cancer Prev.* **2012**, *13*, 6047–6053. [CrossRef]
17. Nassan, M.A.; Soliman, M.M.; Ismail, S.A.; El-Shazly, S. Effect of Taraxacum officinale extract on PI3K/Akt pathway in DMBA-induced breast cancer in albino rats. *Biosci. Rep.* **2018**, *38*, BSR20180334. [CrossRef]

18. Chen, H.; Zhu, T.; Huang, X.; Xu, W.; Di, Z.; Ma, Y.; Xue, M.; Bi, S.; Shen, Y.; Yu, Y.; et al. Xanthatin suppresses proliferation and tumorigenicity of glioma cells through autophagy inhibition via activation of the PI3K-Akt-mTOR pathway. *Pharmacol. Res. Perspect.* **2023**, *11*, e01041. [CrossRef]

19. Chen, W.; Li, Z.; Bai, L.; Lin, Y. NF-kappaB in lung cancer, a carcinogenesis mediator and a prevention and therapy target. *Front. Biosci.* **2011**, *16*, 1172–1185. [CrossRef]

20. Ouhtit, A.; Gaur, R.L.; Abdraboh, M.; Ireland, S.K.; Rao, P.N.; Raj, S.G.; Al-Riyami, H.; Shanmuganathan, S.; Gupta, I.; Murthy, S.N.; et al. Simultaneous Inhibition of Cell-Cycle, Proliferation, Survival, Metastatic Pathways and Induction of Apoptosis in Breast Cancer Cells by a Phytochemical Super-Cocktail: Genes That Underpin Its Mode of Action. *J. Cancer* **2014**, *4*, 703–715. [CrossRef]

21. Kuete, V.; Wiench, B.; Alsaid, M.S.; Alyahya, M.A.; Fankam, A.G.; Shahat, A.A.; Efferth, T. Cytotoxicity, Mode of Action and Antibacterial Activities of Selected Saudi Arabian Medicinal Plants. *BMC Complement. Altern. Med.* **2013**, *13*, 354. [CrossRef]

22. Eissa, T.F.; González-Burgos, E.; Carretero, M.E.; Gómez-Serranillos, M.P. Biological Activity of HPLC-Characterized Ethanol Extract from the Aerial Parts of *Haplophyllum tuberculatum*. *Pharm. Biol.* **2014**, *52*, 151–156. [CrossRef]

23. Aïssaoui, H.; Mencherini, T.; Esposito, T.; De Tommasi, N.; Gazzerro, P.; Benayache, S.; Benayache, F.; Mekkiou, R. *Heliotropium Bacciferum* Forssk. (Boraginaceae) Extracts: Chemical Constituents, Antioxidant Activity and Cytotoxic Effect in Human Cancer Cell Lines. *Nat. Prod. Res.* **2019**, *33*, 1813–1818. [CrossRef]

24. Abdallah, B.M.; Ali, E.M. Therapeutic Effect of Green Synthesized Silver Nanoparticles Using *Erodium glaucophyllum* Extract against Oral Candidiasis: In Vitro and In Vivo Study. *Molecules* **2022**, *27*, 4221. [CrossRef]

25. Barba, F.J.; Alcántara, C.; Abdelkebir, R.; Bäuerl, C.; Pérez-Martínez, G.; Lorenzo, J.M.; Collado, M.C.; García-Pérez, J.V. Ultrasonically-Assisted and Conventional Extraction from *Erodium glaucophyllum* Roots Using Ethanol:Water Mixtures: Phenolic Characterization, Antioxidant, and Anti-Inflammatory Activities. *Molecules* **2020**, *25*, 1759. [CrossRef]

26. Hamza, G.; Emna, B.H.; Yeddes, W.; Dhouafli, Z.; Moufida, T.S.; El Akrem, H. Chemical Composition, Antimicrobial and Antioxidant Activities Data of Three Plants from Tunisia Region: *Erodium glaucophyllum*, *Erodium hirtum* and *Erodium guttatum*. *Data Brief* **2018**, *19*, 2352–2355. [CrossRef]

27. Abdel Bary, E.S.S. The Flora of Qatar: The Dicotyledons (Volume 1) and The Monocotyledons (Volume 2). Available online: https://www.chrome-extension://efaidnbmnnnibpcajpcglclefindmkaj/https://www.qu.edu.qa/static_file/qu/research/ESC/Books/The-Flora-of-Qatar-The-Dicotyledons.pdf (accessed on 23 December 2022).

28. Stanković, M.S.; Petrović, M.; Godjevac, D.; Stevanović, Z.D. Screening Inland Halophytes from the Central Balkan for Their Antioxidant Activity in Relation to Total Phenolic Compounds and Flavonoids: Are There Any Prospective Medicinal Plants? *J. Arid Environ.* **2015**, *120*, 26–32. [CrossRef]

29. Singleton, V.L.; Rossi, J.A., Jr. Colorimetry of Total Phenolics with Phosphomolybdic-Phosphotungstic Acid Reagents. *Am. J. Enol. Vitic.* **1965**, *16*, 144–158. [CrossRef]

30. Chang, C.-C.; Yang, M.-H.; Wen, H.-M.; Chern, J.-C. Estimation of Total Flavonoid Content in Propolis by Two Complementary Colometric Methods. *J. Food Drug Anal.* **2002**, *10*, 3. [CrossRef]

31. Alateyah, N.; Ahmad, S.M.S.; Gupta, I.; Fouzat, A.; Thaher, M.I.; Das, P.; Al Moustafa, A.E.; Ouhtit, A. Haematococcus Pluvialis Microalgae Extract Inhibits Proliferation, Invasion, and Induces Apoptosis in Breast Cancer Cells. *Front. Nutr.* **2022**, *9*, 882956. [CrossRef]

32. Shariati, S.R.P.; Shokrgozar, M.A.; Vossoughi, M.; Eslamifar, A. In vitro Co-Culture of Human Skin Keratinocytes and Fibroblasts on a Biocompatible and Biodegradable Scaffold. *IBJ Iran. Biomed. J.* **2009**, *13*, 169–177. Available online: http://ibj.pasteur.ac.ir/article-1-70-en.htmL (accessed on 23 November 2022).

33. Tsao, R. Chemistry and biochemistry of dietary polyphenols. *Nutrients* **2010**, *2*, 1231–1246. [CrossRef]

34. Ghasemzadeh, A.; Ghasemzadeh, N. Flavonoids and Phenolic Acids: Role and Biochemical Activity in Plants and Human. *J. Med. Plant Res.* **2011**, *5*, 6697–6703. [CrossRef]

35. John, B.; Sulaiman, C.T.; George, S.; Reddy, V.R.K. Total Phenolics and Flavonoids in Selected Medicinal Plants from Kerala. *Int. J. Pharm. Pharm. Sci.* **2014**, *6*, 406–408. Available online: https://www.researchgate.net/publication/267031901 (accessed on 1 February 2023).

36. Khasawneh, M.; Hamza, A.; Fawzi, N. Antioxidant Activity and Phenolic content of Some Emirates Medicinal Plants. *Adv. food Sci.* **2010**, *32*, 62–66. Available online: https://www.researchgate.net/publication/215449624 (accessed on 1 February 2023).

37. Al-Brashdi, A.S.; Al-Ariymi, H.; Al Hashmi, M.; Khan, S.A. Evaluation of Antioxidant Potential, Total Phenolic Content and Phytochemical Screening of Aerial Parts of a Folkloric Medicine, *Haplophyllum tuberculatum* (Forssk) A. Juss. *J. Coast. Life Med.* **2016**, *4*, 315–319. [CrossRef]

38. Gadhoumi, H.; Martinez-Rojas, E.; Tounsi, M.S.; Hayouni, E.A. Phenolics Composition and Biological Activities Assessment of Leaves, Flowers and Roots Extracts from *Erodium glaucophyllum*, *Erodium hirtum* and *Erodium guttatum*. *Biol. Bull.* **2021**, *48*, 667–672. [CrossRef]

39. Wang, Y.; Wan, D.; Zhou, R.; Zhong, W.; Lu, S.; Chai, Y. Geraniin Inhibits Migration and Invasion of Human Osteosarcoma Cancer Cells through Regulation of PI3K/Akt and ERK1/2 Signaling Pathways. *Anticancer Drugs* **2017**, *28*, 959–966. [CrossRef]

40. Li, J.; Wang, S.; Yin, J.; Pan, L. Geraniin Induces Apoptotic Cell Death in Human Lung Adenocarcinoma A549 Cells in Vitro and in Vivo. *Can. J. Physiol. Pharmacol.* **2013**, *91*, 1016–1024. [CrossRef]

41. Gohar, A.A.; Lahloub, M.F.; Niwa, M. Antibacterial Polyphenol from *Erodium glaucophyllum*. *Z. Naturforsch. C J. Biosci.* **2003**, *58*, 670–674. [CrossRef]

42. Bakari, S.; Hajlaoui, H.; Daoud, A.; Mighri, H.; Ross-Garcia, J.M.; Gharsallah, N.; Kadri, A. Phytochemicals, Antioxidant and Antimicrobial Potentials and LC-MS Analysis of Hydroalcoholic Extracts of Leaves and Flowers of *Erodium glaucophyllum* Collected from Tunisian Sahara. *Food Sci. Technol.* **2018**, *38*, 310–317. [CrossRef]

43. Vizetto-Duarte, C.; Custódio, L.; Gangadhar, K.N.; Lago, J.H.G.; Dias, C.; Matos, A.M.; Neng, N.; Nogueira, J.M.F.; Barreira, L.; Albericio, F.; et al. Isololiolide, a Carotenoid Metabolite Isolated from the Brown Alga Cystoseira Tamariscifolia, Is Cytotoxic and Able to Induce Apoptosis in Hepatocarcinoma Cells through Caspase-3 Activation, Decreased BCL-2 Levels, Increased P53 Expression and PARP Cleavage. *Phytomedicine* **2016**, *23*, 550–557. [CrossRef] [PubMed]

44. Al-Rehaily, A.J.; Al-Howiriny, T.A.; Ahmad, M.S.; Al-Yahya, M.A.; El-Feraly, F.S.; Hufford, C.D.; McPhail, A.T. Alkaloids from *Haplophyllum tuberculatum*. *Phytochemistry* **2001**, *57*, 597–602. [CrossRef] [PubMed]

45. Hamdi, A.; Viane, J.; Mahjoub, M.A.; Majouli, K.; Gad, M.H.H.; Kharbach, M.; Demeyer, K.; Marzouk, Z.; Heyden, Y.V. Polyphenolic Contents, Antioxidant Activities and UPLC–ESI–MS Analysis of *Haplophyllum tuberculatum*, A. Juss Leaves Extracts. *Int. J. Biol. Macromol.* **2018**, *106*, 1071–1079. [CrossRef]

46. Mahmoud, A.B.; Danton, O.; Kaiser, M.; Han, S.; Moreno, A.; Algaffar, S.A.; Khalid, S.; Oh, W.K.; Hamburger, M.; Mäser, P. Lignans, Amides, and Saponins from *Haplophyllum tuberculatum* and Their Antiprotozoal Activity. *Molecules* **2020**, *25*, 2825. [CrossRef] [PubMed]

47. Abdelkhalek, A.; Salem, M.Z.M.; Hafez, E.; Behiry, S.I.; Qari, S.H. The Phytochemical, Antifungal, and First Report of the Antiviral Properties of Egyptian *Haplophyllum tuberculatum* Extract. *Biology* **2020**, *9*, 248. [CrossRef]

48. Yang, M.H.; Ha, I.J.; Ahn, J.; Kim, C.-K.; Lee, M.; Ahn, K.S. Potential Function of Loliolide as a Novel Blocker of Epithelial-Mesenchymal Transition in Colorectal and Breast Cancer Cells. *Cell. Signal.* **2023**, *105*, 110610. [CrossRef]

49. Mondal, A.; Gandhi, A.; Fimognari, C.; Atanasov, A.G.; Bishayee, A. Alkaloids for Cancer Prevention and Therapy: Current Progress and Future Perspectives. *Eur. J. Pharmacol.* **2019**, *858*, 172472. [CrossRef]

50. Khan, H.; Alam, W.; Alsharif, K.F.; Aschner, M.; Pervez, S.; Saso, L. Alkaloids and Colon Cancer: Molecular Mechanisms and Therapeutic Implications for Cell Cycle Arrest. *Molecules* **2022**, *27*, 920. [CrossRef]

51. Cháirez-Ramírez, M.H.; de la Cruz-López, K.G.; García-Carrancá, A. Polyphenols as Antitumor Agents Targeting Key Players in Cancer-Driving Signaling Pathways. *Front. Pharmacol.* **2021**, *12*, 710304. [CrossRef]

52. Hui, L.; Zheng, Y.; Yan, Y.; Bargonetti, J.; Foster, D.A. Mutant P53 in MDA-MB-231 Breast Cancer Cells Is Stabilized by Elevated Phospholipase D Activity and Contributes to Survival Signals Generated by Phospholipase D. *Oncogene* **2006**, *25*, 7305–7310. [CrossRef] [PubMed]

53. Basu, A.; Haldar, S. The Relationship between BCL-2, Bax and P53: Consequences for Cell Cycle Progression and Cell Death. *Mol. Hum. Reprod.* **1998**, *4*, 1099–1109. [CrossRef] [PubMed]

54. Levine, A.J.; Momand, J.; Finlay, C.A. The P53 Tumour Suppressor Gene. *Nature* **1991**, *351*, 453–456. [CrossRef] [PubMed]

55. Zambetti, G.P.; Levine, A.J. A Comparison of the Biological Activities of Wild-type and Mutant P53. *FASEB J.* **1993**, *7*, 855–865. [CrossRef] [PubMed]

56. Blandino, G.; Levine, A.J.; Oren, M. Mutant P53 Gain of Function: Differential Effects of Different P53 Mutants on Resistance of Cultured Cells to Chemotherapy. *Oncogene* **1999**, *18*, 477–485. [CrossRef]

57. Cadwell, C.; Zambetti, G.P. The Effects of Wild-Type P53 Tumor Suppressor Activity and Mutant P53 Gain-of-Function on Cell Growth. *Gene* **2001**, *277*, 15–30. [CrossRef]

Article

Drastic Synergy of Lovastatin and *Antrodia camphorata* Extract Combination against PC3 Androgen-Refractory Prostate Cancer Cells, Accompanied by AXL and Stemness Molecules Inhibition

Chih-Jung Yao [1,2,†], Chia-Lun Chang [2,3,4,†], Ming-Hung Hu [3,4], Chien-Huang Liao [4], Gi-Ming Lai [3,4], Tzeon-Jye Chiou [3,4], Hsien-Ling Ho [4], Hui-Ching Kuo [4], Ya-Yu Yang [5], Jacqueline Whang-Peng [3,4] and Shuang-En Chuang [5,*]

[1] Department of Medical Education and Research, Wan Fang Hospital, Taipei Medical University, Taipei 11696, Taiwan; yao0928@tmu.edu.tw
[2] Department of Internal Medicine, School of Medicine, College of Medicine, Taipei Medical University, Taipei 11031, Taiwan; 101255@w.tmu.edu.tw
[3] Division of Hematology and Medical Oncology, Department of Internal Medicine, Wan Fang Hospital, Taipei Medical University, Taipei 11696, Taiwan; lancehu7@gmail.com (M.-H.H.); gminlai@canceraway.org.tw (G.-M.L.); 108178@w.tmu.edu.tw (T.-J.C.); jqwpeng@nhri.org.tw (J.W.-P.)
[4] Cancer Center, Wan Fang Hospital, Taipei Medical University, Taipei 11696, Taiwan; a2639264@ms25.hinet.net (C.-H.L.); injection63@gmail.com (H.-L.H.); ghost1132@yahoo.com (H.-C.K.)
[5] National Institute of Cancer Research, National Health Research Institutes, Miaoli 35053, Taiwan; irelandfish@nhri.edu.tw
* Correspondence: sechuang@nhri.edu.tw
† These authors contributed equally to this work.

Citation: Yao, C.-J.; Chang, C.-L.; Hu, M.-H.; Liao, C.-H.; Lai, G.-M.; Chiou, T.-J.; Ho, H.-L.; Kuo, H.-C.; Yang, Y.-Y.; Whang-Peng, J.; et al. Drastic Synergy of Lovastatin and *Antrodia camphorata* Extract Combination against PC3 Androgen-Refractory Prostate Cancer Cells, Accompanied by AXL and Stemness Molecules Inhibition. *Nutrients* **2023**, *15*, 4493. https://doi.org/10.3390/nu15214493

Academic Editors: Yi-Wen Liu and Ching-Hsein Chen

Received: 28 September 2023
Revised: 20 October 2023
Accepted: 21 October 2023
Published: 24 October 2023

Abstract: Prostate cancer (PC) is the second most frequently diagnosed cancer and the fifth leading cause of cancer-related death in males worldwide. Early-stage PC patients can benefit from surgical, radiation, and hormonal therapies; however, once the tumor transitions to an androgen-refractory state, the efficacy of treatments diminishes considerably. Recently, the exploration of natural products, particularly dietary phytochemicals, has intensified in response to addressing this prevailing medical challenge. In this study, we uncovered a synergistic effect from combinatorial treatment with lovastatin (an active component in red yeast rice) and *Antrodia camphorata* (AC, a folk mushroom) extract against PC3 human androgen-refractory PC cells. This combinatorial modality resulted in cell cycle arrest at the G0/G1 phase and induced apoptosis, accompanied by a marked reduction in molecules responsible for cellular proliferation (p-Rb/Rb, Cyclin A, Cyclin D1, and CDK1), aggressiveness (AXL, p-AKT, and survivin), and stemness (SIRT1, Notch1, and c-Myc). In contrast, treatment with either AC or lovastatin alone only exerted limited impacts on the cell cycle, apoptosis, and the aforementioned signaling molecules. Notably, significant reductions in canonical PC stemness markers (CD44 and CD133) were observed in lovastatin/AC-treated PC3 cells. Furthermore, lovastatin and AC have been individually examined for their anti-PC properties. Our findings elucidate a pioneering discovery in the synergistic combinatorial efficacy of AC and clinically viable concentrations of lovastatin on PC3 PC cells, offering novel insights into improving the therapeutic effects of dietary natural products for future strategic design of therapeutics against androgen-refractory prostate cancer.

Keywords: lovastatin; *Antrodia camphorata*; prostate cancer; PC3; AXL

1. Introduction

Prostate cancer (PC) has been the second-most commonly diagnosed cancer and the fifth leading cause of cancer death in males worldwide [1]. Although androgen deprivation therapy remains the mainstay first-line treatment for PC patients, most patients eventually become "androgen refractory" (resistant to androgen-ablation therapy) within a few years after the initial response [2,3]. Due to the scarcity of efficacious therapeutic options, addressing androgen-refractory PC remains a formidable clinical challenge [4]. Despite significant advances in currently approved treatment options, only a marginal extension of patient survival has been achieved [3,5]. More effective treatment strategies to combat this lethal disease are urgently needed.

Statins, the 3-hydroxy-3-methyl-glutaryl-CoA (HMG-CoA) reductase inhibitors such as FDA-approved simvastatin, atorvastatin, pravastatin, and lovastatin, are the most commonly prescribed drugs for cardiovascular diseases due to their abilities to disrupt cholesterol synthesis and to mitigate arterial cholesterol plaque accumulation [6–9]. Recently, increasing studies have been initiated to investigate the anti-cancer properties of statins, either alone or in combination with radiotherapy or chemotherapy [10–12]. A recent meta-analysis study indicates that using statins in cancer patients can significantly lower cancer-related risk and mortality [13]. Another study by John Hopkins Hospital shows that statins may attenuate PC progression and protect patients from relapse after prostatectomy [14]. Among the statins, lovastatin (Mevacor) is a naturally occurring active ingredient (monacolin K) contained in red yeast rice, a Chinese fermented rice product (*Monascus purpureus*) [15]. This small natural compound (MW 404 Da) is a widely renowned medicine for cholesterol reduction with staggering safety profiles [16]. An investigation carried out by Park et al. [17] demonstrates that lovastatin-induced death of PC3 human androgen-refractory prostate cancer cells occurs via abating transcription factor E2F-1 and its downstream signaling molecules such as c-Myc, cyclin D1, cyclin A, and cyclin B1. In line with this study, Hoque et al. [18] also report that lovastatin can induce apoptosis and G1 phase arrest while reducing cellular levels of phospho-Rb (p-Rb), cyclin D1, cyclin D3, CDK4, and CDK6 in PC3 cells. Owing to the constraints of clinically attainable concentrations of lovastatin [19], it is recommended that further research explore the combinatorial effects of statins and other apoptosis-inducing agents to evaluate their potency against prostate cancer [18].

Antrodia camphorata (AC), also known as *Antrodia cinnamomea* or by its Chinese nomenclature "Niu-Chang-Chih", is an edible Taiwanese mushroom that has been traditionally employed in folk medicine to treat various health disorders [20]. Extracts of AC are reported to possess various biological activities, including anti-inflammatory, antioxidant, hepatoprotective, antihypertensive, antihyperlipidemic, immunomodulatory, and anti-cancer activities [20]. In fact, the anticancer effects of AC extract have been shown to induce apoptosis and arrest the cell cycle at a concentration of around 150 μg/mL in PC cells [21,22]. Although cyclin B1 and CDK1 (CDC2) reductions are observed in 150 μg/mL of AC extract-treated PC cells, only a limited degree of apoptosis can be induced [22]. AC thus holds promise as a potential adjuvant anticancer agent for treating prostate cancers. Nevertheless, whether its therapeutic potency can be augmented with other dietary components still requires further investigation before optimal integration into clinical practice.

In this study, therefore, we systemically examined whether combining AC extract with clinically achievable concentrations of lovastatin could lead to synergistic efficacy against human androgen-refractory PC cells. The compelling therapeutic effects exerted by AC and lovastatin were molecularly dissected by investigating changes in the protein expression of genes liable for cellular proliferation (p-Rb/Rb, Cyclin A, Cyclin D1, and CDK1), aggressiveness (AXL, p-AKT, and survivin), and stemness (SIRT1, Notch1, and c-Myc) (CD44 and CD133) in PC cells.

2. Materials and Methods

2.1. Cell Culture

Human androgen-refractory prostate cancer cell lines PC3 (bone-metastatic) [23] and DU145 (brain-metastatic) [24] were maintained in RPMI-1640 (Gibco, Carlsbad, CA, USA) and DMEM (Gibco) medium, respectively. HS68 primary human foreskin fibroblast cells were cultured in DMEM (Gibco) medium. The mediums were supplemented with 10% fetal bovine serum (Gibco) and 1 × penicillin streptomycin-glutamine (Gibco). All cells were cultured at 37 °C in a water-jacketed, 5% CO_2 incubator.

2.2. Preparation of Lovastatin and AC Extract

Lovastatin was purchased from Calbiochem (San Diego, CA, USA) and was dissolved in dimethyl sulfoxide (DMSO) to make a 10 mM stock solution. The crude material of *Antrodia camphorata* (AC) was provided by Well Shine Biotechnology Development Co. (Taipei, Taiwan) and was extracted by 95% ethanol in a 1:20 (w/v) ratio for 24 h at room temperature with shaking. The supernatant of extracts was centrifuged at 3000 rpm for 30 min to remove the precipitate, further filtered by a 0.22 micrometer pore size filter (Merck Millipore, Carrigtwohill, Cork, Ireland), and was then lyophilized and stored at −20 °C before use. The final product of the AC extract was reconstituted in 50% dimethyl sulfoxide (DMSO) and 50% EtOH to make a 100 mg/mL stock solution. The stock solutions of lovastatin and AC extract were diluted with culture medium before being added to cultured cells.

2.3. Sulforhodamine B (SRB) Assay for Cell Viability

PC3 and DU145 prostate cancer cells were seeded in a 96-well plate at a density of 2×10^3 cells/well. The primary human foreskin fibroblast HS68 cells were seeded in a 96-well plate at a density of 8×10^3 cells/well. After 24 h of incubation at 37 °C, cells were treated with various agents as indicated in the figure legends for a further 72 h. Cells were then harvested and fixed in 10% trichloroacetic acid (TCA). The fixed cells were washed with distilled water and stained with 0.4% (w/v) SRB dye dissolved in 1% acetic acid. The unbound dye was washed away by 1% acetic acid, and the plates were then air-dried. The cell-bound SRB dye was dissolved by adding 200 μL per well of 10 mM Tris-base, and the absorbance was measured at 570 nm. The absorbance is directly proportional to the cell number over a wide range.

2.4. Photograph of the Cells

The images of cells were photographed using a digital microscope camera PAXcam2+ (Midwest Information Systems, Inc., Villa Park, IL, USA) adapted to an inverted microscope CKX31 (Olympus Co., Tokyo, Japan).

2.5. Cell Cycle Analysis

A total of 2×10^5 PC3 cells per well was plated on 6-well plates and incubated for 24 h, and then treated with lovastatin (2 μM) and AC extract (40 μg/mL) individually or in combination for 24, 48, and 72 h. At harvest, cells were trypsinized using Trypsin-EDTA (Invitrogen, Carlsbad, CA, USA), washed twice with ice-cold phosphate-buffered saline (PBS), and fixed in cold 70% ethanol overnight at 4 °C. The cells were washed twice with PBS and incubated with 100 μg/mL of propidium iodide (PI) containing 100 μg/mL of RNase at 37 °C for 30 min, and then stored on ice protected from light. Cell cycle analysis was performed by flow cytometry (Beckman Coulter EPICS XL, Fullerton, CA, USA). To evaluate the changes more accurately in distribution at G0/G1, S, and G2/M phases, the percentage of cell-cycle distribution was re-calculated after excluding the sub-G1 proportion.

2.6. Western Blotting

PC3 cells were seeded in 10 cm dishes at a density of 4×10^5 cells/dish for 24 h and then treated with agents as described in the figure legends. On the day of harvest, total protein extracts (10 μg each) were size-fractionated electrophoretically by a 10% polyacrylamide SDS-PAGE gel and transferred onto a PVDF membrane using the BioRad Mini Protean electrotransfer system. The blots were incubated with 5% milk in PBST (phosphate buffered saline with 0.05% Tween-20) for 1 h to block nonspecific binding and incubated with individual primary antibodies overnight at 4 °C, respectively. The primary antibodies purchased from Cell Signaling Technology, Inc. (Danvers, MA, USA) include phospho-Rb (p-Rb) (#9308), SIRT1 (#8469), p-AKT (#9271), and survivin (#2808); antibodies obtained from Abcam, Inc. (Cambridge, MA, USA) include GAPDH (ab8245),

CDK4 (ab68266), CD44 (ab51037), and Notch1 (ab52627); those purchased from Santa Cruz Biotechnology (San Diego, CA, USA) include Rb (sc-102), Cyclin A (sc239), CDK-2 (sc-6248), Cyclin D1 (SC-246), AXL (sc-1096), and c-Myc (sc-40); while CDK1 (CDC2) (#1161-1) is obtained from Epitomics (Burlingame, CA, USA); CD133 (#18470-1-AP) is obtained from Proteintech (Chicago, IL, USA). The membranes were incubated with an appropriate peroxidase-conjugated secondary antibody at room temperature for 1 h and washed intensively with PBS. The immune complexes (protein bands) were visualized using an enhanced chemiluminescence detection system (ECL, Perkin Elmer, Waltham, MA, USA) according to the manufacturer's instructions. The intensities of Western blot bands were quantified using ImageJ software (ImageJ Version 1.52, National Institutes of Health, Bethesda, MD, USA) downloaded from https://imagej.nih.gov/ij/download.html (accessed on 25 September 2019).

3. Results

3.1. Lovastatin and AC Extract Synergistically Suppress Proliferation and Induce Apoptosis in PC3 Cells

Firstly, a bone-metastatic androgen-refractory human prostate cancer cell line, PC3, was employed to assess the combinatorial effects of lovastatin and AC extract on cellular proliferation. As shown in Figure 1, lovastatin (2 µM) alone (Figure 1B) and AC extract (40 µg/mL) alone (Figure 1C) showed only marginal impacts on the growth of PC3 cells as compared to the untreated control (Figure 1A). Intriguingly, the combination treatment of both agents dramatically reduced the growth of PC3 cells (Figure 1D). The sulforhodamine B colorimetric (SRB) assay was performed to further demonstrate the synergistic anticancer effect of this combination. After 72 h of treatment, AC extract alone exerted only a minimal inhibitory effect at concentrations of 10 to 40 µg/mL, as cell viability largely remained above 90% (Figure 2A). By contrast, lovastatin alone dose-dependently reduced PC3 cell viability to 48.2% of control at a concentration of 4 µM (Figure 2A). Consistent with this observation, the survival inhibition curve of lovastatin was markedly shifted downward in an AC dose-dependent manner (Figure 2A). A similar phenomenon was also observed in another androgen-refractory and bone-metastatic human prostate cancer cell line, DU145 (Figure 2B). Although higher concentrations of lovastatin alone (8 µM and above) were required to suppress the cell viability of DU145 cells compared to PC3 cells, a similar downward shift of the survival inhibition curve was observed when treated in combination with AC extract (Figure 2B). Furthermore, lovastatin and AC extract did not exhibit inhibitory or synergistic effects on the viability of HS68 primary human foreskin fibroblast cells (Figure 2C), suggesting the proliferation-inhibitor effects of the combinatorial treatment modality using lovastatin and AC extract were exclusive to the growth of malignant cells but not normal cells.

The combined effect of lovastatin and AC extract was further investigated by assessing alterations in the cell cycle of PC3 cells. As shown in Figure 3, apoptosis induction and G1 phase arrest were found in the combined treatment group after 72 h, while the effects of lovastatin (2 µM) and AC extract (40 µg/mL) alone on apoptosis or G1 arrest were insignificant (Figure 3). The cell cycle distributions of these lovastatin and/or AC-treated cells are shown in Table 1. In accordance with the findings shown in Figures 1 and 2A, the sub-G1 arrest population was significantly augmented when PC3 cells were treated with lovastatin (2 µM) and AC (40 µg/mL) at both 48 and 72 h, indicating potent combinatorial efficacy from lovastatin and AC extract.

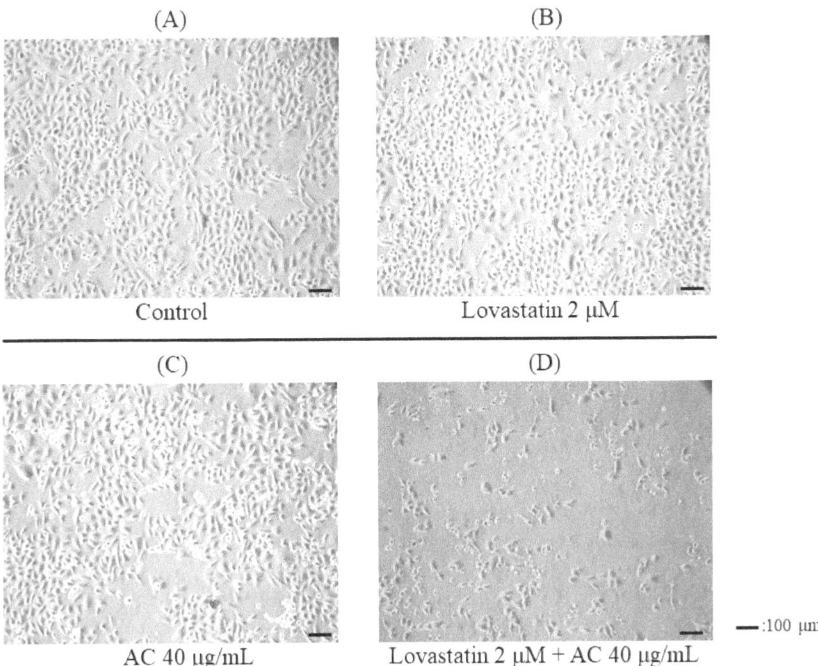

Figure 1. The androgen-refractory prostate cancer PC3 cells were treated with lovastatin and AC extract, individually or in combination, for 72 h. PC3 cells were treated with (**A**) vehicle as the control, (**B**) 2 μM lovastatin, (**C**) 40 μg/mL AC extract, and (**D**) 2 μM lovastatin plus 40 μg/mL AC extract for 72 h. Bar = 100 μm.

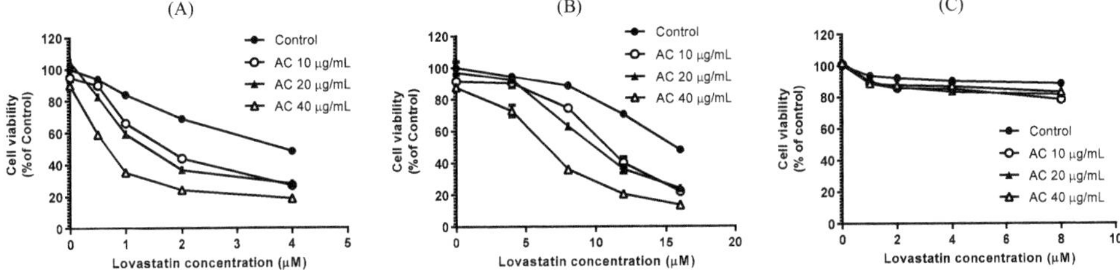

Figure 2. Combination effects of lovastatin and AC extract on cell viability. (**A**) PC3, (**B**) DU145 androgen-refractory prostate cancer cells, and (**C**) HS68 primary human foreskin fibroblast cells were treated as indicated (lovastatin and AC extract, individually or in combination) for 72 h, and then the cell viability was determined by SRB assay as described in Materials and methods.

Table 1. Tabulated cell-cycle distribution analysis (by percentages from Figure 3) of PC3 cells treated with lovastatin and/or AC extract after 72 h.

Lovastatin (μM)	0	2	0	2
AC (μg/mL)	0	0	40	40
Sub-G1 (%)	2.7	6.2	1.2	23.5
G0/G1 (%)	62.2	63.3	59.7	81.0
S (%)	21.7	19.0	18.5	10.4
G2/M (%)	16.1	17.7	21.8	8.6

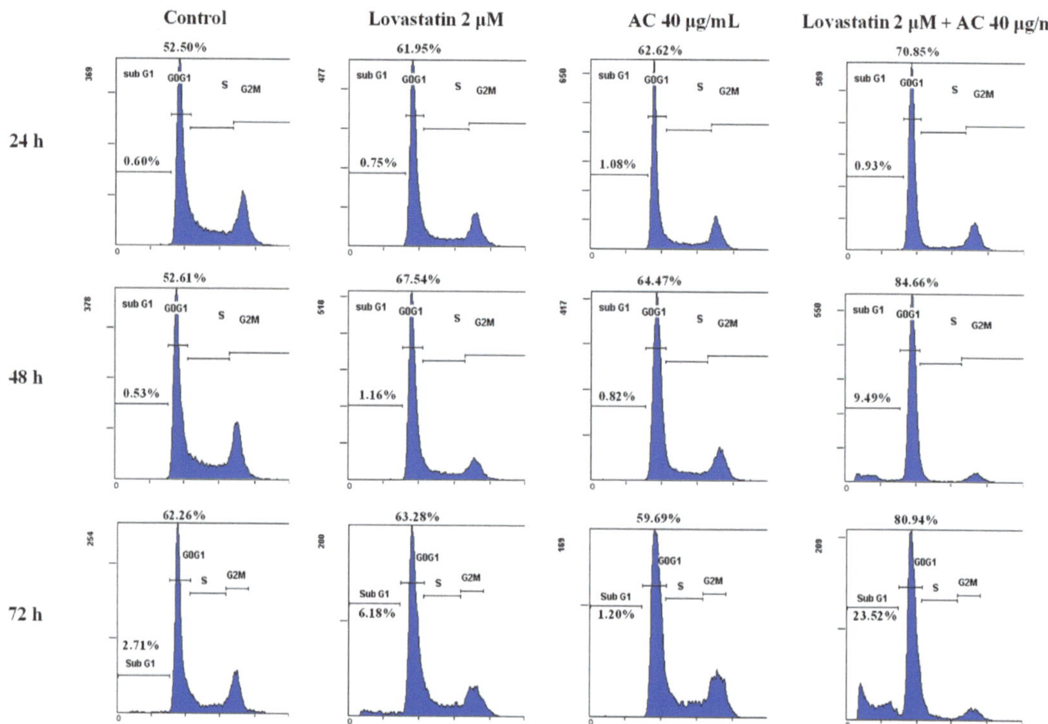

Figure 3. Combination effects of lovastatin and AC extract on the distribution of cell-cycle phase and induction of apoptotic sub-G1 fraction in PC3 cells. Cells were treated as indicated for 24, 48, and 72 h. The percentages of the G0/G1 and sub-G1 fractions are shown in the respective flow cytometric histograms.

3.2. Lovastatin and AC Extract Synergistically Reduce the Proteins Crucial for Cell Cycle Progression in PC3 Cells

To extrapolate the potential mechanism that underlines the novel therapeutic potency exerted by the combination modality of lovastatin and AC through cell cycle regulation in PC3 cells, we next assessed modulations in the expression of proteins crucial for cell cycle progression such as Rb (retinoblastoma protein), p-Rb (phospho-Rb), cyclin A, cyclin D1, CDK1, CDK2, and CDK4. Rb protein is known to play a central role in cell cycle regulation, and its inactivation by phosphorylation, which triggers uncontrolled proliferation, is most common in human sporadic cancers [25]. As expected, the phosphorylation level of Rb was greatly reduced in cells treated with AC extract (40 µg/mL) combined with lovastatin (1 or 2 µM), while individual lovastatin showed only limited (1 or 2 µM) or moderate (4 µM) suppression (Figure 4). Similar changes were also observed in the super-shifted (low-mobility) bands detected in the total Rb blot panel (Figure 4), correlating to the inhibition of Rb phosphorylation. Further, we also revealed marked suppression in cyclin A, cyclin D1, and CDK1 protein levels in a combinatorial group compared to those treated individually (Figure 4). In contrast, protein levels of CDK2 and CDK4 were not markedly affected even in the combined treatment groups, suggesting cyclin A, cyclin D1, and CDK1 as the primary molecules accountable for Rb-mediated cell cycle regulation by the lovastatin/AC combinatorial modality.

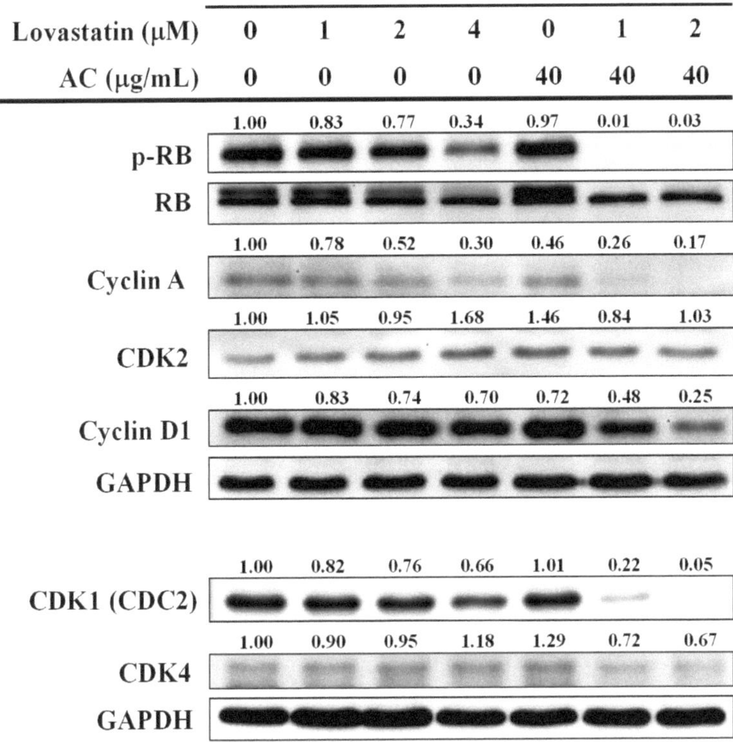

Figure 4. Combination effects of lovastatin and AC extract on cell cycle-regulating molecules in PC3 cells. Cells were treated as indicated for 72 h. Whole-cell lysates were analyzed by Western blot for the proteins of interest. The numbers above the bands indicate the relative densitometric ratios to the bands of loading control (GAPDH). Protein size (kDa): RB (110), Cyclin A (54), CDK2 (33), Cyclin D1 (37), CDK1 (34), CDK4 (34), and GAPDH (35).

3.3. Lovastatin and AC Extract Synergistically Diminish AXL and Survivin in PC3 Cells

In addition to malignant cellular proliferation, aggressive behavior of prostate cancer cells has also been a critical factor contributing to the disease progression [26]. Receptor tyrosine kinase AXL (from the Greek word anexelekto, or uncontrolled) and its downstream phospho-AKT (p-AKT), for instance, are reportedly associated with the aggressiveness and progression of PC [27,28]. Our data demonstrated that expression levels of AXL and p-AKT were dramatically decreased only in PC3 cells treated with a combination of lovastatin and AC extract in a time-dependent manner at 48 and 72 h (Figure 5, upper and middle panels). Apart from AXL, survivin is also frequently associated with biologically aggressive prostate carcinoma [26]. In line with the findings from Figure 4 and AXL, p-ATK levels from Figure 5 and combinatorial lovastatin/AC treatment also predominantly suppressed protein levels of survivin in PC3 cells (Figure 5, lower panel). The aforementioned inhibition of AXL and survivin might profoundly constrain the biologically aggressive behavior of PC3 cells.

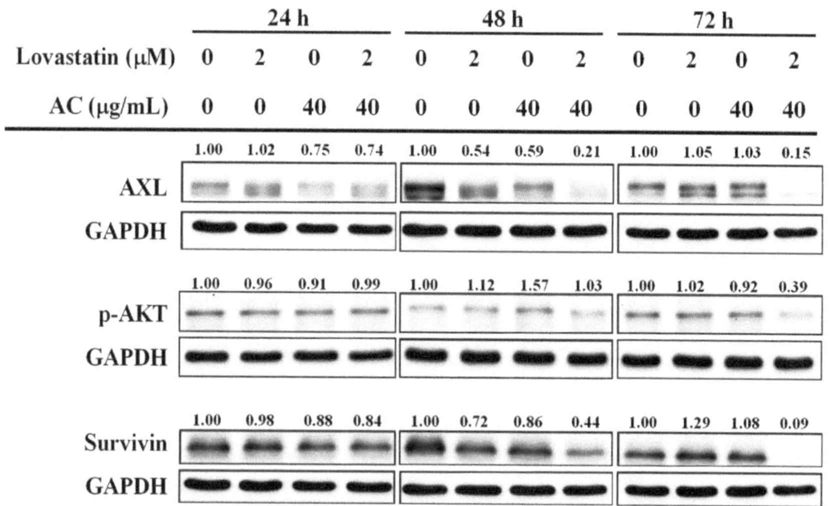

Figure 5. Combination effects of lovastatin and AC extract on aggressiveness-related molecules in PC3 cells. Cells were treated as indicated for 24, 48, and 72 h. Whole-cell lysates were analyzed by Western blot for the proteins of interest. The numbers above the bands indicate the relative densitometric ratios to the bands of loading control (GAPDH). Protein size (kDa): AXL (140), AKT (60), Survivin (16), and GAPDH (35).

3.4. Lovastatin and AC Extract Synergistically Suppress Stemness Molecules and Markers in PC3 Cells

Recent research hypothesizes that the primary contributors to distant metastasis and treatment failure in PC are the prostate cancer stem cells, which possess self-renewal properties and elicit resistance to conventional anticancer therapies [29]. Previous studies have shown that SIRT1, Notch1, c-Myc, and downstream targets of miR-34a are positively associated with prostate cancer stem cell traits [30,31]. Next, we analyzed the protein levels of SIRT1, Notch1, and c-Myc in PC3 cells after combinatorial treatment with lovastatin and AC extract to substantiate its potential merits in treating PC. In agreement with their combination effect on the aggressiveness of PC3 cells (Figure 5), lovastatin and AC extract consistently exhibited synergistic suppression in protein levels of SIRT1, Notch1, and c-Myc as compared with individual treatment (Figure 6A). Moreover, in addition to CD44, a direct target of miR-34a [30,31], CD133 is also reported to be an important marker for prostate cancer stem cells [29]. Our data in Figure 6B revealed substantial downregulation of CD133 and CD44 levels in the combinatorial groups but not in the individual treatment groups. Reductions of these stemness molecules and markers might implicate that the synergistically reduced PC proliferation and aggressiveness are attributed to diminished cancer stem subpopulation in lovastatin/AC-treated PC3 cells.

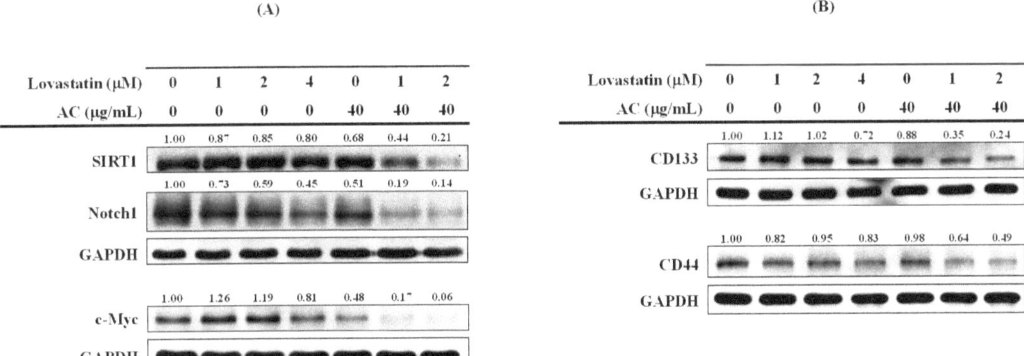

Figure 6. Combination effects of lovastatin and AC extract on the (**A**) stemness molecules and (**B**) stemness markers of PC3 cells. Cells were treated as indicated for 72 h. Whole-cell lysates were analyzed by Western blot for the proteins of interest. The numbers above the bands indicate the relative densitometric ratios to the bands of the loading control (GAPDH). Protein size (kDa): SIRT1 (120), Notch1 (125), c-Myc (67), and GAPDH (35).

4. Discussion

Among the male population in the US, prostate cancer remains the predominant malignancy [30]. Due to the marginal survival benefits that can be offered by current standard-of-care therapies for metastatic androgen-refractory PC patients [30], alternative treatment options like drug repurposing, medicinal plants, and traditional medicine have also been explored in order to fulfill the unmet medical need [5]. The naturally occurring lovastatin from red yeast rice and the edible Taiwanese mushroom *Antrodia camphorata* [15,16,20] have been individually investigated for their effects against prostate cancer, but disappointingly, no clinically viable effects of either compound on prostate cancer have been conclusively reported till now. Results from this study successfully demonstrated synergistic efficacies from a combinatorial strategy using both lovastatin and AC extract, eliciting phenomenal suppression effects on proliferation, aggressiveness, and stemness in PC3 androgen-refractory prostate cancer cells (Figures 1, 2, 5 and 6). Based on the findings presented in our study, lovastatin and AC extract could jointly target molecules responsible for cellular proliferation, aggressiveness, and stemness in PC, providing a plausible explanation for the previous research failure to utilize lovastatin or AC extract alone. This synergism might promote its potential for clinical applications by reducing the required effective concentrations.

Canonically, the tumor suppressor protein retinoblastoma (RB) crucially regulates cell-cycle progression and proliferation by repressing the E2F1-mediated transcriptional program [32–34]. In parallel with our results on the dephosphorylation and thus activation of RB (Figure 4), lovastatin was reported to repress E2F1 and induce apoptosis of PC3 cells at a concentration of 10 µM by Park et al. [17]. In fact, higher concentrations (above 10 µM) required for inducing lovastatin-mediated cytotoxicity were also reported by two previous studies in PC cells [35,36]. Considering that 3.92 µM is the maximum plasma lovastatin concentration detected in clinical trials [37], lovastatin alone seems unlikely to effectively modulate E2F1 in prostate cancer patients. Likewise, the most effective concentrations of statins in other research papers are far from the clinically viable range [38]. Although the anticancer effects of statins against prostate cancer had been intensively studied [39,40], the hurdle remained, thus becoming the choice of a combinatorial compound that could potentiate the therapeutic efficacy of lovastatin to achieve a clinically viable concentration. In our study, an edible compound like AC extract and its renowned safety profile [16] were chosen to investigate its combinatorial anticancer effect with lovastatin against PC.

Multiple oncogenic transcription factors, including but not limited to E2F1, will be rewired after the loss of RB function [34,41]. In addition to regulating the cell cycle, RB plays a pleiotropic role in cancer restriction [34]. Recent studies have declared that RB loss-of-function is tightly associated with aggressive disease and poor outcomes in PC [33,34,41]. In combination with AC extract, lovastatin dephosphorylated and thus activated RB in PC3 cells at concentrations below 3.92 μM. Our Figure 5 showed a reduction of AXL and survivin, two known aggressiveness-regulating molecules of prostate cancer [26–28,42], by combinatorial lovastatin and AC extract, which echoed the roles of RB in cancer control. Survivin is originally known as an IAP (inhibitor of apoptosis) protein [26]. The induction of the apoptotic sub-G1 fraction in lovastatin/AC extract-treated PC3 cells might largely be attributed to the profound decrease of survivin in these cells.

Further, it is noteworthy that AXL is reportedly overexpressed in prostate cancer cell lines and human prostate tumors [42], and its expression is considerably higher in more aggressive androgen-refractory PC3 and DU145 cells compared with the androgen-dependent cell line LNCaP [43]. Consequently, the crucial role of AXL in the progression and metastasis of prostate cancer has made it an attractive therapeutic target [42], as various synthetic inhibitors have been conducted in clinical trials for AXL-targeted therapies [44]. Thus, the drastic AXL-diminishing effect of combining lovastatin with AC extract opened a new avenue to future strategic integration with currently available AXL inhibitors for better treatment against aggressive prostate cancers.

Furthermore, AXL is also closely associated with epithelial-mesenchymal transition (EMT), drug resistance, and cancer stemness [44]. In lovastatin/AC extract-treated PC3 cells, the greatly reduced AXL levels from Figure 5 thus consistently reflected another observation in ablating stemness molecules (SIRT1, Notch1, and c-Myc) and markers (CD44 and CD133) in Figure 6. Contrary to the expected oncogenic roles of Notch1 and c-Myc, the biological functions of SIRT1 in cancer remain controversial [45]. It is reported that the NAD+-dependent deacetylase, i.e., Sirtuin 1 (SIRT1), possesses both oncogenic and tumor-suppressive functions in PC, possibly depending on the stage of tumor progression in context-dependent manners [46]. Compared with normal prostate epithelial PrEC cells and normal prostate cells obtained from patients, SIRT1 expression is markedly up-regulated in human PC cell lines such as LNCap, PC3, and DU145 cells [47,48]. Regarding the roles of the SIRT1-c-Myc axis in supporting cancer stem cell maintenance [30,31,45,49]. Moreover, the reduction of SIRT1 in lovastatin/AC extract-treated PC3 cells might contribute to not only growth inhibition but also mitigated cancer stemness. Since it has been reported that CD133 is a robust biomarker for prostate cancer stem cells, a combination of CD133+ and CD44+ markers, with or without integrin 21, may further improve the isolation of prostate cancer stem cells from clinical specimens [29]. Statins have been strategically proposed as an anti-cancer stem cell compound. Nonetheless, clinical studies conducted to date have yet to deliver conclusive evidence [50]. Given the aforementioned combination effects of lovastatin and AC extract, the CD133+/CD44+ population might be substantially eliminated in the combinatorial-treated PC3 cells.

5. Conclusions

Despite recent advances, androgen-refractory prostate cancer remains without a definitive cure [3,5]. Our current study exploited a novel therapeutic intervention targeting AXL [44], Notch1 [51], Myc [52], or SIRT1 [46] in an attempt to improve the management of androgen-refractory prostate cancers. The findings from this study underscore the potential of integrating lovastatin with AC extract, providing a synergistic or adjunctive pharmacological strategy to amplify the clinical effectiveness of lovastatin or AC extract alone. Since edible dietary products typically exhibit negligible physiological toxicity, a comprehensive preclinical evaluation is also imperative to optimize the dosage, initiation, sequence, and duration of this novel combinatorial treatment modality warranted by lovastatin and AC extract prior to their clinical use.

6. Patents

An invention patent (NO. I 311912, TW) resulted from this work.

Author Contributions: Conceptualization, C.-J.Y., C.-L.C. and S.-E.C.; methodology, H.-L.H., Y.-Y.Y. and H.-C.K.; investigation, C.-H.L., M.-H.H., G.-M.L., T.-J.C. and J.W.-P.; execution of experiments, H.-L.H.; resources, S.-E.C. and G.-M.L.; writing—original draft preparation, C.-J.Y. and H.-C.K.; writing—review and editing, C.-J.Y. and S.-E.C.; supervision, S.-E.C., T.-J.C. and J.W.-P.; project administration, C.-J.Y.; funding acquisition, S.-E.C., G.-M.L. and C.-J.Y. All authors have read and agreed to the published version of the manuscript.

Funding: This work was supported by the Health and Welfare Surcharge of Tobacco Products (MOHW103-TD-B-111-01 and MOHW104-TDU-B-212-124-001), Taiwan, and a research grant from Wan Fang Hospital, Taipei Medical University, Taipei, Taiwan (110-wf-eva-11), and the National Health Research Institutes, Miaoli, Taiwan (CA-111-PP-16).

Institutional Review Board Statement: Not applicable.

Informed Consent Statement: Not applicable.

Data Availability Statement: Not applicable.

Acknowledgments: The authors would like to thank Well Shine Biotechnology Development Co. (Taipei, Taiwan) and Ching-Hua Su (Graduate Institute of Biomedical Material, Taipei Medical University, Taipei, Taiwan) for providing and preparing the *Antrodia camphorata* extract.

Conflicts of Interest: The authors declare no conflict of interest.

References

1. Wang, L.; Lu, B.; He, M.; Wang, Y.; Wang, Z.; Du, L. Prostate Cancer Incidence and Mortality: Global Status and Temporal Trends in 89 Countries from 2000 to 2019. *Front. Public Health* **2022**, *10*, 811044. [CrossRef]
2. Hara, T.; Miyazaki, H.; Lee, A.; Tran, C.P.; Reiter, R.E. Androgen receptor and invasion in prostate cancer. *Cancer Res.* **2008**, *68*, 1128–1135. [CrossRef]
3. Dong, L.; Zieren, R.C.; Xue, W.; de Reijke, T.M.; Pienta, K.J. Metastatic prostate cancer remains incurable, why? *Asian J. Urol.* **2019**, *6*, 26–41. [CrossRef] [PubMed]
4. Kita, Y.; Goto, T.; Akamatsu, S.; Yamasaki, T.; Inoue, T.; Ogawa, O.; Kobayashi, T. Castration-Resistant Prostate Cancer Refractory to Second-Generation Androgen Receptor Axis-Targeted Agents: Opportunities and Challenges. *Cancers* **2018**, *10*, 345. [CrossRef] [PubMed]
5. Sekhoacha, M.; Riet, K.; Motloung, P.; Gumenku, L.; Adegoke, A.; Mashele, S. Prostate Cancer Review: Genetics, Diagnosis, Treatment Options, and Alternative Approaches. *Molecules* **2022**, *27*, 5730. [CrossRef]
6. De Pinieux, G.; Chariot, P.; Ammi-Said, M.; Louarn, F.; Lejonc, J.L.; Astier, A.; Jacotot, B.; Gherardi, R. Lipid-lowering drugs and mitochondrial function: Effects of HMG-CoA reductase inhibitors on serum ubiquinone and blood lactate/pyruvate ratio. *Br. J. Clin. Pharmacol.* **1996**, *42*, 333–337. [CrossRef] [PubMed]
7. Prospective Studies, C.; Lewington, S.; Whitlock, G.; Clarke, R.; Sherliker, P.; Emberson, J.; Halsey, J.; Qizilbash, N.; Peto, R.; Collins, R. Blood cholesterol and vascular mortality by age, sex, and blood pressure: A meta-analysis of individual data from 61 prospective studies with 55,000 vascular deaths. *Lancet* **2007**, *370*, 1829–1839. [CrossRef]
8. Taylor, F.C.; Huffman, M.; Ebrahim, S. Statin therapy for primary prevention of cardiovascular disease. *JAMA* **2013**, *310*, 2451–2452. [CrossRef]
9. Barrios-Gonzalez, J.; Miranda, R.U. Biotechnological production and applications of statins. *Appl. Microbiol. Biotechnol.* **2010**, *85*, 869–883. [CrossRef]
10. Chae, Y.K.; Yousaf, M.; Malecek, M.K.; Carneiro, B.; Chandra, S.; Kaplan, J.; Kalyan, A.; Sassano, A.; Platanias, L.C.; Giles, F. Statins as anti-cancer therapy; Can we translate preclinical and epidemiologic data into clinical benefit? *Discov. Med.* **2015**, *20*, 413–427. [PubMed]
11. Lim, T.; Lee, I.; Kim, J.; Kang, W.K. Synergistic Effect of Simvastatin Plus Radiation in Gastric Cancer and Colorectal Cancer: Implications of BIRC5 and Connective Tissue Growth Factor. *Int. J. Radiat. Oncol. Biol. Phys.* **2015**, *93*, 316–325. [CrossRef]
12. Keskivali, T.; Kujala, P.; Visakorpi, T.; Tammela, T.L.; Murtola, T.J. Statin use and risk of disease recurrence and death after radical prostatectomy. *Prostate* **2016**, *76*, 469–478. [CrossRef]
13. Graaf, M.R.; Beiderbeck, A.B.; Egberts, A.C.; Richel, D.J.; Guchelaar, H.J. The risk of cancer in users of statins. *J. Clin. Oncol.* **2004**, *22*, 2388–2394. [CrossRef]
14. Mondul, A.M.; Han, M.; Humphreys, E.B.; Meinhold, C.L.; Walsh, P.C.; Platz, E.A. Association of statin use with pathological tumor characteristics and prostate cancer recurrence after surgery. *J. Urol.* **2011**, *185*, 1268–1273. [CrossRef]
15. Klimek, M.; Wang, S.; Ogunkanmi, A. Safety and efficacy of red yeast rice (*Monascus purpureus*) as an alternative therapy for hyperlipidemia. *Pharm. Ther.* **2009**, *34*, 313–327.

16. Takwi, A.A.; Li, Y.; Becker Buscaglia, L.E.; Zhang, J.; Choudhury, S.; Park, A.K.; Liu, M.; Young, K.H.; Park, W.Y.; Martin, R.C.; et al. A statin-regulated microRNA represses human c-Myc expression and function. *EMBO Mol. Med.* **2012**, *4*, 896–909. [CrossRef] [PubMed]

17. Park, C.; Lee, I.; Kang, W.K. Lovastatin-induced E2F-1 modulation and its effect on prostate cancer cell death. *Carcinogenesis* **2001**, *22*, 1727–1731. [CrossRef] [PubMed]

18. Hoque, A.; Chen, H.; Xu, X.C. Statin induces apoptosis and cell growth arrest in prostate cancer cells. *Cancer Epidemiol. Biomarkers Prev.* **2008**, *17*, 88–94. [CrossRef] [PubMed]

19. Knox, J.J.; Siu, L.L.; Chen, E.; Dimitroulakos, J.; Kamel-Reid, S.; Moore, M.J.; Chin, S.; Irish, J.; LaFramboise, S.; Oza, A.M. A Phase I trial of prolonged administration of lovastatin in patients with recurrent or metastatic squamous cell carcinoma of the head and neck or of the cervix. *Eur. J. Cancer* **2005**, *41*, 523–530. [CrossRef] [PubMed]

20. Ganesan, N.; Baskaran, R.; Velmurugan, B.K.; Thanh, N.C. Antrodia cinnamomea-An updated minireview of its bioactive components and biological activity. *J. Food Biochem.* **2019**, *43*, e12936. [CrossRef] [PubMed]

21. Ho, C.M.; Huang, C.C.; Huang, C.J.; Cheng, J.S.; Chen, I.S.; Tsai, J.Y.; Jiann, B.P.; Tseng, P.L.; Kuo, S.J.; Jan, C.R. Effects of antrodia camphorata on viability, apoptosis, and [Ca^{2+}]i in PC3 human prostate cancer cells. *Chin. J. Physiol.* **2008**, *51*, 78–84. [PubMed]

22. Chen, K.C.; Peng, C.C.; Peng, R.Y.; Su, C.H.; Chiang, H.S.; Yan, J.H.; Hsieh-Li, H.M. Unique formosan mushroom Antrodia camphorata differentially inhibits androgen-responsive LNCaP and -independent PC-3 prostate cancer cells. *Nutr. Cancer* **2007**, *57*, 111–121. [CrossRef]

23. Kaighn, M.E.; Narayan, K.S.; Ohnuki, Y.; Lechner, J.F.; Jones, L.W. Establishment and characterization of a human prostatic carcinoma cell line (PC-3). *Investig. Urol.* **1979**, *17*, 16–23.

24. Stone, K.R.; Mickey, D.D.; Wunderli, H.; Mickey, G.H.; Paulson, D.F. Isolation of a human prostate carcinoma cell line (DU 145). *Int. J. Cancer* **1978**, *21*, 274–281. [CrossRef] [PubMed]

25. Rizzolio, F.; Lucchetti, C.; Caligiuri, I.; Marchesi, I.; Caputo, M.; Klein-Szanto, A.J.; Bagella, L.; Castronovo, M.; Giordano, A. Retinoblastoma tumor-suppressor protein phosphorylation and inactivation depend on direct interaction with Pin1. *Cell Death Differ.* **2012**, *19*, 1152–1161. [CrossRef]

26. Shariat, S.F.; Lotan, Y.; Saboorian, H.; Khoddami, S.M.; Roehrborn, C.G.; Slawin, K.M.; Ashfaq, R. Survivin expression is associated with features of biologically aggressive prostate carcinoma. *Cancer* **2004**, *100*, 751–757. [CrossRef] [PubMed]

27. Shiozawa, Y.; Pedersen, E.A.; Patel, L.R.; Ziegler, A.M.; Havens, A.M.; Jung, Y.; Wang, J.; Zalucha, S.; Loberg, R.D.; Pienta, K.J.; et al. GAS6/AXL axis regulates prostate cancer invasion, proliferation, and survival in the bone marrow niche. *Neoplasia* **2010**, *12*, 116–127. [CrossRef] [PubMed]

28. Vanli, N.; Sheng, J.; Li, S.; Xu, Z.; Hu, G.F. Ribonuclease 4 is associated with aggressiveness and progression of prostate cancer. *Commun. Biol.* **2022**, *5*, 625. [CrossRef]

29. Yang, J.; Aljitawi, O.; Van Veldhuizen, P. Prostate Cancer Stem Cells: The Role of CD133. *Cancers* **2022**, *14*, 5448. [CrossRef]

30. Li, W.J.; Liu, X.; Dougherty, E.M.; Tang, D.G. MicroRNA-34a, Prostate Cancer Stem Cells, and Therapeutic Development. *Cancers* **2022**, *14*, 4538. [CrossRef]

31. Liu, C.; Kelnar, K.; Liu, B.; Chen, X.; Calhoun-Davis, T.; Li, H.; Patrawala, L.; Yan, H.; Jeter, C.; Honorio, S.; et al. The microRNA miR-34a inhibits prostate cancer stem cells and metastasis by directly repressing CD44. *Nat. Med.* **2011**, *17*, 211–215. [CrossRef] [PubMed]

32. Thangavel, C.; Boopathi, E.; Liu, Y.; Haber, A.; Ertel, A.; Bhardwaj, A.; Addya, S.; Williams, N.; Ciment, S.J.; Cotzia, P.; et al. RB Loss Promotes Prostate Cancer Metastasis. *Cancer Res.* **2017**, *77*, 982–995. [CrossRef]

33. McNair, C.; Xu, K.; Mandigo, A.C.; Benelli, M.; Leiby, B.; Rodrigues, D.; Lindberg, J.; Gronberg, H.; Crespo, M.; De Laere, B.; et al. Differential impact of RB status on E2F1 reprogramming in human cancer. *J. Clin. Investig.* **2018**, *128*, 341–358. [CrossRef] [PubMed]

34. Mandigo, A.C.; Shafi, A.A.; McCann, J.J.; Yuan, W.; Laufer, T.S.; Bogdan, D.; Gallagher, L.; Dylgjeri, E.; Semenova, G.; Vasilevskaya, I.A.; et al. Novel Oncogenic Transcription Factor Cooperation in RB-Deficient Cancer. *Cancer Res.* **2022**, *82*, 221–234. [CrossRef] [PubMed]

35. Niknejad, N.; Gorn-Hondermann, I.; Ma, L.; Zahr, S.; Johnson-Obeseki, S.; Corsten, M.; Dimitroulakos, J. Lovastatin-induced apoptosis is mediated by activating transcription factor 3 and enhanced in combination with salubrinal. *Int. J. Cancer* **2014**, *134*, 268–279. [CrossRef] [PubMed]

36. Dimitroulakos, J.; Ye, L.Y.; Benzaquen, M.; Moore, M.J.; Kamel-Reid, S.; Freedman, M.H.; Yeger, H.; Penn, L.Z. Differential sensitivity of various pediatric cancers and squamous cell carcinomas to lovastatin-induced apoptosis: Therapeutic implications. *Clin. Cancer Res.* **2001**, *7*, 158–167.

37. Thibault, A.; Samid, D.; Tompkins, A.C.; Figg, W.D.; Cooper, M.R.; Hohl, R.J.; Trepel, J.; Liang, B.; Patronas, N.; Venzon, D.J.; et al. Phase I study of lovastatin, an inhibitor of the mevalonate pathway, in patients with cancer. *Clin. Cancer Res.* **1996**, *2*, 483–491.

38. Dulak, J.; Jozkowicz, A. Anti-angiogenic and anti-inflammatory effects of statins: Relevance to anti-cancer therapy. *Curr. Cancer Drug Targets* **2005**, *5*, 579–594. [CrossRef]

39. Craig, E.L.; Stopsack, K.H.; Evergren, E.; Penn, L.Z.; Freedland, S.J.; Hamilton, R.J.; Allott, E.H. Statins and prostate cancer-hype or hope? The epidemiological perspective. *Prostate Cancer Prostatic Dis.* **2022**, *25*, 641–649. [CrossRef]

40. Longo, J.; Freedland, S.J.; Penn, L.Z.; Hamilton, R.J. Statins and prostate cancer-hype or hope? The biological perspective. *Prostate Cancer Prostatic Dis.* **2022**, *25*, 650–656. [CrossRef]

41. Mandigo, A.C.; Yuan, W.; Xu, K.; Gallagher, P.; Pang, A.; Guan, Y.F.; Shafi, A.A.; Thangavel, C.; Sheehan, B.; Bogdan, D.; et al. RB/E2F1 as a Master Regulator of Cancer Cell Metabolism in Advanced Disease. *Cancer Discov.* **2021**, *11*, 2334–2353. [CrossRef] [PubMed]
42. Paccez, J.D.; Vasques, G.J.; Correa, R.G.; Vasconcellos, J.F.; Duncan, K.; Gu, X.; Bhasin, M.; Libermann, T.A.; Zerbini, L.F. The receptor tyrosine kinase Axl is an essential regulator of prostate cancer proliferation and tumor growth and represents a new therapeutic target. *Oncogene* **2013**, *32*, 689–698. [CrossRef] [PubMed]
43. Xiao, Z.; Hammes, S.R. AXL cooperates with EGFR to mediate neutrophil elastase-induced migration of prostate cancer cells. *iScience* **2021**, *24*, 103270. [CrossRef]
44. Zhu, C.; Wei, Y.; Wei, X. AXL receptor tyrosine kinase as a promising anti-cancer approach: Functions, molecular mechanisms and clinical applications. *Mol. Cancer* **2019**, *18*, 153. [CrossRef]
45. Garcia-Peterson, L.M.; Li, X. Trending topics of SIRT1 in tumorigenicity. *Biochim. Biophys. Acta Gen. Subj.* **2021**, *1865*, 129952. [CrossRef]
46. Huang, S.B.; Thapa, D.; Munoz, A.R.; Hussain, S.S.; Yang, X.; Bedolla, R.G.; Osmulski, P.; Gaczynska, M.E.; Lai, Z.; Chiu, Y.C.; et al. Androgen deprivation-induced elevated nuclear SIRT1 promotes prostate tumor cell survival by reactivation of AR signaling. *Cancer Lett.* **2021**, *505*, 24–36. [CrossRef] [PubMed]
47. Jung-Hynes, B.; Nihal, M.; Zhong, W.; Ahmad, N. Role of sirtuin histone deacetylase SIRT1 in prostate cancer. A target for prostate cancer management via its inhibition? *J. Biol. Chem.* **2009**, *284*, 3823–3832. [CrossRef] [PubMed]
48. Kojima, K.; Ohhashi, R.; Fujita, Y.; Hamada, N.; Akao, Y.; Nozawa, Y.; Deguchi, T.; Ito, M. A role for SIRT1 in cell growth and chemoresistance in prostate cancer PC3 and DU145 cells. *Biochem. Biophys. Res. Commun.* **2008**, *373*, 423–428. [CrossRef] [PubMed]
49. Fan, W.; Li, X. The SIRT1-c-Myc axis in regulation of stem cells. *Front. Cell Dev. Biol.* **2023**, *11*, 1236968. [CrossRef]
50. Shin, S.W. Clinical and Therapeutic Implications of Cancer Stem Cells. *N. Engl. J. Med.* **2019**, *381*, e19. [CrossRef] [PubMed]
51. Stoyanova, T.; Riedinger, M.; Lin, S.; Faltermeier, C.M.; Smith, B.A.; Zhang, K.X.; Going, C.C.; Goldstein, A.S.; Lee, J.K.; Drake, J.M.; et al. Activation of Notch1 synergizes with multiple pathways in promoting castration-resistant prostate cancer. *Proc. Natl. Acad. Sci. USA* **2016**, *113*, E6457–E6466. [CrossRef] [PubMed]
52. Shukla, S.; Fletcher, S.; Chauhan, J.; Chalfant, V.; Riveros, C.; Mackeyev, Y.; Singh, P.K.; Krishnan, S.; Osumi, T.; Balaji, K.C. 3JC48-3 (methyl 4′-methyl-5-(7-nitrobenzo[c][1,2,5]oxadiazol-4-yl)-[1,1′-biphenyl]-3-carboxylate): A novel MYC/MAX dimerization inhibitor reduces prostate cancer growth. *Cancer Gene Ther.* **2022**, *29*, 1550–1557. [CrossRef] [PubMed]

Article

Amurensin G Sensitized Cholangiocarcinoma to the Anti-Cancer Effect of Gemcitabine via the Downregulation of Cancer Stem-like Properties

Yun-Jung Na, Hong Kyu Lee and Kyung-Chul Choi *

Laboratory of Biochemistry and Immunology, College of Veterinary Medicine, Chungbuk National University, Cheongju 28644, Chungbuk, Republic of Korea; sky1105a@naver.com (Y.-J.N.); hklee83@cbnu.ac.kr (H.K.L.)
* Correspondence: kchoi@cbu.ac.kr; Tel.: +82-43-261-3664; Fax: +82-43-267-3150

Abstract: Cholangiocarcinoma (CCA) is a malignant biliary tract tumor with a high mortality rate and refractoriness to chemotherapy. Gemcitabine is an anti-cancer chemotherapeutic agent used for CCA, but the efficacy of gemcitabine in CCA treatment is limited, due to the acquisition of chemoresistance. The present study evaluated the chemosensitizing effects of Amurensin G (AMG), a natural sirtuin-1 inhibitor derived from *Vitis amurensis*, in the SNU-478 CCA cells. Treatment with AMG decreased the SNU-478 cell viability and the colony formation ability. Annexin V/ Propidium iodide staining showed that the AMG increased apoptotic death. In addition, AMG downregulated anti-apoptotic Bcl-2 expression, while upregulating pro-apoptotic cleaved caspase-3 expression. Treatment with AMG decreased the migratory ability of the cells in a wound healing assay and transwell migration assay. It was observed that AMG decreased the gemcitabine-induced increase in $CD44^{high}CD24^{high}CD133^{high}$ cell populations, and the expression of the Sox-2 protein was decreased by AMG treatment. Co-treatment of AMG with gemcitabine significantly enhanced the production of reactive oxygen species, as observed through mitochondrial superoxide staining, which might be associated with the downregulation of the Sirt1/Nrf2 pathway by AMG. These results indicate that AMG enhances the chemotherapeutic ability of gemcitabine by downregulating cancer stem-like properties in CCA cells. Hence, a combination therapy of AMG with gemcitabine may be an attractive therapeutic strategy for cholangiocarcinoma.

Keywords: cholangiocarcinoma; amurensin G; gemcitabine; cancer stem cell; chemoresistance

Citation: Na, Y.-J.; Lee, H.K.; Choi, K.-C. Amurensin G Sensitized Cholangiocarcinoma to the Anti-Cancer Effect of Gemcitabine via the Downregulation of Cancer Stem-like Properties. *Nutrients* **2024**, *16*, 73. https://doi.org/10.3390/nu16010073

Academic Editors: Debasis Bagchi, Yi-Wen Liu and Ching-Hsein Chen

Received: 20 October 2023
Revised: 8 December 2023
Accepted: 20 December 2023
Published: 25 December 2023

1. Introduction

Cholangiocarcinoma (CCA) is a malignant biliary tract tumor with a poor prognosis. Despite significant advances in diagnosis and therapy, the 5-year survival rate of CCA is still less than 20%. Because CCA has no clinical symptoms and no specific diagnostic indicators in the early stage, CCA is commonly diagnosed in the advanced stages. Surgical resection remains the mainstay of therapy, but recurrences are common, due to incomplete tumor removal [1,2]. The adjuvant chemotherapy with gemcitabine and cisplatin after CCA resection may reduce recurrence risk and improve survival [3,4]. However, the efficacy of these adjuvant chemotherapies for CCA treatment is limited, due to the acquisition of resistance, and relapse occurs frequently.

Gemcitabine inhibits cell division and the growth of cancer cells by interfering with DNA synthesis [5]. However, persistent treatment of gemcitabine, in turn, exerts drug resistance and toxicity through its effects on the tumor microenvironment and epithelial-mesenchymal transition (EMT) in cancer cells [5,6]. Drug resistance induced by gemcitabine is caused by various factors, including the high expression of drug efflux pumps and the upregulation of cancer stem-like cells (CSCs) via the activation of the Hedgehog, Wnt, and Notch pathways [7]. CSCs are subpopulations of undifferentiated cancer cells within the tumor that are responsible for tumor initiation, progression, and metastasis [8]. In addition,

CSCs mediate EMT, stemness, and chemoresistance, thereby causing relapse in tumors [9]. Therefore, various approaches have been tested to sensitize cancer cells to conventional chemotherapeutics, through CSC downregulation.

Amurensin G (AMG), a member of the amurensis family, is a resveratrol trimer and is known to derive from specific plants, including *Phellodendron amurense*, some *Papaver* species, and *Vitis amurensis* [10,11]. AMG has been reported as a potent natural sirtuin 1 (Sirt1) inhibitor, and has shown neuroprotective effects in animal models [10,12,13]. In addition, AMG has anti-oxidant activity, induces apoptosis, and has anti-angiogenesis effects in cancer cells [14–16]. Previous studies have shown that AMG inhibits drug-resistant cell growth and reduces CSCs [14,15]. However, the anti-cancer effect of AMG in CCA and its related mechanisms have not been elucidated, to date.

In the present study, we investigated the effects of AMG on CCA sensitization to the gemcitabine by confirming cancer stem-like cell properties, including apoptosis, colony formation, migratory ability, and ROS formation in the SNU-478 CCA cell line. This study suggests that AMG increases the anti-cancer effect mediated by gemcitabine, and combination treatment of AMG with gemcitabine may be a promising therapeutic strategy for CCA.

2. Materials and Methods

2.1. Cell Culture and Reagents

CCA (SNU-478) cell line was obtained from the Korean Cell Line Bank (KCLB; Seoul, Republic of Korea). The SNU-478 cells were maintained in Roswell Park Memorial Institute (RPMI) 1640 (Welgene, Gyeongsan, Republic of Korea), supplemented with 10% fetal bovine serum (FBS; R&D systems, Minneapolis, MN, USA), 1% antibiotic-antimycotic solution (Welgene) and 1% HEPES buffer solution (Welgene) in a humidified incubator with 5% CO_2 at 37 °C. When the cell density was 70–80%, trypsinization was performed using 0.25% trypsin (Welgene). AMG, isolated from the stem of *Vitis amurensis*, according to the procedure previously described [13], was kindly provided by Prof. Dr. Yeon Hee Seong (Laboratory of Pharmacology, College of Veterinary Medicine, Chungbuk National University, Cheongju, Republic of Korea). Gemcitabine was purchased from Sigma-Aldrich (St. Louis, MO, USA).

2.2. Cell Viability Assay

Cell viability was measured using a Quanti-MAX water-soluble tetrazolium salt (WST) assay kit (Biomax, Seoul, Republic of Korea), according to the manufacturer's instructions, with slight modifications. SNU-478 cells (4×10^3 cells/well) were plated in a 96-well cell culture plate (Sarstedt, Nümbrecht, Germany). The cells were pre-incubated for 24 h, and AMG and/or gemcitabine were treated. After a further 72 h of incubation, the absorbance was measured using a Neo2 Hybrid Multimode Reader (Agilent Technologies, Inc., Santa Clara, CA, USA).

2.3. Colony Formation Assay

The ability of colony formation was evaluated using a colony formation assay, as described previously, with slight modifications [17]. CCA (SNU-478) cells (5×10^2 cells/well) were seeded in 6-well cell culture plates. After 24 h of pre-incubation, the cells were treated with AMG (5 µM) and/or gemcitabine (1 nM). After 8 days of incubation, the cells were fixed with 4% paraformaldehyde and stained with 0.5% crystal violet (Sigma-Aldrich). The colony areas were measured using Image J software v.1.49.

2.4. Cell Apoptosis Assay

The types of cell death were evaluated using Annexin V/Propidium iodide (PI) staining, according to the manufacturer's instructions. SNU-478 cells (2×10^5 cells/well) were plated in 6-well cell culture plates. After 24 h of incubation, the cells were co-incubated with AMG (5 µM) and/or gemcitabine (500 nM) for 48 h. After collecting the cells, they were reacted with Annexin V (BioLegend, San Diego, CA, USA) and propidium iodide (PI) (Invitrogen, Waltham, MA, USA). The samples were acquired using fluorescence-activated cell sorting (FACS) Symphony A3 (BD Biosciences, San Diego, CA, USA), and results were analyzed with the FlowJo Software v. 10.8.1 (TreeStar, San Carlos, CA, USA).

2.5. Wound Healing Assay

The cell migratory ability was evaluated using the wound healing assay, as described previously, with slight modifications [18]. SNU-478 cells (5×10^5 cells/well) were seeded in 6-well plates (SPL, Gyeonggi, Republic of Korea). After 24 h, cells were treated with 3.5 μg/mL Mitomycin C (Sigma-Aldrich) for 1.5 h, and scratched using a sterile plastic tip. After washing with Dulbecco's Phosphate-Buffered Saline (DPBS, Welgene), the cells were treated with AMG (5 μM) and/or gemcitabine (100 nM) for 48 h. Images were acquired using the IX-73 inverted microscope (Olympus, Tokyo, Japan), and the scratch area was measured by CellSens Dimension 1.13 version (Olympus).

2.6. Transwell Migration Assay

The cell migratory ability was evaluated using the transwell migration assay, as described previously, with slight modifications [19]. SNU-478 cells (1×10^5 cells/well) were seeded in the upper chamber (24-well, 8.0 μm pore membrane, Corning, Somerville, MA, USA). The upper chamber was treated with serum-free media with AMG (5 μM) and/or gemcitabine (100 nM), and the lower chamber was treated with media (10% FBS) with AMG (5 μM) and/or gemcitabine (100 nM). After 24 h, the surface of the upper chamber was fixed with 4% paraformaldehyde (Geneall, Seoul, Republic of Korea) and stained with 0.5% crystal violet (Sigma-Aldrich). The images of migrated cells were photographed using an (Olympus) IX-73 inverted microscope, and the number of migrated cells was counted using CellSens Dimension 1.13 version (Olympus).

2.7. Flow Cytometry Analysis

The CD44, CD24, and CD133 populations were detected by FACS analysis. SNU-478 cells (2×10^5 cells/well) were plated in 6-well cell culture plates (Sarstedt). After 24 h, the cells were treated with AMG (5 μM) and/or gemcitabine (500 nM) for 48 h. After harvesting, cells were stained with allophycocyanin (APC)-labeled anti-human CD133 (BioLegend), phycoerythrin (PE)-labeled anti-human CD24 (BioLegend), and fluorescein isothiocyanate (FITC)-labeled anti-human CD44 (BioLegend) (Table 1). Stained cells were acquired using FACS Symphony A3 (BD Life Sciences Bioscience, San Diego, CA, USA) and results were analyzed using the FlowJo Software v. 10.8.1 (TreeStar, San Carlos, CA, USA).

Table 1. List of antibodies used in Western blot assay.

Protein	Manufacturer	Molecular Weight (kDa)	Source	Clone	Dilution Ratio
Bcl-2	BioLegend	22, 26	Mouse	100	1:100
c-cas3	Cell Signaling Technology	17, 19	Rabbit	Polyclonal	1:100
Sox-2	Cell Signaling Technology	35	Rabbit	D6D9	1:100
Sirt1	Cell Signaling Technology	120	Mouse	1F3	1:100
Nrf2	Cell Signaling Technology	97–100	Rabbit	D1Z9C	1:100
Keap1	Cell Signaling Technology	60–64	Rabbit	D6B12	1:100
GAPDH	Abcam	40.2	Mouse	6C5	1:200
β-actin	Cell Signaling Technology	45	Rabbit	13E5	1:200

2.8. Mitochondrial Superoxide (MitoSOX) Assay

MitoSOX^{TM} staining was performed according to the manufacturer's instructions, with slight modifications. SNU-478 cells (6×10^3 cells/well) were seeded in a 96-well cell culture plate. After 24 h, cells were treated with AMG (5 μM) and/or gemcitabine (500 nM) for 48 h. Then, staining was performed with MitoSOX^{TM} (5 μM) (Invitrogen) and Hoechst 33342 (10 μg/mL) (Sigma-Aldrich). Images of stained cells were captured by Lionheart^{TM}

FX Automated Microscope (BioTek Instruments Inc., Winooski, VT, USA), and analyzed using Gen5 v 3.14.03 (Agilent Technologies, Inc.).

2.9. Western Blot Assay

The proteins from SNU-478 cells were extracted from cells using PRO-PREP™ (iNtRON Biotechnology, Inc., Seongnam, Republic of Korea). Protein concentration was determined by bicinchoninic acid assay (Sigma-Aldrich), according to the manufacturer's instructions. Western blot was performed using the JESS™ Simple Western automated nano-immunoassay system (ProteinSimple, San Jose, CA, USA). The antibodies used were the following: Sirt1, Kelch-like ECH-associated protein 1 (Keap1), nuclear factor erythroid-2 related factor 2 (Nrf2), Sex determining region Y-box 2 (Sox-2), cleaved caspase-3 (c-cas3) (Cell Signaling Technology, Danvers, MA, USA), B-cell lymphoma-2 (Bcl-2; Biolegend), β-actin and glyceraldehyde-3-phosphate dehydrogenase (GAPDH) (Abcam, Cambridge, UK) (Table 1). The band images were captured by Compass Simple Western software (v 6.0.0, ProteinSimple), and the intensities of the target protein were measured using Image J software v.1.49.

2.10. Statistical Analysis

All data were presented as means $\pm$ standard deviation (S.D.). Statistical significances of all the data were analyzed using one-way analysis of variance (ANOVA), followed by a post hoc Dunnett's test, using the GraphPad Prism 5.01 software (GraphPad Software Inc., San Diego, CA, USA). A p-value < 0.05 is considered statistically significant.

3. Results

3.1. AMG Decreases Cell Growth in SNU-478 Cells

To evaluate the effects of AMG on cell viability, a WST assay was performed on SNU-478 cells. Based on preliminary experimental results, gemcitabine was chosen at an appropriate concentration of 500 nM. AMG decreased cell viability, with half-maximal inhibitory concentration (IC_{50}) values of 4.57 μM in the SNU-478 cells (Figure 1A). AMG or gemcitabine significantly decreased cell viability. Combination treatment of AMG with gemcitabine presented additively reduced cell viability (Figure 1B). A colony formation assay was performed to confirm the effect of AMG and gemcitabine on cell proliferation. The use of gemcitabine 500 nM caused almost all cell death, preventing the continuation of the experiment. Therefore, we used an appropriate concentration of gemcitabine 1 nM. AMG (5 μM) significantly reduced the colony-forming area compared with the control, while gemcitabine (1 nM) treatment showed no significant difference in the colony-forming area. Combination treatment of AMG (5 μM) with gemcitabine (1 nM) significantly decreased the colony area compared with the gemcitabine-alone treatment (Figure 1C,D). These results indicated that the AMG reduced cell survival and proliferation, and co-treatment of AMG with gemcitabine additively inhibits cell survival and proliferation.

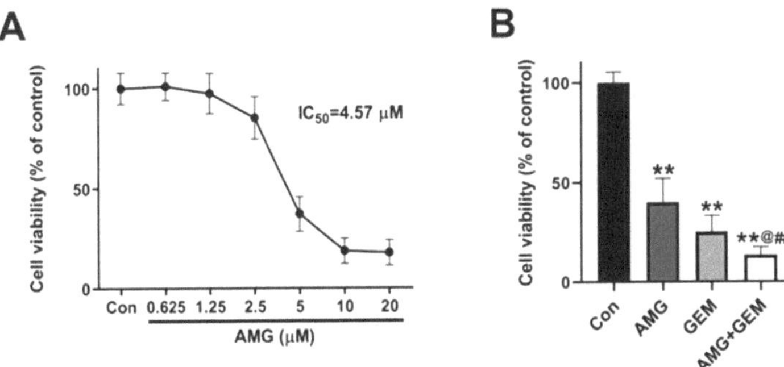

Figure 1. *Cont.*

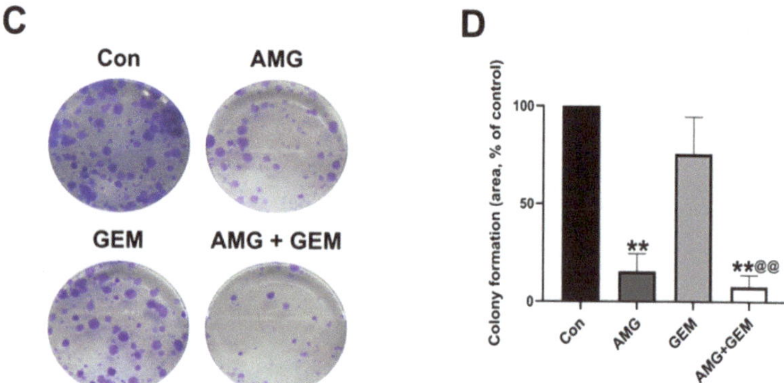

Figure 1. Effects of co-treatment of AMG and gemcitabine on cell viability and proliferation of SNU-478 cells. (**A**) The representative graph shows the changes in cell viability by AMG, for 72 h. (**B**) The graph shows the change in cell viability by AMG and/or gemcitabine, for 72 h. (**C**) The representative images show colony formation in SNU-478 cells. (**D**) Quantification of colony-forming area. Data are presented as means ± S.D. from at least three independent experiments. ** $p < 0.01$ vs. Con; @ $p < 0.05$, @@ $p < 0.01$ vs. GEM; ## $p < 0.01$ vs. AMG (Dunnet's test). Con, control; AMG, amurensin G; GEM, gemcitabine.

3.2. Combination Treatment of AMG with Gemcitabine Induces Apoptosis in SNU-478 Cells

Annexin V/PI staining was used to investigate whether AMG induces apoptosis. AMG (5 μM) and gemcitabine (500 nM) significantly increased the Annexin V^+ population, compared with the control. Combination treatment of AMG (5 μM) with gemcitabine (500 nM) significantly increased the Annexin V^+ population, compared with all other groups. (Figure 2A,B). In addition, the expression of apoptosis-associated protein was examined using Western blot (Figure 2C). AMG treatment significantly reduced the level of Bcl-2, compared with the control. The expression level of Bcl-2 in the co-treatment of AMG with the gemcitabine group was lower than that of the control group. The level of c-cas3 in the combination treatment of AMG with the gemcitabine group was significantly higher than that of the control group (Figure 2D,E). These results indicated that AMG and gemcitabine co-treatment increased apoptotic cell death.

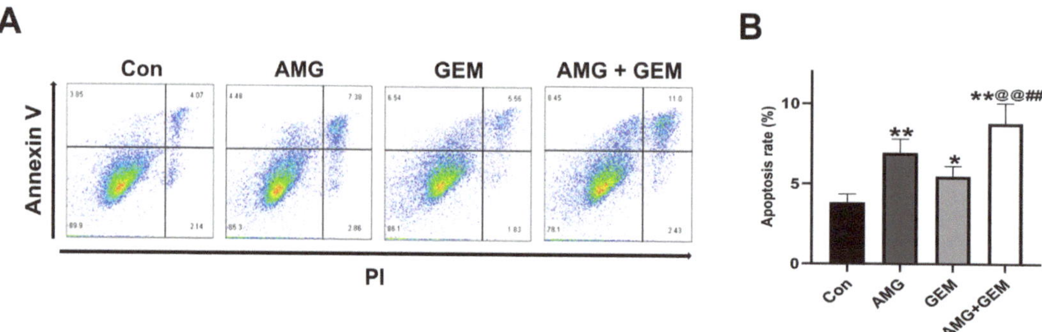

Figure 2. *Cont.*

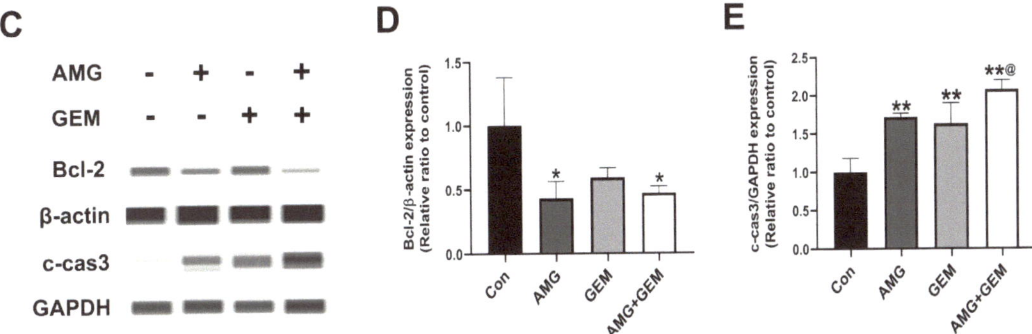

Figure 2. Effects of co-treatment of AMG and gemcitabine on types of cell death in SNU-478 cells. (**A**) Representative flow cytometry plots with Annexin V/PI staining for apoptosis. (**B**) Percentage of apoptotic cells. (**C**) The protein expression of Bcl-2 and c-cas3 was detected by Western blotting after the treatment for 48 h. (**D**) The protein expression of Bcl-2 was normalized with β-actin. (**E**) The protein expression of c-cas3 was normalized with GAPDH. Data are presented as means ± S.D. from at least three independent experiments. * $p < 0.05$, ** $p < 0.01$ vs. Con; @ $p < 0.05$, @@ $p < 0.01$ vs. GEM; ## $p < 0.01$ vs. AMG (Dunnet's test). Con, control; AMG, amurensin G; GEM, gemcitabine.

3.3. AMG Suppresses the Migratory Ability in SNU-478 Cells

To evaluate the effect of AMG on the migratory ability of SNU-478 cells, a wound healing assay and transwell migration assay were performed. The use of gemcitabine 500 nM caused almost all cell death, preventing the continuation of the experiment. Therefore, we used an appropriate concentration of gemcitabine 100 nM. At 24 h, AMG (5 μM) significantly decreased wound closure, compared with the control, while gemcitabine (100 nM) significantly increased wound closure, compared with the control. Combination treatment of AMG (5 μM) with gemcitabine (100 nM) significantly decreased wound closure, compared with gemcitabine-alone treatment. At 48 h, AMG (5 μM) decreased wound closure compared with control, and wound closure of gemcitabine (100 nM) showed no significant difference. Combination treatment of gemcitabine (100 nM) with AMG (5 μM) significantly reduced wound closure, compared with gemcitabine-alone treatment (Figure 3A,B). In the transwell migration assay, AMG (5 μM) decreased the number of migrated cells when compared with the control, while gemcitabine (100 nM) showed no significant difference. Combination treatment of AMG (5 μM) with gemcitabine (100 nM) significantly decreased the number of migrated cells, compared with gemcitabine-alone treatment (Figure 3C,D). These results indicated that the migratory ability is inhibited by the treatment of AMG.

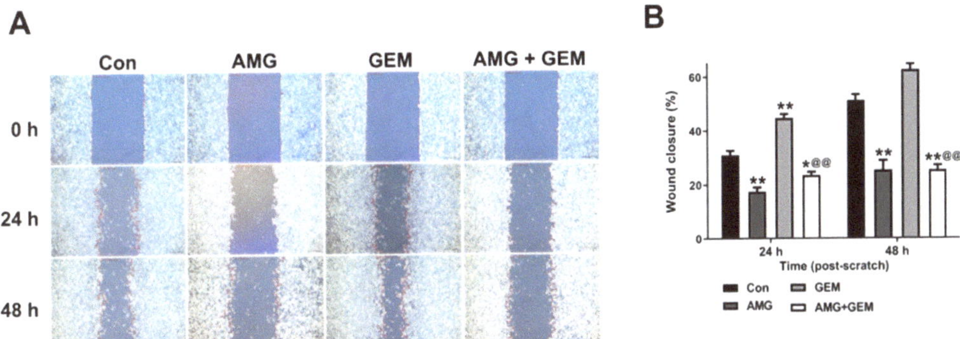

Figure 3. *Cont.*

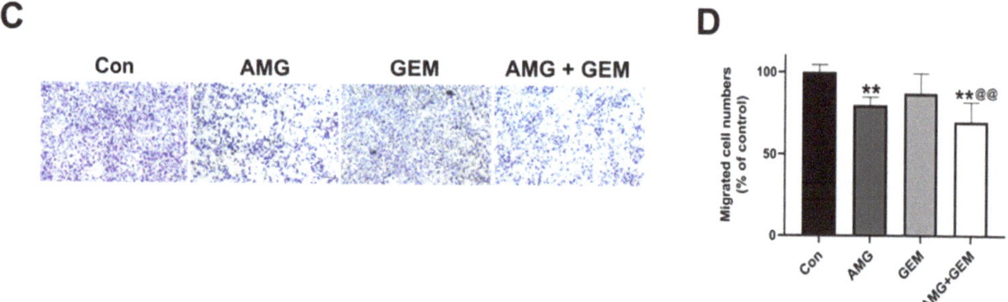

Figure 3. Effect of co-treatment of AMG and gemcitabine on cell migratory ability in SNU-478 cells. (**A**) The representative images of a wound healing assay in SNU-478 cells. Bar = 500 μm. (**B**) Quantification of wound-healing scratch area. (**C**) The representative images of a transwell migration assay in SNU-478 cells. Bar = 200 μm. (**D**) Quantification of transwell migrated cells. Data are presented as means ± S.D. from at least three independent experiments. * $p < 0.05$, ** $p < 0.01$ vs. Con; @@ $p < 0.01$ vs. GEM (Dunnet's test). Con, control; AMG, amurensin G; GEM, gemcitabine.

3.4. AMG Inhibits Cancer Stem-like Cell Properties in SNU-478 Cells

To confirm the effect of AMG treatment in inhibiting CSCs, cancer stem-like cell subpopulations were assessed using FACS. AMG (5 μM) or gemcitabine (500 nM) decreased the CD44^{high} population, compared with the control. Combination treatment of AMG (5 μM) with gemcitabine (500 nM) significantly decreased the CD44^{high} population, compared with gemcitabine-alone treatment (Figure 4A,B). AMG (5 μM) significantly decreased the CD44^{high}CD24^{high}CD133^{high} populations compared with the control, while gemcitabine (500 nM) single treatment significantly increased CD44^{high}CD24^{high}CD133^{high} populations, compared with the control. Combination treatment of AMG (5 μM) with gemcitabine (500 nM) significantly decreased CD44^{high}CD24^{high}CD133^{high} populations when compared with other groups (Figure 4C,D). In addition, the expression of Sox-2 was examined using Western blot (Figure 4E). The expression level of Sox-2 in the AMG and gemcitabine co-treatment group was lower than that of the control group (Figure 4F). These results indicated that co-treatment of AMG with gemcitabine downregulates the cancer stem-like cell population.

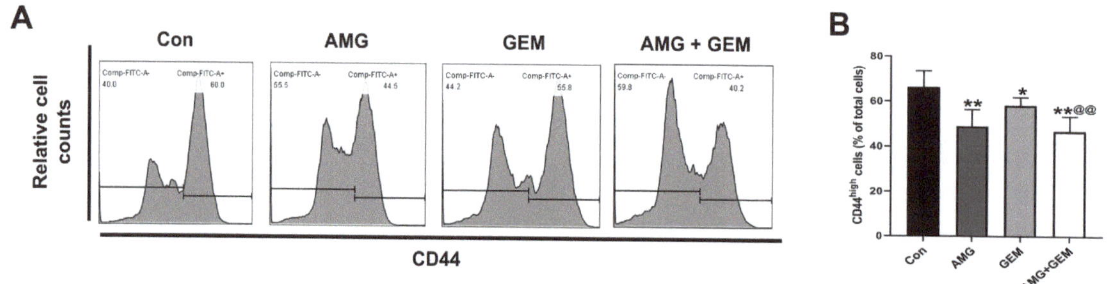

Figure 4. *Cont.*

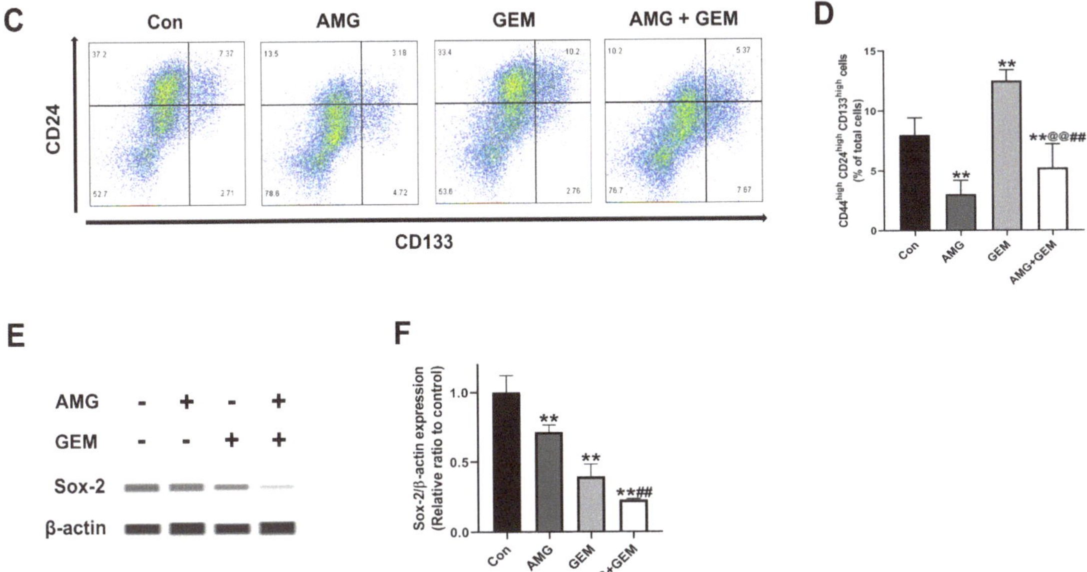

Figure 4. Effect of co-treatment of AMG and gemcitabine on cancer stem-like cell properties in SNU-478 cells. (**A**) Representative histogram of CD44. (**B**) The graph shows the percentage of CD44^{high} populations. (**C**) Representative plot of CD24 and CD133 among CD44^{high} cells. (**D**) The graph shows the percentage of CD44^{high}CD24^{high}CD133^{high} populations. (**E**) The protein expression of Sox-2 was detected using Western blotting. (**F**) The Sox-2 protein expression was normalized with β-actin. Data are presented as means ± S.D. from at least three independent experiments. * $p < 0.05$, ** $p < 0.01$ vs. Con; @@ $p < 0.01$ vs. GEM; ## $p < 0.01$ vs. AMG (Dunnet's test). Con, control; AMG, amurensin G; GEM, gemcitabine.

3.5. Combination Treatment of AMG with Gemcitabine Increases Mitochondrial ROS Production in SNU-478 Cells

The Sirt1/Nrf2/Keap1 signaling pathway plays a crucial role in cellular signal transduction related to oxidative stress. This pathway is activated to help cells respond to oxidative stress and maintain survival. Sirt1 inhibits the activation of Keap1, activating Nrf2, which increases anti-oxidant molecules, reducing the accumulation of ROS in the cell. To investigate the effect of AMG on ROS production in SNU-478 cells, MitoSOX^{TM} staining was performed. Combination treatment of AMG (5 μM) with gemcitabine (500 nM) significantly increased MitoSOX fluorescence intensity, compared with all other groups (Figure 5A,B). In addition, the expression of proteins related to the Sirt1/ Nrf2/ Keap1 signaling was examined, using Western blot (Figure 5C). Treatment of AMG significantly decreased the level of Sirt1, compared with the control (Figure 5D). Treatment of AMG significantly increased the Keap1/Nrf2 expression ratio, compared with the control (Figure 5E). These results indicated that co-treatment of AMG with gemcitabine increased mitochondrial ROS production in CCA, partly mediated by the downregulation of Sirt1/Nrf2 signaling by AMG.

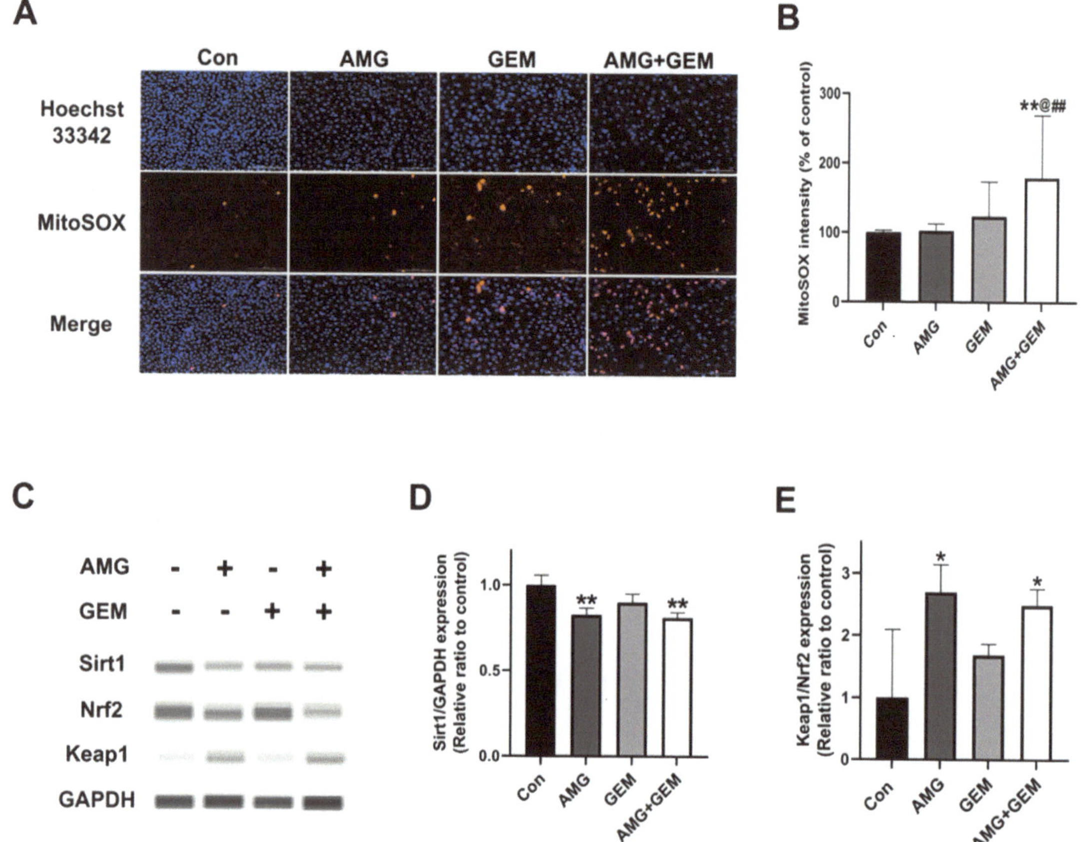

Figure 5. Effects of co-treatment of AMG and gemcitabine on mitochondrial ROS production. (**A**) The representative images show ROS formation using MitoSOX™ and Hoechst 33342 in SNU-478 cells. Bar = 200 μm. (**B**) Quantification of MitoSOX fluorescence intensity. (**C**) The protein expression of Sirt1, Nrf2, and Keap1 was detected using Western blotting. (**D**) The Sirt1 protein expression was normalized with GAPDH. (**E**) The Keap1/Nrf2 protein expression ratio was assessed. Data are presented as means ± S.D. from at least three independent experiments. * $p < 0.05$, ** $p < 0.01$ vs. Con; @ $p < 0.05$ vs. GEM; ## $p < 0.01$ vs. AMG (Dunnet's test). Con, control; AMG, amurensin G; GEM, gemcitabine.

4. Discussion

The CSCs are a small subset of tumor cells that have the capability of self-renewal, differentiation, and tumor recurrence. Due to these characteristics, CSCs are responsible for tumor initiation, metastasis, relapse, and chemoresistance in CCA [2,20]. Commonly, CSCs are known to comprise approximately <3% of total cells in most solid cancers, including breast cancer and colon cancer [21]. However, 20–30% of total tumor mass in CCA exhibits expression of CSC markers, which can increase the risk of tumor progression and recurrence associated with poor prognosis of CCA [1]. CSCs have unique cell surface markers in each cancer type. In CCA, CSCs exhibit specific cell markers such as CD44, CD24, CD133, and the epithelial cell adhesion molecule (EpCAM) [2]. Current data indicate that gemcitabine treatment induced the CSC population with an increased CD44[high] CD24[high]CD133[high] cell population. These results are partially consistent with previous studies, which show that low doses of gemcitabine increase CSCs, and increased CSCs cause higher resistance and tumor growth [22]. During the co-treatment of gemcitabine with AMG, AMG reduced CD44^{high}CD24^{high}CD133^{high} populations induced by gemcitabine, and significantly downregulated Sox-2 expression, in the current study. These results suggest that AMG downregulates the cancer stem-like properties mediated by gemcitabine.

In CCA, the migratory ability is one of the most distinctive characteristics associated with invasiveness, metastasis, and chemoresistance [2]. A previous study has shown that CD44 reduction is associated with proliferation, migration, and invasion in most solid tumors [23]. In addition, as a primary regulator of stemness, Sox-2 accelerates the acquisition of resistance characteristics, including tumor aggressiveness, chemoresistance, and EMT [24]. Our data showed that AMG inhibits migratory ability in CCA cells, which might be associated with the reduction in CD44 and Sox-2 expression by AMG. Moreover, CD44 is associated with ROS production, by regulating the redox status [25]. Low levels of ROS can promote tumor formation and progression by stimulating cell proliferation, angiogenesis, survival, and invasion [26]. However, high amounts of ROS can trigger cell death by inducing oxidative stress and apoptosis. In the present study, the combination treatment of AMG with gemcitabine increased mitochondrial ROS production, resulting in apoptotic death of SNU-478 CCA cells. Previous studies have reported that a high level of CD44 expression enhances the defense against ROS in gastrointestinal cancer, and the deletion of CD44+ CSC populations suppresses tumor growth [27]. Gemcitabine was reported to eliminate CSCs by increasing ROS formation [28], but the current results showed that gemcitabine treatment did not result in a significant increase compared with the control, a fact which may be related to the increase in CD133. CD133 is a pentaspan transmembrane protein associated with stem cell regeneration, differentiation, and metabolism [29]. There is evidence that gemcitabine increases the CD133 expression in pancreatic cancer, and the increasing CD133 expression inhibits the accumulation of ROS [30]. Therefore, the downregulation of the CSC subpopulation by AMG might contribute to increases in ROS formation in the present study.

AMG is known as a potent natural Sirt1 inhibitor [31]. Sirt1 is commonly found in the nucleus of almost all cells, and regulates various cellular functions, including differentiation, proliferation, and stemness [32–34]. In particular, activation of Sirt1 regulates the redox system via the upregulating of the Nrf2-mediated anti-oxidant system [35]. Activation of Sirt1 activates Nrf2; this activated Nrf2 enters the nucleus without binding to Keap1, and starts transcription, which increases anti-oxidants within the cell [35]. In the current study, AMG significantly inhibits Sirt1 and Nrf2 expression and increases Keap1, which is associated with a significant increase in mitochondrial ROS by the co-treatment of AMG with gemcitabine (Figure 6). Previous studies have shown that an excessive accumulation of ROS induces a decrease in the mitochondrial membrane potential, which leads to a reduction in the Bcl-2 expression, and an increase in Bcl-2-associated X protein (BAX) and c-cas3 expression, thus resulting in apoptosis in CCA [36]. In the present study, AMG increases apoptosis in CCA cells by decreasing Bcl-2 and through an increase in c-cas3 expressions. These results suggest that the inhibition of Sirt1 by AMG might contribute to increases in apoptotic death, through an increase in mitochondrial ROS production.

Phytochemicals act through multiple pathways, making it difficult to identify the specific pathway of interest. This study was conducted at a level that suggests the possibility that AMG can reduce CCA by enhancing the anti-cancer properties of gemcitabine. Therefore, further studies will be needed to determine a clear signaling pathway, and in vivo studies are necessary to confirm the therapeutic effects of AMG and to select the optimum dose, based on clinical correlation.

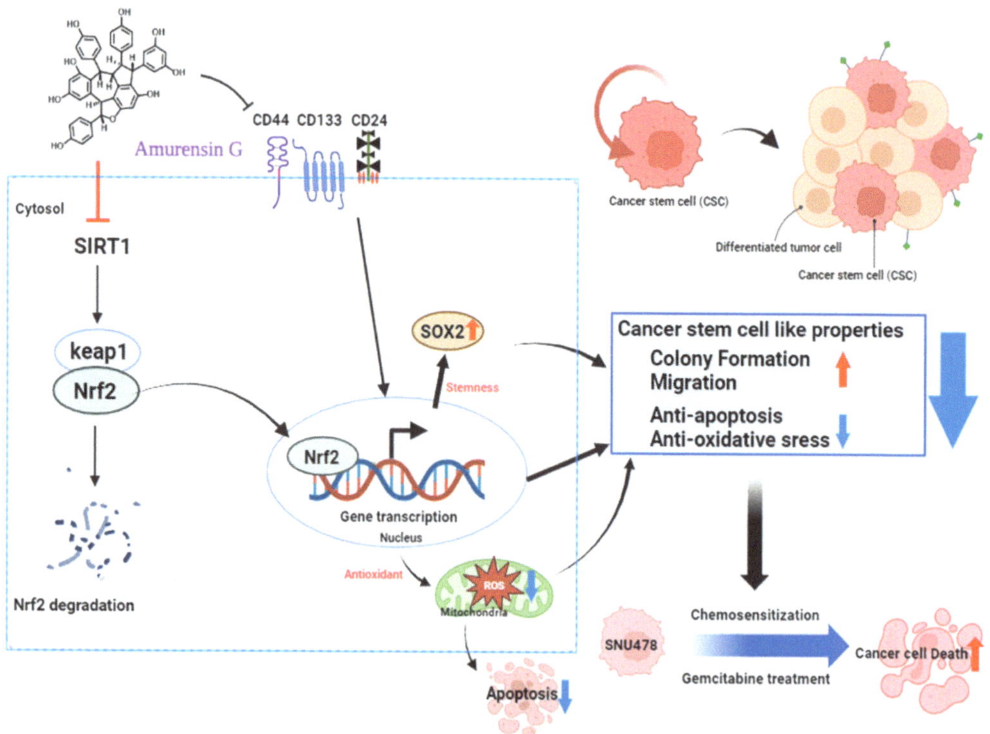

Figure 6. Schematic model of the mechanism of action upon treatment with AMG in SNU-478 cells (created with BioRender.com). Treatment of AMG reduces colony formation and cell migration, by inhibiting cancer stem-like properties. Inhibition of Sirt1 reduces the expression of Nrf2, leading to apoptosis through increased ROS formation. Additionally, AMG enhances cell death by sensitizing CCA to gemcitabine.

5. Conclusions

The current study demonstrated that AMG enhanced the chemotherapeutic ability of gemcitabine by targeting CSCs; this is associated with the upregulation of apoptosis and oxidative stress and the downregulation of migratory potential in CCA cells. Based on these results, it can be concluded that AMG may be a promising agent for CCA therapy.

Author Contributions: Conceptualization, H.K.L.; Methodology, Y.-J.N. and H.K.L.; Validation, Y.-J.N., H.K.L. and K.-C.C.; Formal analysis, Y.-J.N.; Investigation, Y.-J.N.; Data curation, H.K.L. and K.-C.C.; Writing—original draft, Y.-J.N.; Writing—review & editing, H.K.L. and K.-C.C.; Visualization, H.K.L.; Supervision, K.-C.C.; Project administration, H.K.L. and K.-C.C.; Funding acquisition, K.-C.C. All authors have read and agreed to the published version of the manuscript.

Funding: This work was supported by the Basic Research Lab Program (2022R1A4A1025557) through the National Research Foundation (NRF) of Korea, funded by the Ministry of Science and ICT. In addition, this study was also supported by "Regional Innovation Strategy (RIS)" through the National Research Foundation of Korea (NRF), funded by the Ministry of Education (MOE; 2021RIS-001) in 2023.

Institutional Review Board Statement: Not applicable.

Informed Consent Statement: Not applicable.

Data Availability Statement: Data are contained within the article.

Conflicts of Interest: The authors do not have any conflicts of interest to declare.

Abbreviations

AMG, Amurensin G; ANOVA, One-way analysis of variance; APC, Allophycocyanin; BAX, Bcl-2 associated X; Bcl-2, B-cell lymphoma 2; CCA, Cholangiocarcinoma; c-cas3, cleaved caspase 3; CD, Cluster of differentiation; CSC, Cancer stem cell; DPBS, Dulbecco's phosphate-buffered saline; EMT, Epithelial-mesenchymal transition; EpCAM, Epithelial cell adhesion molecule; FACS, Fluorescence-activated cell sorting; FBS, Fetal bovine serum; FITC, Fluorescein isothiocyanate; GAPDH, Glyceraldehyde-3-phosphate dehydrogenase; HEPES, 4-(2-hydroxyethyl)-1-piperazineethanesulfonic acid; IC_{50}, Half-maximal inhibitory concentration; KCLB, Korean cell line bank; Keap1, Kelch-like ECH-associated protein 1; MitoSox, Mitochondrial superoxide; Notch, Neurogenic locus notch homolog protein1; Nrf2, Nuclear factor-like 2; PE, Phycoerythrin; PI, Propidium iodide; ROS, Reactive oxygen species; RPMI, Roswell Park Memorial Institute; S.D., Standard deviation; Sirt1, Sirtuin1; SNU, Seoul National University; Sox-2, SRY-BOX transcription factor; Wnt, Wingless-related integration site; WST, Water-soluble tetrazolium salt.

References

1. Banales, J.M.; Marin, J.J.G.; Lamarca, A.; Rodrigues, P.M.; Khan, S.A.; Roberts, L.R.; Cardinale, V.; Carpino, G.; Andersen, J.B.; Braconi, C.; et al. Cholangiocarcinoma 2020: The next horizon in mechanisms and management. *Nat. Rev. Gastroenterol. Hepatol.* **2020**, *17*, 557–588. [CrossRef] [PubMed]
2. Wu, H.J.; Chu, P.Y. Role of cancer stem cells in cholangiocarcinoma and therapeutic implications. *Int. J. Mol. Sci.* **2019**, *20*, 4154. [CrossRef] [PubMed]
3. Valle, J.; Wasan, H.; Palmer, D.H.; Cunningham, D.; Anthoney, A.; Maraveyas, A.; Madhusudan, S.; Iveson, T.; Hughes, S.; Pereira, S.P.; et al. Cisplatin plus gemcitabine versus gemcitabine for biliary tract cancer. *N. Engl. J. Med.* **2010**, *362*, 1273–1281. [CrossRef] [PubMed]
4. Elvevi, A.; Laffusa, A.; Scaravaglio, M.; Rossi, R.E.; Longarini, R.; Stagno, A.M.; Cristoferi, L.; Ciaccio, A.; Cortinovis, D.L.; Invernizzi, P.; et al. Clinical treatment of cholangiocarcinoma: An updated comprehensive review. *Ann. Hepatol.* **2022**, *27*, 100737. [CrossRef] [PubMed]
5. Pandit, B.; Royzen, M. Recent development of prodrugs of gemcitabine. *Genes* **2022**, *13*, 466. [CrossRef] [PubMed]
6. Miao, H.; Chen, X.; Luan, Y. Small molecular gemcitabine prodrugs for cancer therapy. *Curr. Med. Chem.* **2020**, *27*, 5562–5582. [CrossRef] [PubMed]
7. Jia, Y.; Xie, J. Promising molecular mechanisms responsible for gemcitabine resistance in cancer. *Genes. Dis.* **2015**, *2*, 299–306. [CrossRef]
8. Lei, M.M.L.; Lee, T.K.W. Cancer stem cells: Emerging key players in immune evasion of cancers. *Front. Cell Dev. Biol.* **2021**, *9*, 692940. [CrossRef]
9. Yang, L.; Shi, P.; Zhao, G.; Xu, J.; Peng, W.; Zhang, J.; Zhang, G.; Wang, X.; Dong, Z.; Chen, F.; et al. Targeting cancer stem cell pathways for cancer therapy. *Signal Transduct. Target. Ther.* **2020**, *5*, 8. [CrossRef]
10. Ryu, H.W.; Oh, W.K.; Jang, I.S.; Park, J. Amurensin g induces autophagy and attenuates cellular toxicities in a rotenone model of parkinson's disease. *Biochem. Biophys. Res. Commun.* **2013**, *433*, 121–126. [CrossRef]
11. Butnariu, M.; Quispe, C.; Herrera-Bravo, J.; Pentea, M.; Sarac, I.; Kusumler, A.S.; Ozcelik, B.; Painuli, S.; Semwal, P.; Imran, M.; et al. Papaver plants: Current insights on phytochemical and nutritional composition along with biotechnological applications. *Oxid. Med. Cell Longev.* **2022**, *2022*, 2041769. [CrossRef] [PubMed]
12. Hong, K.S.; Park, J.I.; Kim, M.J.; Kim, H.B.; Lee, J.W.; Dao, T.T.; Oh, W.K.; Kang, C.D.; Kim, S.H. Involvement of sirt1 in hypoxic down-regulation of c-myc and beta-catenin and hypoxic preconditioning effect of polyphenols. *Toxicol. Appl. Pharmacol.* **2012**, *259*, 210–218. [CrossRef] [PubMed]
13. Jeong, H.Y.; Kim, J.Y.; Lee, H.K.; Ha do, T.; Song, K.S.; Bae, K.; Seong, Y.H. Leaf and stem of vitis amurensis and its active components protect against amyloid beta protein (25-35)-induced neurotoxicity. *Arch. Pharm. Res.* **2010**, *33*, 1655–1664. [CrossRef] [PubMed]
14. Kim, J.A.; Kim, M.R.; Kim, O.; Phuong, N.T.; Yun, J.; Oh, W.K.; Bae, K.; Kang, K.W. Amurensin g inhibits angiogenesis and tumor growth of tamoxifen-resistant breast cancer via pin1 inhibition. *Food Chem. Toxicol.* **2012**, *50*, 3625–3634. [CrossRef] [PubMed]
15. Lee, S.H.; Kim, M.J.; Kim, D.W.; Kang, C.D.; Kim, S.H. Amurensin g enhances the susceptibility to tumor necrosis factor-related apoptosis-inducing ligand-mediated cytotoxicity of cancer stem-like cells of hct-15 cells. *Cancer Sci.* **2013**, *104*, 1632–1639. [CrossRef] [PubMed]
16. Oh, W.K.; Cho, K.B.; Hien, T.T.; Kim, T.H.; Kim, H.S.; Dao, T.T.; Han, H.K.; Kwon, S.M.; Ahn, S.G.; Yoon, J.H.; et al. Amurensin g, a potent natural sirt1 inhibitor, rescues doxorubicin responsiveness via down-regulation of multidrug resistance 1. *Mol. Pharmacol.* **2010**, *78*, 855–864. [CrossRef] [PubMed]

17. Kim, C.W.; Lee, H.K.; Nam, M.W.; Lee, G.; Choi, K.C. The role of kiss1 gene on the growth and migration of prostate cancer and the underlying molecular mechanisms. *Life Sci.* **2022**, *310*, 121009. [CrossRef]

18. Nam, M.W.; Lee, H.K.; Kim, C.W.; Choi, Y.; Ahn, D.; Go, R.E.; Choi, K.C. Effects of ccn6 overexpression on the cell motility and activation of p38/bone morphogenetic protein signaling pathways in pancreatic cancer cells. *Biomed. Pharmacother.* **2023**, *163*, 114780. [CrossRef]

19. Kim, C.W.; Lee, H.K.; Nam, M.W.; Choi, Y.; Choi, K.C. Overexpression of kiss1 induces the proliferation of hepatocarcinoma and increases metastatic potential by increasing migratory ability and angiogenic capacity. *Mol. Cells* **2022**, *45*, 935–949. [CrossRef]

20. Kokuryo, T.; Yokoyama, Y.; Nagino, M. Recent advances in cancer stem cell research for cholangiocarcinoma. *J. Hepatobiliary Pancreat. Sci.* **2012**, *19*, 606–613. [CrossRef]

21. Banales, J.M.; Cardinale, V.; Carpino, G.; Marzioni, M.; Andersen, J.B.; Invernizzi, P.; Lind, G.E.; Folseraas, T.; Forbes, S.J.; Fouassier, L.; et al. Expert consensus document: Cholangiocarcinoma: Current knowledge and future perspectives consensus statement from the european network for the study of cholangiocarcinoma (ens-cca). *Nat. Rev. Gastroenterol. Hepatol.* **2016**, *13*, 261–280. [CrossRef] [PubMed]

22. Zhang, Z.; Duan, Q.; Zhao, H.; Liu, T.; Wu, H.; Shen, Q.; Wang, C.; Yin, T. Gemcitabine treatment promotes pancreatic cancer stemness through the nox/ros/nf-kappab/stat3 signaling cascade. *Cancer Lett.* **2016**, *382*, 53–63. [CrossRef] [PubMed]

23. Kong, L.; Ji, H.; Gan, X.; Cao, S.; Li, Z.; Jin, Y. Knockdown of cd44 inhibits proliferation, migration and invasion of osteosarcoma cells accompanied by downregulation of cathepsin s. *J. Orthop. Surg. Res.* **2022**, *17*, 154. [CrossRef] [PubMed]

24. Zhu, Y.; Huang, S.; Chen, S.; Chen, J.; Wang, Z.; Wang, Y.; Zheng, H. Sox2 promotes chemoresistance, cancer stem cells properties, and epithelial-mesenchymal transition by beta-catenin and beclin1/autophagy signaling in colorectal cancer. *Cell Death Dis.* **2021**, *12*, 449. [CrossRef] [PubMed]

25. Thanee, M.; Dokduang, H.; Kittirat, Y.; Phetcharaburanin, J.; Klanrit, P.; Titapun, A.; Namwat, N.; Khuntikeo, N.; Wangwiwatsin, A.; Saya, H.; et al. Cd44 modulates metabolic pathways and altered ros-mediated akt signal promoting cholangiocarcinoma progression. *PLoS ONE* **2021**, *16*, e0245871. [CrossRef] [PubMed]

26. Moloney, J.N.; Cotter, T.G. Ros signalling in the biology of cancer. *Semin. Cell Dev. Biol.* **2018**, *80*, 50–64. [CrossRef] [PubMed]

27. Ishimoto, T.; Nagano, O.; Yae, T.; Tamada, M.; Motohara, T.; Oshima, H.; Oshima, M.; Ikeda, T.; Asaba, R.; Yagi, H.; et al. Cd44 variant regulates redox status in cancer cells by stabilizing the xct subunit of system xc($-$) and thereby promotes tumor growth. *Cancer Cell* **2011**, *19*, 387–400. [CrossRef] [PubMed]

28. Zhao, H.; Wu, S.; Li, H.; Duan, Q.; Zhang, Z.; Shen, Q.; Wang, C.; Yin, T. Ros/kras/ampk signaling contributes to gemcitabine-induced stem-like cell properties in pancreatic cancer. *Mol. Ther. Oncolytics* **2019**, *14*, 299–312. [CrossRef]

29. Li, Z. Cd133: A stem cell biomarker and beyond. *Exp. Hematol. Oncol.* **2013**, *2*, 17. [CrossRef]

30. Nomura, A.; Dauer, P.; Gupta, V.; McGinn, O.; Arora, N.; Majumdar, K.; Uhlrich, C., 3rd; Dalluge, J.; Dudeja, V.; Saluja, A.; et al. Microenvironment mediated alterations to metabolic pathways confer increased chemo-resistance in cd133+ tumor initiating cells. *Oncotarget* **2016**, *7*, 56324–56337. [CrossRef]

31. Kim, H.B.; Kim, M.J.; Lee, S.H.; Lee, J.W.; Bae, J.H.; Kim, D.W.; Dao, T.T.; Oh, W.K.; Kang, C.D.; Kim, S.H. Amurensin g, a novel sirt1 inhibitor, sensitizes trail-resistant human leukemic k562 cells to trail-induced apoptosis. *Biochem. Pharmacol.* **2012**, *84*, 402–410. [CrossRef] [PubMed]

32. An, Y.; Wang, B.; Wang, X.; Dong, G.; Jia, J.; Yang, Q. Sirt1 inhibits chemoresistance and cancer stemness of gastric cancer by initiating an ampk/foxo3 positive feedback loop. *Cell Death Dis.* **2020**, *11*, 115. [CrossRef] [PubMed]

33. Wang, Z.; Chen, W. Emerging roles of sirt1 in cancer drug resistance. *Genes Cancer* **2013**, *4*, 82–90. [CrossRef] [PubMed]

34. Alves-Fernandes, D.K.; Jasiulionis, M.G. The role of sirt1 on DNA damage response and epigenetic alterations in cancer. *Int. J. Mol. Sci.* **2019**, *20*, 3153. [CrossRef]

35. Singh, C.K.; Chhabra, G.; Ndiaye, M.A.; Garcia-Peterson, L.M.; Mack, N.J.; Ahmad, N. The role of sirtuins in antioxidant and redox signaling. *Antioxid. Redox Signal.* **2018**, *28*, 643–661. [CrossRef]

36. Zhang, C.; Yang, H.Y.; Gao, L.; Bai, M.Z.; Fu, W.K.; Huang, C.F.; Mi, N.N.; Ma, H.D.; Lu, Y.W.; Jiang, N.Z.; et al. Lanatoside c decelerates proliferation and induces apoptosis through inhibition of stat3 and ros-mediated mitochondrial membrane potential transformation in cholangiocarcinoma. *Front. Pharmacol.* **2023**, *14*, 1098915. [CrossRef]

Article

Antitumor Effect and Gut Microbiota Modulation by Quercetin, Luteolin, and Xanthohumol in a Rat Model for Colorectal Cancer Prevention

Álvaro Pérez-Valero [1,2,3], Patricia Magadán-Corpas [1,2,3], Suhui Ye [1,2,3], Juan Serna-Diestro [1,2,3], Sandra Sordon [4], Ewa Huszcza [4], Jarosław Popłoński [4], Claudio J. Villar [1,2,3] and Felipe Lombó [1,2,3,*]

[1] Research Group BIONUC (Biotechnology of Nutraceuticals and Bioactive Compounds), Departamento de Biología Funcional, Área de Microbiología, Universidad de Oviedo, 33006 Oviedo, Spain; sernadjuan@uniovi.es (J.S.-D.); cjvg@uniovi.es (C.J.V.)
[2] Instituto Universitario de Oncología del Principado de Asturias (IUOPA), 33006 Oviedo, Spain
[3] Instituto de Investigación Sanitaria del Principado de Asturias (ISPA), 33006 Oviedo, Spain
[4] Department of Food Chemistry and Biocatalysis, Wrocław University of Environmental and Life Sciences, Norwida 25, 50-375 Wrocław, Poland; sandra.sordon@upwr.edu.pl (S.S.); ewa.huszcza@upwr.edu.pl (E.H.); jaroslaw.poplonski@upwr.edu.pl (J.P.)
* Correspondence: lombofelipe@uniovi.es

Abstract: Colorectal cancer stands as the third most prevalent form of cancer worldwide, with a notable increase in incidence in Western countries, mainly attributable to unhealthy dietary habits and other factors, such as smoking or reduced physical activity. Greater consumption of vegetables and fruits has been associated with a lower incidence of colorectal cancer, which is attributed to their high content of fiber and bioactive compounds, such as flavonoids. In this study, we have tested the flavonoids quercetin, luteolin, and xanthohumol as potential antitumor agents in an animal model of colorectal cancer induced by azoxymethane and dodecyl sodium sulphate. Forty rats were divided into four cohorts: Cohort 1 (control cohort), Cohort 2 (quercetin cohort), Cohort 3 (luteolin cohort), and Cohort 4 (xanthohumol cohort). These flavonoids were administered intraperitoneally to evaluate their antitumor potential as pharmaceutical agents. At the end of the experiment, after euthanasia, different physical parameters and the intestinal microbiota populations were analyzed. Luteolin was effective in significantly reducing the number of tumors compared to the control cohort. Furthermore, the main significant differences at the microbiota level were observed between the control cohort and the cohort treated with luteolin, which experienced a significant reduction in the abundance of genera associated with disease or inflammatory conditions, such as *Clostridia UCG-014* or *Turicibacter*. On the other hand, genera associated with a healthy state, such as *Muribaculum*, showed a significant increase in the luteolin cohort. These results underline the anti-colorectal cancer potential of luteolin, manifested through a modulation of the intestinal microbiota and a reduction in the number of tumors.

Keywords: flavonoid; intraperitoneal; colorectal cancer; rat model; gut microbiota

Citation: Pérez-Valero, Á.; Magadán-Corpas, P.; Ye, S.; Serna-Diestro, J.; Sordon, S.; Huszcza, E.; Popłoński, J.; Villar, C.J.; Lombó, F. Antitumor Effect and Gut Microbiota Modulation by Quercetin, Luteolin, and Xanthohumol in a Rat Model for Colorectal Cancer Prevention. *Nutrients* **2024**, *16*, 1161. https://doi.org/10.3390/nu16081161

Academic Editors: Yi-Wen Liu and Ching-Hsein Chen

Received: 13 March 2024
Revised: 9 April 2024
Accepted: 11 April 2024
Published: 13 April 2024

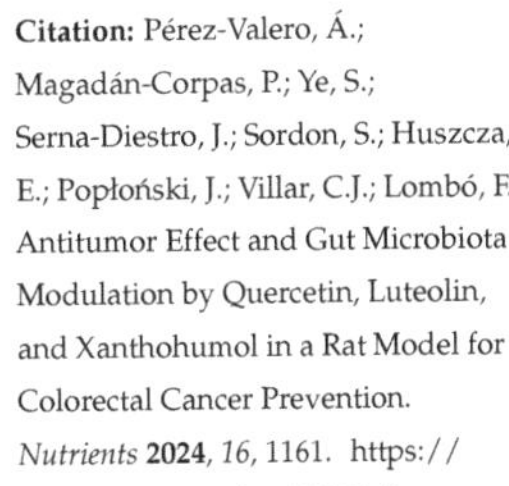

1. Introduction

Cancer, a global health challenge, encompasses a group of disorders arising from abnormal and uncontrolled cell proliferation with invasive characteristics [1,2]. Colorectal cancer (CRC) stands as the third most prevalent malignancy worldwide, with over 1.93 million newly reported cases and 935,173 deaths in 2020, securing its place as the third leading cause of mortality in both, males and females, globally [3–5]. Notably, CRC has emerged as a predominant cancer in Western countries, contributing to 10% of worldwide cancer incidence and 9.4% of cancer-related deaths [3].

This higher prevalence of CRC in Western countries is attributed to the aging population and the prevalence of unhealthy dietary practices, including high intakes of saturated fats and nitrosamines, along with a low consumption of fruits and vegetables. Additional contributing factors include smoking, low physical activity, and obesity, which together affect colonic mucosal health [6,7]. Another factor affecting the prevalence of this disease is the time of its onset. CRC typically appears late in life, as its progression requires several genetic mutations, resulting in a higher incidence in the adult population over 55–60 years old [8].

The initiation of CRC involves genetic (such as loss of *APC*, *TP53*, or *KRAS* genes or in the mismatch repair genes *MLH1*, *MSH2*, *MSH6*, or *PMS2*, due to direct mutations, or the *BRAF* gene due to microsatellite instability-associated point mutations), and epigenetic alterations (such as silencing of *MLH1* gene by promoter hypermethylation, CpG islands hypermethylation silencing at promoter regions in several genes, or miRNA methylation) in the stem cells located at the base of the colon crypts, which globally generate different subtypes of CRC cases (based on chromosomal instability, microsatellite instability, or CpG-island methylator phenotype) as those following the adenoma/carcinoma pathway or the serrated pathway [9,10]. These modifications affect tumor suppressor genes and oncogenes, leading to the transformation of normal stem cells into neoplastic stem cells [11]. The resulting genetic and epigenetic changes contribute to the loss of genomic and/or epigenomic stability, leading to tumor lesions in the colon, including aberrant crypt foci (ACF), adenomas, and serrated polyps. This process is a pivotal event in the initiation and development of CRC, both at the pathophysiological and molecular levels [12,13].

In CRC, most tumors originate from a polyp formed from an ACF. This polyp progresses to an early adenoma of less than 1 cm, which then evolves to an advanced adenoma exceeding 1 cm in size. Ultimately, the advanced adenoma becomes transformed into a malignant tumor (adenocarcinoma), which can acquire metastatic capabilities [14].

The neoplastic progression caused by ACF within the colonic mucosa can be modulated to attenuate or prevent its progression via the presence of a variety of nutraceutical compounds in the colon lumen, particularly polyphenols and short-chain fatty acids from prebiotic fiber fermentation [8]. Polyphenols are the largest group of plant bioactive chemicals known in nature. They have been associated with numerous health benefits, such as cancer prevention [15]. Within polyphenols, flavonoids are a large family of nutraceuticals widely distributed in plants, including edible plants [15–19]. Although flavonoids are primarily known for their antioxidant attributes, findings from in vitro and in vivo studies underline their ability as anti-inflammatory and immunomodulatory bioactives [20], as well as robust anticancer compounds [21–24]. The molecular mechanisms underlying the anticancer effects of flavonoids are not yet fully elucidated [24]. Nevertheless, their recognized functions include the modulation of reactive oxygen species (ROS)-scavenging enzymatic activities. Furthermore, flavonoids are involved in the regulation of critical cellular processes, such as cell cycle arrest, apoptosis induction, autophagy, as well as suppression of cancer cell proliferation and invasiveness [20–26].

Within the extensive family of flavonoids, quercetin, luteolin, and xanthohumol possess anti-CRC activity [1,27,28]. Quercetin, which has been shown to be safe for addition to foods [29], shows an inhibition of cell viability in CT26 and MC38 colon cancer cells. It induces apoptosis through the mitogen-activated protein kinases (MAPKs) pathway and modulates the expression of epithelial-mesenchymal transition (EMT) markers, including E-cadherin, N-cadherin, β-catenin, and SNAI1 [27].

Luteolin, a safe chemopreventive agent [30] characterized by its potent antioxidant and anti-inflammatory effects, is remarkably effective in CRC and its associated complications. It attenuates the expression of nitric oxide synthase (iNOS) and cyclooxygenase-2 (COX-2). Additionally, luteolin suppresses the expression of matrix metalloproteinase-2 (MMP-2) and matrix metalloproteinase-9 (MMP-9) to address CRC-related issues [1]. The antitumor activity of luteolin shows synergy in combination with current antitumor drugs, such as 5-fluorouracil (5-FU), against human CRC cell lines in vitro, which makes this flavonoid highly interesting for future in vivo co-therapy studies [31].

On the other hand, xanthohumol, whose administration in mice has been proved as safe [32], is able to reduce proliferation in HT-29 CRC cells [28,33]. Moreover, xanthohumol decreases the expression of several drug efflux genes. This property makes it an interesting candidate for combination therapy with other anticancer chemotherapeutic agents, presenting a potential strategy to mitigate drug resistance by inhibiting drug efflux transporters [28,34]. Luteolin and xanthohumol have also demonstrated good in vitro antitumor activity against several human colon cancer cell lines [35].

Most studies available in the literature have tested flavonoids against CRC cell lines or in murine CRC models through oral administration as nutraceuticals. However, oral administration affects the bioavailability of flavonoids mainly due to their limited solubility and changes carried out by gut microbiota in their chemical skeletons, such as hydrolysis at ring C. In contrasts, intraperitoneal administration allows the final molecule to be actually absorbed to the portal circulation directly from the peritoneal cavity to the plasma via a network of capillaries in the peritoneum that arrive at the liver, where they undergo phase II metabolism (sulfation and methylation) and finally reach the digestive tract via bile component [36–38]. In contrast, the use of flavonoids as therapeutic drugs, intraperitoneally (or via intravenous administration), instead of nutraceuticals to combat CRC, has rarely been studied [39].

The present investigation was designed to examine the possible antitumor effects of quercetin, luteolin, and xanthohumol administered intraperitoneally in an animal model of CRC, specifically *Rattus norvegicus* F344. In this study, CRC was chemically induced by a combination regime including azoxymethane (AOM) and dextran sodium sulphate (DSS). The animals were systematically analyzed for several parameters, including body weight, caecum weight, hyperplastic Peyer's patches count, colon length, and tumor number. Additionally, the composition of the gut microbiota was examined within the four cohorts (control, quercetin, luteolin, and xanthohumol), revealing notable distinctions among them.

2. Materials and Methods

2.1. Animals and Experimental Design

A total of 40 male Fischer 344 rats were maintained in the Animal Facilities at the University of Oviedo (authorized facility No. ES330440003591). All rat experiments were approved by the ethics committee of the Principality of Asturias (authorization code PROAE 14/2022).

The rats (five weeks old) were randomly divided into four cohorts of ten individuals each and fed ad libitum (2014 Teklad Global 14% Protein Rodent Maintenance Harlan diet feed, Harlan Laboratories, Barcelona, Spain). This feed contained 14.3% protein, 4% fat, 48% carbohydrates, 22.1% fiber, and 4.7% ashes. The rats were placed in a room with constant temperature (21 °C) and humidity on a 12:12 h dark/light cycle throughout the experiment.

Cohort 1 was injected intraperitoneally with 200 µL of phosphate buffered saline (PBS) (VWR International, Radnor, PA, USA) as control. Cohort 2 was injected with 25 mg per kg body weight (mg/kg/bw) of quercetin (Apollo Scientific, Bredbury, UK). Cohort 3 was injected with 25 mg/kg/bw of luteolin (Fluorochem, Hadfield, UK). Cohort 4 was injected with 25 mg/kg/bw of xanthohumol. All the flavonoids were dissolved in PBS and injected into the animals 3 days a week during the 18 weeks of the experiment.

Xanthohumol was purified following a modified procedure described previously [40]. The same batch of spent hops, stored in high-density polyethylene (HDPE) industrial barrels and closed under a nitrogen atmosphere, was used. The purification modification involved only the initial extract preparation step, as it was fully completed at the Department of Food Chemistry and Biocatalysis, Wrocław University of Environmental and Life Sciences laboratories. Eighteen kilograms of spent hops were extracted with ninety L of acetone in 0.2 kg:1.4 L batches, each made in a 2 L Erlenmeyer flask shaken for 3 h on a rotary shaker (120 rpm). The formed pulp was vacuum-filtered on Whatman filter paper no. 4 and concentrated using a laboratory rotary evaporator. The combined extracts were further subjected to polyphenol precipitation and Sephadex LH 20 column chromatography steps, resulting in 20.233 g of Xanthohumol (>98% purity with HPLC).

2.2. Colorectal Cancer Induction and Monitoring

One week after the animals arrived at the animal facility, the injections started. After one week receiving the flavonoids, CRC was induced in eight rats from each cohort. The two other rats were kept free of CRC induction as absolute control animals, used as sentinels for detecting any potential side effect of the treatments. CRC induction was carried out in those eight rats of each cohort using AOM (Sigma-Aldrich, Madrid, Spain) dissolved in sterile saline (0.9% NaCl) at a concentration of 2 mg/mL. This AOM solution was injected intraperitoneally at a final concentration of 10 mg/kg/bw. The AOM treatment was repeated seven days after the first injection (weeks 2 and 3). The two absolute control animals received sterile saline solution in both injections.

In weeks 4 and 15, the eight rats of each cohort (those treated with AOM) received drinking water ad libitum for 7 days containing 3% and 2% DSS (40.000 g/Mol, VWR), respectively. This ulcerative colitis step was repeated twice because it enhances the pro-carcinogenic effect caused by AOM administration.

The rats were weighed once a week during the 18 experimental weeks. At the end of the experiment, the rats were sacrificed using pneumothorax.

2.3. Tissue Samples

The rats were anesthetized (isoflurane) and submitted to euthanasia (pneumothorax) at week 18. The small intestine was removed fresh and the hyperplastic Peyer's patches were counted.

The caecums were weighed immediately after sacrifice using a precision scale and then frozen at $-20\,^{\circ}$C.

Finally, the colon was opened longitudinally and washed with PBS before keeping it in 4% formaldehyde at $4\,^{\circ}$C. Fixed colons were meticulously examined in order to count the number of tumors.

2.4. Genomic DNA Extraction and 16S Ribosomal RNA Sequencing for Gut Microbiota Analysis

A metagenomics analysis of caecal stool specimens was conducted employing the Pathogen Detection Protocol facilitated by the E. Z. N. A.® Stool DNA Kit (Ref. D4015-02, VWR, Madrid, Spain). The caecal specimens, after being thawed in ice for a duration of 30 min, underwent the extraction of 200 mg of feces from the midsection of the caecum, which was then placed in a 25 mL tube for subsequent protocol adherence. Subsequently, 200 μL of genomic DNA was isolated and quantified using a BioPhotometer® (Eppendorf, Madrid, Spain).

The processed DNA samples were frozen at $-20\,^{\circ}$C for subsequent analysis via the amplification and sequencing of the variable regions V3 and V4 of the 16S ribosomal RNA gene using Illumina MiSeq (Microomics Systems, Barcelona, Spain). The Illumina Miseq sequencing 300×2 methodology was employed, with amplification conducted after 25 PCR cycles. Quality control measures included the incorporation of a negative control for DNA extraction and a positive mock community control. This approach facilitated the characterization and quantification of microbial alpha and beta diversity, along with the examination of taxonomic profiles spanning from phylum to species levels.

2.5. Bioinformatics Analysis

The phylotype data served as the basis for calculating alpha diversity metrics, facilitating the analysis of the microbial community diversity. The assessment encompassed community richness, denoted by the count of observed operating taxonomic units (OTUs), representing distinct phylotypes within a community. Additionally, community evenness was evaluated using Pielou's evenness index, quantifying the numerical equality of phylotypes within a community, considering both their abundance and number. The determination of alpha diversity metrics also included indices such as Chao1 (indicative of species richness), Simpson (reflecting biodiversity levels), and Shannon (representing species diversity).

Beta diversity metrics, comparing microbial community structure among cohorts, were computed based on both phylotype and phylogenetic data. A principal coordinate analysis (PCoA) employing unweighted Unifrac distance, a phylogenetic qualitative measure, was executed to discern differences in beta diversity among the microbial communities.

2.6. Statistical Methods

In the context of the metagenomics analysis, comparisons of alpha diversity were executed utilizing a linear model with an appropriate distribution, specifically the negative binomial model for Chao1, beta regression for Simpson, and a linear model for Shannon diversity. Beta diversity distance matrices were employed to compute PCoA and construct ordination plots using the R software version 4.2.0. The assessment of community structure significance among groups was conducted through the permutational multivariate analysis of variance (Permanova) test.

The differential relative abundance of taxa was scrutinized using a linear model based on the negative binomial distribution or ANOVA. The statistical analyses involved the utilization of BiodiversityR version 2.14-1, PMCMRplus version 1.9.4, RVAideMemoire version 0.9-8, and vegan version 2.5-6 packages.

For additional comparisons, group data were expressed as mean $\pm$ standard error of the mean (SEM). The Shapiro–Wilk test was used for calculating the Gaussian distribution of the different variables. One-way ANOVA (Holm–Šídák multiple comparison test) was used for comparisons between flavonoid-treated cohorts and the control cohort following a Gaussian distribution. In the case of no Gaussian distribution, one-way ANOVA (Kruskal–Wallis test) was used for determining the statistical differences among cohorts. The graphical representation of the data was executed using GraphPad Prism software (version 9.0.2, GraphPad Software, San Diego, CA, USA).

In the case of the number of tumors, the ROUT method ($Q = 5\%$) was used to identify outliers. Rat number 8 in the Cohort 3 was identified as an outlier (31 tumors) and removed for the statistical analysis.

In each case, a p-value < 0.05 was considered statistically significant (* $p < 0.05$; ** $p < 0.005$; *** $p < 0.0005$; **** $p < 0.0001$).

3. Results

3.1. Effect of Quercetin, Luteolin, and Xanthohumol Administration on Body Weight

Animals in all four cohorts showed constant weight gain over the 18 experimental weeks. CRC was induced in rats 1–8 of each cohort. The two AOM challenges for CRC induction took place at weeks 2 and 3, and both DSS events were carried out at weeks 4 and 15. Figure 1 shows a slightly slowdown in weight gain after the second DSS event, with a notably recovery in weight gain from week 16 until the end of the experiment. However, no obvious slowdown in weight gain was observed after the first DSS event during week 4. Furthermore, rats numbered 3 and 7 in Cohort 2 (quercetin) died after the first DSS event, in week 5, as a consequence of the episode of transitional ulcerative colitis, a pro-inflammatory step necessary to increase the final number of tumors generated by AOM.

On the other hand, in the absolute control animals (rats 9 and 10 in each cohort), the increase in body weight was constant throughout the experiment since these animals did not receive AOM or DSS. In summary, none of the treatments caused a strong increase or reduction in body weight.

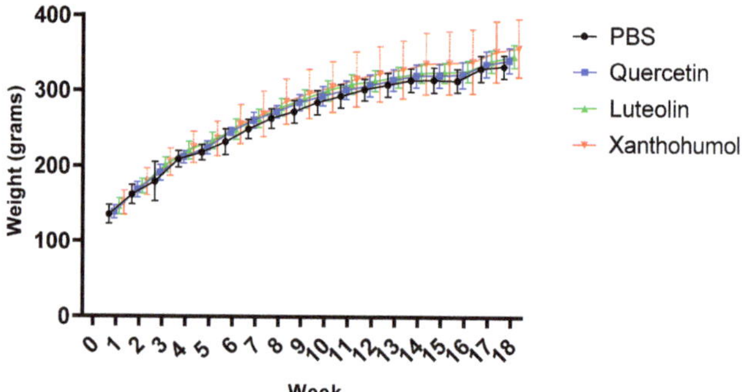

Figure 1. Evolution of body weight throughout the entire experiment (18 weeks, X axis) of the eight rats with CRC induction in the four cohorts. Body weight was measured every week. When the animals were sacrificed, the mean value for Cohort 1 was 332.9 ± 15.3 g, for Cohort 2 it was 341.9 ± 16.6 g, for Cohort 3 it was 347.8 ± 15.2 g, and for Cohort 4 it was 358.5 ± 38.8 g.

3.2. Effect of Quercetin, Luteolin, and Xanthohumol Administration on Hyperplastic Peyer's Patches

The number of Peyer's patches in the small intestine was quantified when the animals were sacrificed. Peyer's patches contain large numbers of lymphocytes and can become hyperplastic. These lymphoid nodules are easily observable as elongated thickenings in the intestinal mucosa, measuring a few millimeters in length.

In this study, the differences in the mean values of Peyer's patches were statistically significant between Cohort 3 (luteolin) and Cohort 1 (PBS), and between Cohort 4 (xanthohumol) and Cohort 1, where a drastic decrease of 40.2% and 54.2% in the number of Peyer's patches, respectively, was observed (Figure 2), a sign of anti-inflammatory protective bioactivities due to luteolin and xanthohumol administration.

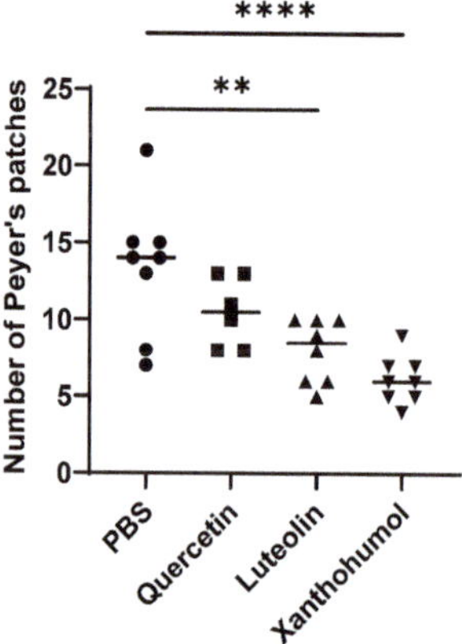

Figure 2. Mean number of hyperplastic Peyer's patches in the small intestine of animals with induced CRC from each cohort (rats 1 to 8 in the 4 groups). The median value of each cohort is represented as a horizontal line within the corresponding box plots. Asterisks indicate statistically significant differences (** $p < 0.005$; **** $p < 0.0001$) (ANOVA test).

3.3. Effect of Quercetin, Luteolin, and Xanthohumol Administration on Caecum Weight

No statistically significant differences were observed in the mean values of caecum weight when comparing the animals induced for CRC within the control cohort, with those subjected to treatments with quercetin, luteolin, or xanthohumol (Figure S1). Likewise, comparable non-significant distinctions were observed between the absolute control animals of the control cohort and the flavonoid-treated cohorts.

3.4. Effect of Quercetin, Luteolin, and Xanthohumol Administration on Colon Length

No statistically significant differences were found in mean colonic length measurements between the control cohort and the cohorts treated with quercetin, luteolin, or xanthohumol (Figure S2). Similarly, no significant differences were observed between the absolute control animals of the control cohort and those of the flavonoid-treated cohorts.

3.5. Effect of Quercetin, Luteolin, and Xanthohumol Administration on the Number of Tumors

The colon mucosa in each of the rats was analyzed to determine the number of tumors. A statistically significant difference was only observed between rats in Cohort 3 (luteolin) and Cohort 1 (control). Cohort 3 showed a drastic 63.9% reduction in the number of tumors (Figure 3), indicating a potent antitumor potential for luteolin. The number of tumors was also reduced in Cohort 2 (quercetin) and Cohort 4 (xanthohumol), but these reductions were not statistically significant (Figure 3). No tumors were found in the colonic mucosa of the absolute control animals (rats 9 and 10) of each cohort, as expected, as those animals were not submitted to AOM tumor induction, and the flavonoid treatments were not carcinogenic.

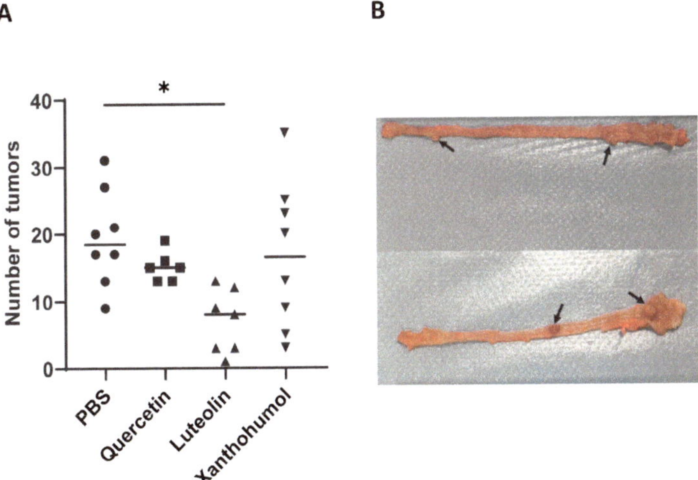

Figure 3. (**A**) Average number of colon tumors in each cohort (rats 1 to 8 in the 4 cohorts). The average number of tumors showed a statistically significant decrease in the case of Cohort 3 (luteolin) compared to the control Cohort 1. The median value of each cohort is represented as a horizontal line within the corresponding box plots. (**B**) Image showing two representative colons of rats in which CRC was induced. Some tumors are highlighted with black arrows. Asterisks indicate statistically significant differences (* $p < 0.05$) (ANOVA test).

3.6. Effect of Quercetin, Luteolin, and Xanthohumol Administration on the Gut Microbiota
3.6.1. Alpha and Beta Diversity

Alpha diversity metrics, which encompass diversity within each animal sample, were assessed by measuring richness (observed operational taxonomic units or OTUs) and evenness within microbial communities. Additionally, alpha diversity was comprehensively assessed using statistical indexes such as Chao1, Simpson, and Shannon. Figure S3 illustrates box plot representations of these diversity measures. When comparing CRC-induced animals from the PBS cohort with the corresponding animals from each of the flavonoid treatment cohorts, no statistically significant differences were discerned for any of the alpha diversity measures. This absence of significant differences indicates a lack of alterations in microbial alpha diversity in CRC-induced animals under the influence of the flavonoid treatments.

The unweighted Unifrac beta diversity index, representing diversity among samples, was calculated to assess differences between groups with respect to species complexity. The PCoA plot, represented in Figure 4, visually shows the structural variations within the microbial communities of CRC-induced rats. Differences in beta diversity were evident in the comparison between the PBS control group and the quercetin-, luteolin-, or xanthohumol-treated cohorts, as illustrated in Table 1. Moreover, discernible differences in beta diversity were observed between the quercetin and both the luteolin and xanthohumol cohorts, as well as between the two cohorts undergoing treatments with luteolin or xanthohumol (Table 1), indicating that these two treatments caused more similar alterations in colon microbiota populations, far from PBS and quercetin conditions.

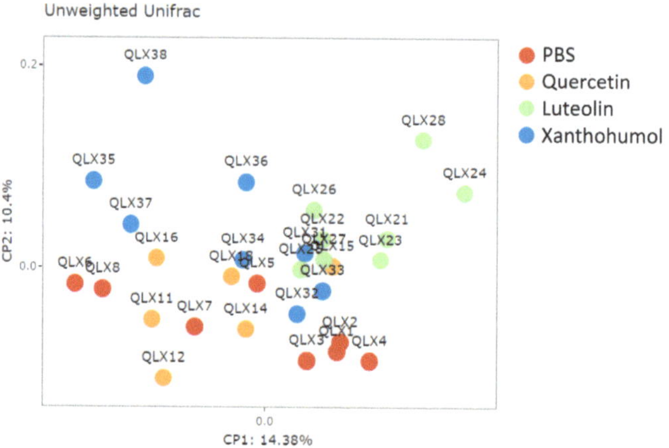

Figure 4. PCoA plot showing structural variations within microbial communities measured using the unweighted Unifrac beta diversity index. Each dot labeled with "QLX" and a numerical identifier indicates an individual participant in the experiment. Specifically, red dots represent CRC-induced animals in the PBS-treated cohort, orange dots belong to CRC-induced quercetin-treated animals, green dots correspond to CRC-induced luteolin-treated animals, and blue dots represent CRC-induced individuals of the xanthohumol-treated cohort.

Table 1. Significance (*p*-value) between cohorts in terms of beta diversity using the Permanova test. Cohort 1: PBS administration; Cohort 2: quercetin administration; Cohort 3: luteolin administration; Cohort 4: xanthohumol administration.

	Cohort 1	**Cohort 2**	**Cohort 3**
Cohort 2	0.037		
Cohort 3	0.002	0.002	
Cohort 4	0.0084	0.0075	0.002

3.6.2. Taxonomic Profile

Metagenomic analysis of the cecal microbiota in CRC-induced animals revealed significant differences between various taxonomic levels and between different experimental cohorts. In general, *Bacillota* (formerly *Firmicutes*) and *Bacteroidota* (formerly *Bacteroidetes*) emerge as the predominant phyla, being consistently observed in all cohorts and comparisons. The relative abundance of the different phyla in the CRC-induced animals of the four cohorts shows variations that depend on the specific treatment and the comparisons made (Table 2).

Table 2. Average percentage composition of the gut microbiota at the phylum level in CRC-induced animals for the four cohorts studied. Underlined phyla indicate significant differences (ANOVA test) between at least one of the cohorts treated with flavonoids (Cohorts 2, 3, and 4) and the control cohort (Cohort 1). Percentages marked by asterisks indicate that the cohort shows a statistically significant difference compared to the control within a specific phylum, along with the corresponding level of statistical significance. Asterisks indicate statistically significant differences (* $p < 0.05$; ** $p < 0.005$; **** $p < 0.0001$).

	Cohort 1 (%)	Cohort 2 (%)	Cohort 3 (%)	Cohort 4 (%)
Actinomycetota	0.33	0.30	0.10 ****	0.26
Bacteroidota	32.01	32.40	36.08 **	35.80 **
Cyanobacteria	0.27	0.34	0.69	1.06 *
Deferribacterota	0.31	0.34	0.43	0.19
Thermodesulfobacteriota	0.53	0.43	0.95	0.60
Bacillota	64.78	63.33	59.87 *	59.41 *
Patescibacteria	0.35	2.06 **	1.07	1.28 *
Pseudomonadota	0.80	0.15 **	0.34	0.87
Verrucomicrobiota	0.62	0.67	0.48	0.52

The quercetin cohort showed a statistically significant increase compared to the PBS cohort in the abundance of the phylum *Patescibacteria* (0.35% vs. 2.06% in the PBS cohort, p-value 0.00581) and a significant decrease in the phylum *Pseudomonadota* (0.80% vs. 0.15% in the PBS cohort, p-value 0.00913). In the case of the luteolin cohort, compared to the PBS cohort, notable variations were observed between different phyla. Specifically, the phyla *Actinomycetota* and *Bacillota* experienced reductions after luteolin treatment (0.33% vs. 0.10% in the luteolin cohort and 64.78% vs. 59.87% in the luteolin cohort, respectively) with associated p-values of 0.0001 and 0.02865, respectively. In contrast, the abundance of the phylum *Bacteroidota* exceeded that of the PBS cohort (31.01% vs. 36.08% in the luteolin cohort (p-value 0.00179). Finally, the xanthohumol cohort demonstrated an increase in the average percentage of the phyla *Bacteroidota* (32.01% in the PBS cohort vs. 35.80%), *Cyanobacteria* (0.27% in the PBS cohort vs. 1.06%), and *Patescibacteria* (0.35% in the PBS cohort vs. 1.28%), with associated p-values of 0.00415, 0.00167, and 0.04696, respectively. On the contrary, the percentage of the phylum *Bacillota* decreased from 64.78% in the PBS cohort to 59.41% in the xanthohumol cohort (p-value 0.01227). These variations are graphically represented in Figure 5.

At the taxonomic family level, the variations in abundance between families of CRC-induced animals in the four cohorts depend on the specific treatment administered and the comparisons made. The predominant families in all cohorts include *Lachnospiraceae*, *Muribaculaceae*, *Oscillospiraceae*, and *Ruminococcaceae*, as shown in Figure 6. Substantial differences were seen between the cohorts subjected to flavonoid treatments and the control cohort. Families exhibiting statistically significant variations are described in Table 3.

Significant differences were identified between PBS and quercetin cohorts in the families *Saccharimonadaceae*, which presented an increase from 0.35% to 2.06%, and *Enterobacteriaceae*, which decreased from 0.69% to 0.07%. The most pronounced disparities were observed between the PBS cohort and the luteolin cohort, where a significant decrease was observed in the families *Eggerthellaceae* (0.24% vs. 0.05%), *Erysipelotrichaceae* (0.82% vs. 0.13%), *Christensenellaceae* (0.77 vs. 0.09%), and *Anaerovoracaceae* (0.26% vs. 0.10%). On the contrary, an increase was observed in the families *Clostridia* vadinBB60 group (0.19% vs. 0.88%) and *Butyricicoccaceae* (0.23% vs. 0.61%).

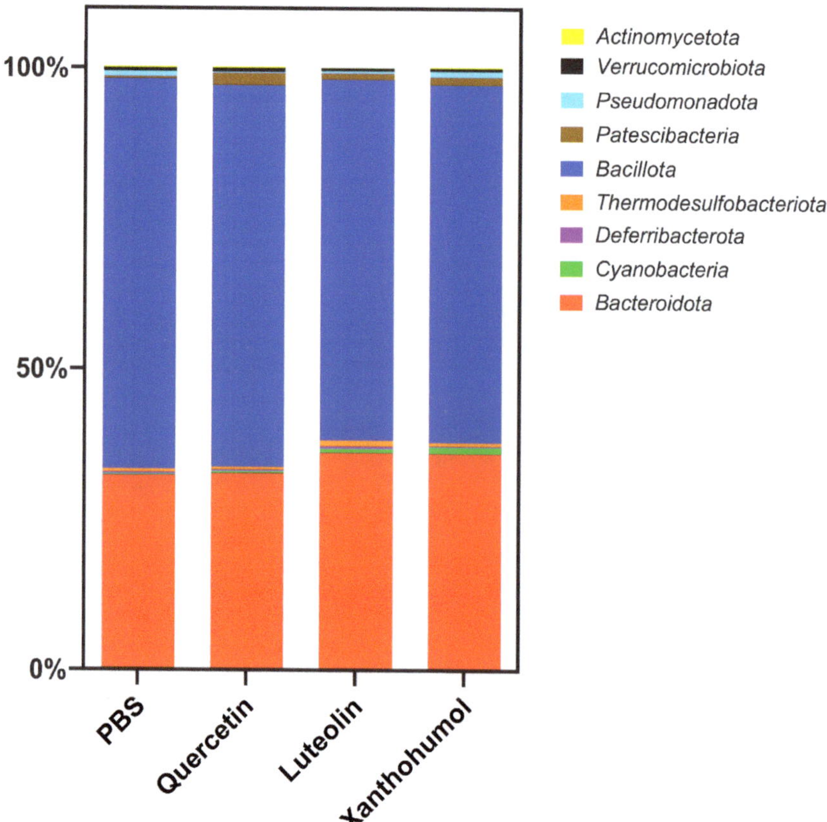

Figure 5. Average gut microbiota composition at the phylum level for the CRC-induced rats belonging to the four cohorts.

Significant differences were also found between the PBS cohort and the xanthohumol cohort, revealing an increase in the abundance of the families *Gastranaerophilales*, *Saccharimonadaceae*, and *Sutterellaceae* (0.27% vs. 1.06%, 0.35% vs. 1.28%, and 0.01% vs. 0.16%, respectively). Finally, a significant increase in the abundance of the *Prevotellaceae* familly was observed between the PBS cohort and the luteolin and xanthohumol cohorts (6.83% vs. 11.09% and 6.83% vs. 10.80%, respectively).

At the taxonomic genus level, statistically significant differences were also observed through the comparative analysis of the PBS cohort and the cohorts subjected to flavonoid treatments, as shown in Figure 7A. The main differences were found between PBS (Cohort 1) and luteolin (Cohort 3). Genera such as *Clostridia UCG-014*, *Turicibacter*, and *Christensenellaceae R-7* showed significant reductions exclusively in luteolin (Cohort 3) (from 4.37%, 0.77%, and 0.76% in the PBS cohort vs. 1.88%, 0.07%, and 0.08% in the luteolin cohort, respectively). Furthermore, the *Eubacterium xylanophilum* group experienced a significant decrease in all flavonoid-treated cohorts (from 1.40% in the PBS cohort to 0.71%, 0.31%, and 0.69% in quercetin, luteolin, and xanthohumol, respectively). In contrast, the relative abundance of *Muribaculum* increased significantly in all cohorts (0.24% in the PBS cohort vs. 0.37% in the quercetin cohort, 0.56% in the luteolin cohort, and 0.44% in the xanthohumol cohort). Similarly, an uncultured genus of the family *Ruminococcaceae* increased significantly in luteolin and xanthohumol cohorts (from 0.45% in the PBS cohort to 1.27% and 0.98% in the luteolin and xanthohumol cohorts, respectively), while the genera *Bilophila* and another uncultured genus of the family *Oscillospiraceae* showed significant increases only in the luteolin cohort (from 0.18% and 3.99% in the PBS cohort to 0.88% and 6.10% in the luteolin cohort, respectively). Finally, the genera *Parasutterella* and *Gastranaerophilales* experienced a significant increase in the xanthohumol cohort (from 0.01% and 0.20% in the PBS cohort to 0.16% and 0.71 in the xanthohumol cohort, respectively), and an uncultured genus of the family *Erysipelotrichaceae* and a Candidatus *Saccharimonas* experienced an increase only in the quercetin cohort (from 0.02% and 0.35% in the PBS cohort to 0.13% and 2.06% in the quercetin cohort, respectively). All of these notable variations are described in Table 4.

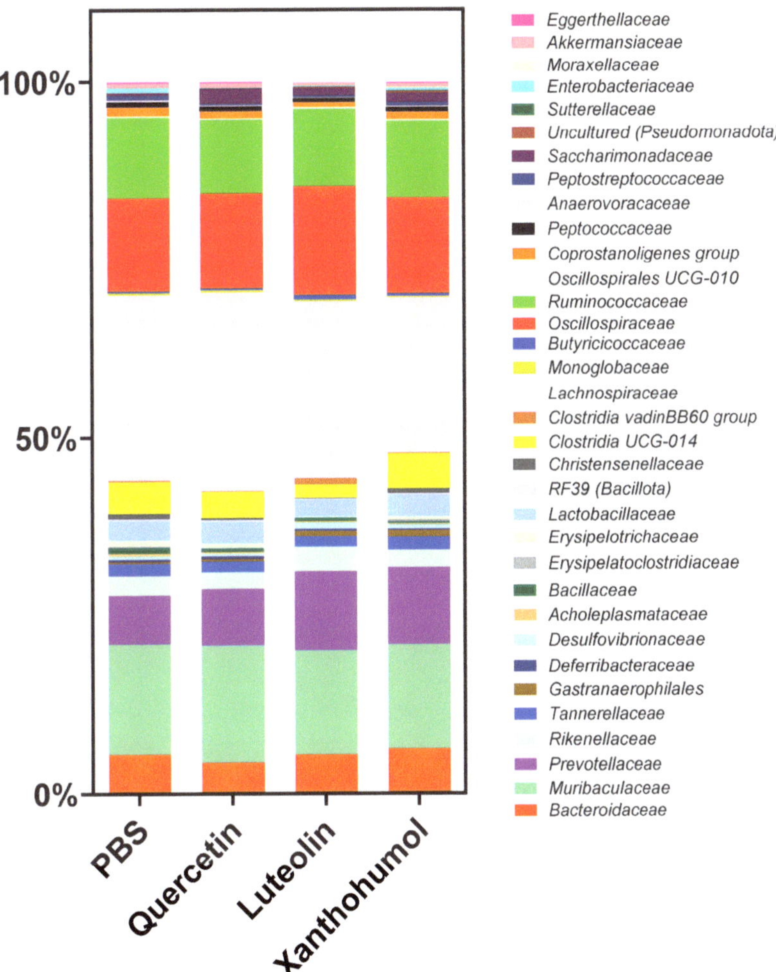

Figure 6. Differences in the average composition of the gut microbiota at the family level for the CRC-induced rats belonging to the four cohorts.

At the species level, notable variations in relative abundance were also observed. However, the majority of these species remain unidentified (Figure 7B). Three species were identified that exhibit notable differences between the PBS control cohort and the flavonoid-treated cohorts, none of which belong to the aforementioned genera. It is worth highlighting the abundance of *Bacteroides* sp., which exhibited an increase in the luteolin cohort compared to the control cohort (0.23% vs. 1.45%). On the contrary, *Eubacterium* sp. *coprostanoligenes* group showed a higher abundance in the xanthohumol cohort in relation to the control cohort (0.06% vs. 0.58%). Finally, a species identified as *UCG-005 metagenome* showed an increase in its abundance in the quercetin cohort compared to the PBS cohort (0.05% vs. 0.13%).

Table 3. Average percentage composition of statistically significant intestinal microbiota families in CRC-induced animals for the four cohorts studied. Underlined families indicate significant differences (ANOVA test) between at least one of the cohorts treated with flavonoids (Cohorts 2, 3, and 4) and the control cohort (Cohort 1). Percentages marked with asterisks indicate that the cohort shows a statistically significant difference compared to the control cohort within a specific family, along with the corresponding level of statistical significance. Cohort 1: PBS administration; Cohort 2: quercetin administration; Cohort 3: luteolin administration; Cohort 4: xanthohumol administration. Asterisks indicate statistically significant differences (* $p < 0.05$; ** $p < 0.005$; *** $p < 0.0005$; **** $p < 0.0001$).

	Cohort 1 (%)	Cohort 2 (%)	Cohort 3 (%)	Cohort 4 (%)
Eggerthellaceae	0.24	0.16	0.05 ****	0.18
Bacteroidaceae	5.26	4.21	5.42	6.12
Muribaculaceae	15.50	16.42	14.61	14.61
Prevotellaceae	6.83	7.93	11.09 ****	10.80 ****
Rikenellaceae	2.76	2.34	3.40	2.38
Tannerellaceae	1.63	1.49	1.49	1.88
Gastranaerophilales	0.27	0.34	0.69	1.06 **
Deferribacteraceae	0.31	0.34	0.43	0.19
Desulfovibrionaceae	0.53	0.43	0.95	0.60
Acholeplasmataceae	0.30	0.17	0.15	0.14
Bacillaceae	0.85	0.53	0.49	0.30
Erysipelatoclostridiaceae	0.17	0.05	0.04	0.24
Erysipelotrichaceae	0.82	0.60	0.13 ***	0.43
Lactobacillaceae	2.93	3.20	2.35	2.97
RF39 (Bacillota)	0.25	0.29	0.11	0.24
Christensenellaceae	0.77	0.28	0.09 ****	0.67
Clostridia UCG-014	4.37	3.64	1.88	4.94
Clostridia vadin BB60 group	0.19	0.09	0.88 *	0.11
Lachnospiraceae	26.07	27.94	24.93	21.84
Monoglobaceae	0.25	0.23	0.20	0.13
Butyricicoccaceae	0.23	0.26	0.61 **	0.37
Oscillospiraceae	13.11	13.43	15.38	13.48
Ruminococcaceae	11.36	10.31	10.86	10.80
Oscillospirales UCG-010	0.21	0.18	0.25	0.22
Coprostanoligenes group	1.26	1.06	0.67	1.09
Peptococcaceae	0.72	0.65	0.52	0.62
Anaerovoracaceae	0.26	0.16	0.10 ****	0.16
Peptostreptococcaceae	0.58	0.21	0.25	0.53
Saccharimonadaceae	0.35	2.06 *	1.07	1.28 *
Uncultured (*Rhodospirillales*)	0.03	0.02	0.17	0.22
Sutterellaceae	0.01	0.02	0.03	0.16 ***
Enterobacteriaceae	0.69	0.07 *	0.08 *	0.31
Moraxellaceae	0.00	0.00	0.00	0.12
Akkermansiaceae	0.62	0.67	0.48	0.52

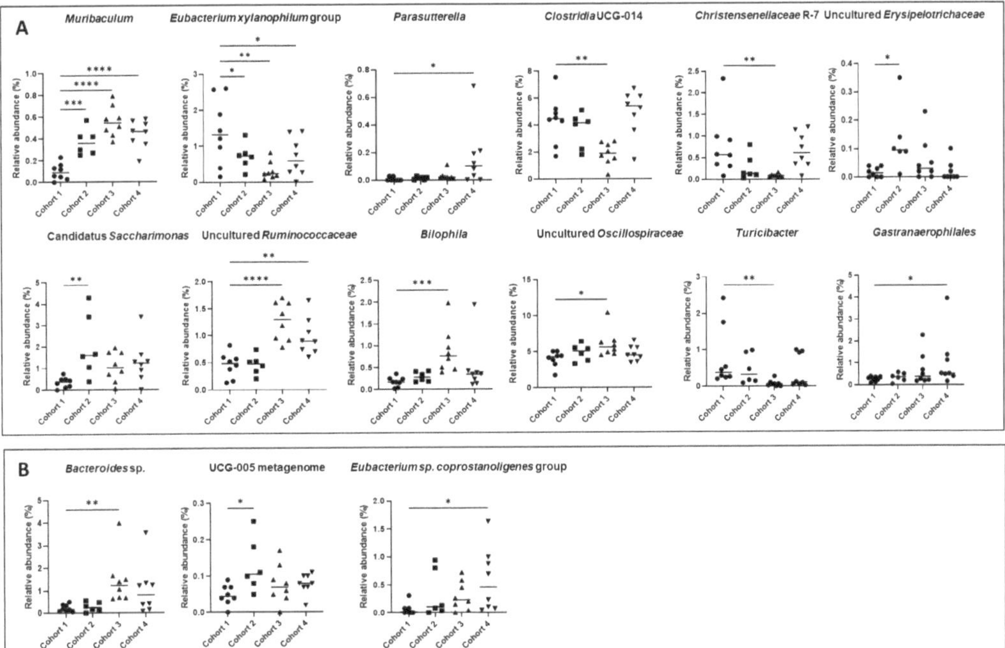

Figure 7. (**A**) Significant taxonomic variations at the genus level observed between rats with CRC-induced within the PBS cohort (Cohort 1) and the cohorts subjected to treatment with quercetin, luteolin, or xanthohumol (Cohorts 2, 3, and 4, respectively). (**B**) Significant taxonomic differences at the species level observed between rats with CRC induced within the PBS cohort (Cohort 1) and the cohorts subjected to treatment with quercetin, luteolin, or xanthohumol (Cohorts 2, 3, and 4, respectively). The median value of each cohort is represented as a horizontal line within each cohort. Asterisks indicate statistically significant differences (* $p < 0.05$; ** $p < 0.005$; *** $p < 0.0005$; **** $p < 0.0001$) (ANOVA test).

Table 4. Average percentage composition of statistically significant (ANOVA test) intestinal microbiota genera and species in CRC-induced animals for the four cohorts studied. Asterisks indicate the level of significant difference compared to the control cohort. Cohort 1: PBS administration; Cohort 2: quercetin administration; Cohort 3: luteolin administration; Cohort 4: xanthohumol administration. Asterisks indicate statistically significant differences (* $p < 0.05$; ** $p < 0.005$; *** $p < 0.0005$; **** $p < 0.0001$).

Genus	Cohort 1 (%)	Cohort 2 (%)	Cohort 3 (%)	Cohort 4 (%)
Muribaculum	0.24	0.37 ***	0.56 ****	0.44 ****
Bilophila	0.18	0.29	0.80 ***	0.49
Christensenellaceae R-7	0.76	0.27	0.08 **	0.66
Clostridia UCG-014	4.37	3.64	1.88 **	4.94
Eubacterium xylanophilum group	1.40	0.71 *	0.31 **	0.69 *
Uncultured (*Oscillospiraceae*)	3.99	4.84	6.06 *	4.72
Uncultured (*Ruminococcaceae*)	0.45	0.46	1.27 ****	0.98 **
"*Candidatus* Saccharimonas"	0.35	2.06 **	1.07	1.28
Parasutterella	0.01	0.02	0.03	0.16 *
Turicibacter	0.77	0.46	0.07 **	0.40
Gastranaerophilales	0.20	0.25	0.22	0.71 *
Uncultured (*Erysipelotrichaceae*)	0.02	0.13 *	0.06	0.02
Species				
Bacteroides sp.	0.23	0.27	1.45 **	1.04
Eubacterium sp. Coprostanoligenes group	0.06	0.33	0.30	0.58 *
UCG-005 metagenome	0.05	0.13 *	0.08	0.08

4. Discussion

This work has evaluated the antitumor potential of the flavonoids quercetin, luteolin, and xanthohumol in an animal model where CRC was induced using AOM and DSS. These three flavonoids have already been shown to have antitumor effects in the treatment of CRC through oral administration as nutraceuticals [1,27,28,33,34,41]. Here, we performed intraperitoneal administration of the compounds to study their effects as pharmaceutical compounds against CRC. A total of forty rats participated in this study, divided into four cohorts: Cohort 1, which received PBS instead of flavonoids (used as a control), Cohort 2 (treated with quercetin), Cohort 3 (treated with luteolin), and Cohort 4 (treated with xanthohumol). Within each cohort, eight rats were induced for CRC, while two rats did not receive AOM nor DSS (absolute controls, which were used as sentinels for detecting potential side effects of the flavonoid treatments). The CRC-induced animals from the flavonoid-treated cohorts were compared with the corresponding animals from the control cohort at different levels (physical and metagenomics parameters).

Animals from the four cohorts showed a constant weight gain throughout the 18 experimental weeks and no significant differences were found between the flavonoid-treated cohorts and the control cohort. However, after the first DSS event (week 4), rats numbered 3 and 7 in Cohort 2 (quercetin) died as a result of intense colitis. Furthermore, a slowdown in weight gain was observed after the second DSS event (week 15) in all cohorts, with rapid recovery from week 16 until the end of the experiment (Figure 1). Once the experiment was completed at week 18, the 38 surviving animals were sacrificed.

In order to evaluate the anti-inflammatory potential of these compounds, quantification of hyperplastic Peyer's patches was performed in all rats. These protuberances within the mucosal structures of the small intestine have an abundance of lymphocytes and undergo hyperplasia in response to alterations in the digestive tract. These hyperplastic changes imply a pro-inflammatory immune status in the animals [42]. Luteolin and xanthohumol demonstrated the potential to reduce the number of hyperplastic Peyer's patches after intraperitoneal administration, with this second compound being the most effective (Figure 2). This result can be easily explained since luteolin and xanthohumol are well-known anti-inflammatory compounds [43,44]. In the case of quercetin, although a reduction in the average number of hyperplastic Peyer's patches was observed, no statistically significant variations were found when compared with the control cohort (Figure 2). Like many other flavonoids, quercetin has also been shown to be an anti-inflammatory compound [45]; however, an in vitro study showed that a significant amount of quercetin can remain adhered to the wall of the small intestine of the rat, reducing its availability [46], which could explain, in our case, its lower anti-inflammatory effect.

Regarding caecum weight (Figure S1), as expected, no significant variations were observed between cohorts since flavonoids, although they are prebiotics, are not fermentable compounds like inulin or other polysaccharide-type prebiotics [47–49] and therefore do not contribute to the gain of microbial mass in this organ. However, flavonoids and their derived metabolites do have the ability to modulate the composition of the gut microbiota. This modulation occurs by inhibiting certain undesirable bacterial strains or increasing concentrations of beneficial genera [50].

At the colon level, no significant differences were found in terms of their lengths (Figure S2). In contrast, the number of colon tumors showed a significant decrease in the luteolin cohort (average of 7 tumors) when compared to the PBS control cohort (average of 19 tumors) (Figure 3), demonstrating the antitumor potential of luteolin when administered intraperitoneally at the tested dose (25 mg/kg/bw). These results are in accordance with another study carried out in mice to which luteolin was administered intraperitoneally in AOM-induced animals, where the average number of tumors was reduced from 9.4 in the control cohort to 4.2 in the group that received luteolin [51]. In vitro, luteolin has shown high antitumor activity against the human CRC cell lines HCT116, HT-29, and T84, even with a synergistic effect when administered together with the commercial antitumor 5-FU [35]

Regarding quercetin, it has been shown to exert an anti-CRC effect when administered orally at a concentration of 2% in the diet [52]. However, lower doses of quercetin were ineffective in reducing tumor numbers [53]. In this study, quercetin was administered intraperitoneally at 25 mg/kg/bw, but no significant reduction in tumor count was observed (Figure 3). This inconsistency may be attributed to the different bioavailability profile associated with intraperitoneal administration compared to oral administration, which would make the dose ineffective in achieving the desired antitumor activity. Finally, in previous works, xanthohumol has demonstrated antitumor potential in CRC cell lines [35,54,55], displaying a better antitumor effect than that of current pharmacological drugs in use for chemotherapy of this cancer, such as 5-FU [35]. In contrast to the antitumor properties exhibited by xanthohumol in CRC cell lines, the performance of this compound in this rat CRC model closely reflects (at the level of tumor numbers) that of the control group that did not receive any flavonoid treatment (Figure 3). These observations suggest that xanthohumol may not be a potent antitumor agent in this context, despite its notable anti-inflammatory activity evidenced by the reduction in the number of hyperplastic Peyer's patches.

Regarding the transport of intraperitoneally administered flavonoids, the sub-mesothelial stratum of the peritoneum supports an intricate but effective vascular network comprising blood and lymphatic vessels [56]. Compounds within the visceral peritoneum traverse the venous system, gaining entry into the portal vein. The peritoneum is richly perfused with blood capillaries, thus providing an optimal surface for the bidirectional exchange of pharmaceutical agents between the peritoneal cavity and plasma. Molecules introduced via portal circulation undergo integration with systemic circulation after hepatic transit, leading to rapid first-pass metabolism of the administered substances [37]. The liver functions as a pivotal organ in phase II metabolism, facilitating the conjugation of flavonoids through processes such as sulfation and methylation. Subsequently, the resulting metabolites are excreted as bile components back to the small intestine and reach the intestinal microbiota, undergoing deconjugation and subsequent reabsorption [38].

A fundamental disparity arises in flavonoid processing between oral and intraperitoneal intake. Orally administered flavonoids undergo alterations catalyzed by intestinal phase II enzymes, producing conjugated metabolites. These metabolites enter the portal circulation and subsequently reach the liver where additional modifications take place [38]. In contrast, flavonoids administered intraperitoneally directly access the portal system on their route to the liver [37]. This difference makes intraperitoneal administration a faster route for flavonoid absorption. Furthermore, the complexity of the flavonoid absorption process, whether by oral or intraperitoneal administration, could explain the differences observed in the effect of flavonoids against CRC between CRC cell lines and CRC animal models due to the absence of this entire circulatory network in the in vitro tests on cell lines, where flavonoids are absorbed directly by cells with a smaller amount (or absence) of enzymatic modifications in their structure.

The processing of flavonoids by the intestinal microbiota, after deconjugation, leads to the production of various hydroxyphenylacetic acids [57–59]. For example, quercetin undergoes metabolic transformations that lead to the formation of 2-(3,4-dihydroxyphenyl) acetic acid, 2-(3-hydroxyphenyl) acetic acid, and 3,4-dihydroxybenzoic acid from its B ring. Simultaneously, the A ring produces phloroglucinol, 3-(3,4-dihydroxyphenyl) propionic acid, and 3-(3-hydroxyphenyl) propionic acid [60]. Within the intestinal environment, these compounds can also undergo initial alterations through the fission of the C ring, involving various metabolic pathways, followed by subsequent dihydroxylation reactions [61]. In the case of luteolin, the compounds generated due to microbiota metabolism are 3-(4'-hydroxyphenyl) propionic acid and 3-(3',4'-dihydroxyphenyl) propionic acid, which are formed by cleavage of the C ring. In both cases, the release of phloroglucinol occurs as a byproduct of this process [59,62]. For xanthohumol, the metabolite resulting from the action of the gut microbiota is 8-prenylnaringenin, a well-known potent phytoestrogen [62].

All of these compounds have the ability to shape the intestinal microbiota in different ways, which will affect their antitumor activity. To study the taxonomic variations between the different cohorts of this work, the composition of the intestinal microbiota was determined by metagenomic 16S rRNA sequencing of cecal content. In terms of alpha diversity, richness and evenness remained similar in the different cohorts (Figure S3), while interesting differences were found with respect to beta diversity between the groups (Figure 4). The beta diversity measure suggests that the PBS cohort and the quercetin cohort have a similar community structure although significant differences were found between them (Table 1). In the case of the luteolin cohort, large differences were found when compared with the rest of the cohorts (p-value 0.002 for all comparisons) (Table 1), indicating a less related community structure with the rest of the cohorts. Finally, the xanthohumol cohort also showed significant differences with all the other cohorts, especially with luteolin, which is in accordance with the good antitumor performance observed in the case of luteolin but not with xanthohumol. These findings suggest that quercetin, luteolin, and xanthohumol exhibit the ability to modulate the community structure of the gut microbiota in the CRC-induced animals, with special emphasis on the notable differences observed in luteolin treatment. These distinctions are evident when clustering and plotting samples on a heatmap using genus abundance data (Figure S4).

At the phylum level, *Bacillota* was significantly reduced in both luteolin and xanthohumol cohorts, while *Bacteroidota* was significantly increased in both of them. It has been previously shown that the abundance of the phylum *Bacillota* decreased in the lumen of a CRC rat model compared to healthy rats [63], supporting our observations in the case of the luteolin treatment, which has been shown to possess the best antitumor effect among the flavonoids analyzed in this work. However, this dysbiosis was also observed in the xanthohumol cohort, which showed a tumor count similar to that of the control cohort. In contrast, the phylum *Actinomycetota* was reduced only in the luteolin cohort (Table 2), and this lower abundance has been associated with healthy rats [63].

At the family level, *Prevotellaceae* was the most abundant family in all cohorts, being extremely lower in the CRC-induced rats of PBS cohort (control) compared to the luteolin and xanthohumol cohorts, especially in the case of the luteolin cohort, where the abundance was almost the double (Table 3). Other authors have observed a similar result, where the *Prevotellaceae* family has been significantly increased in healthy rats compared to CRC rats [63]. In this study, based on the results obtained, the luteolin cohort is the healthiest group, which supports that the abundance of the *Prevotellaceae* family may be related to a better state of health. Paradoxically, this family was also overrepresented in the xanthohumol cohort, where the number of tumors was similar to the one observed in the PBS cohort. However, xanthohumol demonstrated a great anti-inflammatory effect, which may explain the high abundance of this family. The *Erysipelotrichaceae* family was also overrepresented in the control cohort and significantly reduced in the luteolin cohort (Table 3). The high abundance of this family has been linked to CRC status [63]. Regarding the family *Christensenellaceae*, a notable reduction was observed in the luteolin cohort (Table 3). A decrease in the abundance of this family has been postulated to be advantageous to health, based on a study among African American patients with colorectal cancer (CRC) [64], which is consistent with our findings. However, a different study reported elevated levels of *Christenellaceae* in healthy controls compared to individuals with CRC [65]. These incongruent trends can be explained since it has been observed that the association of *Christensenellaceae* with CRC depends on the type of specific mutation present [66,67]. On the other hand, as shown in Table 3, we observed a significant decrease in the abundance of the family *Enterobacteriaceae* in the luteolin cohort, which may be related to an unhealthy state (due to the presence of lipopolysaccharide in this family, as well as other virulence factors), since this family is associated with CRC due to the production of the organic compound trimethylamine n-oxide [68]. Other families, such as *Eggerthellaceae* and *Anaerovoracaceae*, showed a reduction in the luteolin cohort, while the *Clostridia vadinBB60* group and *Butyricicoccaceae* experienced an increase. This dysbiosis observed at the family level can be correlated with an improvement in health status (due to a higher production of anti-inflammatory short-chain fatty acids, among other factors such as lower production or presence of virulence factors) after the administration of luteolin as a therapeutic intervention [69].

At the genus level, *Muribaculum* increased significantly in all the flavonoid-treated cohorts, especially in the luteolin cohort (Table 4 and Figure 7). A significant abundance in this genus has been positively associated in CRC mouse models, as bacteria of this genus have demonstrated the ability to maintain intestinal homeostasis by utilizing mucin monosaccharides [70]. The genus *Bilophila* was overrepresented in the luteolin cohort (Table 4 and Figure 7). It is well known that *Bilophila wadsworthia* converts taurine to the toxic metabolite hydrogen sulfide, an activity associated with CRC [71], and this pathogenic gut population has been shown to be inhibited after administration of functional meats enriched in flavonoids [8]. However, this species did not experience dysbiosis and the differences observed in *Bilophila* may be associated with other species in this genus. In contrast, a notable reduction in the abundance of the genus *Christensenellaceae R-7* was observed exclusively within the luteolin cohort (Table 4 and Figure 7). This reduction is notable, particularly considering that the prevalence of this genus has previously been correlated with healthy conditions [72]. In the case of *Clostridia UCG-014*, which has been commonly reported as pro-inflammatory bacteria [73], a significant reduction in the luteolin cohort was observed, compared to the control cohort (Table 4 and Figure 7). This is supported by a study conducted in a colitis-associated CRC mouse, where a decrease in this genus was observed within the CRC cohort that received natural shikonin, which was found to be a preventive agent for this neoplasia [74]. The genus *Eubacterium xylanophilum* group was reduced in all flavonoid-treated cohorts, especially in the luteolin cohort (Table 4 and Figure 7). Controversially, this genus was found to be increased in a CRC-mice cohort fed with rice bran, which improved the CRC condition [75]. Regarding the *Oscillospiraceae* and *Ruminococcaceae* families, a notable increase in the abundance of an uncultured genus within each family was found, which was particularly evident in the luteolin cohort (Table 4 and Figure 7). The taxon "*Candidatus* Saccharimonas" was significantly increased in the quercetin cohort, while the beneficial genus *Parasutterella* [76] remained more abundant in the xanthohumol cohort. Regarding the genus *Turicibacter*, it has been associated with various diseases, such as acute appendicitis [77,78]. Moreover, *Turicibacter* was increased in other studies with CRC-induced mice [78,79], which supports the observed reduction in its abundance here in the luteolin cohort.

5. Conclusions

In summary, this study has elucidated the antitumor potential of the flavonoid luteolin in a CRC rat model when administered intraperitoneally. Both luteolin and xanthohumol have shown the ability to significantly reduce the number of hyperplastic Peyer's patches in the small intestine, which is an inflammation biomarker. The luteolin cohort experienced a significant reduction in the number of colon tumors compared to the control cohort. In addition, a metagenomic study has been carried out to analyze the possible differences in the microbiota of the different cohorts, finding the main differences (with respect to the control cohort) in the luteolin cohort, where some bacterial families and genera associated with a good prognosis (such as those ones generating antitumor short-chain fatty acids like propionate or butyrate, which inhibit histone deacetylases, inducing apoptosis only in tumor colonocytes) experienced an increase, while other groups of harmful bacterial decreased (such as those ones involved in inflammation onset, stimulation of colonocytes proliferation via β-catenin activation, generation of reactive oxygen or nitrogen species (RONS) able to cause DNA mutations on colonocytes, or in the activation of procarcinogens) [69,80,81]. These results show the ability of flavonoids, particularly luteolin, to modulate the intestinal microbiota in an animal model for CRC to contribute to an improvement in the health of individuals. In addition, it confirms the effectiveness of the intraperitoneal administration of flavonoids, as drugs.

Supplementary Materials: The following supporting information can be downloaded at: https://www.mdpi.com/article/10.3390/nu16081161/s1. Figure S1. Mean number of caecum weight in grams for each cohort in CRC-induced animals (rats 1 to 8 in each cohort). The median value of each cohort is represented as a horizontal line within the corresponding box plots. Figure S2. Mean numbers of colon length in centimeters for each cohort in CRC-induced animals (rats 1 to 8 in each cohort). The median value of each cohort is represented as a horizontal line within the corresponding box plots. Figure S3. Box plots indicating alpha diversity measures of CRC-induced animals using the Chao1 (A), Simpson (B), and Shannon (C) indices. The median value of each cohort is represented as a horizontal line within the corresponding box plots. Figure S4. Heatmap with cluster samples by genus abundance. A dendrogram is used to show how samples are grouped based on genus abundance. Row labels add phylum information. Colors represent standardized abundances (red means high abundance of the given genus, while blue means low abundance).

Author Contributions: Funding acquisition, F.L.; investigation, Á.P.-V., P.M.-C., S.Y., J.S.-D., S.S., E.H. and J.P.; supervision, C.J.V. and F.L.; writing—original draft preparation, Á.P.-V.; writing—review and editing, Á.P.-V., C.J.V. and F.L. All authors have read and agreed to the published version of the manuscript.

Funding: This research was funded by Principado de Asturias (Spain) through the program Ayudas a organismos públicos para apoyar las actividades de I + D + I de sus grupos de investigación (grant AYUD/2021/51347) and through "Programa Severo Ochoa de Ayudas Predoctorales para la investigación y docencia" from Principado de Asturias (PhD grant PA-21-PF-BP20-150 to Á.P.-V. and grant PA-20-PF-BP19-058 to P.M.-C.), Programa de Ayudas FPI from MICINN (PhD grant PRE2022-102792 to J.S.-D.), the research project PID2021-127812OB-I00 from MICINN (Spanish Ministry of Science and Innovation), and the European Union's Horizon 2020 Research and Innovation Program under grant agreement no. 814650 for the project SynBio4Flav.

Institutional Review Board Statement: The animal study protocol used in this study was approved by the University of Oviedo Ethics Committee (protocol code PROAE 14/2022 from 22 June 2022).

Informed Consent Statement: Not applicable.

Data Availability Statement: Data (numbers of tumors, hyperplastic Peyer patches, etc.) and materials (tissues maintained in paraformaldehyde) can be obtained from the research group upon request. Publicly available datasets (metagenome sequences) were analyzed in this study, and these data can be found at the NCBI SRA database with access number PRJNA1083865.

Conflicts of Interest: The authors declare no conflict of interest.

References

1. Imran, M.; Rauf, A.; Abu-Izneid, T.; Nadeem, M.; Shariati, M.A.; Khan, I.A.; Imran, A.; Orhan, I.E.; Rizwan, M.; Atif, M.; et al. Luteolin, a Flavonoid, as an Anticancer Agent: A Review. *Biomed. Pharmacother.* **2019**, *112*, 108612. [CrossRef] [PubMed]
2. Devi, K.P.; Rajavel, T.; Nabavi, S.F.; Setzer, W.N.; Ahmadi, A.; Mansouri, K.; Nabavi, S.M. Hesperidin: A Promising Anticancer Agent from Nature. *Ind. Crops Prod.* **2015**, *76*, 582–589. [CrossRef]
3. Xi, Y.; Xu, P. Global Colorectal Cancer Burden in 2020 and Projections to 2040. *Transl. Oncol.* **2021**, *14*, 101174. [CrossRef] [PubMed]
4. Siegel, R.L.; Miller, K.D.; Fuchs, H.E.; Jemal, A. Cancer Statistics, 2021. *CA Cancer J. Clin.* **2021**, *71*, 7–33. [CrossRef] [PubMed]
5. Constantinou, V.; Constantinou, C. Focusing on Colorectal Cancer in Young Adults (Review). *Mol. Clin. Oncol.* **2024**, *20*, 8. [CrossRef] [PubMed]
6. Kuipers, E.J.; Grady, W.M.; Lieberman, D.; Seufferlein, T.; Sung, J.J.; Boelens, P.G.; van de Velde, C.J.H.; Watanabe, T. Colorectal Cancer. *Nat. Rev. Dis. Primers* **2015**, *1*, 15065. [CrossRef] [PubMed]
7. Jemal, A.; Bray, F.; Center, M.M.; Ferlay, J.; Ward, E.; Forman, D. Global Cancer Statistics. *CA Cancer J. Clin.* **2011**, *61*, 69–90. [CrossRef] [PubMed]
8. Fernández, J.; García, L.; Monte, J.; Villar, C.J.; Lombó, F. Functional Anthocyanin-Rich Sausages Diminish Colorectal Cancer in an Animal Model and Reduce pro-Inflammatory Bacteria in the Intestinal Microbiota. *Genes* **2018**, *9*, 133. [CrossRef] [PubMed]
9. Parmar, S.; Easwaran, H. Genetic and Epigenetic Dependencies in Colorectal Cancer Development. *Gastroenterol. Rep.* **2022**, *10*, goac035. [CrossRef]
10. Kim, U.; Lee, D.S. Epigenetic Regulations in Mammalian Cells: Roles and Profiling Techniques. *Mol. Cells* **2023**, *46*, 86–98. [CrossRef]
11. Zeki, S.S.; Graham, T.A.; Wright, N.A. Stem Cells and Their Implications for Colorectal Cancer. *Nat. Rev. Gastroenterol. Hepatol.* **2011**, *8*, 90–100. [CrossRef] [PubMed]

12. Colussi, D.; Brandi, G.; Bazzoli, F.; Ricciardiello, L. Molecular Pathways Involved in Colorectal Cancer: Implications for Disease Behavior and Prevention. *Int. J. Mol. Sci.* **2013**, *14*, 16365–16385. [CrossRef] [PubMed]

13. Grady, W.M.; Carethers, J.M. Genomic and Epigenetic Instability in Colorectal Cancer Pathogenesis. *Gastroenterology* **2008**, *135*, 1079–1099. [CrossRef] [PubMed]

14. Jones, S.; Chen, W.D.; Parmigiani, G.; Diehl, F.; Beerenwinkel, N.; Antal, T.; Traulsen, A.; Nowak, M.A.; Siegel, C.; Velculescu, V.E.; et al. Comparative Lesion Sequencing Provides Insights into Tumor Evolution. *Proc. Natl. Acad. Sci. USA* **2008**, *105*, 4283–4288. [CrossRef] [PubMed]

15. Tsao, R. Chemistry and Biochemistry of Dietary Polyphenols. *Nutrients* **2010**, *2*, 1231–1246. [CrossRef] [PubMed]

16. Manach, C.; Scalbert, A.; Morand, C.; Rémésy, C.; Jiménez, L. Polyphenols: Food Sources and Bioavailability. *Am. J. Clin. Nutr.* **2004**, *79*, 727–747. [CrossRef] [PubMed]

17. Chaudhuri, S.; Sengupta, B.; Taylor, J.; Pahari, B.P.; Sengupta, P.K. Interactions of Dietary Flavonoids with Proteins: Insights from Fluorescence Spectroscopy and Other Related Biophysical Studies. *Curr. Drug Metab.* **2013**, *14*, 491–503. [CrossRef]

18. González-Vallinas, M.; González-Castejón, M.; Rodríguez-Casado, A.; Ramírez de Molina, A. Dietary Phytochemicals in Cancer Prevention and Therapy: A Complementary Approach with Promising Perspectives. *Nutr. Rev.* **2013**, *71*, 585–599. [CrossRef]

19. Li, A.-N.; Li, S.; Zhang, Y.-J.; Xu, X.-R.; Chen, Y.-M.; Li, H.-B. Resources and Biological Activities of Natural Polyphenols. *Nutrients* **2014**, *6*, 6020–6047. [CrossRef]

20. Yahfoufi, N.; Alsadi, N.; Jambi, M.; Matar, C. The Immunomodulatory and Anti-Inflammatory Role of Polyphenols. *Nutrients* **2018**, *10*, 1618. [CrossRef]

21. Rodríguez-García, C.; Sánchez-Quesada, C.; Gaforio, J.J. Dietary Flavonoids as Cancer Chemopreventive Agents: An Updated Review of Human Studies. *Antioxidants* **2019**, *8*, 137. [CrossRef] [PubMed]

22. Abotaleb, M.; Samuel, S.; Varghese, E.; Varghese, S.; Kubatka, P.; Liskova, A.; Büsselberg, D. Flavonoids in Cancer and Apoptosis. *Cancers* **2018**, *11*, 28. [CrossRef] [PubMed]

23. Chirumbolo, S.; Bjørklund, G.; Lysiuk, R.; Vella, A.; Lenchyk, L.; Upyr, T. Targeting Cancer with Phytochemicals via Their Fine Tuning of the Cell Survival Signaling Pathways. *Int. J. Mol. Sci.* **2018**, *19*, 3568. [CrossRef] [PubMed]

24. Kopustinskiene, D.M.; Jakstas, V.; Savickas, A.; Bernatoniene, J. Flavonoids as Anticancer Agents. *Nutrients* **2020**, *12*, 457. [CrossRef] [PubMed]

25. Gorlach, S.; Fichna, J.; Lewandowska, U. Polyphenols as Mitochondria-Targeted Anticancer Drugs. *Cancer Lett.* **2015**, *366*, 141–149. [CrossRef] [PubMed]

26. Perez-Vizcaino, F.; Fraga, C.G. Research Trends in Flavonoids and Health. *Arch. Biochem. Biophys.* **2018**, *646*, 107–112. [CrossRef] [PubMed]

27. Rauf, A.; Imran, M.; Khan, I.A.; ur-Rehman, M.; Gilani, S.A.; Mehmood, Z.; Mubarak, M.S. Anticancer Potential of Quercetin: A Comprehensive Review. *Phytother. Res.* **2018**, *32*, 2109–2130. [CrossRef] [PubMed]

28. Jiang, C.H.; Sun, T.L.; Xiang, D.X.; Wei, S.S.; Li, W.Q. Anticancer Activity and Mechanism of Xanthohumol: A Prenylated Flavonoid from Hops (*Humulus lupulus* L.). *Front. Pharmacol.* **2018**, *9*, 530. [CrossRef]

29. Harwood, M.; Danielewska-Nikiel, B.; Borzelleca, J.F.; Flamm, G.W.; Williams, G.M.; Lines, T.C. A Critical Review of the Data Related to the Safety of Quercetin and Lack of Evidence of in Vivo Toxicity, Including Lack of Genotoxic/Carcinogenic Properties. *Food Chem. Toxicol.* **2007**, *45*, 2179–2205. [CrossRef]

30. Kawanishi, S.; Oikawa, S.; Murata, M. Evaluation for Safety of Antioxidant Chemopreventive Agents. *Antioxid. Redox Signal.* **2005**, *7*, 1728–1739. [CrossRef]

31. Liu, H.; Zhang, L.; Li, G.; Gao, Z. Xanthohumol Protects against Azoxymethane-Induced Colorectal Cancer in Sprague-Dawley Rats. *Environ. Toxicol.* **2020**, *35*, 136–144. [CrossRef] [PubMed]

32. Vanhoecke, B.W.; Delporte, F.; Van Braeckel, E.; Heyerick, A.; Depypere, H.T.; Nuytinck, M.; De Keukeleire, D.; Bracke, M.E. A Safety Study of Oral Tangeretin and Xanthohumol Administration to Laboratory Mice. *In Vivo* **2005**, *19*, 103–107. [PubMed]

33. Miranda, C.L.; Stevens, J.F.; Helmrich, A.; Henderson, M.C.; Rodriguez, R.J.; Yang, Y.-H.; Deinzer, M.L.; Barnes, D.W.; Buhler, D.R. Antiproliferative and Cytotoxic Effects of Prenylated Flavonoids from Hops (*Humulus lupulus*) in Human Cancer Cell Lines. *Food Chem. Toxicol.* **1999**, *37*, 271–285. [CrossRef]

34. Lee, S.H.; Kim, H.J.; Lee, J.S.; Lee, I.-S.; Kang, B.Y. Inhibition of Topoisomerase I Activity and Efflux Drug Transporters' Expression by Xanthohumol from Hops. *Arch. Pharm. Res.* **2007**, *30*, 1435–1439. [CrossRef]

35. Fernández, J.; Silván, B.; Entrialgo-Cadierno, R.; Villar, C.J.; Capasso, R.; Uranga, J.A.; Lombó, F.; Abalo, R. Antiproliferative and Palliative Activity of Flavonoids in Colorectal Cancer. *Biomed. Pharmacother.* **2021**, *143*, 112241. [CrossRef]

36. Nejabati, H.R.; Roshangar, L. Kaempferol: A Potential Agent in the Prevention of Colorectal Cancer. *Physiol. Rep.* **2022**, *10*, e15488. [CrossRef]

37. Al Shoyaib, A.; Archie, S.R.; Karamyan, V.T. Intraperitoneal Route of Drug Administration: Should It Be Used in Experimental Animal Studies? *Pharm. Res.* **2020**, *37*, 12. [CrossRef]

38. Murota, K.; Nakamura, Y.; Uehara, M. Flavonoid Metabolism: The Interaction of Metabolites and Gut Microbiota. *Biosci. Biotechnol. Biochem.* **2018**, *82*, 600–610. [CrossRef] [PubMed]

39. Thangaraj, K.; Natesan, K.; Settu, K.; Palani, M.; Govindarasu, M.; Subborayan, V.; Vaiyapuri, M. Orientin Mitigates 1, 2-Dimethylhydrazine Induced Lipid Peroxidation, Antioxidant and Biotransforming Bacterial Enzyme Alterations in Experimental Rats. *J. Cancer Res. Ther.* **2018**, *14*, 1379–1388. [CrossRef]

40. Nowak, B.; Poźniak, B.; Popłoński, J.; Bobak, Ł.; Matuszewska, A.; Kwiatkowska, J.; Dziewiszek, W.; Huszcza, E.; Szeląg, A. Pharmacokinetics of Xanthohumol in Rats of Both Sexes after Oral and Intravenous Administration of Pure Xanthohumol and Prenylflavonoid Extract. *Adv. Clin. Exp. Med.* **2020**, *29*, 1101–1109. [CrossRef]

41. Neamtu, A.; Maghiar, T.; Alaya, A.; Olah, N.; Turcus, V.; Pelea, D.; Totolici, B.D.; Neamtu, C.; Maghiar, A.M.; Mathe, E. A Comprehensive View on the Quercetin Impact on Colorectal Cancer. *Molecules* **2022**, *27*, 1873. [CrossRef] [PubMed]

42. Mishra, S.; Tripathi, A.; Chaudhari, B.P.; Dwivedi, P.D.; Pandey, H.P.; Das, M. Deoxynivalenol Induced Mouse Skin Cell Proliferation and Inflammation via MAPK Pathway. *Toxicol. Appl. Pharmacol.* **2014**, *279*, 186–197. [CrossRef] [PubMed]

43. Aziz, N.; Kim, M.Y.; Cho, J.Y. Anti-Inflammatory Effects of Luteolin: A Review of in Vitro, in Vivo, and in Silico Studies. *J. Ethnopharmacol.* **2018**, *225*, 342–358. [CrossRef] [PubMed]

44. Cho, Y.C.; Kim, H.J.; Kim, Y.J.; Lee, K.Y.; Choi, H.J.; Lee, I.S.; Kang, B.Y. Differential Anti-Inflammatory Pathway by Xanthohumol in IFN-γ and LPS-Activated Macrophages. *Int. Immunopharmacol.* **2008**, *8*, 567–573. [CrossRef] [PubMed]

45. Chiang, M.C.; Tsai, T.Y.; Wang, C.J. The Potential Benefits of Quercetin for Brain Health: A Review of Anti-Inflammatory and Neuroprotective Mechanisms. *Int. J. Mol. Sci.* **2023**, *24*, 6328. [CrossRef] [PubMed]

46. Carbonaro, M.; Grant, G. Absorption of Quercetin and Rutin in Rat Small Intestine. *Ann. Nutr. Metab.* **2005**, *49*, 178–182. [CrossRef] [PubMed]

47. Hughes, R.L.; Alvarado, D.A.; Swanson, K.S.; Holscher, H.D. The Prebiotic Potential of Inulin-Type Fructans: A Systematic Review. *Adv. Nutr.* **2022**, *13*, 492–529. [CrossRef] [PubMed]

48. Fernández, J.; Ledesma, E.; Monte, J.; Millán, E.; Costa, P.; de la Fuente, V.G.; García, M.T.F.; Martínez-Camblor, P.; Villar, C.J.; Lombó, F. Traditional Processed Meat Products Re-Designed Towards Inulin-Rich Functional Foods Reduce Polyps in Two Colorectal Cancer Animal Models. *Sci. Rep.* **2019**, *9*, 14783. [CrossRef] [PubMed]

49. Xiao, J.; Metzler-Zebeli, B.U.; Zebeli, Q. Gut Function-Enhancing Properties and Metabolic Effects of Dietary Indigestible Sugars in Rodents and Rabbits. *Nutrients* **2015**, *7*, 8348–8365. [CrossRef]

50. Oteiza, P.I.; Fraga, C.G.; Mills, D.A.; Taft, D.H. Flavonoids and the Gastrointestinal Tract: Local and Systemic Effects. *Mol. Asp. Med.* **2018**, *61*, 41–49. [CrossRef]

51. Ashokkumar, P.; Sudhandiran, G. Luteolin Inhibits Cell Proliferation during Azoxymethane-Induced Experimental Colon Carcinogenesis via Wnt/ β-Catenin Pathway. *Investig. New Drugs* **2011**, *29*, 273–284. [CrossRef] [PubMed]

52. Matsukawa, Y.; Nishino, H.; Okuyama, Y.; Matsui, T.; Matsumoto, T.; Matsumura, S.; Shimizu, Y.; Sowa, Y.; Sakai, T. Effects of Quercetin and/or Restraint Stress on Formation of Aberrant Crypt Foci Induced by Azoxymethane in Rat Colons. *Oncology* **1997**, *54*, 118–121. [CrossRef] [PubMed]

53. Deschner, E.E.; Ruperto, J.; Wong, G.; Newmark, H.L. Quercetin and Rutin as Inhibitors of Azoxymethanol-Induced Colonic Neoplasia. *Carcinogenesis* **1991**, *12*, 1193–1196. [CrossRef] [PubMed]

54. Torrens-Mas, M.; Alorda-Clara, M.; Martínez-Vigara, M.; Roca, P.; Sastre-Serra, J.; Oliver, J.; Pons, D.G. Xanthohumol Reduces Inflammation and Cell Metabolism in HT29 Primary Colon Cancer Cells. *Int. J. Food Sci. Nutr.* **2022**, *73*, 471–479. [CrossRef] [PubMed]

55. Sastre-Serra, J.; Ahmiane, Y.; Roca, P.; Oliver, J.; Pons, D.G. Xanthohumol, a Hop-Derived Prenylflavonoid Present in Beer, Impairs Mitochondrial Functionality of SW620 Colon Cancer Cells. *Int. J. Food Sci. Nutr.* **2019**, *70*, 396–404. [CrossRef] [PubMed]

56. Aguirre, A.R.; Abensur, H. Physiology of Fluid and Solute Transport across the Peritoneal Membrane. *J. Bras. Nefrol.* **2014**, *36*, 74–79. [CrossRef] [PubMed]

57. Landete, J.M. Ellagitannins, Ellagic Acid and Their Derived Metabolites: A Review about Source, Metabolism, Functions and Health. *Food Res. Int.* **2011**, *44*, 1150–1160. [CrossRef]

58. Jaganath, I.B.; Jaganath, I.B.; Mullen, W.; Edwards, C.A.; Crozier, A. The Relative Contribution of the Small and Large Intestine to the Absorption and Metabolism of Rutin in Man. *Free Radic. Res.* **2006**, *40*, 1035–1046. [CrossRef]

59. Marín, L.; Miguélez, E.M.; Villar, C.J.; Lombó, F. Bioavailability of Dietary Polyphenols and Gut Microbiota Metabolism: Antimicrobial Properties. *BioMed Res. Int.* **2015**, *2015*, 905215. [CrossRef]

60. Selma, M.V.; Espín, J.C.; Tomás-Barberán, F.A. Interaction between Phenolics and Gut Microbiota: Role in Human Health. *J. Agric. Food Chem.* **2009**, *57*, 6485–6501. [CrossRef]

61. Nazzaro, F.; Fratianni, F.; De Feo, V.; Battistelli, A.; Da Cruz, A.G.; Coppola, R. *Polyphenols, the New Frontiers of Prebiotics*, 1st ed.; Elsevier Inc.: Amsterdam, The Netherlands, 2020; Volume 94, ISBN 9780128202180.

62. Makarewicz, M.; Drożdż, I.; Tarko, T.; Duda-Chodak, A. The Interactions between Polyphenols and Microorganisms, Especially Gut Microbiota. *Antioxidants* **2021**, *10*, 188. [CrossRef] [PubMed]

63. Zhu, Q.; Jin, Z.; Wu, W.; Gao, R.; Guo, B.; Gao, Z.; Yang, Y.; Qin, H. Analysis of the Intestinal Lumen Microbiota in an Animal Model of Colorectal Cancer. *PLoS ONE* **2014**, *9*, e90849. [CrossRef]

64. Yazici, C.; Wolf, P.G.; Kim, H.; Cross, T.-W.L.; Vermillion, K.; Carroll, T.; Augustus, G.J.; Mutlu, E.; Tussing-Humphreys, L.; Braunschweig, C.; et al. Race-Dependent Association of Sulfidogenic Bacteria with Colorectal Cancer. *Gut* **2017**, *66*, 1983–1994. [CrossRef] [PubMed]

65. Le Gall, G.; Guttula, K.; Kellingray, L.; Tett, A.J.; ten Hoopen, R.; Kemsley, E.K.; Savva, G.M.; Ibrahim, A.; Narbad, A. Metabolite Quantification of Faecal Extracts from Colorectal Cancer Patients and Healthy Controls. *Oncotarget* **2018**, *9*, 33278–33289. [CrossRef]

66. Burns, M.B.; Montassier, E.; Abrahante, J.; Priya, S.; Niccum, D.E.; Khoruts, A.; Starr, T.K.; Knights, D.; Blekhman, R. Colorectal Cancer Mutational Profiles Correlate with Defined Microbial Communities in the Tumor Microenvironment. *PLoS Genet.* **2018**, *14*, e1007376. [CrossRef]
67. Waters, J.L.; Ley, R.E. The Human Gut Bacteria Christensenellaceae Are Widespread, Heritable, and Associated with Health. *BMC Biol.* **2019**, *17*, 83. [CrossRef]
68. Qu, R.; Zhang, Y.; Ma, Y.; Zhou, X.; Sun, L.; Jiang, C.; Zhang, Z.; Fu, W. Role of the Gut Microbiota and Its Metabolites in Tumorigenesis or Development of Colorectal Cancer. *Adv. Sci.* **2023**, *10*, e2205563. [CrossRef] [PubMed]
69. Fernández, J.; Redondo-Blanco, S.; Gutiérrez-del-Río, I.; Miguélez, E.M.; Villar, C.J.; Lombó, F. Colon Microbiota Fermentation of Dietary Prebiotics towards Short-Chain Fatty Acids and Their Roles as Anti-Inflammatory and Antitumour Agents: A Review. *J. Funct. Foods* **2016**, *25*, 511–522. [CrossRef]
70. Liu, N.; Zou, S.; Xie, C.; Meng, Y.; Xu, X. Effect of the β-Glucan from Lentinus Edodes on Colitis-Associated Colorectal Cancer and Gut Microbiota. *Carbohydr. Polym.* **2023**, *316*, 121069. [CrossRef]
71. Peck, S.C.; Denger, K.; Burrichter, A.; Irwin, S.M.; Balskus, E.P.; Schleheck, D. A Glycyl Radical Enzyme Enables Hydrogen Sulfide Production by the Human Intestinal Bacterium Bilophila Wadsworthia. *Proc. Natl. Acad. Sci. USA* **2019**, *116*, 3171–3176. [CrossRef]
72. Mancabelli, L.; Milani, C.; Lugli, G.A.; Turroni, F.; Cocconi, D.; van Sinderen, D.; Ventura, M. Identification of Universal Gut Microbial Biomarkers of Common Human Intestinal Diseases by Meta-Analysis. *FEMS Microbiol. Ecol.* **2017**, *93*, fix153. [CrossRef] [PubMed]
73. Wang, Y.; Nan, X.; Zhao, Y.; Jiang, L.; Wang, H.; Zhang, F.; Hua, D.; Liu, J.; Yao, J.; Yang, L.; et al. Dietary Supplementation of Inulin Ameliorates Subclinical Mastitis via Regulation of Rumen Microbial Community and Metabolites in Dairy Cows. *Microbiol. Spectr.* **2021**, *9*, e0010521. [CrossRef]
74. Lin, H.; Ma, X.; Yang, X.; Chen, Q.; Wen, Z.; Yang, M.; Fu, J.; Yin, T.; Lu, G.; Qi, J.; et al. Natural Shikonin and Acetyl-Shikonin Improve Intestinal Microbial and Protein Composition to Alleviate Colitis-Associated Colorectal Cancer. *Int. Immunopharmacol.* **2022**, *111*, 109097. [CrossRef]
75. Weber, A.M.; Ibrahim, H.; Baxter, B.A.; Kumar, R.; Maurya, A.K.; Kumar, D.; Agarwal, R.; Raina, K.; Ryan, E.P. Integrated Microbiota and Metabolite Changes Following Rice Bran Intake during Murine Inflammatory Colitis-Associated Colon Cancer and in Colorectal Cancer Survivors. *Cancers* **2023**, *15*, 2231. [CrossRef] [PubMed]
76. Zhang, J.; Chen, Z.; Lu, Y.; Tu, D.; Zou, F.; Lin, S.; Yu, W.; Miao, M.; Shi, H. A Functional Food Inhibits Azoxymethane/Dextran Sulfate Sodium-Induced Inflammatory Colorectal Cancer in Mice. *Onco. Targets Ther.* **2021**, *14*, 1465–1477. [CrossRef]
77. Bosshard, P.P.; Zbinden, R.; Altwegg, M. Turicibacter Sanguinis Gen. Nov., Sp. Nov., a Novel Anaerobic, Gram-Positive Bacterium. *Int. J. Syst. Evol. Microbiol.* **2002**, *52*, 1263–1266. [CrossRef] [PubMed]
78. Chung, Y.; Ryu, Y.; An, B.C.; Yoon, Y.S.; Choi, O.; Kim, T.Y.; Yoon, J.; Ahn, J.Y.; Park, H.J.; Kwon, S.K.; et al. A Synthetic Probiotic Engineered for Colorectal Cancer Therapy Modulates Gut Microbiota. *Microbiome* **2021**, *9*, 122. [CrossRef] [PubMed]
79. Zackular, J.P.; Baxter, N.T.; Iverson, K.D.; Sadler, W.D.; Petrosino, J.F.; Chen, G.Y.; Schloss, P.D. The Gut Microbiome Modulates Colon Tumorigenesis. *mBio* **2013**, *4*, e00692-13. [CrossRef] [PubMed]
80. Jahani-Sherafat, S.; Alebouyeh, M.; Moghim, S.; Amoli, H.A.; Ghasemian-Safaei, H. Role of Gut Microbiota in the Pathogenesis of Colorectal Cancer: A Review Article. *Gastroenterol. Hepatol. Bed Bench* **2018**, *11*, 101.
81. Cipe, G.; Idiz, U.O.; Firat, D.; Bektasoglu, H. Relationship Between Intestinal Microbiota and Colorectal Cancer. *World J. Gastrointest. Oncol.* **2015**, *7*, 233–240. [CrossRef]

Article

Fractionated Leaf Extracts of *Ocimum gratissimum* Inhibit the Proliferation and Induce Apoptosis of A549 Lung Adenocarcinoma Cells

Rachael M. Curtis [1,*], Heng-Shan Wang [2], Xuan Luo [3], Erika B. Dugo [4], Jacqueline J. Stevens [1] and Paul B. Tchounwou [1,5]

[1] College of Science, Engineering, and Technology, Jackson State University, 1400 JR Lynch Street, Jackson, MS 39217, USA; jacqueline.j.stevens@jsums.edu (J.J.S.); paul.tchounwou@morgan.edu (P.B.T.)
[2] School of Chemistry and Pharmacy, Guangxi Normal University, No. 15 Yu Cai Road, Guilin 541004, China
[3] Department of Applied Chemistry, School of Chemistry and Chemical Engineering, Guangxi University, No. 100 East Daxue Road, Nanning 530004, China
[4] Department of Natural and Behavioral Sciences, College of Science, Technology, Engineering, and Mathematics, Johnson C. Smith University, 100 Beatties Ford Road, Charlotte, NC 28216, USA
[5] RCMI Center for Urban Health Disparities Research and Innovation, Morgan State University, 1700 E. Cold Spring Lane, Baltimore, MD 21252, USA
* Correspondence: rachael.curtis@jsums.edu; Tel.: +1-601-376-9531

Abstract: Previous in vitro studies in our laboratory demonstrated that ethyl acetate (P_2) and water-soluble (PS/PT1) fractionated leaf extracts of *Ocimum gratissimum* inhibit the proliferation of prostate cancer cells. It has been reported that the crude aqueous extract induces apoptosis in lung adenocarcinoma cells; however, the efficacy of the fractionated extracts against these cells remains unclear. In the present study, we hypothesized that the ability of the fractionated extracts to inhibit proliferation and induce apoptosis is associated with the activation of pro-apoptotic proteins and induction of DNA condensation in A549 cells. *Ocimum gratissimum* was cultivated and its leaves were harvested, extracted, and fractionated to produce fractions P_2 and PS/PT1. Anti-proliferative activity was assessed by direct cell count. For morphological characterization of apoptosis, 4′,6-diamidino-2-phenylindole staining was employed. Western blot analysis was performed to evaluate the apoptotic activity of the fractionated extracts. In data generated from anti-proliferation studies, P_2 significantly inhibited cell proliferation in a concentration-dependent manner; PS/PT1 elicited a decrease in the viability of cells, occurring at 500 µg/mL. 4′,6-diamidino-2-phenylindole staining revealed the induction of apoptosis, as evidenced by the formation of apoptotic bodies. Increased levels of pro-apoptotic proteins were observed as the concentrations of the fractionated extracts increased. These results suggest that fractionated leaf extracts of *Ocimum gratissimum* inhibit the proliferation and induce apoptosis of A549 cells.

Keywords: *Ocimum gratissimum*; A549 cells; apoptosis

Citation: Curtis, R.M.; Wang, H.-S.; Luo, X.; Dugo, E.B.; Stevens, J.J.; Tchounwou, P.B. Fractionated Leaf Extracts of *Ocimum gratissimum* Inhibit the Proliferation and Induce Apoptosis of A549 Lung Adenocarcinoma Cells. *Nutrients* **2024**, *16*, 2737. https://doi.org/10.3390/nu16162737

Academic Editors: Ching-Hsein Chen and Yi-Wen Liu

Received: 11 July 2024
Revised: 13 August 2024
Accepted: 14 August 2024
Published: 16 August 2024

1. Introduction

Lung cancer is the leading cause of cancer-related death worldwide [1]. Mainstream treatment options include surgery, radiation therapy, chemotherapy, or a combination of these treatments [2]. These expensive, harsh treatment options often produce toxic side effects in patients. Therefore, a critical need exists for the application of efficacious cancer treatments with low toxicity, which is commonly found in natural products.

Due to their abundant medicinal benefits, natural products are considered excellent sources for the discovery of novel anticancer agents. Cancer research has increasingly focused on using natural products such as vegetables, fruits, spices, and other plants to destroy cancer cells and improve overall health. Increasing evidence suggests that myricetin, a flavonoid found in fruits and vegetables, inhibits the migration of human lung adenocarcinoma cells [3]. Additionally, curcumin, the principal bioactive compound of the popular Indian spice, turmeric, has been shown to induce autophagy, and apoptosis, and inhibit invasion and proliferation in lung adenocarcinoma cells [4–6]. Furthermore, crocin, the major constituent of saffron, a spice derived from the flower of *Crocus sativus*, has been found to stimulate apoptosis and cell cycle arrest in lung adenocarcinoma cells [7]. These studies, amongst many, suggest that some compounds obtained from medicinal plants and herbs are excellent agents for the prevention and treatment of lung cancer.

The *Ocimum* genus belongs to the Lamiaceae family of aromatic mints, herbs, and shrubs that are native to tropical climates. The plants of this family possess unique pharmacological properties and they are often used as food and medicine. The extracts and essential oils of three well-known species in this family, *Ocimum basilicum*, *Ocimum sanctum*, and *Ocimum gratissimum*, are known to possess medicinal properties. Essential oils of *Ocimum basilicum* have demonstrated anti-fungal, anti-bacterial, and anti-oxidant capacity [8]. Specifically, when tested against liver carcinoma cells, ethanol leaf extracts of *O. bacilicum* were shown to possess free-radical scavenging and DNA protective properties, which may prevent carcinogenesis [9]. It has been reported that essential oils and ethanolic extracts from *O. sanctum* inhibit proliferation, trigger apoptosis, and impede angiogenesis in breast, prostate, and lung cancer cell lines [10–12].

The species of interest for this project, *Ocimum gratissimum* (*Og*), is a basil indigenous to West Africa [13]. Known as Clove Basil or Scent Leaf, *Og* has been used extensively in African traditional medicinal systems for centuries. In coastal Nigeria, the plant is used to treat respiratory infections, conjunctivitis, epilepsy, high fever, and diarrhea. *Og* has been shown to have anti-oxidant, anti-microbial, anti-fungal, anti-proliferative, anti-diarrheal, antiviral, anti-hyperglycemic, anti-anemic, anti-hyperlipidemic, dermal-protective, and cytotoxic properties that contribute to its pharmacological value [14–21]. There is anecdotal evidence that the leaf extracts of *Og* shrink hemorrhoids while obviating associated bleeding and itching. It is, therefore, not surprising that crude aqueous extracts of *Og* leaves are reported to also possess anti-inflammatory properties [22]. Terpenoids, alkaloids, flavonoids, saponins, steroids, tannins, anthraquinones, and cardiac glycosides are among several biologically active constituents said to be responsible for the activity of *Og* [23].

Previous studies conducted in our laboratory have shown that aqueous and fractionated leaf extracts of *Og* inhibit the proliferation of prostate cancer cells [24,25]. Within the past decade, the anti-cancer activity of extracts of *Og* has also been reported in liver, breast, osteosarcoma, and cervical cell lines [26–29]. Most notably, the aqueous extract of *Og* has been shown to activate both intrinsic and extrinsic apoptosis pathways in (A549) lung adenocarcinoma cells [30]; however, the effects of fractionated extracts of *Og* leaves against these cells have not been reported. Therefore, this study evaluated the efficacy of fractionated leaf extracts of *Ocimum gratissimum* (OGFEs) against human lung adenocarcinoma (A549) cells in vitro.

2. Materials and Methods

2.1. Plant Preparation

Og was cultivated in the Jackson State University greenhouse. The plants were grown for approximately three to four months and were then harvested by cutting off the stems. The leaves were picked from the stems and air-dried at room temperature for ten days. The leaves were turned daily to prevent mold growth and to increase drying capacity. A conventional blender was used to grind the leaves into a fine powder which was then sifted using a sieve to remove unwanted woody fibers. The powdered leaves were stored at 4 °C until ready for use.

2.2. Extract Preparation and Preparation of Stock Solution

Og was fractionated by Dr. Heng-Shan Wang (Guangxi Normal University, Guilin, China). Ethyl acetate soluble extract P_2 was prepared as follows: 3 sets of 100 g of powdered *Og* leaves were extracted in 500 mL round-bottom flasks fitted with reflux condensers using 450 mL aliquots of 95% ethanol (EtOH) in a 50 °C water bath for 2 h. Extracts from all three flasks were vacuum-filtered, combined, diluted to 70% EtOH, and then incubated at −20 °C for 2 h to precipitate chlorophyll from the extract. The extract was vacuum-dried using a rotary evaporator. The extract was then completely re-dissolved in distilled water and acetone. The acetone was volatilized using the rotary evaporator at 50 °C with occasional rotation of the flask, yielding a dark brown crude ethanol extract. Using a 1 L separatory funnel, the crude ethanol extract was re-extracted with 250 mL ethyl acetate, collected, and then dried in a rotary evaporator, resulting in fraction P_2. This was then stored at −20 °C until further use.

Dr. Xuan Luo (Guangxi University, Nanning, China) provided the water-soluble fraction, PS/PT1. The fraction was obtained by first extracting 500 g of powdered *Og* with distilled water three times at 50 °C. The water solutions were combined and dried using a freeze dryer (FD-1A-50, Beijing Boyikang Experiment Instrument Co. Ltd., Liangxiang Industrial Development Zone, Beijing, China). The freeze-dried extract was re-extracted with anhydrous methanol three times. Any remaining solid residue was then dissolved in distilled water and the solution was vacuum-filtered. Anhydrous ethanol was added to the filtrate to a final concentration of 40%. The solution was centrifuged at 4500 rpm for five minutes to yield the extract PS/PT1.

The stock solutions were prepared as follows: P_2 was dissolved in 0.8% EtOH, and PS/PT1 was dissolved in sterile distilled water (Figure 1).

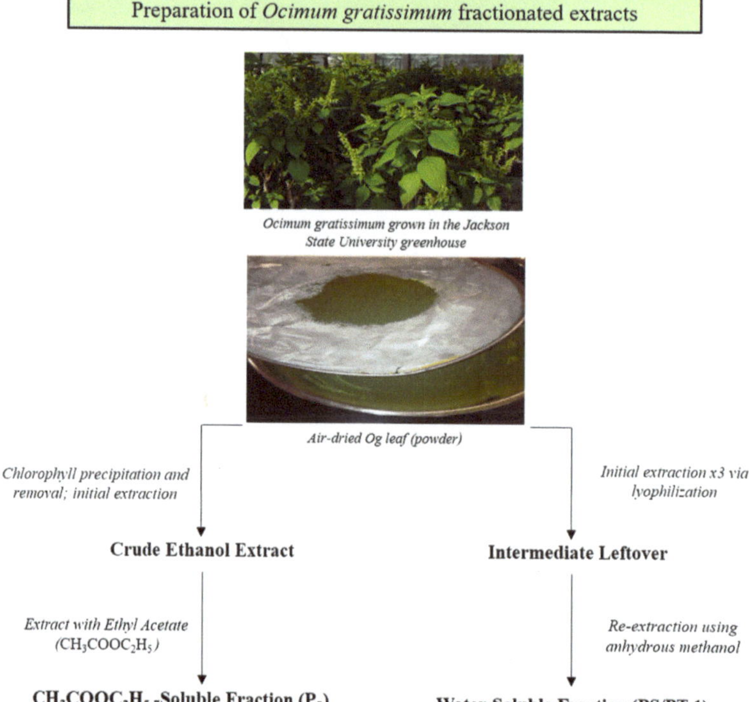

Figure 1. Preparation of *Ocimum gratissimum* fractionated extracts, P_2 and PS/PT1.

2.3. Cell Culture

A549 lung adenocarcinoma cells (Catalog #CCL-185) and F-12K medium were purchased from the American Type Culture Collection (Rockville, MD, USA). F-12K medium was supplemented with 10% fetal bovine serum (FBS) and 1% Penicillin/Streptomycin (Thermo Scientific, Waltham, MA, USA) to produce the complete growth medium (CGM).

A549 cells were cultured in Primaria™ tissue culture dishes, with dimensions of 100×20 mm, and maintained in a 5% CO_2 humidified incubator at 37 °C. CGM was aspirated and refreshed every 48 h and the cells were grown until they were 100% confluent. Cell detachment was performed by incubating the cells with 0.25% trypsin (Thermo Scientific, Waltham, MA, USA). Detached cells were collected into a sterile centrifuge tube and cell density was calculated using a hemocytometer for subsequent analysis.

2.4. Anti-Proliferation Studies

Two milliliters of A549 cell suspension were seeded in Primaria™ 6-well tissue culture plates (BD Biosciences, San Jose, CA, USA) at a density of 5×10^4 cells/mL. The cells were grown until they reached 65% confluency. Cells were treated with OGFEs at concentrations of 200, 300, 400, and 500 µg/mL and incubated again for 24 h. Cells grown in serum-free media served as the negative control and cells grown in CGM served as the positive control for testing with each fraction. For activity testing using P_2, CGM was supplemented with the vehicle, 0.8% EtOH, which served as a second positive control. Cells were washed with $1 \times$ PBS, harvested using a cell scraper, and collected into a sterile centrifuge tube. The density of viable cells was then calculated directly, using a hemocytometer.

2.5. 4′,6-Diamidino-2-Phenylindole (DAPI) Staining

DAPI staining was performed to observe chromatin condensation and nuclear morphological changes in A549 cells treated with *Og* extracts. Three milliliters of A549 cells were seeded in NUNC™ glass chamber slides (Lab Tek, Nunc Products, Naperville, IL, USA) at a density of 5×10^4 cells/mL and grown to 65% confluency. The chamber slides were separated into two groups and were treated with 200 and 400 µg/mL OGFEs; one chamber slide remained untreated and served as the control. After 24 h of treatment, cells were washed with $1 \times$ PBS and fixed to each slide by incubating with 10% trichloroacetic acid (MP Biomedicals, Solon, OH, USA) (10 min, 4 °C). The cells were washed thrice with ice-cold $1 \times$ PBS before detaching the slide from the media chamber. Three drops of Prolong™ Gold Antifade reagent with DAPI solution (Invitrogen Corporation, Carlsbad, CA, USA) were added to the slide and allowed to cure for 24 h in the dark. Cells were visualized using an Olympus Epifluorescence Microscope (Center Valley, PA, USA) between 350 and 461 nm.

2.6. Western Immunoblot Analysis

For protein analysis, A549 cells were grown until approximately 90% confluent. Tissue culture dishes were separated into two groups and treated with OGFEs at concentrations of 200 and 400 µg/mL for 24 h; one dish remained untreated and served as the control. After incubation, cells were harvested and a pellet was obtained by centrifugation (2500 rpm, 5 min, 4 °C). The pellet was washed twice in ice-cold PBS, lysed in RIPA buffer, (20 mM Tris-HCl, pH 8.0, 0.5% (w/v) Nonidet P-40, 1 µg/mL leupeptin, 1.0 µg/mL pepstatin, 1 mM dithiothreitol, 1 mL PMSF, and 0.1 NaCl) vortexed, and then incubated on ice. The lysates were then clarified by centrifugation (13,500 rpm, 15 min, 4 °C). The supernatant was collected into sterile microcentrifuge tubes and protein concentration was determined using the Bicinchoninic Acid method. Cell lysates containing 100 µg of proteins were subjected to electrophoresis in a 10% SDS–polyacrylamide gel, using a BioRad® Mini-PROTEAN Tetra Gel Electrophoresis System at 100 V (BioRad Corporation, Hercules, CA, USA). Protein was then slowly transferred to a polyvinylidene difluoride membrane by electroblotting overnight. The membrane was blocked with 5% (w/v) biotin-free non-fat dry milk and probed with monoclonal mouse anti-Caspase-8 (97446 s), anti-BH3 interacting-domain death agonist (BID) (8762), anti-Cytochrome C (12963), anti-Caspase-9 (9508 s), and anti-Caspase 3 (9668 s) (Cell Signaling, Technology, Danvers, MA) at a 1:1000 dilution overnight at 4 °C. The secondary antibody consisted of HRP-conjugated anti-mouse whole IgG used at a 1:10,000 dilution for 1 h at room temperature. Protein bands were then visualized using an enhanced chemiluminescence (ECL) detection system (GE Healthcare, Little Chalfont, Buckinghamshire, England).

2.7. Statistical Analysis

Quantitative data obtained from growth inhibition and DAPI staining experiments were presented as means $\pm$ SDs of triplicate counts. A p value < 0.05 was considered significantly different from the control according to Dunnett's test.

3. Results

3.1. Determination of Anti-Proliferative Activity of OG Extracts on A549 Cells

Figure 2 shows the results of the anti-proliferative activity of P_2 and PS/PT1 OGFEs against A549 lung adenocarcinoma cells. The results represent the mean of triplicate counts $\pm$ SD. Exposure to all concentrations of P_2 reduced cell proliferation in a concentration-dependent manner. Treatment with PS/PT1 resulted in a significant decrease in cell viability at 500 µg/mL only.

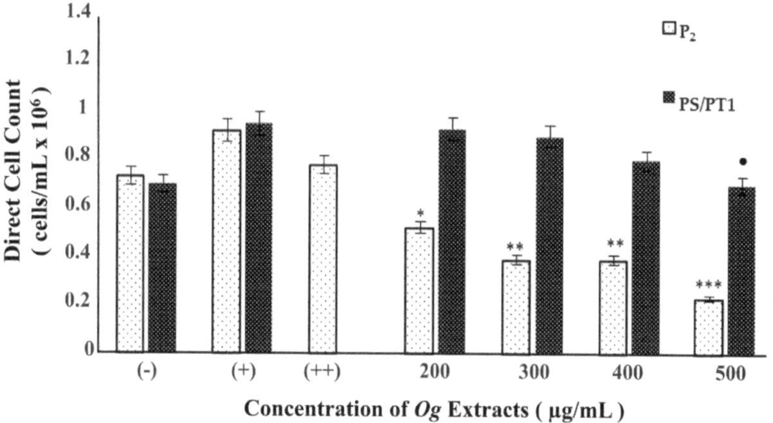

Figure 2. Determination of anti-proliferative activity of fractions P_2 and PS/PT1 on A549 cells after 24 h of treatment. p values denote the level of significance of each treatment, when compared to (++, P2) and (+, PS/PT1), using Dunnett's test. Corresponding p values: * = 0.003, ** = 0.0003, *** = 0.0007, • = 0.01.

3.2. Determination of DNA Condensation and Morphological Changes in A549 Cells by Og Extracts, through DAPI Staining

To evaluate the ability of OGFEs to induce chromatin condensation and nuclear morphological changes in A549 cells, DAPI staining was performed. As shown in Figure 3a, viable cells displayed a normal nuclear size and fluorescence. Cells treated with OGFEs displayed hyper-fluorescence, characterized by the cell nuclei undergoing fragmentation and forming apoptotic bodies. In the treatment with both fractions, the number of apoptotic bodies was proportional to the treatment concentration (Table 1).

Table 1. Quantification of apoptotic bodies in A549 cells exposed to fractions P_2 and PS/PT1. Data are expressed as means $\pm$ standard deviations.

Treatment	Mean Number of Apoptotic Bodies $\pm$ SD, Control	Mean Number of Apoptotic Bodies $\pm$ SD, 200 µg/mL	Mean Number of Apoptotic Bodies $\pm$ SD, 400 µg/mL
P_2	27 $\pm$ 3.06	55 $\pm$ 4.04	107 $\pm$ 11.06
PS/PT1	27 $\pm$ 3.06	34 $\pm$ 4.16	57 $\pm$ 2.89

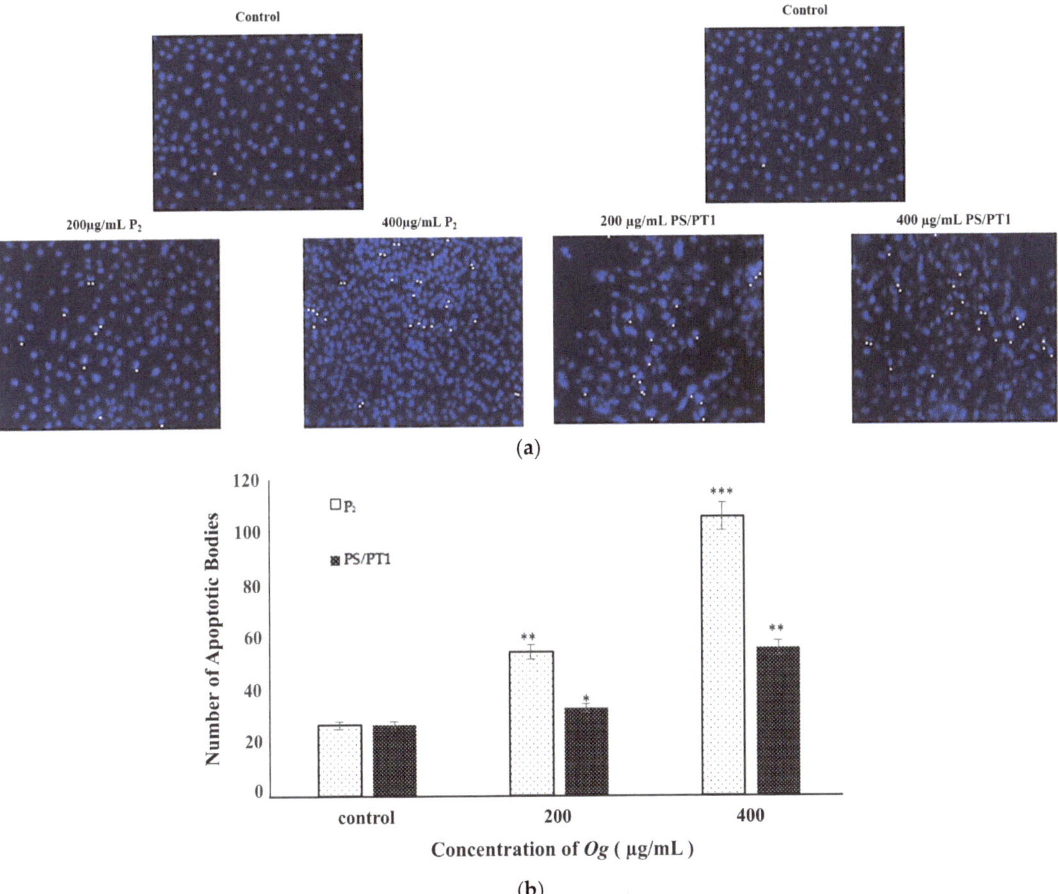

Figure 3. Determination of DNA condensation and nuclear morphological changes in A549 cells by fractions P$_2$ and PS/PT1 through 4′,6-diamidino-2-phenylindole staining. (**a**) Represents photographed cells after treatment. Hyperfluorescence is indicated by white arrowheads. (**b**) Graphical representation of apoptotic body formation in cells from treatment with OGFEs. Corresponding *p* values: * = 0.05, ** = 0.003, *** = 0.0002.

3.3. Assessment of Apoptotic Effects of Og Fractions by Western Immunoblot Analysis

To evaluate the apoptotic activity of OGFEs, Western blot analysis was performed. A549 cells were treated with P$_2$ and PS/PT1 extracts at concentrations of 200 µg/mL and 400 µg/mL for 24 h. Figure 4 displays the Western blots used to examine the levels of Caspase-8, BID, Cytochrome C, Caspase-9, and Caspase-3 proteins expressed in A549 cells treated with P$_2$ and PS/PT1 fractions, respectively. Beta-actin was used as the loading control for all Western blot experiments.

A concentration-dependent increase in Caspase-8 expression was shown in cells treated with P$_2$ and PS/PT1. Although P$_2$ was found to have a stronger effect on the upregulation of Caspase-8 than PS/PT1 at 200 µg/mL and 400 µg/mL concentrations, both extracts were found to induce the production of cleaved Caspase-8. Decreased levels of full-length BID after exposure to P$_2$ and PS/PT1 were also observed. Treatment with both extracts showed a noticeable degradation in this protein at 400 µg/mL. Both fractions were also found to increase the expression level of cytosolic Cytochrome C at 200 µg/mL and 400 µg/mL. Both P$_2$ and PS/PT1 also increased the expression levels of Caspase 3 and Caspase-9.

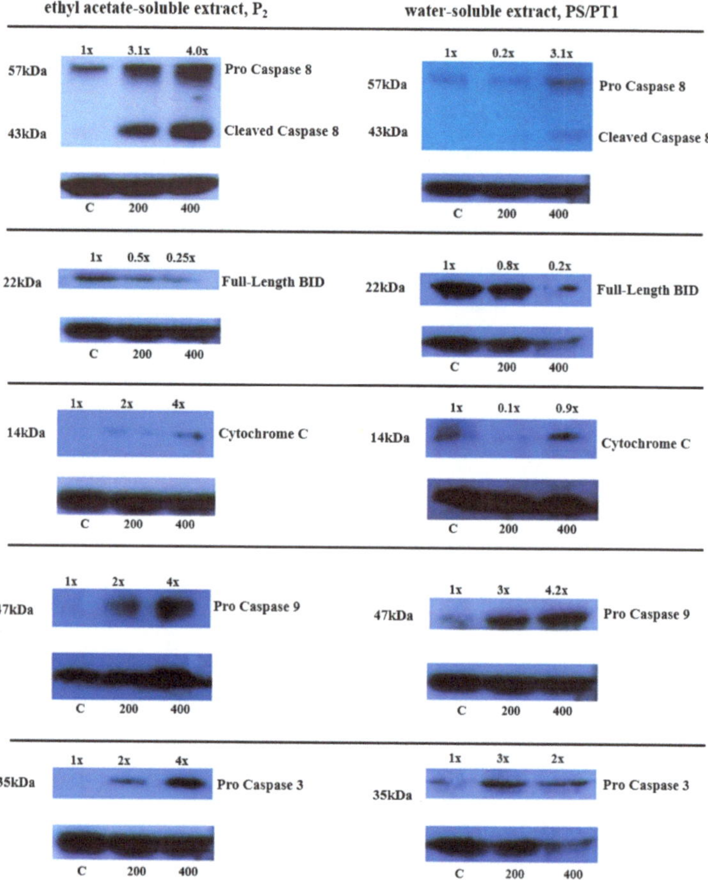

Figure 4. Evaluation of Caspase-8, BID, Cytochrome C, Caspase-9, and Caspase-3 proteins in A549 cells treated with fractions P$_2$ and PS/PT1 for 24 h. C = untreated cells. 200 and 400 = cells treated with 200 and 400 µg/mL, respectively.

4. Discussion

Lung cancer is the leading cause of global cancer incidence and mortality [31]. Although several studies in recent years have targeted potential therapies for the treatment of lung cancer, it remains a major health issue. The side effects that are associated with the established treatments for this disease can be toxic and debilitating for the patient, presenting a lack of selectivity for tumor cells, thereby negatively affecting healthy cells and tissues. Hence, there is a great need for less harsh, fortifying treatment agents, such as those harbored by plants and herbs.

Studies have shown that the extracts of various medicinal plants, bushes, and shrubs have profound effects on A549 cells [32,33]. It is believed that species within the *Ocimum* genus may qualify as potential candidates for the treatment of lung cancer, based on previous studies with various cancer cell lines [34–37]. Our laboratory has previously reported that fractionated extracts of *Ocimum gratssimum* leaf extract impede the growth of prostate cancer [24,25]. Furthermore, in A549 cells specifically, apoptotic signaling was induced by the crude aqueous extracts alone [30]. We previously determined the antiproliferative activity of OGFEs to be significantly greater than that of the crude aqueous extract on prostate cancer cell lines [24]; therefore, we sought to determine if the same effect would occur in A549 cells. This study has now demonstrated that fractionated extracts more effectively decrease the proliferation of A549 cells than the aqueous extract previously reported [30]. We have also shown that the fractionated extracts of *Og* also activate apoptosis pathways to suppress cell viability in lung adenocarcinoma cells.

A549 cells were exposed to OGFESs to determine their effectiveness for growth inhibition. We observed that OGFEs inhibit the proliferation of A549 cells in a concentration-dependent manner. Although PS/PT1 hindered the growth of A549 cells at 500 ug/mL only, P_2 had the strongest effect on growth inhibition, with a significant decrease in proliferation at all concentrations. Proliferation studies also determined the fractionated extracts to be more potent than the aqueous extract used in previous lung cancer studies [30], where significant cell death was observed at all concentrations using fraction P_2 and at 500 µg/mL using fraction PS/PT1.

During apoptosis, there are morphological characteristics exhibited by cells. These may include detachment, cell shrinkage, chromatin condensation, and membrane blebbing [38]. DAPI permeates the cell membrane and yields blue fluorescence in viable cells; however, this fluorescence is heightened in the presence of apoptotic cells, as the dye penetrates the compromised membrane [39]. We examined the apoptotic features of A549 cells exposed to OGFEs. While viable cells displayed a normal nuclear size and fluorescence (Figure 3a), those treated with P_2 and PS/PT1 exhibited hyperfluorescence, morphological alterations, and chromatin condensation in a concentration-dependent manner. Of all treatments, 400 µg/mL of fraction P_2 produced the most apoptotic bodies.

The two best-known pathways of apoptosis are the intrinsic (mitochondrial) and extrinsic (death receptor-dependent) pathways. Both pathways are activated in response to stress and may lead to cell death through cytoskeleton cleavage [40]. The extrinsic apoptotic pathway is initiated by the binding of extracellular ligands to cell-surface death receptors. Through this pathway, Caspase-8 is activated and may also activate the downstream executioner, Caspase-3 [41]. Evidence suggests that Caspase 8 also plays a role in the intrinsic pathway through its interaction with BID [42]. Once cleaved by Caspase-8, BID trans-locates to the mitochondria and assists with the outpour of Cytochrome C. Once released into the cytosol, Cytochrome C binds with APAF-1 and Caspase-9 to form an apoptosome trimer. Caspase-9 is then cleaved from the complex to activate Caspase-3. Once activated, Caspase-3 cleaves the actin cytoskeleton, causing the cell to collapse [43,44]. The Western blot results of the present study indicate that exposure to OGFEs induces the upregulation of cleaved Caspase-8 expression. Whether this expression was stimulated via the intrinsic or extrinsic pathway is inconclusive; however, our findings confirm that BID was indeed stimulated by Caspase-8 cleavage. A concentration-dependent decrease in full-length BID was also observed. This decrease is attributable to it changing into its cleaved form, which translocates to the mitochondrial membrane to amplify the signal in the aforementioned apoptotic cascade. Ultimately, treatment with OGFEs leads to the activation of the Caspase-3 protein. Taken together, these findings indicate that fractions P_2 and PS/PT1 upregulate the expression of pro-apoptotic proteins as a molecular mechanism for the inhibition of the proliferation of A549 cells.

5. Conclusions

The present study concludes that fractionated leaf extracts of *Ocimum gratissimum* inhibit the proliferation and strongly induce programmed cell death, as evidenced by DNA condensation, nuclear morphological changes, and upregulation of the expression of apoptotic proteins, in lung adenocarcinoma (A549) cells in a concentration-dependent manner. This is the first time that the effects of fractionated *Og* extracts have been evaluated on A549 cells. These results provide new insights into the therapeutic potential of *Og* and underscore the need for further pharmacological studies on the fractionation, isolation, purification, and testing of the anti-cancer activity of its active compounds.

Author Contributions: Conceptualization, R.M.C.; methodology, R.M.C., X.L., H.-S.W. and E.B.D.; formal analysis, R.M.C.; data curation, R.M.C. and E.B.D.; writing—review and editing, R.M.C., J.J.S. and P.B.T.; funding acquisition, P.B.T. All authors have read and agreed to the published version of the manuscript.

Funding: This work was supported by the National Institutes of Health (NIH) Grant No. 5U54MD015 929-05 through the RCMI Center for Health Disparities Research at Jackson State University, Jackson, Mississippi, USA; NIH Grant No. 5U54MD013376-05 through the RCMI Center for Health Disparities Research and Innovation at Morgan State University, Baltimore, Maryland, USA; NIH Grant No. 5 R25 GM067122 through the Research Initiative for Scientific Enhancement (RISE) Program at Jackson State University, Jackson, Mississippi, USA, and Grant No. XBZ 100817 through the Scientific Research Foundation of Guangxi University, Guangxi, China.

Institutional Review Board Statement: Not applicable.

Informed Consent Statement: Not applicable.

Data Availability Statement: The original contributions presented in the study are included in the article; further inquiries can be directed to the corresponding authors.

Acknowledgments: This manuscript is submitted in remembrance of the late Stephen I. N. Ekunwe and Gregorio B. Begonia. May their souls rest in peace.

Conflicts of Interest: The authors declare no conflicts of interest.

References

1. Lung Source: Globocan 2023 Number of New Cases in 2023, Both Sexes, All Ages. Available online: https://gco.iarc.fr/today (accessed on 2 July 2024).
2. Miller, K.D.; Nogueira, L.; Mariotto, A.B.; Rowland, J.H.; Yabroff, K.R.; Alfano, C.M.; Jemal, A.; Kramer, J.L.; Siegel, R.L. Cancer treatment and survivorship statistics, 2019. *CA Cancer J. Clin.* **2019**, *69*, 363–385. [CrossRef]
3. Kang, H.R.; Moon, J.Y.; Ediriweera, M.K.; Song, Y.W.; Cho, M.; Dharanibalan, K.; Cho, S.K. Dietary flavonoid myricetin inhibits invasion and migration of radioresistant lung cancer cells (A549-IR) by suppressing MMP-2 and MMP-9 expressions through inhibition of the FAK-ERK signaling pathway. *Food Sci. Nutr.* **2020**, *8*, 2059–2067. [CrossRef] [PubMed]
4. Lee, M.; Kim, K.S.; Fukushi, A.; Kim, D.H.; Kim, C.H.; Lee, Y.C. Transcriptional activation of human GD3 synthase (hST8sia i) gene in curcumin-induced autophagy in A549 human lung carcinoma cells. *Int. J. Mol. Sci.* **2018**, *19*, 1943. [CrossRef]
5. Jin, H.; Qiao, F.; Wang, Y.; Xu, Y.; Shang, Y. Curcumin inhibits cell proliferation and induces apoptosis of human non-small cell lung cancer cells through the upregulation of miR-192-5p and suppression of PI3K/Akt signaling pathway. *Oncol. Rep.* **2015**, *34*, 2782–2789. [CrossRef] [PubMed]
6. Lu, Y.; Wei, C.; Xi, Z. Curcumin suppresses proliferation and invasion in non-small cell lung cancer by modulation of MTA1-mediated Wnt/β-catenin pathway. *Vitr. Cell. Dev. Biol.—Anim.* **2014**, *50*, 840–850. [CrossRef] [PubMed]
7. Chen, S.; Zhao, S.; Wang, X.; Zhang, L.; Jiang, E.; Gu, Y.; Shangguan, A.J.; Zhao, H.; Lv, T.; Yu, Z. Crocin inhibits cell proliferation and enhances cisplatin and pemetrexed chemosensitivity in lung cancer cells. *Transl. Lung. Cancer Res.* **2015**, *4*, 775–783. [CrossRef]
8. Avetisyan, A.; Markosian, A.; Petrosyan, M.; Sahakyan, N.; Babayan, A.; Aloyan, S.; Trchounian, A. Chemical composition and some biological activities of the essential oils from basil *Ocimum* different cultivars. *BMC Complement. Altern. Med.* **2017**, *17*, 60. [CrossRef]
9. Thirugnanasampandan, R.; Jayakumar, R. Protection of cadmium chloride induced DNA damage by *Lamiaceae* plants. *Asian Pac. J. Trop. Biomed.* **2011**, *1*, 391–394. [CrossRef]
10. Manaharan, T.; Thirugnanasampandan, R.; Jayakumar, R.; Kanthimathi, M.S.; Ramya, G.; Ramnath, M.G. Purified essential oil from *Ocimum sanctum* Linn. triggers the apoptotic mechanism in human breast cancer cells. *Pharmacogn. Mag.* **2016**, *12*, S327–S331. [CrossRef]
11. Dhandayuthapani, S.; Azad, H.; Rathinavelu, A. Apoptosis Induction by *Ocimum sanctum* Extract in LNCaP Prostate Cancer Cells. *J. Med. Food.* **2015**, *18*, 776–785. [CrossRef]
12. Wihadmadyatami, H.; Hening, P.; Kustiati, U.; Kusindarta, D.L.; Triyono, T.; Supriatno, S. *Ocimum sanctum* Linn. Ethanolic extract inhibits angiogenesis in human lung adenocarcinoma (a549) cells. *Vet. World.* **2020**, *13*, 2028–2032. [CrossRef] [PubMed]
13. Paton, A. A Synopsis of *Ocimum* L. (*Labiatae*) in Africa. *Kew. Bull.* **1992**, *47*, 403–437. [CrossRef]
14. Onyebuchi, C.; Kavaz, D. Chitosan and N, N, N-trimethyl chitosan nanoparticle encapsulation of *Ocimum gratissimum* essential oil: Optimised synthesis, in vitro release and bioactivity. *Int. J. Nanomed.* **2019**, *14*, 7707–7727. [CrossRef] [PubMed]
15. Omodamiro, O.D.; Jimoh, M.A. Antioxidant and Antibacterial Activities of *Ocimum gratissimum*. *Am. J. Phytomedicine Clin. Ther.* **2015**, *3*, 10–19.
16. Akara, E.U.; Emmanuel, O.; Ude, V.C.; Uche-Ikonne, C.; Eke, G.; Ugbogu, E.A. *Ocimum gratissimum* leaf extract ameliorates phenylhydrazine-induced anaemia and toxicity in Wistar rats. *Drug. Metab. Pers. Ther.* **2021**, *36*, 311–320. [CrossRef]

17. Okoduwa, S.; Umar, I.; James, D.; Inuwa, H. Anti-Diabetic Potential of *Ocimum gratissimum* Leaf Fractions in Fortified Diet-Fed Streptozotocin Treated Rat Model of Type-2 Diabetes. *Medicines* **2017**, *4*, 73. [CrossRef]

18. Casanova, L.M.; DaSilva, D.; Sola-Penna, M.; Camargo, L.M.; Celestrini, D.; Tinoco, L.W.; Costa, S.S. Identification of chicoric acid as a hypoglycemic agent from *Ocimum gratissimum* leaf extract in a biomonitoring in vivo study. *Fitoterapia* **2014**, *93*, 132–141. [CrossRef] [PubMed]

19. Chang, S.H.; Liu, J.Y.; Hsiao, M.W.; Yang, H.L.; Wang, G.W.; Ye, J.C. Protective effects of *Ocimum gratissimum aqueous* extracts on hacat cells against uvc-induced inhibition of cell viability and migration. *Int. J. Med. Sci.* **2021**, *18*, 2086–2092. [CrossRef]

20. Chao, P.Y.; Lin, J.A.; Ting, W.J.; Lee, H.H.; Hsieh, K.; Chiu, Y.W.; Lai, T.J.; Hwang, J.M.; Liu, J.; Huang, C.Y. *Ocimum gratissmum aqueous* extract reduces plasma lipid in hypercholesterol-fed hamsters. *Int. J. Med. Sci.* **2016**, *13*, 819–824. [CrossRef]

21. Chao, P.Y.; Lin, J.A.; Ye, J.C.; Hwang, J.M.; Ting, W.J.; Huang, C.Y.; Liu, J.Y. Attenuation of oxidative stress-induced cell apoptosis in schwann RSC96 cells by *Ocimum gratissimum aqueous* extract. *Int. J. Med. Sci.* **2017**, *14*, 764–771. [CrossRef]

22. Ajayi, A.M.; Martins, D.T.; Balogun, S.O.; de Oliveira, R.G.; Ascencio, S.D.; Soares, I.M.; Barbosa, R.D.S.; Ademowo, O.D. *Ocimum gratissimum* L. leaf flavonoid-rich fraction suppress LPS-induced inflammatory response in RAW 264.7 macrophages and peritonitis in mice. *J. Ethnopharmacol.* **2017**, *204*, 169–178. [CrossRef]

23. Ajayi, A.M.; Tanayen, J.K.; Ezeonwumelu, J.; Dare, S.; Okwanachi, A.; Adzu, B. Ademowo. O.G. Anti-inflammatory, Anti-nociceptive and Total polyphenolic Content of Hydroethanolic Extract of *Ocimum gratissimum* L. *Afr. J. Med. Med. Sci.* **2014**, *43*, 215–224. [PubMed]

24. Ekunwe, S.I.N.; Thomas, M.; Luo, X.; Wang, H.S.; Chen, Y.; Zhang, X.; Begonia, G.B. Potential cancer-fighting *Ocimum gratissimum* (*Og*) leaf extracts: Increased anti-proliferation activity of partially purified fractions and their spectral fingerprints. *Ethn. Dis.* **2010**, *20*, 12–16.

25. Ekunwe, S.I.N.; Hall, S.M.; Luo, X.; Wang, H.; Begonia, G.B. Fractionated *Ocimum gratissimum* leaf extract inhibit prostate cancer (PC3•AR) cells growth by reducing androgen receptor and survivin levels. *J. Health Care Poor Underserved.* **2013**, *24*, 61–69. [CrossRef] [PubMed]

26. Huang, C.C.; Hwang, J.M.; Tsai, J.H.; Chen, J.H.; Lin, H.; Lin, G.J.; Yang, H.L.; Liu, J.Y.; Yang, C.Y.; Ye, J.C. Aqueous *Ocimum. gratissimum* extract induces cell apoptosis in human hepatocellular carcinoma cells. *Int. J. Med. Sci.* **2020**, *17*, 338–346. [CrossRef] [PubMed]

27. Nangia-Makker, P.; Raz, T.; Tait, L.; Shekhar, M.P.V.; Li, H.; Balan, V.; Makker, H.; Fridman, R.; Maddipati, K.; Raz, A. *Ocimum gratissimum* retards breast cancer growth and progression and is a natural inhibitor of matrix metalloproteases. *Cancer Biol. Ther.* **2013**, *14*, 417–427. [CrossRef] [PubMed]

28. Lin, C.C.; Chao, P.Y.; Shen, C.Y.; Shu, J.J.; Yen, S.K.; Huang, C.Y.; Liu, J.Y. Novel target genes responsive to apoptotic activity by *Ocimum gratissimum* in human osteosarcoma cells. *Am. J. Chin. Med.* **2014**, *42*, 743–767. [CrossRef] [PubMed]

29. Chang, W.C.; Hsieh, C.H.; Hsiao, M.W.; Lin, W.C.; Hung, Y.C.; Ye, J.C. Caffeic Acid Induces Apoptosis in Human Cervical Cancer Cells Through the Mitochondrial Pathway. *Taiwan J. Obstet. Gynecol.* **2010**, *49*, 419–424. [CrossRef] [PubMed]

30. Kao, S.H.; Chen, H.M.; Lee, M.J.; Kuo, C.Y.; Tsai, P.L.; Liu, J.Y. *Ocimum gratissimum aqueous* extract induces apoptotic signalling in lung adenocarcinoma cell A549. *Evid.-Based Complement. Altern. Med.* **2011**, *2011*, 739093. [CrossRef]

31. Lung Cancer Statistics. Available online: https://www.wcrf.org/cancer-trends/lung-cancer-statistics/ (accessed on 2 July 2024).

32. Yang, C.; Chen, H.; Chen, H.; Zhong, B.; Luo, X.; Chun, J. Antioxidant and anticancer activities of essential oil from gannan navel orange peel. *Molecules* **2017**, *22*, 1391. [CrossRef]

33. Liu, Y.; Liu, W.; Chen, C.; Xiang, Z.; Liu, H. A Cytotoxic Natural Product from *Patrinia villosa* Juss. *Anti-Cancer Agents Med. Chem.* **2019**, *19*, 1399–1404. [CrossRef] [PubMed]

34. Alkhateeb, M.A.; Al-Otaibi, W.R.; Al-Gabbani, Q.; Alsakran, A.A.; Alnafjan, A.A.; Alotaibi, A.M.; Al-Qahtani, W.S. Low-temperature extracts of Purple blossoms of basil (*Ocimum basilicum* L.) intervened mitochondrial translocation contributes prompted apoptosis in human breast cancer cells. *Biol. Res.* **2021**, *54*, 2. [CrossRef]

35. Luke, A.M.; Patnaik, R.; Kuriadom, S.T.; Jaber, M.; Mathew, S. An in vitro study of *Ocimum sanctum* as a chemotherapeutic agent on oral cancer cell-line. *Saudi. J. Biol. Sci.* **2021**, *28*, 887–890. [CrossRef]

36. Yang, D.; Zhang, X.; Zhang, W.; Rengarajan, T. Vicenin-2 inhibits Wnt/β-catenin signaling and induces apoptosis in HT-29 human colon cancer cell line. *Drug Des. Devel. Ther.* **2018**, *12*, 1303–1310. [CrossRef]

37. Harsha, M.; Mohan Kumar, K.; Kagathur, S.; Amberkar, V. Effect of *Ocimum sanctum* extract on leukemic cell lines:A preliminary in-vitro study. *J. Oral. Maxillofac. Pathol.* **2020**, *24*, 93–98. [CrossRef]

38. Yan, G.; Elbadawi, M.; Efferth, T. Multiple Cell Death Modalities and Their Key Features. *World. Acad. Sci. J.* **2020**, *2*, 39–48. [CrossRef]

39. Kapuscinski, J. DAPI: A DMA-Specific fluorescent probe. *Biotech. Histochem.* **1995**, *70*, 220–233. [CrossRef] [PubMed]

40. Slee, E.A.; Adrain, C.; Martin, S.J. Serial killers: Ordering caspase activation events in apoptosis. *Cell. Death. Differ.* **1999**, *6*, 1067–1074. [CrossRef] [PubMed]

41. Guicciardi, M.E.; Gores, G.J. Life and death by death receptors. *FASEB J.* **2009**, *23*, 1625–1637. [CrossRef]

42. Luo, X.; Budihardjo, I.; Zou, H.; Slaughter, C.; Wang, X. Bid, a Bcl2 interacting protein, mediates cytochrome c release from mitochondria in response to activation of cell surface death receptors. *Cell* **1998**, *94*, 481–490. [CrossRef]

43. Degterev, A.; Boyce, M.; Yuan, J. A decade of caspases. *Oncogene* **2003**, *22*, 8543–8567. [CrossRef] [PubMed]
44. Elmore, S. Apoptosis: A Review of Programmed Cell Death. *Toxicol. Pathol.* **2007**, *35*, 495–516. [CrossRef] [PubMed]

nutrients

MDPI

Article

Parecoxib Enhances Resveratrol against Human Colorectal Cancer Cells through Akt and TXNDC5 Inhibition and MAPK Regulation

Wan-Ling Chang [1], Kai-Chien Yang [2], Jyun-Yu Peng [1], Chain-Lang Hong [1], Pei-Ching Li [1], Soi Moi Chye [3], Fung-Jou Lu [4], Ching-Wei Shih [5] and Ching-Hsein Chen [5,*]

[1] Department of Anesthesiology, Chang Gung Memorial Hospital at Chiayi, No. 8, West Section of Jiapu Road, Chiayi County, Puzi City 613016, Taiwan; chjack1975@yahoo.com.tw (W.-L.C.); pengjyun550@gmail.com (J.-Y.P.); leisure@cgmh.org.tw (C.-L.H.); peiching@cgmh.org.tw (P.-C.L.)
[2] Department and Graduate Institute of Pharmacology, College of Medicine, National Taiwan University, No. 1, Jen Ai Road Section 1, Taipei 100233, Taiwan; kcyang@ntu.edu.tw
[3] School of Health Science, Division of Applied Biomedical Science and Biotechnology, IMU University, Bukit Jalil, Kuala Lumpur 57000, Malaysia; chye_soimoi@imu.edu.my
[4] Institute of Medicine, Chung Shan Medical University, No. 110, Section 1, Jianguo North Road, Taichung City 402306, Taiwan; fjlu@csmu.edu.tw
[5] Department of Microbiology, Immunology and Biopharmaceuticals, College of Life Sciences, National Chiayi University, A25-303 Room, Life Sciences Hall, No. 300, Syuefu Road, Chiayi City 600355, Taiwan; a0958798219@gmail.com
* Correspondence: chench@mail.ncyu.edu.tw

Abstract: In this study, we discovered the mechanisms underlying parecoxib and resveratrol combination's anti-cancer characteristics against human colorectal cancer DLD-1 cells. We studied its anti-proliferation and apoptosis-provoking effect by utilizing cell viability 3-(4,5-Dimethylthiazol-2-yl)-2,5-diphenyltetrazolium bromide (MTT) assay, fluorescence microscope, gene overexpression, Western blot, and flow cytometry analyses. Parecoxib enhanced the ability of resveratrol to inhibit cell viability and increase apoptosis. Parecoxib in combination with resveratrol strongly enhanced apoptosis by inhibiting the expression of thioredoxin domain containing 5 (TXNDC5) and Akt phosphorylation. Parecoxib enhanced resveratrol-provoked c-Jun *N*-terminal kinase (JNK) and p38 phosphorylation. Overexpression of TXNDC5 and repression of JNK and p38 pathways significantly reversed the inhibition of cell viability and stimulation of apoptosis by the parecoxib/resveratrol combination. This study presents evidence that parecoxib enhances the anti-cancer power of resveratrol in DLD-1 colorectal cancer cells via the inhibition of TXNDC5 and Akt signaling and enhancement of JNK/p38 MAPK pathways. Parecoxib may be provided as an efficient drug to sensitize colorectal cancer by resveratrol.

Keywords: parecoxib; resveratrol; TXNDC5; Akt; MAPK; apoptosis; colorectal cancer

Citation: Chang, W.-L.; Yang, K.-C.; Peng, J.-Y.; Hong, C.-L.; Li, P.-C.; Chye, S.M.; Lu, F.-J.; Shih, C.-W.; Chen, C.-H. Parecoxib Enhances Resveratrol against Human Colorectal Cancer Cells through Akt and TXNDC5 Inhibition and MAPK Regulation. *Nutrients* **2024**, *16*, 3020. https://doi.org/10.3390/nu16173020

Academic Editors: Lynnette Ferguson and Manohar Garg

Received: 15 August 2024
Revised: 29 August 2024
Accepted: 4 September 2024
Published: 6 September 2024

1. Introduction

"Dietary phytochemicals" refer to compounds in fruits, vegetables, grains, and other foods. They are different from vitamins and minerals. Phytochemicals are "non-nutrients"—compounds which are not essential to the organism, but may play a very beneficial role to the body, including prevention of non-communicable diseases, like cardiac vascular diseases, cancer, diabetes, etc. These chemicals have become hot topics for medical research. What excites scientists most is that the dietary phytochemicals in plant foods have a remarkably ability to inhibit many cancer cells [1–3]. As a result, scientists are working to uncover the functions of dietary phytochemicals in plant foods to provide evidence for cancer prevention.

Resveratrol (3,4′,5-trihydroxy-*trans*-stilbene) is generated by certain plants including numerous dietary sources such as peanuts, grapes, blueberries, raspberries, apples, plums, and products derived from them (e.g., wine) [4]. Resveratrol can be isolated and purified from these biological sources or synthesized in a few steps with an overall high yield [4]. The beneficial intake of resveratrol is between 30 and 150 mg [5]. A dietary evaluation was performed to show that the daily resveratrol intake by Chinese people through everyday foods was only 0.783 mg, which was significantly lower than the beneficial dose [5]. Among the chief food kinds, fruits appeared as the primary source of resveratrol, contributing to 88.35% of the total intake [5]. Resveratrol-enriched supplements might be appropriate to permit a daily intake of therapeutically related doses (currently presumed to be 1 g) that are not offered by beverages or conventional foods [4]. Extensive metabolism in the liver and intestine causes an oral bioavailability considerably lower than 1% [6]. Metabolic studies, both in urine and in plasma, have uncovered the major metabolites of resveratrol to be glucuronides and sulfates [6]. Other study demonstrates that no phase I metabolites were detected, but the phase II conjugates resveratrol-3-glucuronide and resveratrol-3-sulfate was found based on LC-MS and LC-MS-MS analysis and comparison with synthetic standards. Although these data point to resveratrol diffusing quickly across the intestinal epithelium, broad phase II metabolism during absorption might decrease resveratrol bioavailability [6,7]. The major sites of resveratrol metabolism include the intestine and liver [6]. There is no first passing effect. The resveratrol is absorbed in the small intestine and sent to the liver for phase II enzyme metabolism and to produce glucuronides and sulfates of resveratrol [6,7]. Resveratrol occurs in the form of two geometric isomers—*trans* and *cis*—and only the *trans* geometric isomer shows biological activity [8]. As a natural phytoalexin, resveratrol is a secondary metabolite derived from plant resistance to pathogenic attack and environmental stress [9]. It was first separated from the roots of *Veratrum album* in 1940 and was later isolated from the roots of *Polygonum cuspidatum* in 1963 [10]. Resveratrol is a dietary phytochemical with potential to improve cancer therapy and it indicates advances in cancer therapy. Resveratrol can induce DNA damage-mediated senescence in breast and liver cancer cells [11]. It has gained increasing interest because of its anti-cancer activities, low toxicity, proliferation suppression [12], apoptosis stimulation [13], autophagic death [14], and anti-metastasis effect [15]. The anti-cancer effects of resveratrol have been demonstrated in various cancer types. In colorectal cancer, resveratrol could inhibit the TGF-β-stimulated epithelial to mesenchymal transition and decrease the metastatic rate of lung and liver [16]. In breast cancer, resveratrol has the potential to suppress cell growth, and can induce apoptosis by increasing cytochrome c release, Bax/Bcl-xL ratio, and the cleavage of caspase 3 and PARP [13]. In human pancreatic cancer cells, resveratrol could induce apoptotic cell death by downregulating the anti-apoptotic protein MCL-1 [17]. Research shows that resveratrol has multiple anti-cancer mechanisms. It can promote apoptosis, which causes cancer cells to die on their own, and prevents their growth. In addition, resveratrol can protect white blood cells and endothelial cells from death due to oxidative stress caused by chemotherapeutical drugs or radiation therapy. Hence, resveratrol has potential value in inhibiting cancer cell growth.

The first-generation non-steroid anti-inflammatory drugs (NSAIDs) are mainly non-selective to inhibit both COX-1 and COX-2. Constitutive COX-1 is considered to mediate prostaglandin-dependent gastric protection [18]. COX-1 can assist the secretion of the gastric wall mucosa and prevent gastric acid from eroding the gastric wall mucosa [18]. Therefore, first-generation NSAIDs often have the serious side effect of gastric ulcers [19]. Selective inhibition of COX-2 is so important in comparison to the first generation NSAIDs, because gastric ulcers rarely occur. One of the serious side effects provided by COX-2 inhibitors is an increased risk of suffering myocardial infarction and death [20]. Rofecoxib was withdrawn from the market for this reason, but the similar COX-2 selective etoricoxib has replenished it in Europe but not in the United States [20]. Parecoxib is a generally prescribed analgesic and antipyretic drug and has an excellent selective inhibitory function on cyclooxygenase (COX)-2. Parecoxib can manage opioid-induced hyperalgesia [21]. Parecoxib diminishes postsurgical pain and accelerates movement more than the controlled analgesia of a patient [22]. Parecoxib can multimodal preemptive analgesia in reducing postoperative acute pain in hip and knee replacement patients, and decrease cumulative opioid consumption without increasing the risk of adverse drug events [23]. Parecoxib protects against myocardial ischemia/reperfusion via targeting the PKA-CREB signaling pathway [24]. Our previous investigations demonstrated that parecoxib possesses an anti-metastasis function by inhibiting the epithelial to mesenchymal transition and the Wnt/β-catenin signaling route in DLD-1 human colorectal cancer cells [25]. Parecoxib decreases glioblastoma migration, invasion, and cell proliferation by upregulating miRNA-29c [26]. Moreover, parecoxib combined with sufentanil inhibits the metastasis and proliferation of HER2-positive breast cancer cells by regulating the epithelial to mesenchymal transition [27], representing a potential synergistic antitumor effect. Recently, we demonstrated that parecoxib and 5-fluorouracil synergistically reduce the epithelial to mesenchymal transition and subsequent metastasis in colorectal cancer through targeting PI3K/Akt/NF-κB signaling [28]. These studies illustrate the potential of parecoxib to assist anti-cancer agents in enhancing anti-cancer effects.

TXNDC5 is a disulfide isomerase predominantly expressed in the endoplasmic reticulum [29]. Evidence indicates that TXNDC5 is upregulated by hypoxia in tumor endothelium and endothelial cells [30]. TXNDC5 is upregulated in several cancer types, such as lung, liver, esophageal, stomach, breast, uterine, and cervical carcinoma [31]. Previous studies validated that colorectal cancer tissues overexpress TXNDC5, and this overexpression is associated with poor clinical pathological characters in vivo. Moreover, an in vitro study showed that hypoxia provokes TXNDC5 production by upregulating HIF-1α; this consequence may elevate the survival and proliferation of colorectal cancer cells in a hypoxia environment. Under hypoxia, TXNDC5 performs as a prominent stress survival factor to induce the tumorigenesis of colorectal cancer by adjusting hypoxia-induced ROS/ER stress signaling [32]. However, no studies have evaluated the role of TXNDC5 or its related mechanisms on the combination of parecoxib and resveratrol on human colorectal cancer cells.

Akt, a proto-oncogene, belongs to the serine/threonine kinase family that adjusts downstream mediators and handles crucial metabolic procedures and cell survival [33]. Moreover, Akt stimulates cell cycle progress and prevents apoptosis. About 60–70% of Akt is highly activated in human colon cancers [34]. Targeting Akt signaling, from the viewpoint of discovering innovative molecular targets for cancer therapy, has revealed therapeutic inhibitors and candidates in critical pathways. The important characteristics for Akt kinase have provided an excellent target for acquiring therapeutic drugs for cancer [35].

Mitogen-activated protein kinase (MAPK) pathways are recognized kinase components that are related to cancer and have a critical function in signal transduction from the environment to the cell for controlling essential cellular progression processes, including differentiation, proliferation, apoptosis, and migration [36]. MAPKs contain three subgroups: JNK, p38, and extracellular signal-regulated kinase (ERK) [37]. Therefore, the discovery of new drugs to target MAPK pathways that can increase the capacity of chemotherapy to act on colorectal cancer homeostasis is noteworthy, because it may offer beneficial clinical consequences, enhance life quality during therapy, and diminish the side effects of chemotherapy.

This study investigated the synergistic anti-cancer effects of parecoxib and resveratrol combination on the cellular viability and apoptosis of human colorectal cancer cells at the cellular level and explored the underlying mechanisms. These findings may uncover the anti-cancer potential of resveratrol joined with clinically practicable concentrations of parecoxib in the therapy of colorectal cancer. This work offers a theoretical foundation for the clinical choice of a suitable chemotherapeutic regimen.

2. Materials and Methods

2.1. Reagents and Chemicals

RPMI-1640 medium was purchased from Hyclone (South Logan, UT, USA). Fetal bovine serum (FBS) and penicillin streptomycin–glutamine was obtained from Gibco Inc. (Freehold, NJ, USA). Resveratrol (3,4′,5-Trihydroxy-*trans*-stilbene), MTT, dimethyl sulfoxide (DMSO), crystal violet, trypan blue, 4,6-diamidino-2-phenylindole (DAPI), dichlorodihydrofluorescein diacetate (DCFH-DA), and other chemicals were acquired from Sigma-Aldrich Corp. (St. Louis, MO, USA). Parecoxib was supplied by Pfizer (Sydney, NSW, Australia). The Bio-Rad protein assay kit was bought from Bio-Rad Laboratories (Richmond, CA, USA). X-tremeGENE™ HP DNA transfection reagent was provided by Roche (Raleigh, NC, USA). Primary antibodies against p53, Bax, PARP, p-Akt, GAPDH and tubulin were bought from Santa Cruz Biotechnology, Inc. (Santa Cruz, CA, USA). Primary antibodies against TXNDC5 were purchased from Abcam (Waltham, MA, USA). The pLAS2w.pPuro and pLAS2w.pPuro-hTXNDC5 vectors were obtained from Addgene Company (Watertown, MA, USA).

2.2. Cell Culture

Human colorectal cancer cell line DLD-1 (BCRC No. 60132) was purchased from Bioresource Collection and Research Center (BCRC, Hsinchu, Taiwan). DLD-1 cells were cultured in RPMI-1640 medium. The medium was complemented with 10% fetal bovine serum (FBS), 2 mM L-glutamine, 100 units/mL penicillin G, and 100 µg/mL streptomycin. All cells were kept at 37 °C in a 5% CO_2 incubator. Stock solutions of resveratrol and parecoxib were dissolved in DMSO, and all treated concentrations were adjusted in the culture medium. The concentration of DMSO did not go beyond 0.05%.

2.3. MTT Assay for Cell Viability

MTT assay was conducted to measure cell viability. About 4×10^4 cells/well in 0.5 mL of RPMI-1640 medium were cultured in 24-well plates. After growth overnight, DLD-1 cells were incubated with parecoxib (3 µM), resveratrol (50, 100, and 200 µM), and parecoxib (3 µM) combined with resveratrol (50, 100, and 200 µM) for 48 h. The plates were then added with 0.5 mg/mL MTT solution and incubated at 37 °C for another 2 h. The supernatant was separated, and the formazan crystals were dissolved in 1 mL of DMSO. An aliquot of the DMSO lysed solution (200 µL) was obtained from the 24-well plates and transmitted to 96-well reader plates. Optical density (OD) was assessed with a microplate reader (Bio-Rad, Richmond, CA, USA) at 570 nm.

2.4. Isobologram Analysis for Synergistic Anti-Cancer Effect

Isobologram analysis [38] was carried out to determine the synergistic anti-cancer effect of a combination of resveratrol and parecoxib. Cell viability after treatment with 0, 1, and 3 µM parecoxib and 0, 100, and 200 µM resveratrol treatment was verified after 48 h, and each concentration was then drawn on each axis of the graph. A diagonal line was plotted between the two concentration spots of each single concentration of resveratrol and parecoxib, denoting the line of additivity as a control. Several values set after treatments with various concentrations of resveratrol and parecoxib in combination were drawn as dots on the graph. The results show antagonism, additivity, or synergy when the dots are localized upon, on or under the diagonal line, respectively.

2.5. Overexpression of TXNDC5 in Cancer Cells

The pLAS2w.pPuro and pLAS2w.pPuro-hTXNDC5 vectors were transfected into cells by the X-treme transfection reagent. Approximately 4×10^4 DLD-1 colorectal cancer cells were cultured in a 6-well plate and placed in a 37 °C, 5% CO_2 incubator for 24 h. After that, 2 mL of serum-free cultured medium was replaced. About 0.2 mL of the serum-free medium was pipetted into the Eppendorf tube, added with 2 µg of vectors and 6 µL of X-tremeGENE™ HP DNA transfection reagent, carefully pipetted and mixed evenly. After reacting at room temperature for 15 min, about 400 μL was obtained and added to the dish. The mixture was shaken evenly, placed back into the incubator at 37 °C for 24 h, and added with 1 μg/mL puromycin antibiotics for 14 days to establish a stable TXNDC5 overexpression cell line. Cells were collected, and their total proteins were extracted to verify the expression level of TXNDC5 by Western blot analyses.

2.6. DAPI Staining for Chromatin Condensation and Fragmented Nucleus

DLD-1 colorectal cancer cells (2×10^5 cells/well) were cultured in 6-well plates. After growth overnight, the cells were incubated with parecoxib (3 μM), resveratrol (200 μM), parecoxib (3 μM)/resveratrol (200 μM) combination for 48 h. After drug treatment, the cells were washed with in PBS, fixed with 4% paraformaldehyde for 15 min at room temperature. Afterward, cells were stained with DAPI (1 μg/mL) for 5 min and exposed to three additional PBS washes. The condensed chromatin and fragmented nucleus were detected and photographed under $200\times$ magnification by a fluorescent microscope.

2.7. Western Blotting

DLD-1 colorectal cancer cells were planted in 6 cm dishes at a density of 5×10^5 cells/dish for 24 h and then cultured with various drugs as explained in figure legends. Cells were collected as programmed after treatment under various conditions, and total protein concentrations were evaluated by the Bio-Rad protein assay kit. Equivalent total proteins (20–50 μg) of cell lysates were separated through 12% SDS-PAGE and then transferred onto a PVDF membrane for 50–75 min. The PVDF membranes were maintained with 5% nonfat milk in PBST buffer for 1 h to block nonspecific binding. After blocking, the PVDF membranes were incubated with the following primary antibodies at 4 °C overnight: anti-PARP (1:1000), anti-Bax (1:500), anti-Bcl-2 (1:500), anti-GAPDH (1:500), anti-TXNDC5 (1:2500), anti-p53 (1:1000), anti-p-Akt (1:1000), anti-Akt (1:1000), anti-p-JNK (1:500), anti-JNK (1:500), anti-p-p38 (1:500), anti-p38 (1:500), anti-p-ERK (1:500), anti-ERK (1:500) and anti-Tubulin (1:500) antibodies. It was then immersed with secondary antibodies for 1 h at room temperature. The antigen–antibody complexes were evaluated by the enhanced chemiluminescence (Amersham Pharmacia Biotech, Piscataway, NJ, USA) using a chemiluminescence analyzer.

2.8. Intracellular ROS Analysis

The level of intracellular ROS was measured by DCFH–DA staining and flow cytometry by. DLD-1 cells were cultured in 6 cm dishes with a density of 4×10^5 cells/dish. Parecoxib (3 μM) alone, resveratrol (200 μM) alone, and parecoxib (3 μM) combined with resveratrol (200 μM) were added, and the cells were treated for 1, 3, and 48 h. After treatment, all cells were incubated with DCFH-DA (10 μM) for intracellular ROS level and determined by using a Backman Coulter cytoFLEX flow cytometer. Cells were treated with 2 mM H_2O_2 as the positive control of intracellular ROS. About 10,000 cells were collected and examined per experimental situation via mean fluorescent intensity.

2.9. Statistical Analysis

Statistical analysis was conducted using a Student's t-test with SigmaPlot 10.0 software. Data are presented as the mean $\pm$ standard deviation from at least three independent experiments. A p value < 0.05 was considered statistically significant.

3. Results

3.1. Parecoxib Enhances Resveratrol to Inhibit Cell Viability in DLD-1 Cells

The non-cytotoxic concentration of parecoxib 3 μM was used to combine with three concentrations of resveratrol and to assess cell viability. As shown in Figure 1, parecoxib combined with resveratrol (200 μM) significantly enhanced the suppression of cell viability compared with resveratrol (200 μM) alone at 48 h treatment. The results showed no significant inhibition in parecoxib combined with 50 and 100 μM resveratrol compared with treatment with 50 and 100 μM of resveratrol alone. However, 1 and 3 μM parecoxib were combined with 100 and 200 µM resveratrol, respectively, resulting in the synergistic effect of reducing cell viability in DLD-1 cells.

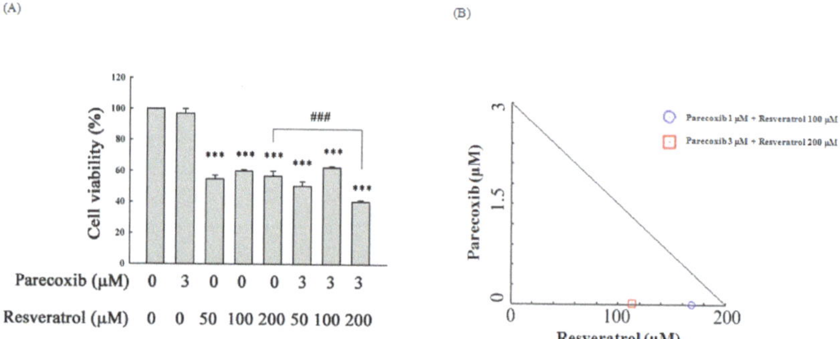

Figure 1. (**A**) Cell viability and (**B**) isobologram analysis in parecoxib and resveratrol treatment at 48 h. After drug treatment, cell viability was measured by MTT analysis. In isobologram analysis, the points under the backslash line show the synergistic effect. Significant differences in the untreated group (UN) and resveratrol are displayed as follows: $p < 0.001$ (***) and $p < 0.001$ (###), respectively.

3.2. Parecoxib Enhances Resveratrol to Induce Apoptosis in DLD-1 Cells

We used DAPI staining to measure nuclear morphological changes and chromatin condensation to demonstrate whether parecoxib can enhance the apoptosis of resveratrol. Parecoxib treatment alone did not cause significant changes in the number of DAPI positive cells in DLD-1 cells. After 48 h of treatment, the resveratrol (200 µM) treatment group had DAPI-positive cells. Combined treatment with 3 µM parecoxib and 200 µM resveratrol significantly increased the numbers of DAPI positive cells (Figure 2A,B). Taken together, these findings demonstrated that, compared with the individual monotherapies, parecoxib and resveratrol combination significantly enlarges the level of apoptosis in DLD-1 colorectal cancer cells.

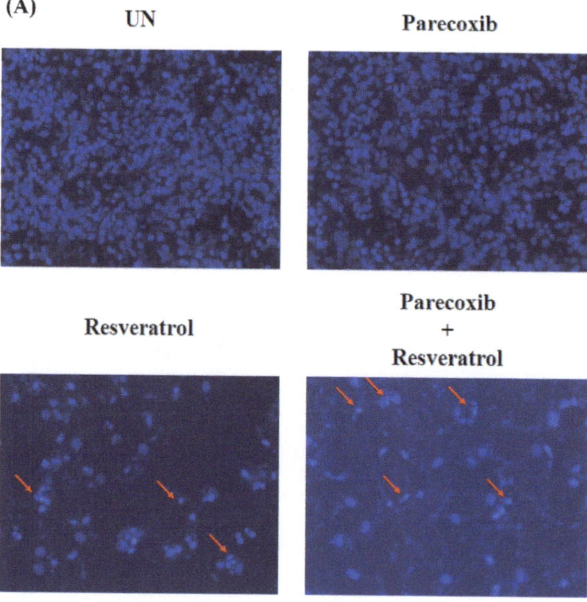

Figure 2. *Cont.*

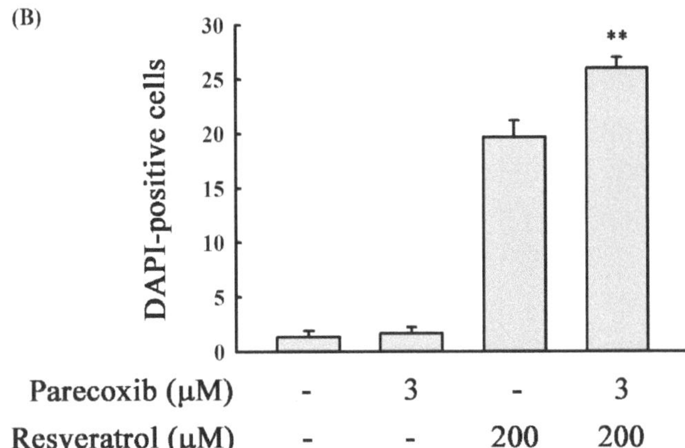

Figure 2. Effect of parecoxib and resveratrol on apoptotic morphology and apoptosis. (**A**) After treatment, chromatin condensation and fragmented nucleus were evaluated using DAPI staining and monitored with fluorescence microscopy (magnification 200×). Red arrows indicate the DAPI positive cells. (**B**) Number of DAPI-positive apoptotic cells per slide was calculated by counting apoptotic cells in five different fields. Each value indicates a mean ± SD (n = 5). Significant difference in the resveratrol is displayed as follows: $p < 0.01$ (**).

3.3. Parecoxib Enhances Resveratrol to Induce Apoptotic Proteins in DLD-1 Cells

To further explore the apoptotic effect of drugs combination, we examined the expression of apoptotic proteins including cleaved PARP, p53, Bax, and Bcl-2 after 48 h treatments. As evaluated by Western blot, the cleaved PARP was markedly enhanced when DLD-1 cells were treated with parecoxib and resveratrol combination compared with resveratrol treatment alone (Figure 3). We next assessed the apoptotic pathway provoked by the combined treatment. Slightly increased levels were detected in the expression of protein p53 and Bax in DLD-1 cells treated with parecoxib alone. Treatment with resveratrol alone resulted in a distinctly increased expression of p53 and Bax and decreased Bcl-2 expression. However, the expression of p53 and Bax noticeably further prominently increased when cells were treated with parecoxib and resveratrol combination, as compared with those treated with each drug alone.

3.4. Role of the PI3K/Akt Signaling Pathway in the Combination Effects of Parecoxib and Resveratrol

To evaluate whether the PI3K/Akt signaling pathway was included in the synergistic effects of parecoxib and resveratrol, the expression of Akt phosphorylation in DLD-1 cells was assessed by Western blotting. As shown in Figure 4, the phosphorylation of Akt was diminished in cells treated with resveratrol and there was no obvious change after parecoxib treatment. However, the phosphorylation of Akt was distinctly lower in cells exposed to combination treatment when compared with resveratrol alone. Hence, parecoxib may exhibit the enhanced repression of resveratrol on the cell viability of colorectal cancer cells by suppressing the PI3K/Akt signaling pathway.

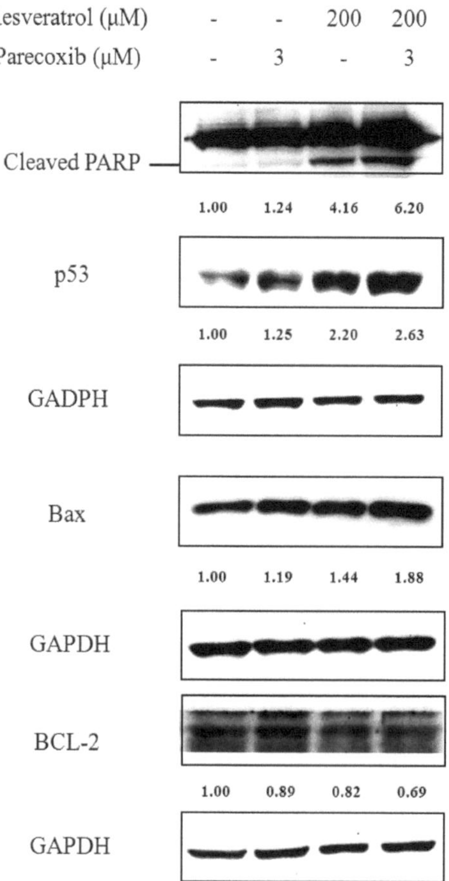

Figure 3. Effect of parecoxib and resveratrol on apoptotic proteins. After treatment, the levels of protein expression were measured using the extracted proteins and determined by Western blot. GAPDH were used as internal control.

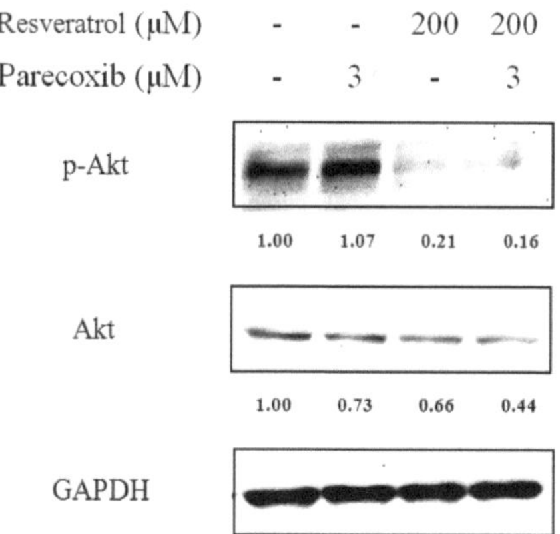

Figure 4. Effect of parecoxib and resveratrol on the expression of p-Akt and Akt. After treatment, the levels of protein expression were measured using the extracted proteins and determined by Western blot. GAPDH were used as internal control.

3.5. Role of TXNDC5 in the Combination Effects of Parecoxib and Resveratrol

To evaluate the role of TXNDC5 in the combination effects of parecoxib and resveratrol, the expression of TXNDC5 in DLD-1 cells were assessed by Western blotting. As shown in Figure 5A, the expression of TXNDC5 did not obviously change between cells treated with parecoxib alone and resveratrol alone. However, the expression of TXNDC5 was reduced to a greater extent after parecoxib and resveratrol combination treatment. Next, we evaluated whether TXNDC5 plays a critical role on the anti-cancer effect in drug combinations; the DLD-1 cells were transfected with an empty vector or TXNDC5 vector, and then we evaluated the expression of TXNDC5 by Western blot. As shown in Figure 5B, the expression of TXNDC5 was exhibited to a distinctly greater extent in TXNDC5 plasmid-transfected cells compared with empty plasmid-transfected cells. After treatment with parecoxib and resveratrol combination, the cell viability in TXNDC5 overexpressed DLD-1 cells was significantly increased compared with empty plasmid-transfected DLD-1 cells (Figure 5C). These results imply that TXNDC5 inhibition is a key mechanism in the anti-cancer effect of parecoxib and resveratrol combination.

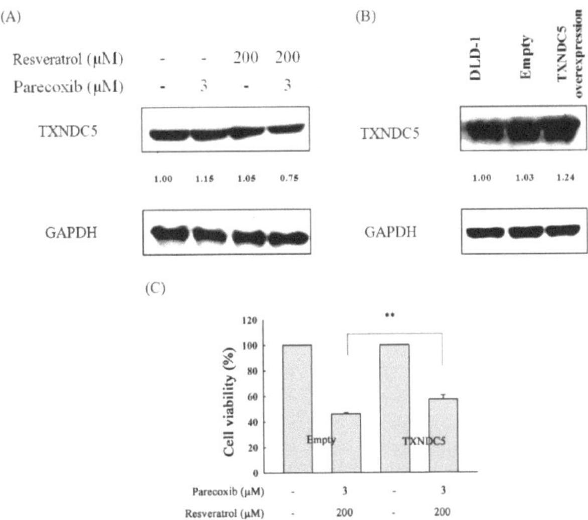

Figure 5. Effect of TXNDC5 overexpression in parecoxib- and resveratrol-treated DLD-1 cells. (**A,B**) Expression of TXNDC5 was measured by Western blot. GAPDH were chosen as loading control. (**C**) The cell viability was measured by MTT assay. These experiments were performed at least three times, and a representative experiment is presented. Data are shown as the mean ± SD of separate tests. Significant differences are expressed as *p* < 0.01 (**).

3.6. Apoptotic Role of TXNDC5 in the Combination Effects of Parecoxib and Resveratrol

We further evaluated the apoptotic role of TXNDC5 in the anti-cancer effect of parecoxib and resveratrol, the TXNDC5 vector-transfected DLD-1 cells and empty vector-transfected cells were treated with parecoxib and resveratrol combination, and then we assessed the expression of cleaved PARP and Bax by Western blotting. As shown in Figure 6, the expression of cleaved PARP and Bax was increased to 2.05-fold and 1.32-fold in empty plasmid-transfected cells treated with parecoxib and resveratrol combination compared with untreated empty plasmid-transfected cells. However, the expression of cleaved PARP and Bax decreased to 0.96-fold and 1.10-fold in TXNDC5 plasmid-transfected cells after parecoxib and resveratrol combination compared with untreated TXNDC5 plasmid-transfected cells. These results suggest that TXNDC5 inhibition is involved in the apoptotic event in the anti-cancer effect of parecoxib and resveratrol combination.

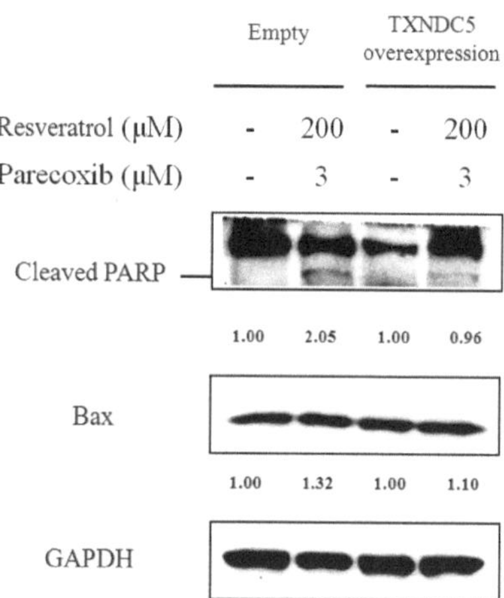

Figure 6. Role of TXNDC5 on apoptosis in parecoxib and resveratrol combination. After treatment, the levels of protein expression were measured using the extracted proteins and determined by Western blot. GAPDH were used as internal control.

3.7. Role of MAPK Signaling in the Combination Effects of Parecoxib and Resveratrol

The MAPK pathway is an important signaling pathway in the regulation of apoptosis. We further explored the expression of p-JNK, p-p38, and p-ERK by Western blot in DLD-1 cells treated with parecoxib alone, resveratrol alone, and parecoxib and resveratrol combination. As shown in Figure 7A, p-JNK expression was slightly increased in parecoxib alone and resveratrol alone treatment. Notably, parecoxib and resveratrol combination induced a large amount of p-JNK expression compared with resveratrol alone. The p-p38 expression was slightly increased in parecoxib alone. In resveratrol alone and drug combination treatment, the p-p38 expression was increased by more than 2.5-fold. The p-p38 expression in parecoxib and resveratrol combination is more than that in resveratrol alone. The p-ERK expression was increased more than 2.5-fold in parecoxib alone compared with the untreated group. Treatment with resveratrol alone and parecoxib and resveratrol combination resulted in a more than 5.5-fold increase in p-ERK expression. However, the p-ERK expression in the drug combination is slightly lower than treatment with resveratrol alone. To evaluate whether JNK and p38 signaling are included in the enhancement of cell viability inhibition and apoptosis of parecoxib and resveratrol combination, SP600125 (a JNK inhibitor) and SB203580 (a p38 inhibitor) were pretreated for 1 h, then treated with parecoxib and resveratrol combination for 48 h. After drug treatment, the cell viability and the expression of cleaved PARP and Bax were assessed by MTT assay and Western blot, respectively. As shown in Figure 7B,C, pretreatment of JNK and p38 inhibitors resulted in significantly recovered cell viability and a lower level of the expression of cleaved PARP and Bax compared with parecoxib and resveratrol combination. These results imply that p38 and JNK signaling participated in the enhancement of cell viability inhibition and apoptosis of parecoxib and resveratrol combination.

3.8. Effects of Parecoxib and Resveratrol on Intracellular ROS in DLD-1 Cells

ROS status is an important factor in cell proliferation. We evaluate whether parecoxib and resveratrol treatment affected the ROS status in DLD-1 cells. As shown in Figure 8A, there was no obvious change in intracellular ROS after treatment with parecoxib and resveratrol for 1 h. The intracellular ROS are significantly decreased to 55% and 60% after 3 h of resveratrol alone and combined treatment, respectively, compared to untreated cells. After 48 h of treatment, the intracellular ROS treated with resveratrol alone and combined treatment decreased below 50% (Figure 8B). There is no difference between resveratrol alone and the combination.

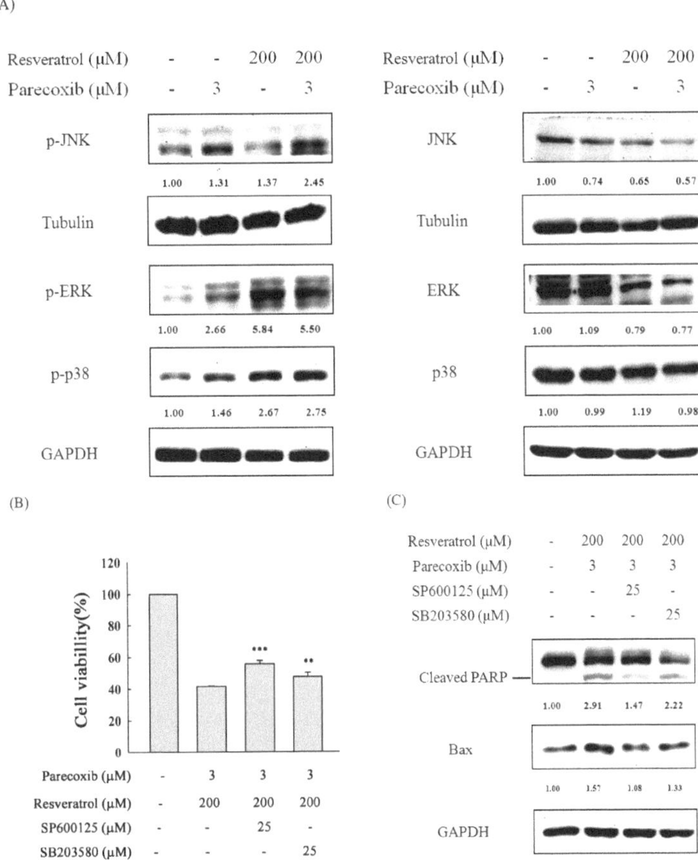

Figure 7. Effect of MAPKs signaling on apoptosis in parecoxib and resveratrol combination. (**A**) Phosphorylation and non-phosphorylation of MAPKs were measured by Western blot. GAPDH and tubulin were selected as loading control. (**B**) The cell viability was analyzed by MTT assay. These experiments were performed at least three times, and a representative experiment is presented. Data are shown as the mean $\pm$ SD of separate tests. Significant differences are expressed as $p < 0.01$ (******) and $p < 0.001$ (*******). (**C**) Cleaved PARP and Bax were detected by Western blot. GAPDH were selected as loading control.

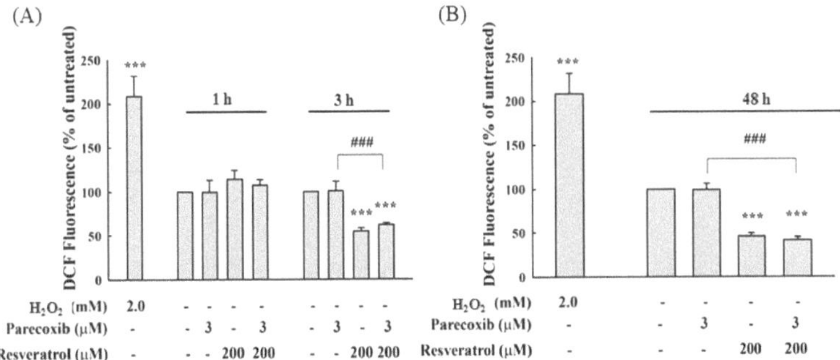

Figure 8. Effect of parecoxib and resveratrol on intracellular ROS in DLD-1 cells. (**A**) Treatment of 1 and 3 h; (**B**) 48 h treatment. After treatment, all cells were stained with DCFH-DA for intracellular

ROS detection and determined by a flow cytometer. H_2O_2 (2.0 mM) treatment was selected as an intracellular ROS positive control. The data are shown as the mean $\pm$ SD (n = 5–8) of individual experiments. Significant differences in the untreated group (UN) and parecoxib are shown as follows: $p < 0.001$ (***) and $p < 0.001$ (###).

4. Discussion

In Figure 1, the lowest concentration (50 μM) of resveratrol is already quite active in cell viability inhibition. However, the lowest concentration (50 μM) of resveratrol combined with a concentration (3 μM) of parecoxib could not exhibit more inhibition of cell viability compared with treatment with resveratrol (50 μM) alone. In contrast, the highest concentration of resveratrol (200 μM) combined with parecoxib (3 μM) appeared to significantly inhibit cell viability compared with treatment with resveratrol (200 μM) alone. For this reason, we used the highest concentration (200 μM) of resveratrol to combine with parecoxib (3 μM) in all subsequent experiments.

TXNDC5 abnormally appeared in several cancers, such as colorectal cancer [32]. Presently, TXNDC5 is thought of as a cancer-enhancing gene [39]. It can stimulate cell proliferation, promote cancer growth, suppress apoptosis, defend cells from oxidative stress, and accelerate the development of disease. Moreover, targeting TXNDC5 in the therapy of diseases has exhibited favorable treatment potentials. TXNDC5 can be exercised as a therapeutic target for cancers. Inhibition of TXNDC5 expression in various cancers such as gastric cancer [39], laryngeal squamous cell carcinoma [40], non-small cell lung carcinoma [41], pancreatic cancer [42], cervical cancer [31], liver cancer [43], and castration-resistant prostate cancer [44] can provoke cell apoptosis and reduce cell proliferation and migration. These results recommend that TXNDC5 can be used as a therapeutic target for cancers. Cetuximab could diminish TXNDC5 expression, thereby augmenting the generation of ROS and developing the endoplasmic reticulum stress-related apoptosis of laryngeal squamous cell carcinoma cells [40]. In addition, knockdown of TXNDC5 can result in clear cell renal cell carcinoma cells sensitive to chemotherapy drugs such as 5-fluorouracil and camptothecin and suppress the growth, migration, and invasion of clear cell renal cell carcinoma cells [45].

The anti-cancer action of resveratrol is to stimulate apoptosis in cancer cells, mediated by the protein p53. Resveratrol can affect the condition of cancer cells through the p53 pathway by enlarging the anti-colorectal cancer force of p53 [46]. In agreement with previous studies validating that resveratrol prompts p53-related apoptosis, resveratrol stimulated p53-related apoptosis in DLD-1 cells in the current study, as verified by the raised levels of PARP cleavage and Bax. TXNDC5 is an endoplasmic reticulum (ER) protein protective against ER stress-associated apoptosis [40]. However, our results showed that resveratrol did not increase TXNDC5 expression in DLD-1 cells; this may be due to the loss of a protective event of DLD-1 cells to provoke apoptosis under resveratrol treatment. The explicit effect of parecoxib was to enhance resveratrol-produced apoptosis, which was demonstrated by the following: (i) the significant augmentation of the cell viability inhibition and the apoptotic effect of resveratrol combined with parecoxib; (ii) parecoxib and resveratrol combination obstructed the expression of TXNDC5, and obviously augmented the apoptotic effect of resveratrol; and (iii) the inhibition of cell viability in the drug combination was recovered by overexpressing TXNDC5. These results suggest that TXNDC5 inhibition is a critical mechanism underlying the apoptotic effect of parecoxib and resveratrol combination in DLD-1 cells.

An important substrate of Akt that induces the mitochondrial apoptotic pathway is caspase-9, which is incapacitated by Akt by the phosphorylation at Ser196 [47]. Caspase-9 deactivation subsequently results in the deactivation of caspase-3 and repression of caspase-dependent apoptosis [48]. Moreover, stimulation of the Akt signaling pathway downregulates the expression of p53 and Bax. Consequently, Akt stimulates cell cycle progression, avoids apoptosis and enhances cell proliferation [49]. Resveratrol reduces the function of Akt and its downstream targets, hence stimulating apoptosis and cell cycle arrest, together with repressing cell proliferation in colon cancer cells [49]. Resveratrol enhances the sensitivity of ovarian cancer cells to cisplatin, causing them to be more sensitive to apoptotic cell killing. The effect of resveratrol was confirmed to initiate from its capacity to trigger p38 MAPK in particular, and diminish Akt activation [37]. Consistent with previous findings, our results found that resveratrol treatment alone diminished p-Akt expression and increased the expression of cleaved PARP, p53, and Bax. Although treatment with parecoxib alone did not affect the expression of p-Akt, it moderately increased the expression of p53 and slightly increased the expression of cleaved PARP and Bax. Parecoxib and resveratrol combination can better inhibit p-Akt expression and increase the expression of cleaved PARP, p53, and Bax compared with resveratrol alone. These results illustrate that parecoxib can enhance resveratrol-induced apoptosis by inhibiting the p-Akt signaling pathway, accompanied by increasing apoptotic-related proteins expression in colorectal cancer calls.

Recently, our study demonstrated that parecoxib can synergistically enhance 5-fluorouracil to repress metastasis in human colorectal cancer [28]. Resveratrol can moderate many cellular pathways correlated with tumorigenesis but it is less effective in colon cancer [50]. In human chronic myelogenous leukemia cells, resveratrol induces apoptosis via activating two MAPK family members, p38 and JNK, and preventing the activation of another MAPK family member, ERK [51]. We postulated that (1) parecoxib may enhance the anti-cancer effect of resveratrol in human colorectal cancer; (2) and resveratrol may employ like signaling pathways to disturb human colorectal cancer cells. Our results from the Western blot evaluation distinctly showed that resveratrol marked the induced phosphorylation of ERK and p38. However, the phosphorylation of JNK only slightly increased in treatment with resveratrol alone. When resveratrol was combined with parecoxib, the phosphorylation of p38 was further slightly increased. We also detected the activation of the ERK upon drug combination but detected no stronger enhancement in the phosphorylation level of the ERK kinase. Parecoxib and resveratrol combination enhanced the phosphorylation of JNK compared with resveratrol alone treatment. Moreover, SP600125 (a JNK inhibitor) and SB203580 (a p38 inhibitor) could alleviate the apoptosis and caused a rebound in the inhibition of cell viability caused by parecoxib and resveratrol combination. These results imply that parecoxib may particularly control the JNK and p38 signal transduction pathways to enhance the anti-cancer effect of resveratrol in human colorectal cancer. Parecoxib could enhance the effectiveness of resveratrol-based chemotherapy against human colorectal cancer. Celecoxib, a COX-2 inhibitor, can enhance apoptosis via upregulating the p-JNK and p-p38 pathway in liver cancer cells [52]. The enhancement of JNK activity under the influence of various anti-cancer compounds provokes the apoptosis of various human cancer cells [53–55]. Resveratrol was described to provoke the activation of JNK, which resulted in induced CHOP-related apoptosis in human colon cancer [56]. Our results indicate the augmented cytotoxic effects of parecoxib on colorectal cancer cell upon resveratrol treatment by the forceful activation of JNK and p38 pathways; as such, colorectal cancer cells become more sensitive to apoptotic cell death stimulation.

Hypoxia causes TXNDC5 expression by increasing hypoxia inducible factor-1α in vivo, thereby suppressing hypoxia-stimulated ROS/ER stress signaling and elevating the reproduction and survival of colorectal cancer cells [57]. These results indicate that there is some special relationship between TXNDC5 and intracellular ROS. Our present study shows that treating cells with resveratrol alone for 3 h and 48 h resulted in reducing intracellular ROS. This result indicates that the antioxidant property of resveratrol leads to the reduction in intracellular ROS. However, parecoxib treatment alone did not change the intracellular ROS levels in DLD-1 cells. Even if parecoxib and resveratrol combination cannot reduce intracellular ROS compared with resveratrol alone treatment, this result shows that the reduction in ROS in cells treated with combined drugs also comes from the antioxidant property of resveratrol. However, treatment with resveratrol alone did not diminish the expression of TXNDC5. Although the combined drugs treatment reduced TXNDC, the amount of intracellular ROS did not decrease. These results illustrate that the reduction in TXNDC5 in the drug combination treatment is not directly related to the change in intracellular ROS in DLD-1 colorectal cancer cells.

Resveratrol can target steroid receptors signaling and result in a potential anti-cancer effect in the treatment of hormone-dependent cancer [58]. We speculate that resveratrol may inhibit cell viability through targeting some unknown receptors in DLD-1 cells. In Figure 1A, the cell viability of cells treated with 100 µM of resveratrol alone is slightly higher than 50 µM of resveratrol alone. It is possible that the concentration of 50 µM resveratrol had reached a saturated state by binding to some unknown receptors, causing the greatest cell viability suppression, while the concentrations of 100 and 200 µM resveratrol might produce a competitive effect with receptors, resulted in an effect of inhibiting cell viability that is slightly less than the resveratrol of 50 µM. This speculation must be further studied in the future.

5. Conclusions

Parecoxib augments the sensitivity of colorectal cancer cells to resveratrol, causing them to become more sensitive to cell viability inhibition and apoptosis. Parecoxib provides a unique key which can not only enhance resveratrol in reducing cancer cell viability, but also enhance apoptosis via the suppression of Akt activation and TXNDC5 expression, upregulating the JNK and p38 MAPK signaling pathway, as well as increasing p53 and Bax expression and decreasing Bcl-2 expression in colorectal cancer, thereby enlarging cleaved PARP and enhancing apoptosis (Figure 9). The significant limitations of this study are that it is only an in vitro study, and that many various body, diet, and drug-dependent factors may have had a significant influence on such interactions. This study only has preliminary significance, showing that parecoxib and resveratrol combined treatment may offer a hopeful future for colorectal cancer patients. Much more research (especially in vivo studies) is needed to fully elucidate the importance of combining parecoxib with resveratrol in the future.

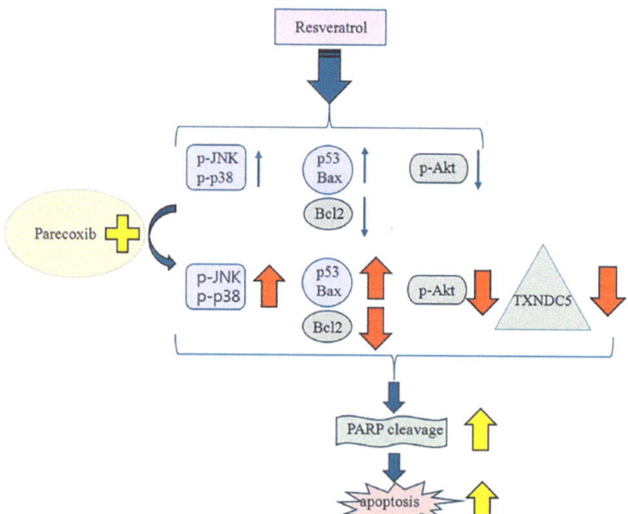

Figure 9. Proposed model of parecoxib and resveratrol combination that enhanced apoptosis in colorectal DLD-1 cancer cells.

Author Contributions: Conceptualization, C.-H.C. and W.-L.C.; methodology, K.-C.Y., J.-Y.P., C.-L.H., P.-C.L., C.-W.S., W.-L.C. and C.-H.C.; software, J.-Y.P., W.-L.C. and F.-J.L.; validation, C.-H.C. and W.-L.C.; resources, C.-L.H., P.-C.L. and W.-L.C.; data curation, W.-L.C., S.M.C. and C.-H.C.; writing—original draft preparation, W.-L.C. and C.-H.C.; writing—review and editing, C.-H.C.; supervision, C.-H.C. and W.-L.C.; project administration, C.-H.C. and W.-L.C.; funding acquisition, W.-L.C. All authors have read and agreed to the published version of the manuscript.

Nutrients 2024, 16, 3020

Funding: This work was supported by the Chang Gung Medical Foundation, Taiwan [grant CM-RPG6M0282 (W.-L.C.)].

Institutional Review Board Statement: Not applicable.

Informed Consent Statement: Not applicable.

Data Availability Statement: The data that support the findings of this study are contained within the article.

Conflicts of Interest: The authors declare no conflicts of interest.

References

1. Ye, Y.; Ma, Y.; Kong, M.; Wang, Z.; Sun, K.; Li, F. Effects of dietary phytochemicals on DNA Damage in cancer cells. *Nutr. Cancer* **2023**, *75*, 761–775. [CrossRef] [PubMed]
2. Tsai, M.C.; Chen, C.C.; Tseng, T.H.; Chang, Y.C.; Lin, Y.J.; Tsai, I.N.; Wang, C.C.; Wang, C.J. Hibiscus anthocyanins extracts induce apoptosis by activating AMP-activated protein kinase in human colorectal cancer cells. *Nutrients* **2023**, *15*, 3972. [CrossRef] [PubMed]
3. Yao, C.J.; Chang, C.L.; Hu, M.H.; Liao, C.H.; Lai, G.M.; Chiou, T.J.; Ho, H.L.; Kio, H.C.; Yang, Y.Y.; Wang-Peng, J.; et al. Drastic synergy of lovastatin and *Antrodia camphorata* extract combination against PC3 androgen-refractory prostatecancer cells, accompanied by AXL and stemness molecules inhibition. *Nutrients* **2023**, *15*, 4493. [CrossRef] [PubMed]
4. Weiskirchen, S.; Weiskirchen, R. Resveratrol: How much wine do you have to drink to stay healthy? *Adv. Nutr.* **2016**, *7*, 706–718. [CrossRef]
5. Xu, Y.; Fang, M.; Li, X.; Wang, D.; Yu, L.; Ma, F.; Jiang, J.; Zhang, L.; Li, P. Contributions of common foods to resveratrol intake in the Chinese diet. *Foods* **2024**, *13*, 1267. [CrossRef]
6. Walle, T. Bioavailability of resveratrol. *Ann. N. Y. Acad. Sci.* **2011**, *1215*, 9–15. [CrossRef]
7. Li, Y.; Shin, Y.G.; Yu, C.; Kosmeder, J.W.; Hirschelman, W.H.; Pezzuto, J.M.; van Breemen, R.B. Increasing the throughput and productivity of Caco-2 cell permeability assays using liquid chromatography-mass spectrometry: Application to resveratrol absorption and metabolism. *Comb. Chem. High. T Scr.* **2003**, *6*, 757–767. [CrossRef]
8. Fiod Riccio, B.V.; Fonseca-Santos, B.; Colerato Ferrari, P.; Chorilli, M. Characteristics, Biological properties and analytical methods of *trans*-resveratrol: A review. *Crit. Rev. Anal. Chem.* **2020**, *50*, 339–358. [CrossRef]
9. Langcake, P.; Pryce, R.J. The production of resveratrol by *Vitis vinifera* and other members of the Vitaceae as a response to infection or injury. *Plant Pathol.* **1976**, *9*, 77–86. [CrossRef]
10. Nonomura, S.; Kanagawa, H.; Makimoto, A. Chemical constituents of polygonaceous plants. I. studies on the components of ko-jo-kon. (*Polygonum cuspidatum* Sieb. et Zucc). *Yakugaku Zasshi* **1963**, *83*, 988–990. [CrossRef]
11. Jin, Y.; Liu, X.; Liang, X.; Liu, J.; Liu, J.; Han, Z.; Lu, Q.; Wang, K.; Meng, B.; Zhang, C.; et al. Resveratrol rescues cutaneous radiation-induced DNA damage via a novel AMPK/SIRT7/HMGB1 regulatory axis. *Cell Death Dis.* **2023**, *13*, 847. [CrossRef] [PubMed]
12. Fukuda, M.; Ogasawara, Y.; Hayashi, H.; Inoue, K.; Sakashita, H. Resveratrol inhibits proliferation and induces autophagy by blocking SREBP1 expression in oral cancer cells. *Molecules.* **2022**, *27*, 8250. [CrossRef] [PubMed]
13. Jang, J.Y.; Im, E.; Kim, N.D. Mechanism of resveratrol-induced programmed cell death and new drug discovery against cancer: A Review. *Int. J. Mol. Sci.* **2022**, *23*, 13689. [CrossRef] [PubMed]
14. Li, J.; Fan, Y.; Zhang, Y.; Liu, Y.; Yu, Y.; Ma, M. Resveratrol induces autophagy and apoptosis in non-small-cell lung cancer cells by activating the NGFR-AMPK-mTOR pathway. *Nutrients* **2022**, *14*, 2413. [CrossRef]
15. Wang, J.; Huang, P.; Pan, X.; Xia, C.; Zhang, H.; Zhao, H.; Yuan, Z.; Liu, J.; Meng, C.; Liu, F. Resveratrol reverses TGF-β1-mediated invasion and metastasis of breast cancer cells via the SIRT3/AMPK/autophagy signal axis. *Phytother. Res.* **2023**, *37*, 211–230. [CrossRef]
16. Ji, Q.; Liu, X.; Han, Z.; Zhou, L.; Sui, H.; Yan, L.; Jiang, H.; Ren, J.; Cai, J.; Li, Q. Resveratrol suppresses epithelial-to-mesenchymal transition in colorectal cancer through TGF-β1/Smads signaling pathway mediated Snail/E-cadherin expression. *BMC Cancer* **2015**, *15*, 97. [CrossRef]
17. Duan, J.J.; Yue, W.; E, J.Y.; Malhotra, J.; Lu, S.; Gu, J.; Xu, F.; Tan, X.-L. In vitro comparative studies of resveratrol and triacetyl-resveratrol on cell proliferation, apoptosis, and STAT3 and NFκB signaling in pancertic cancer cells. *Sci. Rep.* **2016**, *6*, 31672. [CrossRef]
18. Jackson, L.M.; Wu, K.C.; Mahida, Y.R.; Jenkins, D.; Hawkey, C.J. Cyclooxygenase (COX) 1 and 2 in normal, inflamed, and ulcerated human gastric mucosa. *Gut* **2000**, *47*, 762–770. [CrossRef]
19. Lehmann, F.S.; Beglinger, C. Impact of COX-2 inhibitors in common clinical practice a gastroenterologist's perspective. *Curr. Top. Med. Chem.* **2005**, *5*, 449–464. [CrossRef]
20. Stiller, C.O.; Hjemdahl, P. Lessons from 20 years with COX-2 inhibitors: Importance of dose-response considerations and fair play in comparative trials. *J. Intern. Med.* **2022**, *292*, 557–574. [CrossRef]
21. Kellett, E.; Berman, R.; Morgan, H.; Collins, J. Parecoxib for opioid-induced hyperalgesia. *BMJ Support. Palliat. Care.* **2021**, *11*, 126–127. [CrossRef]

22. Chiu, S.C.; Livneh, H.; Chen, J.C.; Chang, C.M.; Hsu, H.; Chiang, T.I.; Tsai, T.Y. Parecoxib reduced postsurgical pain and facilitated movement more than patient controlled analgesia. *Front. Surg.* **2022**, *9*, 799795. [CrossRef] [PubMed]
23. Ge, Z.; Li, M.; Chen, Y.; Sun, Y.; Zhang, R.; Zhang, J.; Bai, X.; Zhang, Y.; Chen, Q. The efficacy and safety of parecoxib multimodal preemptive analgesia in artificial joint replacement: A systematic review and meta-analysis of randomized controlled trials. *Pain Ther.* **2023**, *12*, 1065–1078. [CrossRef]
24. Gao, J.; Zhao, G.; Wang, P.; Li, S.; Wang, X.; Zhao, H. Parecoxib protects against myocardial ischemia/reperfusion via targeting PKA-CREB signaling pathway. *Panminerva Med.* **2022**, *64*, 587–588. [CrossRef] [PubMed]
25. Wong, C.H.; Chang, W.L.; Lu, F.J.; Liu, Y.W.; Peng, J.Y.; Chen, C.H. Parecoxib expresses anti-metastasis effect through inhibition of epithelial-mesenchymal transition and the Wnt/β-catenin signaling pathway in human colon cancer DLD-1 cell line. *Environ. Toxicol.* **2022**, *37*, 2718–2727. [CrossRef] [PubMed]
26. Li, L.Y.; Xiao, J.; Liu, Q.; Xia, K. Parecoxib inhibits glioblastoma cell proliferation, migration and invasion by upregulating miRNA-29c. *Biol. Open.* **2017**, *6*, 311–316. [CrossRef]
27. Xu, S.; Li, X.; Li, W.; Ma, N.; Ma, H.; Cui, J.; You, X.; Chen, X. Sufentanil combined with parecoxib sodium inhibits proliferation and metastasis of HER2-positive breast cancer cells and regulates epithelial-mesenchymal transition. *Clin. Exp. Metastasis.* **2023**, *40*, 149–160. [CrossRef]
28. Chang, W.L.; Peng, J.Y.; Hong, C.L.; Li, P.C.; Lu, F.J.; Chen, C.H. Parecoxib and 5-fluorouracil synergistically inhibit EMT and subsequent metastasis in colorectal cancer by targeting PI3K/Akt/NF-κB signaling. *Biomedicines* **2024**, *12*, 1526. [CrossRef]
29. Horna-Terrón, E.; Pradilla-Dieste, A.; Sanchez-de-Diego, C.; Osada, J. TXNDC5, a newly discovered disulfide isomerase with a key role in cell physiology and pathology. *Int. J. Mol. Sci.* **2014**, *15*, 23501–23518. [CrossRef]
30. Sullivan, D.C.; Huminiecki, L.; Moore, J.W.; Boyle, J.J.; Poulsom, R.; Creamer, D.; Barker, J.; Bicknell, R. EndoPDI. A novel protein-disulfide isomerase-like protein that is preferentially expressed in endothelial cells acts as a stress survival factor. *J. Biol. Chem.* **2003**, *278*, 47079–47088. [CrossRef]
31. Chang, X.; Xu, B.; Wang, L.; Wang, Y.; Wang, Y.; Yan, S. Investigating a pathogenic role for TXNDC5 in tumors. *Int. J. Oncol.* **2013**, *43*, 1871–1884. [CrossRef]
32. Tan, F.; Zhu, H.; He, X.; Yu, N.; Zhang, X.; Xu, H.; Pei, H. Role of TXNDC5 in tumorigenesis of colorectal cancer cells: In vivo and in vitro evidence. *Int. J. Mol. Med.* **2018**, *42*, 935–945. [CrossRef]
33. Cheaib, B.; Auguste, A.; Leary, A. The PI3K/Akt/mTOR pathway in ovarian cancer: Therapeutic opportunities and challenges. *R Chin. J. Cancer* **2015**, *34*, 4–16. [CrossRef] [PubMed]
34. Colakoglu, T.; Yildirim, S.; Kayaselcuk, F.; Nursal, T.Z.; Ezer, A.; Noyan, T.; Karakayali, H.; Haberal, M. Clinicopathological significance of PTEN loss and the phosphoinositide 3-kinase/Akt pathway in sporadic colorectal neoplasms: Is PTEN loss predictor of local recurrence? *Am. J. Surg.* **2008**, *195*, 719–725. [CrossRef]
35. Pal, I.; Mandal, M. PI3K and Akt as molecular targets for cancer therapy: Current clinical outcomes. *Acta Pharmacol. Sin.* **2012**, *33*, 1441–1458. [CrossRef] [PubMed]
36. Dhillon, A.S.; Hagan, S.; Rath, O.; Kolch, W. MAP kinase signalling pathways in cancer. *Oncogene* **2007**, *26*, 3279–3290. [CrossRef]
37. Hankittichai, P.; Thaklaewphan, P.; Wikan, N.; Ruttanapattanakul, J.; Potikanond, S.; Smith, D.R.; Nimlamool, W. Resveratrol enhances cytotoxic effects of cisplatin by inducing cell cycle arrest and apoptosis in ovarian adenocarcinoma SKOV-3 cells through activating the p38 MAPK and suppressing AKT. *Pharmaceuticals* **2023**, *16*, 755. [CrossRef] [PubMed]
38. Huang, R.Y.; Pei, L.; Liu, Q.; Chen, S.; Dou, H.; Shu, G.; Yuan, Z.X.; Lin, J.; Peng, G.; Zhang, W.; et al. Isobologram analysis: A comprehensive review of nethodology and current research. *Front. Pharmacol.* **2019**, *10*, 1222. [CrossRef]
39. Wu, Z.; Zhang, L.; Li, N.; Sha, L.; Zhang, K. An immunohistochemical study of thioredoxin domain-containing 5 expression in gastric adenocarcinoma. *Oncol. Lett.* **2015**, *9*, 1154–1158. [CrossRef]
40. Peng, F.; Zhang, H.; Du, Y.; Tan, P. Cetuximab enhances cisplatin-induced endoplasmic reticulum stress-associated apoptosis in laryngeal squamous cell carcinoma cells by inhibiting expression of TXNDC5. *Mol. Med. Rep.* **2018**, *17*, 4767–4776. [CrossRef]
41. Vincent, E.E.; Elder, D.J.; Phillips, L.; Heesom, K.J.; Pawade, J.; Luckett, M.; Sohail, M.; May, M.T.; Hetzel, M.R.; Tavaré, J.M. Overexpression of the TXNDC5 protein in non-small cell lung carcinoma. *Anticancer Res.* **2011**, *31*, 1577–1582.
42. Lee, S.O.; Jin, U.H.; Kang, J.H.; Kim, S.B.; Guthrie, A.S.; Sreevalsan, S.; Lee, J.S.; Safe, S. The orphan nuclear receptor NR4A1 (Nur77) regulates oxidative and endoplasmic reticulum stress in pancreatic cancer cells. *Mol. Cancer Res.* **2014**, *12*, 527–538. [CrossRef]
43. Zang, H.; Li, Y.; Zhang, X.; Huang, G. Circ_0000517 contributes to hepatocellular carcinoma progression by upregulating TXNDC5 via sponging miR-1296–5p. *Cancer Manag. Res.* **2020**, *12*, 3457–3468. [CrossRef] [PubMed]
44. Wang, L.; Song, G.; Chang, X.; Tan, W.; Pan, J.; Zhu, X.; Liu, Z.; Qi, M.; Yu, J.; Han, B. The role of TXNDC5 in castration-resistant prostate cancer—Involvement of androgen receptor signaling pathway. *Oncogene* **2014**, *34*, 4735–4745. [CrossRef]
45. Mo, R.; Peng, J.; Xiao, J.; Ma, J.; Li, W.; Wang, J.; Ruan, Y.; Ma, S.; Hong, Y.; Wang, C.; et al. High TXNDC5 expression predicts poor prognosis in renal cell carcinoma. *Tumor Biol.* **2016**, *37*, 9797–9806. [CrossRef]
46. Brockmueller, A.; Buhrmann, C.; Moravejolahkami, A.R.; Shakibaei, M. Resveratrol and p53: How are they involved in CRC plasticity and apoptosis? *J. Adv. Res.* **2024**, *1232*, 5. [CrossRef]
47. Sangawa, A.; Shintani, M.; Yamao, N.; Kamoshida, S. Phosphorylation status of Akt and caspase-9 in gastric and colorectal carcinomas. *Int. J. Clin. Exp. Pathol.* **2014**, *7*, 3312–3317.

48. Shen, B.; Chen, H.B.; Zhou, H.G.; Wu, M.H. Celastrol induces caspase-dependent apoptosis of hepatocellular carcinoma cells by suppression of mammalian target of rapamycin. *J. Tradit. Chin. Med.* **2021**, *41*, 381–389. [CrossRef]
49. Li, D.; Wang, G.; Jin, G.; Yao, K.; Zhao, Z.; Bie, L.; Guo, Y.; Li, N.; Deng, W.; Chen, X.; et al. Resveratrol suppresses colon cancer growth by targeting the AKT/STAT3 signaling pathway. *Int. J. Mol. Med.* **2019**, *43*, 630–640. [CrossRef] [PubMed]
50. Mohapatra, P.; Preet, R.; Choudhuri, M.; Choudhuri, T.; Kundu, C.N. 5-fluorouracil increases the chemopreventive potentials of resveratrol through DNA damage and MAPK signaling pathway in human colorectal cancer cells. *Oncol. Res.* **2011**, *19*, 311–321. [CrossRef] [PubMed]
51. Wu, X.; Xiong, M.; Xu, C.; Duan, L.; Dong, Y.; Luo, Y.; Niu, T.; Lu, C. Resveratrol induces apoptosis of human chronic myelogenous leukemia cells in vitro through p38 and JNK-regulated H2AX phosphorylation. *Acta Pharmacol. Sin.* **2015**, *36*, 353–361. [CrossRef] [PubMed]
52. Jia, Z.; Zhang, H.; Ma, C.; Li, N.; Wang, M. Celecoxib enhances apoptosis of the liver cancer cells via regulating ERK/JNK/P38 pathway. *J. BUON.* **2021**, *26*, 875–881. [PubMed]
53. Wu, J.; Wang, D.; Zhou, J.; Li, J.; Xie, R.; Li, Y.; Huang, J.; Liu, B.; Qiu, J. Gambogenic acid induces apoptosis and autophagy through ROS-mediated endoplasmic reticulum stress via JNK pathway in prostate cancer cells. *Phytother. Res.* **2023**, *37*, 310–328. [CrossRef] [PubMed]
54. Wang, Z.; Yu, K.; Hu, Y.; Su, F.; Gao, Z.; Hu, T.; Yang, Y.; Cao, X.; Qian, F. Schisantherin A induces cell apoptosis through ROS/JNK signaling pathway in human gastric cancer cells. *Biochem. Pharmacol.* **2020**, *173*, 113673. [CrossRef]
55. Puckett, D.L.; Alquraishi, M.; Alani, D.; Chahed, S.; Donohoe, D.; Voy, B.; Whelan, J.; Bettaieb, A. Zyflamend induces apoptosis in pancreatic cancer cells via modulation of the JNK pathway. *Cell Commun. Signal.* **2020**, *18*, 126. [CrossRef]
56. Woo, K.J.; Lee, T.J.; Lee, S.H.; Lee, J.M.; Seo, J.H.; Jeong, Y.J.; Park, J.W.; Kwon, T.K. Elevated gadd153/chop expression during resveratrol-induced apoptosis in human colon cancer cells. *Biochem. Pharmacol.* **2007**, *73*, 68–76. [CrossRef]
57. Wang, X.; Li, H.; Chang, X. The role and mechanism of TXNDC5 in diseases. *Eur. J. Med. Res.* **2022**, *27*, 145. [CrossRef]
58. De Amicis, F.; Adele Chimento, A.; Montalto, F.I.; Casaburi, I.; Rosa Sirianni, R.; Vincenzo Pezzi, V. Steroid receptor signallings as targets for resveratrol actions in breast and prostate cancer. *Int. J. Mol. Sci.* **2019**, *20*, 1087. [CrossRef]

Article

Methanolic Extract of *Cimicifuga foetida* Induces G$_1$ Cell Cycle Arrest and Apoptosis and Inhibits Metastasis of Glioma Cells

Chih-Hsuan Chang [1], Hung-Pei Tsai [2], Ming-Hong Yen [1] and Chien-Ju Lin [1,*]

[1] School of Pharmacy, College of Pharmacy, Kaohsiung Medical University, Kaohsiung 80756, Taiwan; sugar0306@gmail.com (C.-H.C.); yen@gap.kmu.edu.tw (M.-H.Y.)

[2] Division of Neurosurgery, Department of Surgery, Kaohsiung Medical University Hospital, Kaohsiung 80708, Taiwan; carbugino@gmail.com

* Correspondence: mistylin@kmu.edu.tw; Tel.: +886-7-312-1101 (ext. 2122)

Citation: Chang, C.-H.; Tsai, H.-P.; Yen, M.-H.; Lin, C.-J. Methanolic Extract of *Cimicifuga foetida* Induces G$_1$ Cell Cycle Arrest and Apoptosis and Inhibits Metastasis of Glioma Cells. *Nutrients* **2024**, *16*, 3254. https://doi.org/10.3390/nu16193254

Academic Editors: Francesca Giampieri and Carmen Mannucci

Received: 24 August 2024
Revised: 20 September 2024
Accepted: 24 September 2024
Published: 26 September 2024

Abstract: Background: Glioblastoma multiforme (GBM) is among the most aggressive and challenging brain tumors, with limited treatment options. *Cimicifuga foetida*, a traditional Chinese medicine, has shown promise due to its bioactive components. This study investigates the anti-glioma effects of a methanolic extract of *C. foetida* (CF-ME) in GBM cell lines. Methods: The effects of CF-ME and its index compounds (caffeic acid, cimifugin, ferulic acid, and isoferulic acid) on GBM cell viability were assessed using MTT assays on U87 MG, A172, and T98G cell lines. The ability of CF-ME to induce cell cycle arrest, apoptosis, and autophagy and inhibit metastasis was evaluated using flow cytometry, Western blotting, and functional assays. Additionally, the synergistic potential of CF-ME with temozolomide (TMZ) was explored. Results: CF-ME significantly reduced GBM cell viability in a dose- and time-dependent manner, induced G1 phase cell cycle arrest, promoted apoptosis via caspase activation, and triggered autophagy. CF-ME also inhibited GBM cell invasion, migration, and adhesion, likely by modulating epithelial–mesenchymal transition (EMT) markers. Combined with TMZ, CF-ME further enhanced reduced GBM cell viability, suggesting a potential synergistic effect. However, the individual index compounds of CF-ME exhibited only modest inhibitory effects, indicating that the full anti-glioma activity may result from the synergistic interactions among its components. Conclusions: CF-ME exhibited potent anti-glioma activity through multiple mechanisms, including cell cycle arrest, apoptosis, autophagy, and the inhibition of metastasis. Combining CF-ME with TMZ further enhanced its therapeutic potential, making it a promising candidate for adjuvant therapy in glioblastoma treatment.

Keywords: *Cimicifuga foetida*; glioma; cell cycle arrest; apoptosis; metastasis inhibition

1. Introduction

Glioblastoma multiforme (GBM) is the most common and aggressive type of malignant glioma in adults, classified as a Grade IV astrocytoma by the World Health Organization (WHO) [1]. Malignant gliomas, including astrocytomas, oligodendrogliomas, and oligoastrocytomas, are the primary malignant brain tumors with the highest incidence rates [2]. GBM accounts for 50.1% of all malignant brain tumors, exhibiting high mobility and invasiveness and contributing to poor prognosis [3]. The 1-year survival rate for GBM patients is approximately 35%, and the 5-year survival rate is <5% [4]. Standard GBM treatment involves a combination of surgery, radiotherapy, and chemotherapy [5]. However, despite advances in treatment, GBM remains a highly lethal cancer due to its aggressive nature and the difficulty of delivering drugs effectively across the blood–brain barrier (BBB). Temozolomide (TMZ), the primary chemotherapeutic agent for GBM, can cross the BBB and has shown effectiveness in treating gliomas [6]. TMZ functions as an alkylating agent, methylating DNA and ultimately inducing cancer cell death [7]. However, despite its widespread use, TMZ is associated with several adverse side effects, including nausea, fatigue, hepatotoxicity, and hematological toxicity, which can severely affect patient quality of life and treatment adherence [8]. Moreover, the development of resistance to both radiotherapy and chemotherapy, including TMZ, is a growing concern in GBM treatment, further complicating the management of the disease. These challenges underscore the need for novel therapeutic strategies that can overcome drug resistance, enhance treatment efficacy, and minimize side effects, highlighting the potential role of natural compounds in cancer therapy.

Cimicifugae rhizoma, known as Sheng Ma in traditional Chinese medicine (TCM), is derived from the dried rhizomes of *Cimicifuga foetida*, *C. heracleifolia*, and *C. dahurica*, all members of the *Ranunculaceae* family [9]. According to the Taiwan Herbal Pharmacopeia, *C. rhizoma* typically appears as irregular masses, measuring approximately 10–20 cm in length and 2–4 cm in diameter. The surface is blackish-brown, with hollow spaces and reticular furrows. It has a hard, light texture and is difficult to break, with an uneven fracture [9]. In TCM, *Cimicifugae rhizoma* is classified as mildly cold, with a pungent and sweet flavor. It targets the meridians of the lungs, spleen, stomach, and large intestine. Traditionally, it is used to relieve exterior disorders and is traditionally used to treat rashes, headaches, and sore throats [10]. Modern pharmacological studies have identified its antiviral [11,12], anti-inflammatory [13], immunomodulatory [14], and anti-tumor activities [15–17]. Research has demonstrated that ethanol extracts of *C. rhizoma*, as well as isolated compounds, such as triterpenoids and actein, can suppress various cancers, including liver, breast, and lung cancers, through mechanisms such as inducing cell cycle arrest, promoting apoptosis, and inhibiting cell metastasis [18–20]. Despite the promising therapeutic potential of Cimicifuga extracts, their application in brain cancer, specifically GBM, remains largely unexplored. Given the urgent need for novel treatments in GBM and the growing interest in natural compounds for cancer therapy, this study aims to investigate the anti-glioma effects of a methanolic extract of *C. foetida* (CF-ME) on GBM cell lines. By examining CF-ME's impact on cell viability, cell cycle progression, apoptosis, autophagy, and metastasis, we seek to elucidate its potential as an adjuvant therapy for GBM. Our findings suggest that CF-ME exhibits potent anti-glioma activity through multiple mechanisms and may enhance the efficacy of current treatments when used in combination.

2. Materials and Methods

2.1. Materials

2.1.1. Source and Extraction of *C. rhizoma*

The *C. rhizoma* material used in this study was purchased from the Hung Chuan Chinese Medicine Store (Sanmin District, Kaohsiung City, Taiwan). The materials were sourced from JET TURN Pharmaceutical Technology (Kaohsiung, Taiwan). The specimens were identified by Dr. Ming-Hong Yen, a specialist in traditional Chinese herbal medicine. The specimens were chopped into small pieces, oven-dried, and weighed. The dried material was soaked in 10 times its weight of water or methanol, subjected to 1 h of ultrasonic shock, and left for 1 d to obtain the first filtrate. This process was repeated with fresh water or methanol to obtain a second filtrate. The filtrates were mixed and concentrated using a rotary evaporator (Vacuum Controller VC-7600, Panchum, Kaohsiung, Taiwan) and freeze dryer (FDU-2000, EYELA, Tokyo, Japan) to obtain a dry powder of the extract, which was stored at $-20\,^{\circ}\text{C}$. Stock solutions were prepared with sterile water (for the water extract) or dimethyl sulfoxide (DMSO) (for the methanolic extract) for subsequent experiments.

2.1.2. HPLC Identification of the *C. rhizoma* Extract

To prepare the samples, methanol was used to prepare 10 mg/mL CF-ME and individual solutions of caffeic acid, cimifugin, ferulic acid, and isoferulic acid at concentrations of 3.125, 6.25, 12.5, 15.625, and 25 μg/mL. These samples were analyzed via high-performance liquid chromatography (HPLC) with ultraviolet detection using a Hitachi L-7100 pump, Hitachi L-7200 autosampler, and Hitachi 5420 UV-vis detector (HITACHI, Tokyo, Japan).

Chromatography was performed on a Mightysil RP-18 GP column (5 μm, 4.6 × 250 mm), with a flow rate of 1.0 mL/min and an injection volume of 20 μL. The UV detection wavelength was set at 254 nm. The mobile phases used were (A) 0.1% H_3PO_4 and (B) acetonitrile. The condition of gradient elution was applied as follows: 0–5 min, 92% → 82% A; 5–20 min, 82% → 76% A; 20–32 min, 76% → 66% A; 32–40 min, 66% → 45% A; 40–48 min, 45% → 30% A; 48–54 min, 30% → 30% A; 54–56 min, 30% → 60% A; 56–58 min, 60% → 92% A; 58–68 min, 92% A. Acetaminophen was used as the internal control.

2.1.3. Cell Lines and Cell Culture

U87 MG, A172, and T98G (GBM cell lines) and SVGp12 (normal human glial cell line) were obtained from the Bioresource Collection and Research Center (BCRC). U87 MG and A172 cells were cultured in Dulbecco's Modified Eagle Medium (DMEM), T98G cells in Minimum Essential Medium (MEM), and SVGp12 cells in Minimum Essential Medium-α (MEM-α). All media were supplemented with 10% fetal bovine serum (FBS), 100 units/mL penicillin, 100 μg/mL streptomycin, and 0.25 μg/mL amphotericin B (Pen-Strep-AmphoB solution, Sartorius, Göttingen, Germany). MEM also contained 1 mM sodium pyruvate and 1 mM non-essential amino acids. Media and supplements were sourced from Hyclone (Vancouver, Canada). All cells were maintained in a cell culture incubator (Thermo Electron Corporation; Waltham, MA, USA) with 5% CO_2 at 37 °C.

2.2. Methods

2.2.1. Cell Viability Assay

Cell viability was determined using the 3-(4,5-dimethylthiazol-2-yl)-2,5-diphenyltetrazolium bromide (MTT) assay. After drug treatment, cells were incubated with 0.5 mg/mL MTT solution for 1 h, and the formazan was dissolved in DMSO. Absorbance was measured with an ELISA reader (Bio-Tek Instruments Inc., Winooski, VT, USA) at a wavelength of 540 nm, and the percentage of cell viability was calculated.

2.2.2. Cell Cycle Analysis

Cells were seeded on 6-well plates (1.5×10^5 cells/well) and treated with 0–150 μg/mL CF-ME for 72 h or 100 μg/mL CF-ME for 24, 48, and 72 h. Cells were collected and fixed in 75% alcohol at 4 °C for 16 h. After centrifugation, cells were washed with phosphate-buffered saline (PBS) and treated with RNase A for 30 min, followed by staining with propidium iodide (PI) for 20 min at 37 °C. The cell cycle was analyzed using flow cytometry and CXP software V2.3 (Beckman Coulter, Brea, CA, USA).

2.2.3. Cell Apoptosis Analysis

Cells were seeded on 6-well plates (1.5×10^5 cells/well) and treated with 0–150 μg/mL CF-ME for 72 h or 100 μg/mL CF-ME for 24, 48, and 72 h. Cells were collected in HEPES solution and incubated with annexin V-FITC and PI for 20 min. Apoptosis was detected via flow cytometry and analyzed using CXP software (Beckman Coulter, Brea, CA, USA). The lower left, upper left, lower right, and upper right quadrants represent the percentages of normal, necrotic, early, and late apoptotic cells, respectively. Total apoptosis was calculated as the sum of early and late apoptosis.

2.2.4. Caspase Activity Assays

After drug treatment, protein samples were collected by lysing the cells with radio-immunoprecipitation assay (RIPA) buffer. Using bovine serum albumin (BSA) as a standard, protein concentrations were determined using the bicinchoninic acid (BCA) assay by measuring the absorbance at a wavelength of 562 nm. Caspase 3, 8, and 9 activities were assayed using Caspase Colorimetric Activity Assay Kits (Millipore, Billerica, MA, USA) according to the manufacturer's instructions. Protein samples were incubated with caspase substrate for 1–2 h at 37 °C. Absorbance was measured at 405 nm using an ELISA reader, and caspase activity was expressed as fold change compared with the control group.

2.2.5. Cell Autophagy Analysis

The formation of acidic vesicular organelles (AVOs), the characteristic of autophagy, could be analyzed by flow cytometry with acridine orange stain. Cells were seeded on 6-well plates (1.5×10^5 cells/well) and treated with 0–150 µg/mL CF-ME for 72 h or 100 µg/mL CF-ME for 24, 48, and 72 h. Cells were collected in PBS solution and incubated with acridine orange for 15 min at 37 °C. The formation of AVOs was detected via flow cytometry and analyzed using CXP software (Beckman Coulter, Brea, CA, USA). The sum of the upper left and upper right quadrants represents the percentages of autophagic cells.

2.2.6. Migration Assay

Cell migration was evaluated using a wound-healing assay. Cells were seeded on 12-well plates (3×10^5 cells/well) and cultured for 24 h. Wounds were created by scratching cells with a sterile 10 µL tip. Cells were treated with CF-ME after washing with PBS. Wound closure was photographed under an inverted microscope (SDPTOP, Ningbo, China) at 0, 6, 12, and 24 h after treatment. Wound width was measured using the MShot Image Analysis System 1.5.2.

2.2.7. Invasion Assay

The in vitro invasion assay was conducted using SPLInsert™ Hanging plates (SPL Life Sciences, Pocheon-si, Gyeonggi-do, Republic of Korea). CF-ME-treated U87 MG, A172, and T98G cells were seeded at a density of 1×10^4 cells/Transwell insert in serum-free medium in the upper chamber, and serum-containing culture medium was added to the lower chamber. After 24 h, the cells were washed with PBS, fixed with 4% formaldehyde for 15 min, and stained with 0.5% crystal violet solution. Cells in the upper layer of the insert were carefully removed using cotton swabs, while cells that invaded through the insert were photographed and counted under an inverted microscope (SDPTOP, Ningbo, China). Cell numbers were counted in three randomly selected fields per insert.

2.2.8. Cell Adhesion Assay

U87 MG, A172, and T98G cells were cultured in 6-well plates at a density of 1.5×10^5 cells/well and treated with 0–150 µg/mL CF-ME for 72 h. After treatment, the cells were harvested via trypsinization and re-seeded in 12-well plates at a density of 1×10^4 cells/well in triplicate wells. After 1 and 24 h of incubation, non-adherent cells were removed by washing with PBS. Adherent cells were fixed with 4% formaldehyde for 15 min, stained with 0.5% crystal violet solution, and photographed using an inverted microscope (SDPTOP; Ningbo, China). Six random regions/well were selected to count the number of cells, and the relative percentage of adherent cells was calculated.

2.2.9. Western Blotting

After drug treatment, the cells were lysed with RIPA buffer, and protein samples were collected. Protein concentration was determined using the BCA assay. An equal amount of protein (50 µg) from each group was separated via sodium dodecyl sulfate-polyacrylamide gel electrophoresis (SDS-PAGE) and transferred to polyvinylidene fluoride (PVDF) membranes (Cytiva/Hyclone, Vancouver, BC, Canada). The membranes were blocked with 5% skim milk in TBS-T buffer (11 mM Tris-base pH 7.4, 154 mM NaCl, 0.1% Tween-20) for 1 h and incubated with specific primary antibodies for 16 h at 4 °C. Subsequently, the membranes were incubated with the appropriate anti-mouse or anti-rabbit secondary antibodies at room temperature for 1–2 h. Immunoreactive bands were detected using enhanced chemiluminescence (ECL) (Millipore, Billerica, MA, USA) and recorded with a MultiGel-21 Image System (TOP BIO CO., New Taipei City, Taiwan). The density of the bands was determined using ImageJ software 1.51j8 and normalized to β-actin. The antibodies used included anti-GADD45A, p21, CDK6, caspase 3, cleaved caspase 3, poly (ADP-ribose) polymerase (PARP), and horseradish peroxidase (HRP)-conjugated goat anti-mouse and anti-rabbit immunoglobulin G (IgG) purchased from Cell Signaling Technology

(Beverly, MA, USA). Antibodies against vimentin, N-cadherin, E-cadherin, and cyclin D1 were purchased from Proteintech (Chicago, IL, USA). Anti-β-actin antibody was purchased from Millipore (Billerica, MA, USA).

2.2.10. Statistical Analysis

At least three independent experiments were conducted for each assay, and the data are expressed as mean $\pm$ standard deviation (SD). Statistical analysis was performed using one-way analysis of variance (ANOVA) to compare three or more groups. A p-value of <0.05 was considered statistically significant.

3. Results

3.1. Investigation of the Effects of C. foetida Extracts on Glioma Cell Growth

This study evaluates the effects of water (CF-WE) and methanol extracts (CF-ME) of *C. foetida* on glioma cell growth and viability. The MTT assay was used to assess short-term cell viability, while the colony formation assay evaluated long-term viability. U87 MG, A172, and T98G glioma cell lines were treated with CF-WE at concentrations ranging from 0 to 800 μg/mL and CF-ME at concentrations from 0 to 150 μg/mL for 24, 48, and 72 h, respectively. Treatment revealed a significant decrease in cell viability, and these changes were more pronounced following CF-ME treatment (Figure 1A,B).

In addition, both CF-WE and CF-ME treatments resulted in dose- and time-dependent reductions in cell viability across all GBM cell lines tested (Figure 1A,B). Specifically, after 72 h of CF-WE treatment, the IC_{50} value was 662.5 ± 70.0 μg/mL for U87 MG cells, while the IC_{50} values for A172 and T98G cells exceeded 800 μg/mL. In contrast, CF-ME demonstrated a stronger inhibitory effect, with IC_{50} values of 94.3 ± 5.2 μg/mL for U87 MG, 96.5 ± 3.7 μg/mL for A172, and 97.3 ± 1.6 μg/mL for T98G cells after 72 h of exposure (Table 1). To assess the selectivity and toxicity of these extracts, the normal human glial cell line SVGp12 was also treated with CF-WE and CF-ME under similar conditions. The results indicated that SVGp12 cells exhibited higher viability and IC_{50} values than the glioma cells (Figure 1C and Table 1). These findings suggest that both CF-WE and CF-ME effectively reduced glioma cell viability while exhibiting lower toxicity toward normal glial cells, with CF-ME showing superior efficacy. Consequently, CF-ME at 100 μg/mL was selected for subsequent experiments. These data support the idea that both water and methanol extracts of *C. foetida* suppress glioma cell growth and reduce cell viability. The methanol extract, in particular, showed greater efficacy and lower toxicity toward normal cells, making it a promising candidate for further investigation in glioma treatment strategies.

Table 1. IC_{50} values of malignant glioma cells and normal glial cells after treatment with extract of *Cimicifuga foetida*. U87 MG, A172, T98G, and SVGp12 cells were treated with different doses of CF-WE and CF-ME for 72 h, respectively. Cell viability was detected using an MTT assay with an ELISA reader to measure absorbance values. IC_{50} values were calculated after plotting the regression lines. Data are presented as means $\pm$ standard deviation.

Cell Lines	IC_{50} (μg/mL)	
	CF-WE	**CF-ME**
U87 MG	662.5 ± 70.0	94.3 ± 5.2
A172	>800	96.5 ± 3.7
T98G	>800	97.3 ± 1.6
SVGp12	>800	>150

U87 MG, A172, and T98G were malignant glioma cell lines; SVGp12 was human normal glial cell line.

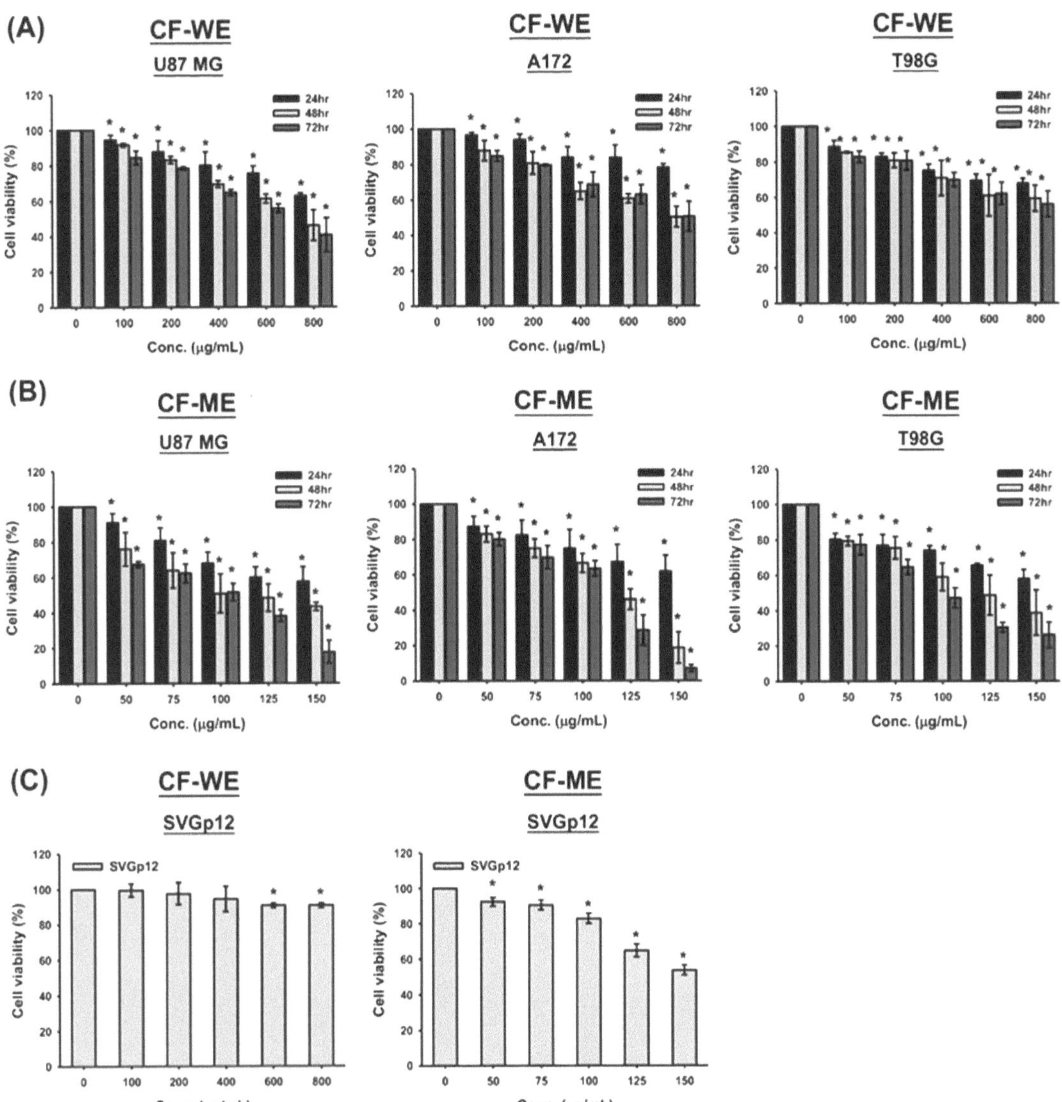

Figure 1. Effects of water (CF-WE) and methanol (CF-ME) extracts of *Cimicifuga foetida* on the viability of glioblastoma (GBM) and glial cells. (**A**) Cell viability of U87 MG, A172, and T98G GBM cell lines treated with varying concentrations (0–800 µg/mL) of CF-WE for 24, 48, and 72 h showed a significant dose- and time-dependent decrease in cell viability, with stronger effects at higher concentrations and longer treatment durations. (**B**) Cell viability of U87 MG, A172, and T98G GBM cell lines treated with CF-ME (0–150 µg/mL) for the same durations exhibited a stronger inhibitory effect, also in a dose- and time-dependent manner. (**C**) Cell viability of human normal glial cell line SVGp12 treated with CF-WE (left) and CF-ME (right) under similar conditions showed reduced cell viability, but SVGp12 cells exhibited higher overall viability and IC_{50} values than GBM cells, indicating lower toxicity toward normal glial cells. * Asterisks indicate statistical significance compared to control (0 µg/mL) ($p < 0.05$). Error bars represent the mean ± SD of three independent experiments.

3.2. CF-ME Induces G_1 Phase Cell Cycle Arrest in Glioma Cells

To determine whether CF-ME induces cell cycle arrest and suppresses cell proliferation, U87 MG, A172, and T98G glioma cells were treated with varying concentrations (0–150 µg/mL) of CF-ME for 72 h (Figure 2A) or with 100 µg/mL CF-ME for 24, 48, and 72 h (Figure 2B). Cell cycle analysis was performed using flow cytometry with PI staining. The results indicated that CF-ME induced G_1 phase cell cycle arrest in glioma cells at lower concentrations and shorter exposure times (Figure 2A,B). In contrast, higher doses and prolonged exposure increased the number of cells in the sub-G1 phase, indicative of apoptosis. Further investigations focused on the molecular mechanisms underlying G_1 phase arrest. Specifically, the study examined the expression of cell cycle-related proteins, such as CDK6 and cyclin D1, as well as the common cell cycle regulators p21 and GADD45A. U87 MG, A172, and T98G cells were treated with 100 µg/mL CF-ME for varying durations (8, 16, 24, 48, and 72 h), and protein levels were analyzed via Western blotting (Figure 2C). The results revealed an increase in GADD45A and p21 protein levels after 8–16 h of treatment, while CDK6 expression decreased in a time-dependent manner. Only minor changes in the cyclin D1 levels were observed (Figure 2C). These findings suggest that CF-ME induces G1 phase cell cycle arrest in glioma cells via the GADD45A/p21/CDK6 signaling pathway.

(A)

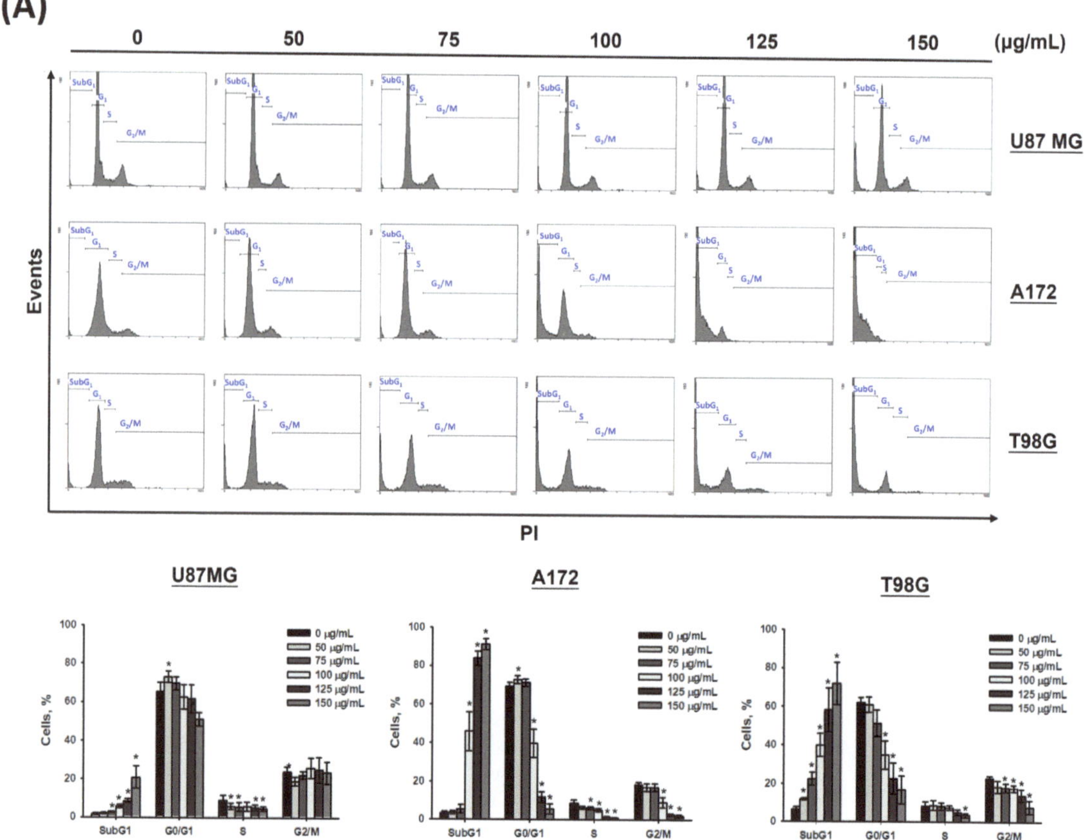

Figure 2. *Cont.*

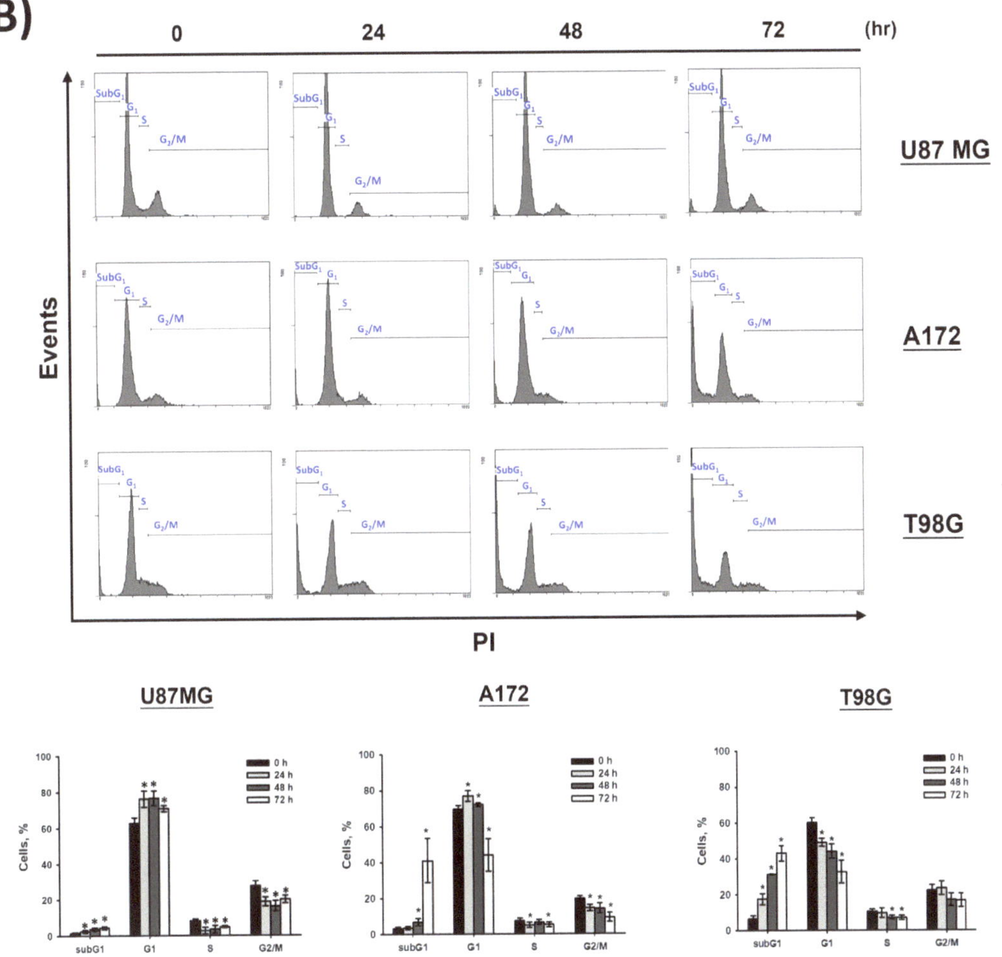

Figure 2. *Cont.*

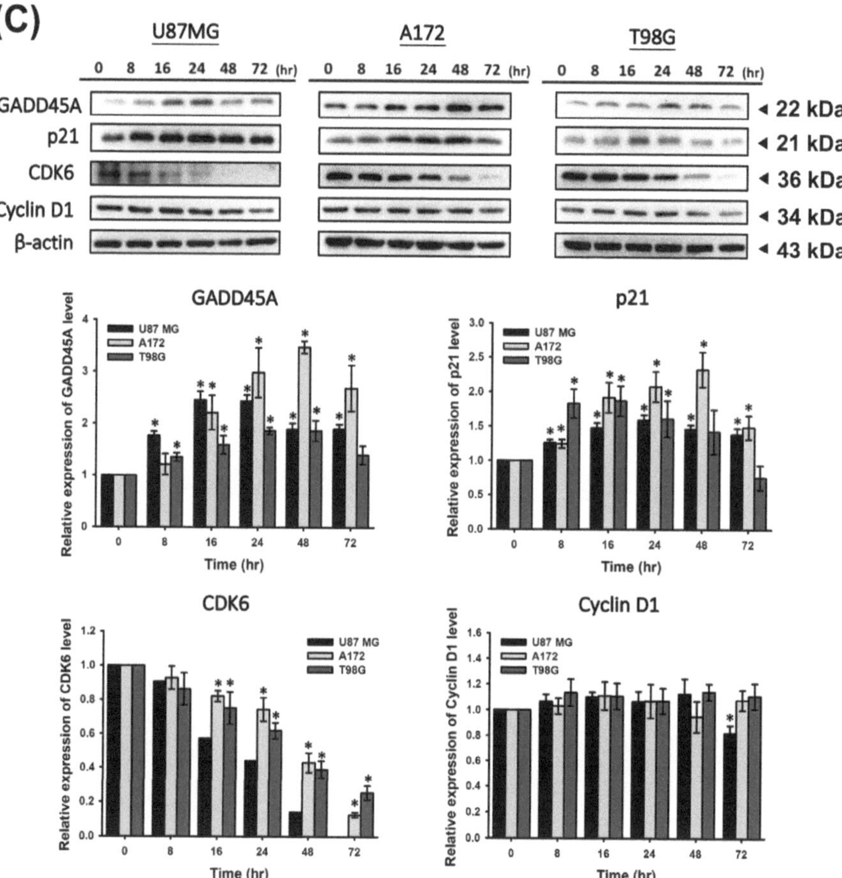

Figure 2. CF-ME induces G_1 phase cell cycle arrest in glioblastoma (GBM) cells. (**A**) Flow cytometry analysis of U87 MG, A172, and T98G GBM cell lines treated with CF-ME (0–150 µg/mL) for 72 h. The histograms show the distribution of cells in different phases of the cell cycle (sub-G_1, G_0/G_1, S, G_2/M), indicating a dose-dependent increase in G_1 phase arrest and a corresponding decrease in S and G_2/M phases across all cell lines. (**B**) Flow cytometry analysis of U87 MG, A172, and T98G cells treated with 100 µg/mL CF-ME for 24, 48, and 72 h display a temporal progression of cell cycle arrest, with a significant accumulation of cells in the G_1 phase over time, and an elevation in the sub-G_1 phase, indicative of apoptosis. (**C**) Western blot analysis of GADD45A, p21, CDK6, and cyclin D1 in U87 MG, A172, and T98G cells treated with 100 µg/mL CF-ME for 0–72 h. The graphs below depict the relative protein expression levels normalized to β-actin. The results show a time-dependent upregulation of GADD45A and p21, downregulation of CDK6, and minimal changes in cyclin D1 levels, suggesting that CF-ME induces G_1 phase cell cycle arrest through the GADD45A/p21/CDK6 signaling pathway. * Asterisks indicate statistical significance compared to control (0 µg/mL or 0 h) ($p < 0.05$). Error bars represent the mean ± SD of three independent experiments.

3.3. CF-ME Induces Apoptosis in Glioma Cells

Given the observed increase in the sub-G_1 phase signal in cell cycle assays under high-dose and prolonged CF-ME treatment (Figure 2A,B), the study further investigated whether CF-ME induces apoptosis in glioma cells. U87 MG, A172, and T98G cells were exposed to different concentrations of CF-ME for 72 h (Figure 3A) or treated with 100 µg/mL CF-ME for 24, 48, and 72 h (Figure 3B).

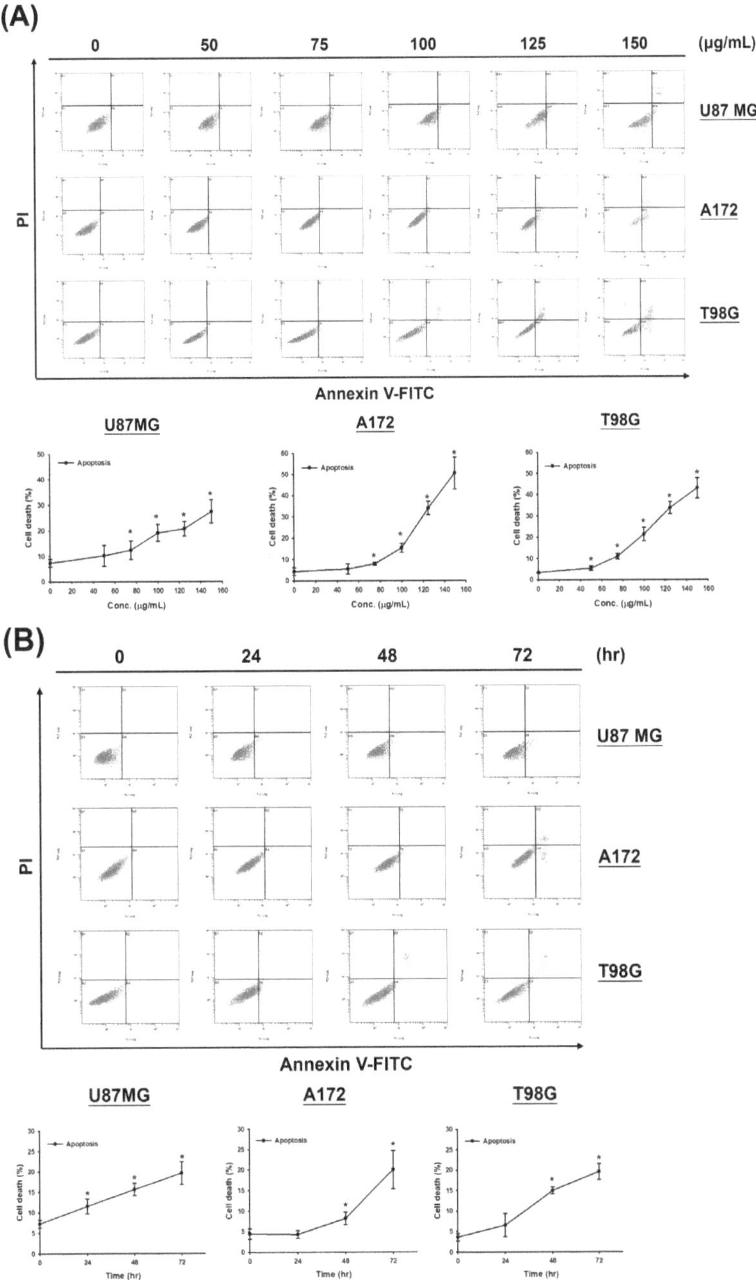

Figure 3. *Cont.*

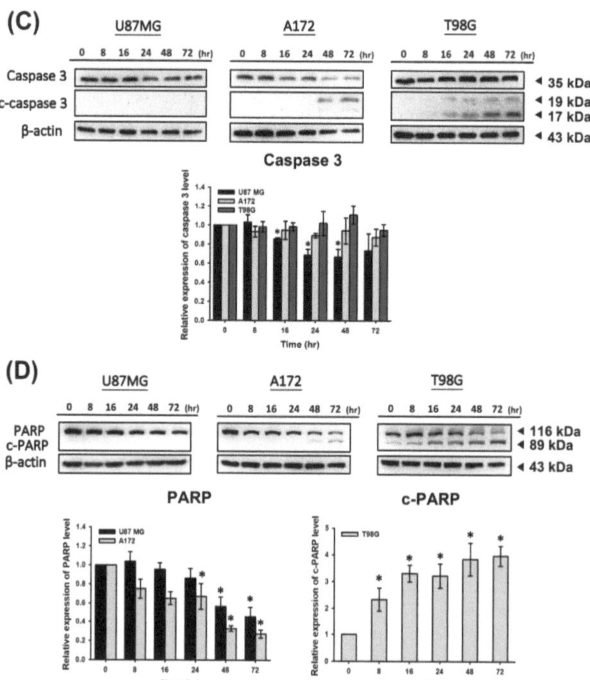

Figure 3. CF-ME induces apoptosis in GBM cells. (**A**) Flow cytometry analysis of apoptosis in U87 MG, A172, and T98G GBM cell lines treated with CF-ME (0–150 µg/mL) for 72 h show a dose-dependent increase in apoptosis. Cells were stained with Annexin V-FITC and PI; the dot plots distinguish early apoptotic (Annexin V-positive, PI-negative) from late apoptotic/necrotic cells (Annexin V-positive, PI-positive). The graphs below show a dose-dependent increase in the percentage of apoptotic cells across all GBM cell lines. (**B**) Time-course analysis of apoptosis in U87 MG, A172, and T98G cells treated with 100 µg/mL CF-ME for 24, 48, and 72 h show a time-dependent increase in both early and late apoptotic cells over time. (**C**) Western blot analysis of caspase 3 and cleaved caspase 3 (c-caspase 3) in U87 MG, A172, and T98G cells treated with 100 µg/mL CF-ME for 0–72 h. The accompanying graph shows the relative expression levels of c-caspase 3 normalized to β-actin. CF-ME treatment showed increased c-caspase 3 levels in A172 and T98G cells, with a decrease in U87 MG cells, suggesting the activation of the caspase pathway in apoptosis. (**D**) Western blot analysis of poly (ADP-ribose) polymerase (PARP) and cleaved PARP (c-PARP) in U87 MG, A172, and T98G cells treated with 100 µg/mL CF-ME for 0–72 h. The graphs below illustrate the relative expression levels of c-PARP and total PARP normalized to β-actin. CF-ME treatment showed increased c-PARP levels, particularly in A172 and T98G cells, with a time-dependent decrease in total PARP in U87 MG and A172 cells indicative of apoptosis progression. * Asterisks indicate statistical significance compared to control (0 µg/mL or 0 h) ($p < 0.05$). Error bars represent the mean ± SD of three independent experiments.

Apoptosis was analyzed via flow cytometry using annexin V-FITC and PI double staining. The results demonstrated that CF-ME induced apoptosis in a dose- and time-dependent manner (Figure 3A,B). To elucidate the molecular mechanisms underlying apoptosis, we examined the expression of apoptosis-related proteins, including cleaved caspase-3 (c-caspase-3) and cleaved PARP (c-PARP), using Western blotting. U87 MG, A172, and T98G cells were treated with 100 µg/mL CF-ME for varying durations (8, 16, 24, 48, and 72 h). CF-ME treatment increased the expression of c-caspase-3 without significant changes in total caspase-3 levels in A172 and T98G cells, while U87 MG cells showed decreased caspase-3 expression (Figure 3C). Similarly, c-PARP levels were increased in

A172 and T98G cells, although no significant changes in pro-PARP levels were observed in T98G cells. In contrast, pro-PARP levels decreased in a time-dependent manner in U87 MG and A172 cells (Figure 3D). To further confirm apoptosis, a Caspase Colorimetric Activity Assay was performed to measure caspase-3 activity after CF-ME treatment (Table 2).

Table 2. Effect of CF-ME on the activities of caspase 3. U87 MG, A172, and T98G cells were treated with 100 µg/mL CF-ME for 0, 24, 48, and 72 h and extracted proteins. Activities of caspase 3 were determined using the Caspase Colorimetric Activity Assay Kit. Absorbance values were measured using an ELISA reader, and fold changes compared to the control group were calculated. Data are presented as means ± standard deviation ($n = 3$). * $p < 0.05$ indicates a statistically significant difference compared to the respective control groups.

Caspase	Cell Line	100 µg/mL CF-ME Treatment (h)			
		0	24	48	72
	U87 MG	1	1.7 ± 1.2	3.3 ± 1.5	4.3 ± 0.6 *
Caspase 3	A172	1	1.3 ± 0.6	3.2 ± 0.8 *	5.0 ± 1.6 *
	T98G	1	3.0 ± 1.0 *	4.7 ± 0.6 *	7.0 ± 1.0 *

These results indicate that CF-ME activates these caspases, suggesting that CF-ME induces apoptosis by activating the caspase cascade and subsequent cleavage of PARP.

3.4. CF-ME Induces Autophagy in Glioma Cells

To determine whether CF-ME induces autophagy in glioma cells, we examined the formation of acidic vesicular organelles (AVOs) and the expression of microtubule-associated protein light chain 3 II (LC3-II). U87 MG, A172, and T98G cells were treated with various concentrations of CF-ME for 72 h (Figure 4A) or with 100 µg/mL CF-ME for 24, 48, and 72 h (Figure 4B). Flow cytometry using acridine orange (AO) staining showed that CF-ME induced AVO formation in a dose- and time-dependent manner (Figure 4A,B). Western blotting showed increased LC3-II levels in U87 MG, A172, and T98G cells treated with 100 µg/mL CF-ME 8, 16, 24, 48, and 72 h (Figure 4C). These results indicate that CF-ME effectively induces autophagy in glioma cells.

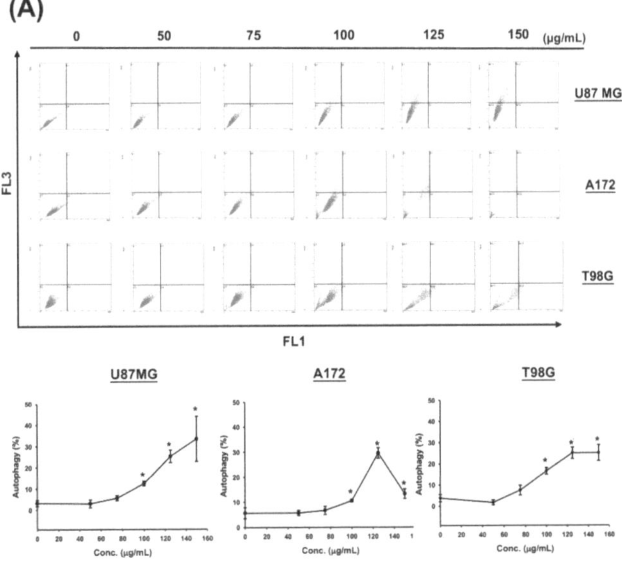

Figure 4. *Cont.*

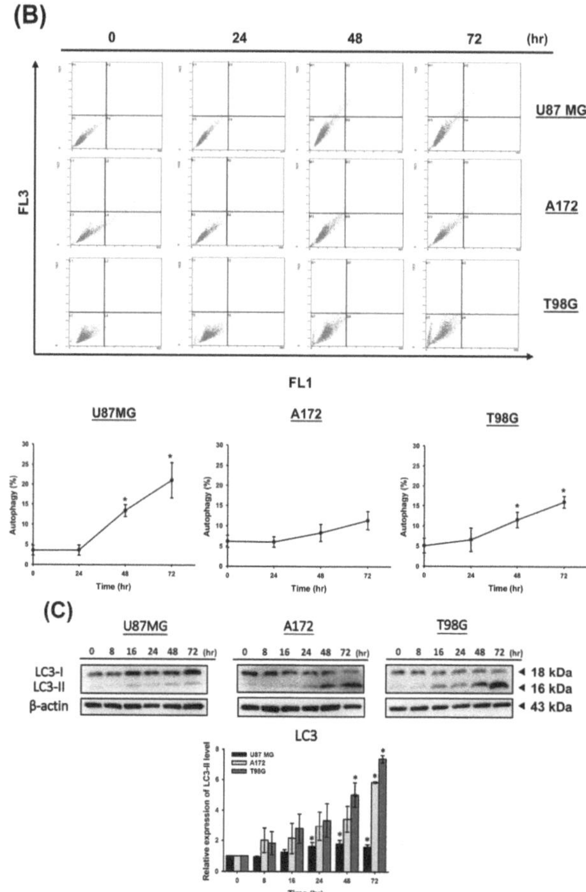

Figure 4. CF-ME induces autophagy in GBM Cells. (**A**) Flow cytometry analysis of autophagy in U87 MG, A172, and T98G GBM cell lines treated with CF-ME (0–150 µg/mL) for 72 h. Cells were stained with acridine orange to detect acidic vesicular organelles (AVOs), a hallmark of autophagy. The dot plots show a dose-dependent increase in AVO formation, indicating that CF-ME induces autophagy across all three cell lines. (**B**) Time-course analysis of autophagy in U87 MG, A172, and T98G cells treated with 100 µg/mL CF-ME for 24, 48, and 72 h reveals a time-dependent increase in AVO formation, with significant autophagic activity observed as early as 24 h, continuing to rise through 72 h. (**C**) Western blot analysis of LC3-I and LC3-II, markers of autophagy, in U87 MG, A172, and T98G cells treated with 100 µg/mL CF-ME for 0–72 h. The graphs below depict the relative expression levels of LC3-II normalized to β-actin. CF-ME treatment led to a time-dependent increase in LC3-II levels, confirming the induction of autophagy in glioma cells by CF-ME. * Asterisks indicate statistical significance compared to control (0 µg/mL or 0 h) ($p < 0.05$). Error bars represent the mean $\pm$ SD of three independent experiments.

3.5. CF-ME Suppresses Metastasis in Glioma Cells

To investigate the CF-ME inhibitory effect on glioma cell metastasis, we conducted Transwell, wound healing, and adhesion assays. First, U87 MG, A172, and T98G cells were treated with 100 µg/mL CF-ME for 72 h, then re-seeded in Transwell chambers (1×10^4 cells/well) and incubated for 24 h. The Transwell assay results (Figure 5A) demonstrated reduced cell invasion in the lower chamber, indicating a CF-ME inhibitory effect. The wound-healing assay assessed cell migration. Scratched U87 MG, A172, and T98G cells treated with 100 µg/mL CF-ME were observed at 6, 12, and 24 h post-treatment. CF-

ME-treated groups exhibited wider wound widths than the control, indicating suppressed glioma cell migration (Figure 5B). Finally, adhesion assays evaluated cells' attachment ability. U87 MG, A172, and T98G cells treated with 0–150 µg/mL CF-ME for 72 h were re-seeded on 12-well plates (1×10^4 cells/well) for 1 h and 24 h. The results (Figure 5C) showed a dose-dependent decrease in attached cells, suggesting CF-ME inhibition of glioma cell adhesion. Given CF-ME's metastasis-suppressing ability, we explored its effect on epithelial–mesenchymal transition (EMT), a crucial process in cancer metastasis. EMT marker proteins—E-cadherin (epithelial), N-cadherin, and vimentin (mesenchymal)—were analyzed via Western blotting following CF-ME treatment (100 µg/mL) for 8, 16, 24, 48, and 72 h (Figure 5D). CF-ME treatment decreased N-cadherin and vimentin expression in glioma cells while increasing E-cadherin levels. However, N-cadherin expression remained unchanged in U87 MG cells (Figure 5D). These results suggest that CF-ME inhibits glioma cell metastasis by modulating EMT through the downregulation of mesenchymal markers and upregulation of epithelial markers.

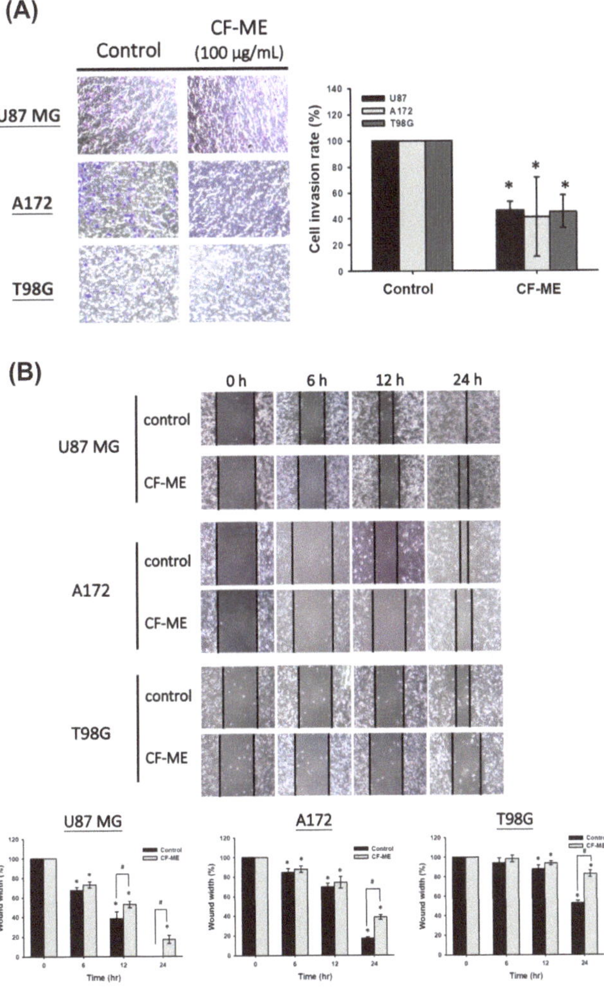

Figure 5. *Cont.*

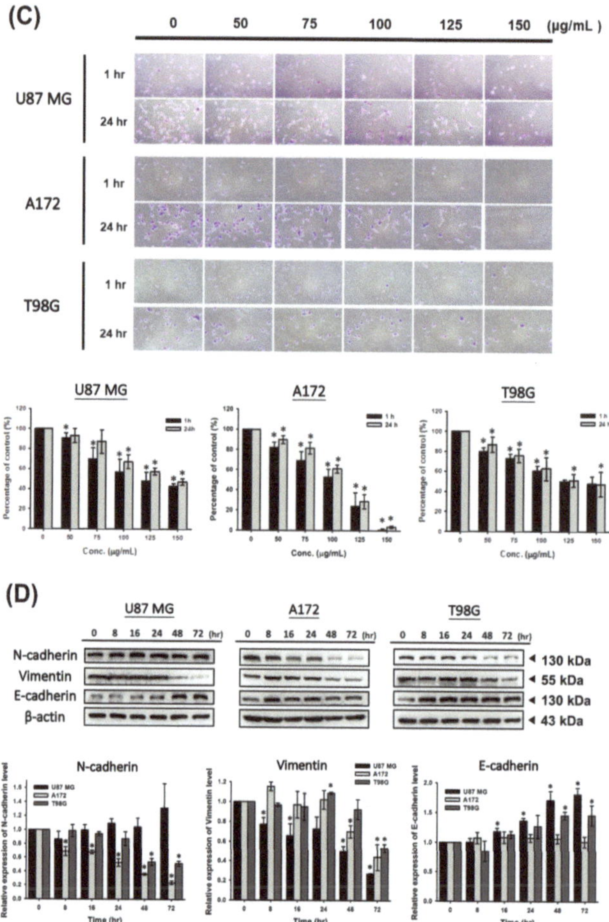

Figure 5. CF-ME inhibits invasion, migration, and adhesion of GBM Cells. (**A**) Invasion assay: U87 MG, A172, and T98G GBM cell lines were treated with 100 µg/mL CF-ME and subjected to a Transwell invasion assay. Representative images show the number of cells that invaded through the Matrigel-coated membrane compared to the control group. The bar graph quantifies the invasion rate as a percentage relative to the control, demonstrating that CF-ME significantly reduces the invasive capability of GBM cells. (**B**) Migration assay: Wound-healing assays were performed on U87 MG, A172, and T98G cells treated with 100 µg/mL CF-ME. Images were captured at 0, 6, 12, and 24 h post-scratch to assess cell migration into the wound area. The graphs quantify wound closure over time, showing that CF-ME significantly inhibits cell migration compared to control. (**C**) Adhesion assay: U87 MG, A172, and T98G cells were treated with CF-ME (0–150 µg/mL) and re-seeded for 1 h and 24 h to assess cell adhesion to the substrate. Representative images display the number of adherent cells at each concentration and time point. The bar graphs show that CF-ME treatment leads to a dose-dependent reduction in cell adhesion, with significant effects observed at higher concentrations and longer exposure times. (**D**) Western blot analysis: N-cadherin, vimentin, and E-cadherin, markers of epithelial–mesenchymal transition (EMT), were analyzed in U87 MG, A172, and T98G cells treated with 100 µg/mL CF-ME for 0–72 h. The graphs below depict the relative expression levels of N-cadherin, vimentin, and E-cadherin normalized to β-actin. * Asterisks indicate statistical significance compared to control is indicated by asterisks ($p < 0.05$), and hashtags indicate statistical significance between control and CF-ME in each treatment time point (# $p < 0.05$). Error bars represent the mean $\pm$ SD of three independent experiments.

3.6. CF-ME as a Potential Adjuvant Therapy for TMZ

Glioma cells were pre-treated with low doses of CF-ME (50 µg/mL and 100 µg/mL) for 1 h, followed by exposure to varying concentrations of TMZ for 72 h. Cell viability was assessed using the MTT assay (Figure 6).

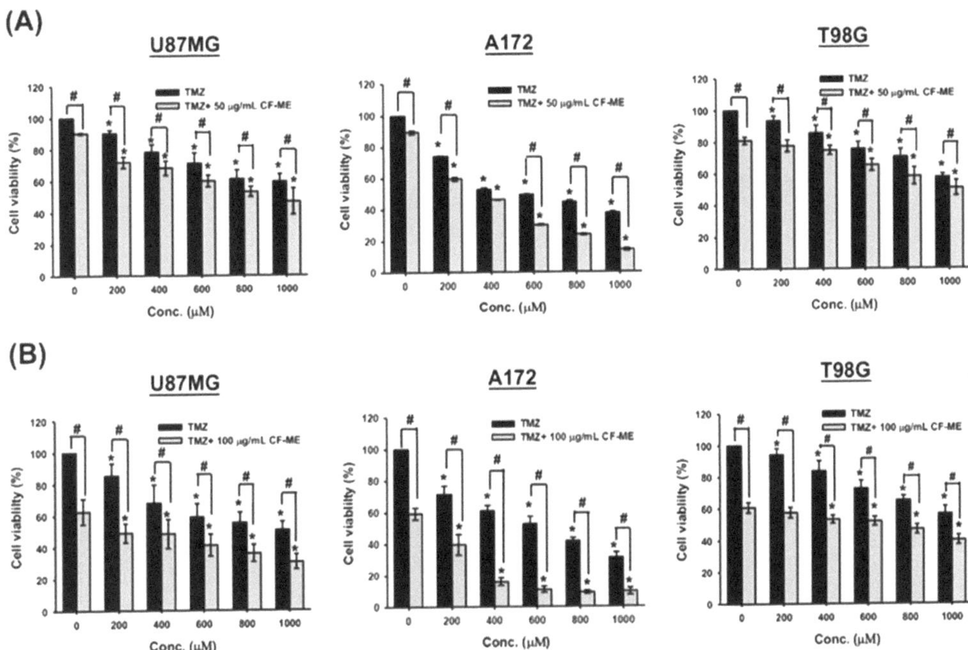

Figure 6. Effects of CF-ME in combination with TMZ on the viability of GBM cells. (**A**) Cell viability of U87 MG, A172, and T98G GBM cell lines treated with temozolomide (TMZ) alone or in combination with 50 µg/mL CF-ME was assessed using the MTT assay. Cells were exposed to varying concentrations of TMZ (0–1000 µM) for 72 h. The combination of CF-ME with TMZ significantly reduced cell viability across all concentrations compared to TMZ treatment alone, suggesting a synergistic effect of CF-ME in enhancing the cytotoxicity of TMZ. (**B**) Cell viability of U87 MG, A172, and T98G GBM cell lines treated with TMZ alone or in combination with 100 µg/mL CF-ME. The combination of CF-ME with TMZ at this higher concentration further reduced cell viability compared to TMZ alone, demonstrating that increasing the dose of CF-ME amplifies its synergistic effect with TMZ. * Asterisks indicate statistical significance compared to 0 µM TMZ is indicated by asterisks ($p < 0.05$), and hashtags indicate statistical significance between TMZ alone and TMZ combined with CF-ME (# $p < 0.05$). Error bars represent the mean ± SD of three independent experiments.

The combination of CF-ME and TMZ further suppressed glioma cell viability compared to TMZ alone (Figure 6). IC_{50} values of TMZ, both alone and in combination with 50 µg/mL and 100 µg/mL CF-ME, were calculated based on cell viability data. For U87 MG cells, the IC_{50} values were 911.1 ± 153.7 µM for TMZ alone, 784.8 ± 81.5 µM when combined with 50 µg/mL CF-ME, and 530.5 ± 87.4 µM when combined with 100 µg/mL CF-ME. For A172 cells, the IC_{50} values were 679.6 ± 13.9 µM for TMZ alone, 488.5 ± 33.8 µM when combined with 50 µg/mL CF-ME, and 259.7 ± 33.5 µM when combined with 100 µg/mL CF-ME. For T98G cells, the IC_{50} values were >1000 µM for TMZ alone, 979.8 ± 70.2 µM when combined with 50 µg/mL CF-ME, and 671.0 ± 45.8 µM when combined with 100 µg/mL CF-ME (Table 3).

Table 3. Combination of CF-ME with temozolomide (TMZ) could decrease the IC_{50} values of TMZ. U87 MG, A172, and T98G cells were treated with different doses of TMZ and combined with low doses (50 μg/mL and 100 μg/mL) of CF-ME for 72 h. Cell viability was detected using an MTT assay with an ELISA reader to measure absorbance values. IC_{50} values were calculated after plotting the regression lines. Data are presented as means $\pm$ standard deviation ($n = 3$). * $p < 0.05$ indicates a statistically significant difference compared to the respective control groups.

Cell Lines	IC_{50} (μM) of TMZ at Different Concentrations of CF-ME		
	0 μg/mL	50 μg/mL	100 μg/mL
U87 MG	911.1 ± 153.7	784.8 ± 81.5	530.5 ± 87.4 *
A172	679.6 ± 13.9	488.5 ± 33.8 *	259.7 ± 33.5 *
T98G	>1000	979.8 ± 70.2 *	671.0 ± 45.8 *

These results suggest that combining CF-ME with TMZ could reduce the required dose of TMZ to achieve the same therapeutic effect, potentially mitigating the side effects of some drugs and overcoming drug resistance in glioma cells. Therefore, CF-ME shows promise as a potential adjuvant therapy to enhance the efficacy of TMZ in the treatment of glioma.

3.7. Content Analysis of Methanol Extract of C. rhizoma

To verify the species of *C. rhizoma* used in this study, key compounds in the methanol extract were analyzed using high-performance liquid chromatography (HPLC). The retention times were 8.12 min for acetaminophen (internal control), 12.23 min for caffeic acid, 16.91 min for cimifugin, 17.85 min for ferulic acid, and 19.13 min for isoferulic acid (Figure 7).

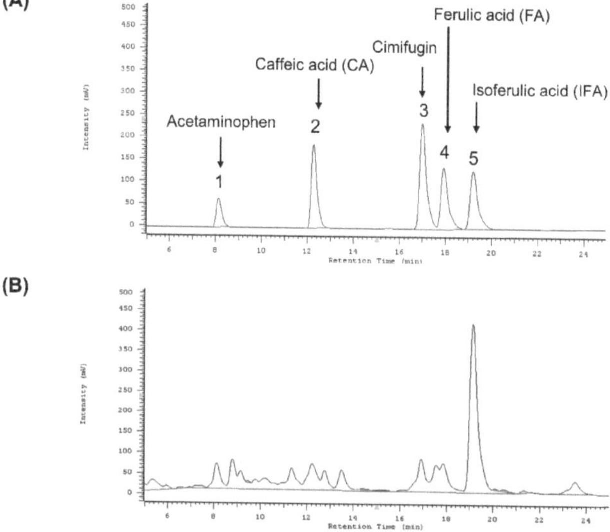

Figure 7. HPLC analysis of standard compounds and methanol extract of *Cimicifuga foetida*. (**A**) HPLC chromatogram showing the retention times of the internal standard acetaminophen and the standard compounds caffeic acid (CA), cimifugin, ferulic acid (FA), and isoferulic acid (IFA). The retention times were as follows: acetaminophen (~8.12 min), caffeic acid (~12.23 min), cimifugin (~16.91 min), ferulic acid (~17.85 min), and isoferulic acid (~19.13 min). (**B**) HPLC chromatogram of the methanol extract of *Cimicifuga foetida*. The chromatogram shows the positions of the identified index compounds (caffeic acid, cimifugin, ferulic acid, and isoferulic acid) after comparison with the retention times of the standard compounds.

These retention times were compared with previously published data to identify the species. Among the three species of *C. rhizoma* listed in the Taiwan Herbal Pharmacopeia (*C. foetida*, *C. heracleifolia*, and *C. dahurica*), only *C. foetida* contained all four index compounds. Thus, the plant material was confirmed to be *C. foetida*. To quantify the content of these compounds in the methanol extract, standard curves (3.125–25 μg/mL) were generated for caffeic acid, cimifugin, ferulic acid, and isoferulic acid (Table 1). The methanolic extract contained 0.11% caffeic acid, 0.08% cimifugin, 0.11% ferulic acid, and 0.78% isoferulic acid (Table 4). These findings validated the plant species and provided a detailed chemical profile essential for experimental reproducibility.

Table 4. Quantitative analysis of four index compounds in the methanol extracts of *Cimicifuga foetida* (CF-ME). 10 mg/mL of CF-ME was detected via HPLC. The data were compared with the standard curves of four index compounds to calculate their respective percentage content in the methanol extract.

Index Compound	Standard Curve	Content (%)
Caffeic acid	y = 0.1221x + 0.1536	0.11
Cimifugin	y = 0.1827x + 0.2071	0.08
Ferulic acid	y = 0.1189x + 0.1466	0.11
Isoferulic acid	y = 0.1144x + 0.1514	0.78
Index compound	Standard curve	Content (%)

3.8. Investigation of the Anti-Glioma Effects of Index Compounds in CF-ME

Previous findings have confirmed the in vitro anti-glioma activity of CF-ME. To identify active components, we assessed a mixture of the four index compounds—caffeic acid, cimifugin, ferulic acid, and isoferulic acid—at a concentration equivalent to 100 μg/mL of CF-ME, using the MTT assay. Unexpectedly, this mixture exhibited minimal anti-glioma activity (Figure 8A). Further investigation using the MTT assay of individual compounds at higher doses (up to 800 μM) showed that caffeic acid at 800 μM reduced the cell viability of U87 MG cells to below 50%, but it had a limited effect on A172 and T98G cells, with their viability remaining above 50%. Ferulic acid, isoferulic acid, and cimifugin at higher doses (1–2 mM) slightly inhibited the viability of all three glioma cell lines (Figure 8B). These results suggested that the individual index compounds did not account for the full anti-glioma effects of CF-ME. Further research is necessary to identify the specific active compounds or potential synergistic interactions within CF-ME that contribute to its anti-glioma activity.

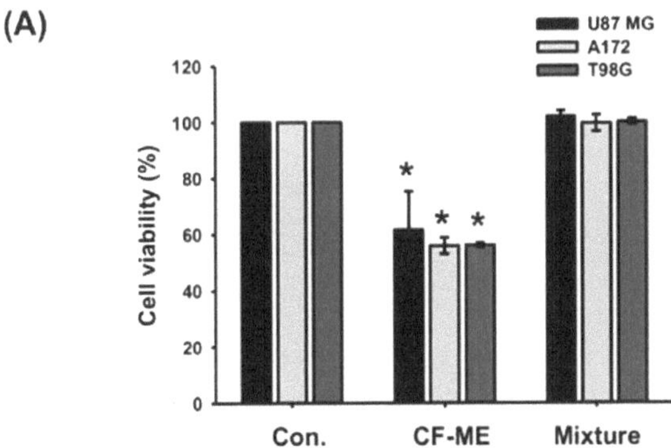

Figure 8. *Cont.*

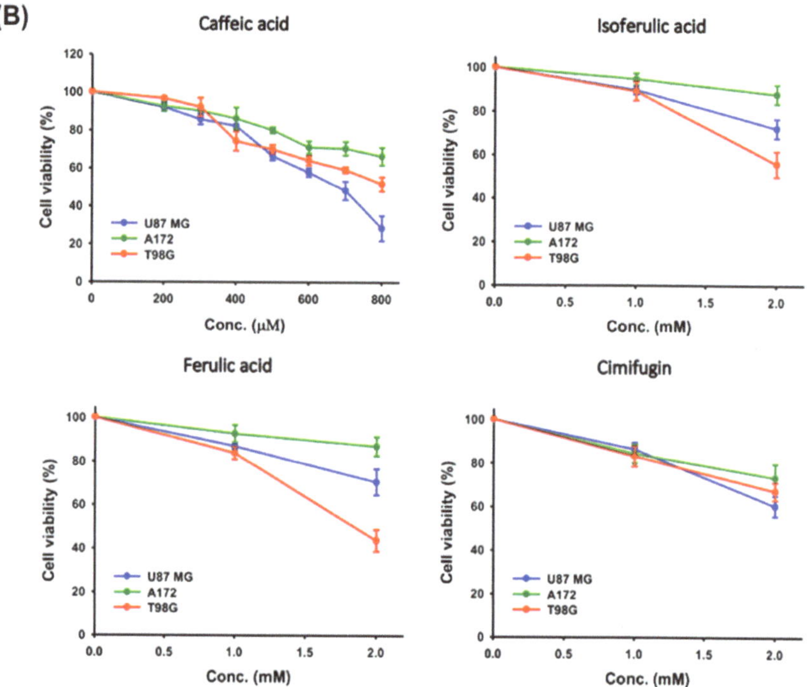

Figure 8. Effect of individual and combined index compounds from CF-ME on the viability of GBM cells. (**A**) Cell viability of U87 MG, A172, and T98G GBM cell lines after 72 h of treatment with CF-ME and a mixture of index compounds (caffeic acid, isoferulic acid, ferulic acid, and cimifugin) at concentrations equivalent to those in 100 μg/mL of CF-ME, as determined via HPLC analysis. The results showed that while CF-ME significantly reduced cell viability in all three GBM cell lines, the mixture of index compounds did not have the same effect. (**B**) Dose-response curves showing the effect of individual index compounds—caffeic acid, isoferulic acid, ferulic acid, and cimifugin—on the viability of U87 MG, A172, and T98G GBM cell lines. Cells were treated for 72 h with increasing concentrations of each compound. The results indicate that each compound has a slight inhibitory effect on cell viability, with caffeic acid being more effective at higher concentrations compared to the other compounds. * Asterisks indicate statistical significance compared to the control group ($p < 0.05$). Error bars represent the mean ± SD of three independent experiments.

4. Discussion

This study systematically evaluated the effects of water (CF-WE) and methanol (CF-ME) extracts on glioma cell growth and viability, focusing on the latter due to its demonstrated efficacy. *Cimicifuga*, a genus encompassing over 18 species, is traditionally used in various cultures [18]. According to the Taiwan Herbal Pharmacopeia (4th edition), Taiwan utilizes *C. foetida*, *C. heracleifolia*, and *C. dahurica*, each with distinct chemical profiles [9]. HPLC fingerprinting is commonly employed to confirm *Cimicifuga* species based on the presence of key compounds, including phenolic acids, triterpenes, and chromones. Specifically, the phenolic acids caffeic acid, ferulic acid, and isoferulic acid, as well as the chromone cimifugin, have been used as marker compounds to identify the botanical origin of *Cimicifuga* herbs [10,21,22]. According to He et al., only *C. foetida* contains all four marker compounds [21]. In the current study, HPLC fingerprinting confirmed that the use of *C. foetida* in our methanol extract (CF-ME) identified caffeic acid, ferulic acid, isoferulic acid, and cimifugin at concentrations of 0.11%, 0.11%, 0.78%, and 0.08%, respectively, with the isoferulic acid content exceeding 0.1%. These findings further revealed that CF-ME significantly inhibited glioma cell growth, inducing G_1 phase cell cycle arrest, apoptosis, and autophagy and suppressing metastasis. These results suggest CF-ME as a potential therapeutic agent for glioma treatment, particularly in combination with TMZ.

CF-ME induced G_1 phase cell cycle arrest in glioma cells, potentially mediated by upregulation of GADD45A and p21 and downregulation of CDK6. GADD45A is known to regulate cell survival and apoptosis by inducing G1 and G2/M cell cycle arrest or by activating the MAPK pathways [23]. Previous studies have shown that overexpression of GADD45A can cause G_2/M cell cycle arrest and reduce proliferation in T24 bladder cancer cells, as well as induce G_0/G_1 and G_2/M phase arrest in EBV$^+$ B lymphoma cells through the activation of the p38 MAPK/TAp73/GADD45A axis [24]. Additionally, GADD45A has been implicated in the regulation of apoptosis via p53 activation and induces apoptosis in pancreatic and rectal cancer cells via MAPK signaling. This mechanism of action is crucial, as it halts cell proliferation, thereby limiting tumor growth. The role of GADD45A in inducing p21 further supports this, as p21 contributes to cell cycle arrest by inhibiting cyclin-dependent kinases like CDK6 [25]. For instance, in liver cancer cells, the activation of GADD45A by isocorydine derivatives upregulates p21, which causes G2/M phase arrest and suppresses tumor growth [25]. Beyond its role in cell cycle arrest, CF-ME has been shown to induce apoptosis in glioma cells, likely through the activation of the caspase cascade and PARP cleavage, both of which are well-established pathways in programmed cell death. The dual action of CF-ME—inducing cell cycle arrest and promoting apoptosis—significantly enhances its anticancer effects, positioning it as a promising candidate for glioma treatment. Additionally, previous research on *Cimicifuga* species has shown that extracts from *C. dahurica* can inhibit breast cancer cell growth. Compounds such as caffeic acid and ferulic acid found in CF-ME have demonstrated anticancer activities in various cancer cell lines by inducing ROS production, disrupting mitochondrial membrane potential, and triggering apoptosis [26–29].

CF-ME induced not only apoptosis but also autophagy in glioma cells, a process that can either promote cell survival or lead to cell death, depending on the context. Apoptosis, characterized by the activation of the caspase cascade and PARP cleavage, is a well-established mechanism of programmed cell death [30]. CF-ME's ability to activate this pathway significantly contributes to its anticancer effects in glioma cells. In addition, the upregulation of LC3-II, a marker of autophagy, suggests that CF-ME also activates the autophagic pathway. While autophagy is typically a protective mechanism under stress, it can also lead to autophagic cell death when excessively activated. The dual induction of apoptosis and autophagy by CF-ME adds complexity to its anti-glioma activity, potentially enhancing its therapeutic efficacy.

This study demonstrated CF-ME's ability to suppress glioma cell metastasis by inhibiting cell invasion, migration, and adhesion. Metastasis, a critical factor in cancer progression, involves the spread of cancer cells from the primary site to distant sites. A key mechanism in metastasis is the EMT, which plays a vital role in tumor formation, malignancy, and cell migration. In GBM, EMT contributes to the cancer's aggressive invasiveness. Cells in an epithelial state exhibit strong adhesion, while those in a mesenchymal state have enhanced migratory abilities. The study found that CF-ME exerts its anti-metastatic effects by modulating EMT markers, specifically decreasing mesenchymal markers (N-cadherin and vimentin) and increasing the epithelial marker (E-cadherin). EMT-inducing transcription factors (EMT-TFs), such as ZEB, Snail, and Twist, facilitate cancer cell invasion by activating mesenchymal and suppressing epithelial markers. Previous studies have shown that drugs can inhibit metastasis by regulating these EMT-related molecules, as seen with polydatin and sotetsuflavone, in other cancer types. These findings suggest that CF-ME may effectively reduce tumor spread by targeting EMT in GBM.

TMZ is a lipophilic small molecule (194 Da) and an oral imidazotetrazine alkylating agent that can cross the blood–brain barrier, making it a key treatment for glioblastoma [7,31]. Its cytotoxic effects stem from O6-methylguanine formation, leading to mismatches with thymine during DNA replication. This mismatch triggers a futile cycle in the mismatch repair system, ultimately resulting in DNA damage and cell death [7]. Although

TMZ treatment often induces G_2-M phase arrest in glioma cells, it leads to apoptosis only in a subset of treated cells [32]. This study found that the combination of CF-ME and TMZ further reduced glioma cell viability, suggesting a synergistic effect. This synergy could potentially reduce TMZ dosage, thereby minimizing its side effects and overcoming drug resistance. These findings suggest that CF-ME is a promising adjuvant therapy for glioma with the potential to improve patient outcomes by enhancing the efficacy of TMZ treatment.

While this study provides valuable insights, the findings have broader implications for cancer treatment and patient outcomes. The ability of CF-ME to enhance the efficacy of TMZ, induce apoptosis and autophagy, and suppress metastasis suggests its potential to improve the therapeutic landscape for glioblastoma. CF-ME as an adjuvant therapy could reduce the required dosage of TMZ, mitigating side effects such as hematological toxicity and liver damage while maintaining or enhancing therapeutic efficacy. This is particularly relevant for patients who develop resistance to TMZ, as CF-ME's multi-targeted approach could offer a new avenue for overcoming treatment barriers. Furthermore, CF-ME's modulation of EMT markers and inhibition of metastasis may hold significant potential for preventing the aggressive spread of glioblastoma cells, thereby improving both treatment efficacy and patient survival rates. To move these findings from in vitro to in vivo applications, several key steps are required. First, studies using animal models are essential to validate CF-ME's efficacy, pharmacokinetics, and toxicity profile. These studies will help determine the optimal dosage, possible side effects, and interactions with existing treatments like TMZ. Upon successful preclinical testing, clinical trials could be initiated to evaluate CF-ME's impact on patient outcomes in a clinical setting, ensuring its effectiveness and safety in humans. This transition is critical for establishing CF-ME as a viable option in glioma therapy. However, this study has several limitations. First, the research was conducted using in vitro glioblastoma cell lines, which, while widely used for preliminary assessments of anticancer activity, do not fully replicate the complex tumor microenvironment found in vivo. As a result, the observed effects of CF-ME and its constituents may vary in animal models or clinical settings. Second, although the study identified several active pathways, including apoptosis, autophagy, and cell cycle arrest, the exact molecular targets and mechanisms remain unclear. Further studies are necessary to elucidate these mechanisms in detail and determine whether the effects are due to specific protein interactions or broader cellular stress responses. Additionally, when tested separately, the individual compounds in CF-ME exhibited only slight inhibitory effects on glioblastoma cell viability, suggesting that CF-ME's anti-glioma activity may result from synergistic interactions among its multiple constituents. However, the nature of these potential synergies was not explored in this study, and future research should focus on identifying the most effective combinations of compounds and testing the BBB permeability of these individual active compounds. Finally, the study did not address the pharmacokinetics, bioavailability, or potential toxicity of CF-ME or its individual components in vivo, which are critical factors in determining the therapeutic potential of any compound. Without these data, the immediate applicability of these findings to clinical contexts remains limited. Future studies should aim to fully characterize CF-ME's therapeutic efficacy and safety, including its effects on normal tissues and its potential interactions with standard glioma treatments like TMZ. Comprehensive research will be essential to translate these promising in vitro results into clinically viable treatments for glioma.

5. Conclusions

CF-ME exhibits potent anti-glioma activity through multiple mechanisms, including G1 phase cell cycle arrest, apoptosis, autophagy, and metastasis inhibition. CF-ME's ability to enhance the efficacy of TMZ further underscores its potential as an adjuvant therapy for gliomas. However, the individual index compounds identified in CF-ME did not fully account for its effects, suggesting potential synergistic actions among its multiple constituents. Future studies are warranted to isolate and identify these active compounds, to elucidate their interactions, and to explore their pharmacokinetics and bioavailability in vivo. Additionally, in-depth investigations into the safety, efficacy, and potential clinical applications of CF-ME are essential to fully realize its therapeutic potential. Overall, CF-ME stands out as a promising candidate for further development in glioma treatment, but comprehensive research is needed to translate these findings into clinical settings.

Author Contributions: Conceptualization, H.-P.T. and C.-J.L.; methodology, H.-P.T., M.-H.Y. and C.-J.L.; software, C.-H.C. and H.-P.T.; validation, C.-H.C. and C.-J.L.; formal analysis, C.-H.C., M.-H.Y. and C.-J.L.; investigation, H.-P.T., M.-H.Y. and C.-J.L.; resources, H.-P.T. and M.-H.Y.; data curation, C.-H.C., H.-P.T. and C.-J.L.; writing—original draft preparation, C.-H.C., H.-P.T. and C.-J.L.; writing—review and editing, C.-H.C., H.-P.T., M.-H.Y. and C.-J.L.; visualization, C.-H.C. and H.-P.T.; supervision, M.-H.Y. and C.-J.L.; project administration, H.-P.T. and C.-J.L.; funding acquisition, H.-P.T. and C.-J.L. All authors have read and agreed to the published version of the manuscript.

Funding: This work was supported by grants from the Kaohsiung Medical University Research Foundation (KMU-Q108022, KMU-M109019, KMU-M111008, and KMU-M112028) and the National Science and Technology Council (MOST109-2320-B-037-020).

Institutional Review Board Statement: This study did not involve human or animal participants.

Informed Consent Statement: Not applicable. This study did not involve human participants; thus, informed consent was not required.

Data Availability Statement: The original contributions presented in the study are included in the article, further inquiries can be directed to the corresponding author.

Conflicts of Interest: The authors declare no conflicts of interest.

References

1. Stoyanov, G.S.; Lyutfi, E.; Georgieva, R.; Georgiev, R.; Dzhenkov, D.L.; Petkova, L.; Ivanov, B.D.; Kaprelyan, A.; Ghenev, P. Reclassification of Glioblastoma Multiforme According to the 2021 World Health Organization Classification of Central Nervous System Tumors: A Single Institution Report and Practical Significance. *Cureus* **2022**, *14*, e21822. [CrossRef] [PubMed]
2. Ohgaki, H.; Kleihues, P. Population-based studies on incidence, survival rates, and genetic alterations in astrocytic and oligodendroglial gliomas. *J. Neuropathol. Exp. Neurol.* **2005**, *64*, 479–489. [CrossRef]
3. Ballestin, A.; Armocida, D.; Ribecco, V.; Seano, G. Peritumoral brain zone in glioblastoma: Biological, clinical and mechanical features. *Front. Immunol.* **2024**, *15*, 1347877. [CrossRef] [PubMed]
4. Samara, K.A.; Al Aghbari, Z.; Abusafia, A. GLIMPSE: A glioblastoma prognostication model using ensemble learning—A surveillance, epidemiology, and end results study. *Health Inf. Sci. Syst.* **2021**, *9*, 5. [CrossRef] [PubMed]
5. Tan, A.C.; Ashley, D.M.; Lopez, G.Y.; Malinzak, M.; Friedman, H.S.; Khasraw, M. Management of glioblastoma: State of the art and future directions. *CA Cancer J. Clin.* **2020**, *70*, 299–312. [CrossRef]
6. Fisher, J.P.; Adamson, D.C. Current FDA-Approved Therapies for High-Grade Malignant Gliomas. *Biomedicines* **2021**, *9*, 324. [CrossRef]
7. Strobel, H.; Baisch, T.; Fitzel, R.; Schilberg, K.; Siegelin, M.D.; Karpel-Massler, G.; Debatin, K.M.; Westhoff, M.A. Temozolomide and Other Alkylating Agents in Glioblastoma Therapy. *Biomedicines* **2019**, *7*, 69. [CrossRef]
8. Czarnywojtek, A.; Dyrka, K.; Borowska, M.; Koscinski, J.; Moskal, J.; Sawicka-Gutaj, N.; Lewandowska, A.; VAN Gool, S.; Fichna, M.; Gut, P.; et al. Temozolomide: A Cytostatic Drug That Is Still Important Today. *Acta Pol. Pharm.* **2022**, *79*, 763–775. [CrossRef]
9. Ministry of Health and Welfare. *Taiwan Herbal Pharmacopeia*, 4th ed.; Ministry of Health and Welfare: Taipei, Taiwan, 2022; p. 96.
10. Pharmaceutical Society of Taiwan. *Concise Traditional Chinese Herbal Medicine*, 2nd ed.; Pharmaceutical Society of Taiwan: Taipei, Taiwan, 2019; p. 145.
11. Wang, K.C.; Chang, J.S.; Lin, L.T.; Chiang, L.C.; Lin, C.C. Antiviral effect of cimicifugin from Cimicifuga foetida against human respiratory syncytial virus. *Am. J. Chin. Med.* **2012**, *40*, 1033–1045. [CrossRef]
12. Dai, X.; Yi, X.; Sun, Z.; Ruan, P. *Cimicifuga foetida* L. plus adefovir effectively inhibits the replication of hepatitis B virus in patients with chronic hepatitis B. *Biomed. Rep.* **2016**, *4*, 493–497. [CrossRef]
13. Thao, N.P.; Lee, Y.S.; Luyen, B.T.T.; Oanh, H.V.; Ali, I.; Arooj, M.; Koh, Y.S.; Yang, S.Y.; Kim, Y.H. Chemicals from Cimicifuga dahurica and Their Inhibitory Effects on Pro-inflammatory Cytokine Production by LPS-stimulated Bone Marrow-derived Dendritic Cells. *Nat. Prod. Sci.* **2018**, *24*, 194–198. [CrossRef]
14. Nishida, M.; Yoshimitsu, H.; Nohara, T. Three cycloartane glycosides from Cimicifuga rhizome and their immunosuppressive activities in mouse allogeneic mixed lymphocyte reaction. *Chem. Pharm. Bull.* **2003**, *51*, 354–356. [CrossRef] [PubMed]
15. Wu, D.; Yao, Q.; Chen, Y.; Hu, X.; Qing, C.; Qiu, M. The In Vitro and In Vivo Antitumor Activities of Tetracyclic Triterpenoids Compounds Actein and 26-Deoxyactein Isolated from Rhizome of *Cimicifuga foetida* L. *Molecules* **2016**, *21*, 1001. [CrossRef] [PubMed]

16. Sun, H.Y.; Liu, B.B.; Hu, J.Y.; Xu, L.J.; Chan, S.W.; Chan, C.O.; Mok, D.K.; Zhang, D.M.; Ye, W.C.; Chen, S.B. Novel cycloartane triterpenoid from *Cimicifuga foetida* (Sheng ma) induces mitochondrial apoptosis via inhibiting Raf/MEK/ERK pathway and Akt phosphorylation in human breast carcinoma MCF-7 cells. *Chin. Med.* **2016**, *11*, 1. [CrossRef]

17. Yue, G.G.; Xie, S.; Lee, J.K.; Kwok, H.F.; Gao, S.; Nian, Y.; Wu, X.X.; Wong, C.K.; Qiu, M.H.; Lau, C.B. New potential beneficial effects of actein, a triterpene glycoside isolated from Cimicifuga species, in breast cancer treatment. *Sci. Rep.* **2016**, *6*, 35263. [CrossRef]

18. Tian, Z.; Pan, R.; Chang, Q.; Si, J.; Xiao, P.; Wu, E. *Cimicifuga foetida* extract inhibits proliferation of hepatocellular cells via induction of cell cycle arrest and apoptosis. *J. Ethnopharmacol.* **2007**, *114*, 227–233. [CrossRef]

19. Kong, Y.; Li, F.; Nian, Y.; Zhou, Z.; Yang, R.; Qiu, M.H.; Chen, C. KHF16 is a Leading Structure from *Cimicifuga foetida* that Suppresses Breast Cancer Partially by Inhibiting the NF-κB Signaling Pathway. *Theranostics* **2016**, *6*, 875–886. [CrossRef]

20. Zhang, Y.; Lian, J.; Wang, X. Actein inhibits cell proliferation and migration and promotes cell apoptosis in human non-small cell lung cancer cells. *Oncol. Lett.* **2018**, *15*, 3155–3160. [CrossRef]

21. He, K.; Pauli, G.F.; Zheng, B.; Wang, H.; Bai, N.; Peng, T.; Roller, M.; Zheng, Q. Cimicifuga species identification by high performance liquid chromatography-photodiode array/mass spectrometric/evaporative light scattering detection for quality control of black cohosh products. *J. Chromatogr. A* **2006**, *1112*, 241–254. [CrossRef]

22. Fam, M.; Ding, F.; Qin, Q.; Yang, B.; Cai, B. Study on HPLC Fingerprints of Different Species of Cimicifuga Rhizoma. *Chin. J. Mod. Appl. Pharm.* **2019**, *36*, 1244–1248. [CrossRef]

23. Liebermann, D.A.; Hoffman, B. Gadd45 in stress signaling. *J. Mol. Signal.* **2008**, *3*, 15. [CrossRef] [PubMed]

24. Park, G.B.; Choi, Y.; Kim, Y.S.; Lee, H.K.; Kim, D.; Hur, D.Y. Silencing of PKCη induces cycle arrest of EBV⁺ B lymphoma cells by upregulating expression of p38-MAPK/TAp73/GADD45α and increases susceptibility to chemotherapeutic agents. *Cancer Lett.* **2014**, *350*, 5–14. [CrossRef] [PubMed]

25. Chen, L.; Tian, H.; Li, M.; Ge, C.; Zhao, F.; Zhang, L.; Li, H.; Liu, J.; Wang, T.; Yao, M.; et al. Derivate isocorydine inhibits cell proliferation in hepatocellular carcinoma cell lines by inducing G2/M cell cycle arrest and apoptosis. *Tumour Biol.* **2016**, *37*, 5951–5961. [CrossRef] [PubMed]

26. Rajendra Prasad, N.; Karthikeyan, A.; Karthikeyan, S.; Reddy, B.V. Inhibitory effect of caffeic acid on cancer cell proliferation by oxidative mechanism in human HT-1080 fibrosarcoma cell line. *Mol. Cell Biochem.* **2011**, *349*, 11–19. [CrossRef]

27. Eroglu, C.; Secme, M.; Bagci, G.; Dodurga, Y. Assessment of the anticancer mechanism of ferulic acid via cell cycle and apoptotic pathways in human prostate cancer cell lines. *Tumour Biol.* **2015**, *36*, 9437–9446. [CrossRef]

28. Ji, L.; Zhong, B.; Jiang, X.; Mao, F.; Liu, G.; Song, B.; Wang, C.Y.; Jiao, Y.; Wang, J.P.; Xu, Z.B.; et al. Actein induces autophagy and apoptosis in human bladder cancer by potentiating ROS/JNK and inhibiting AKT pathways. *Oncotarget* **2017**, *8*, 112498–112515. [CrossRef]

29. Hao, Y.; Huang, Y.; Chen, J.; Li, J.; Yuan, Y.; Wang, M.; Han, L.; Xin, X.; Wang, H.; Lin, D.; et al. Exopolysaccharide from Cryptococcus heimaeyensis S20 induces autophagic cell death in non-small cell lung cancer cells via ROS/p38 and ROS/ERK signalling. *Cell Prolif.* **2020**, *53*, e12869. [CrossRef]

30. Santagostino, S.F.; Assenmacher, C.A.; Tarrant, J.C.; Adedeji, A.O.; Radaelli, E. Mechanisms of Regulated Cell Death: Current Perspectives. *Vet. Pathol.* **2021**, *58*, 596–623. [CrossRef]

31. Ortiz, R.; Perazzoli, G.; Cabeza, L.; Jimenez-Luna, C.; Luque, R.; Prados, J.; Melguizo, C. Temozolomide: An Updated Overview of Resistance Mechanisms, Nanotechnology Advances and Clinical Applications. *Curr. Neuropharmacol.* **2021**, *19*, 513–537. [CrossRef]

32. Yan, Y.; Xu, Z.; Dai, S.; Qian, L.; Sun, L.; Gong, Z. Targeting autophagy to sensitive glioma to temozolomide treatment. *J. Exp. Clin. Cancer Res.* **2016**, *35*, 23. [CrossRef]

Article

High Polygenic Risk Scores Positively Associated with Gastric Cancer Risk Interact with Coffee and Polyphenol Intake and Smoking Status in Korean Adults

Meiling Liu [1,†], **Sang-Shin Song** [2,†] **and Sunmin Park** [2,*]

[1] Department of Chemical Engineering, Shanxi Institute of Science and Technology, Jincheng 048000, China; liumeiling@sxist.edu.cn

[2] Department of Food and Nutrition, Obesity/Diabetes Research Center, Hoseo University, Asan 31499, Republic of Korea; ssscho@empas.com

[*] Correspondence: smpark@hoseo.edu; Tel.: +82-41-540-5345; Fax: +82-41-548-0670

[†] These authors contributed equally to this work.

Abstract: Background/Objectives: This study investigated the relationship between single nucleotide polymorphisms (SNPs) and gastric cancer (GC) risk, while also examining the interaction of genetic factors with lifestyle variables including the nutrient and bioactive compound intake in Korean adults of a large hospital-based cohort. Methods: We conducted a genome-wide association study (GWAS) comparing GC patients (n = 312) with healthy controls without cancers (n = 47,994) to identify relevant genetic variants. Generalized multifactor dimensionality reduction (GMDR) was employed to detect SNP interactions between diets and lifestyles. We utilized polygenic risk scores (PRSs) to assess individuals' GC risk based on multiple SNP loci. Among the selected SNPs, since *SEMA3C*_rs1527482 was a missense mutation, bioactive compounds which decrease the binding energy were found with its wild and mutated proteins by molecular docking analysis. Results: Individuals with high PRSs exhibited a 4.12-fold increased risk of GC compared to those with low PRSs. Additional factors associated with elevated GC risk included a low white blood cell count (OR = 5.13), smoking (OR = 3.83), and low coffee consumption (OR = 6.30). The *SEMA3C*_rs1527482 variant showed a positive correlation with GC risk. Molecular docking analyses suggested that certain polyphenols, including theaflavate, rugosin E, vitisifuran B, and plantacyanin, reduced the binding free energy in both wild-type and mutated *SEMA3C*_rs1527482. However, some polyphenols exhibited differential binding energies between its wild and mutated forms, suggesting they might modulate wild and mutated proteins differently. Conclusion: High PRSs and *SEMA3C*_rs1527482 interact with immune function, coffee intake, polyphenol consumption, and smoking status to influence GC risk. These findings could contribute to developing personalized nutrition and lifestyle interventions to reduce GC risk.

Keywords: gastric cancer; white blood cells; polygenic variants; coffee; smoking status; polyphenols

Citation: Liu, M.; Song, S.-S.; Park, S. High Polygenic Risk Scores Positively Associated with Gastric Cancer Risk Interact with Coffee and Polyphenol Intake and Smoking Status in Korean Adults. *Nutrients* **2024**, *16*, 3263. https://doi.org/10.3390/nu16193263

Academic Editors: Yi-Wen Liu and Ching-Hsein Chen

Received: 12 August 2024
Revised: 20 September 2024
Accepted: 25 September 2024
Published: 27 September 2024

1. Introduction

Gastric cancer remains a significant global health concern, ranking as the fifth most common cancer worldwide (5.6%) according to 2020 estimates from the International Agency for Research on Cancer (IARC) [1]. Its mortality rate ranks fourth (7.7%) among cancers globally [1]. In Korea, the gastric cancer incidence in men was second only to thyroid cancer as of 2018. The disease primarily manifests as adenocarcinomas, originating from the stomach's glandular cells and accounting for 95% of cases, with a notably higher prevalence in men [2].

The etiology of gastric cancer is multifaceted, involving complex interactions between bacterial infections, environmental factors, and genetic predisposition. Modifiable risk factors, including lifestyle and dietary habits, are crucial in gastric cancer development [3].

Geographic variations in the disease's presentation suggest that the interplay between environmental and genetic factors may differ across populations.

Previous genetic studies have identified several single-nucleotide polymorphisms (SNPs) associated with gastric cancer risk. These include variants in tumor suppressor genes (cadherin 1 [*CDH1*], tumor protein 53 [*TP53*]) [4,5], genes involved in mucosal protection against *Helicobacter pylori* (*H. pylori*) (interleukin-1 beta [*IL1B*], interleukin-1 receptor antagonist protein [*IL1RN*], and tumor necrosis factor-α [*TNF-α*]) [6,7], carcinogen metabolism genes (cytochrome P450 family 2 subfamily E member 1 [*CYP2E1*], glutathione S-transferase mu 1 [*GSTM1*]) [8,9], and those related to DNA repair (methylenetetrahydrofolate reductase [*MTHFR*] and X-ray repair cross-complementing group 1 gene [*XRCC1*]) [10,11]. Recent research in Korean populations has uncovered additional genetic markers, including protein kinase AMP-activated catalytic subunit alpha 1 (*PRKAA1*) [12], mucin 1 (*MUC1*), phospholipase C epsilon 1 (*PLCE1*) [13], and prostate stem cell antigen (*PSCA*) [14].

While numerous studies have explored the relationship between dietary factors, environmental influences, and gastric cancer risk, the results have often been inconclusive. Moreover, genetic investigations have typically focused on individual susceptibility genes rather than considering the cumulative effect of multiple genetic variants. The concept of polygenic risk scores (PRSs) offers a more comprehensive approach to assessing genetic predisposition to gastric cancer.

Limited research has examined the interaction between dietary patterns and combinations of genetic variants in identifying high-risk groups for gastric cancer. Previous studies have analyzed the association between PRSs and the age at onset of gastric cancer [15] and predicted gastric cancer risk using PRSs according to *H. pylori* infection status [16]. A meta-analysis has also suggested that a healthy lifestyle can mitigate the genetic risk of gastric cancer [17].

This study aimed to address this gap by investigating the hypothesis that polygenic variants interact with metabolic parameters, dietary intake, and lifestyle factors to influence gastric cancer risk. Utilizing data from the Korean Genome and Epidemiology Study (KoGES), we evaluated the interplay between PRSs for gastric cancer and various factors, including immune function, lifestyle choices, and environmental exposures in the Korean adult population. Our findings suggested significant interactions between genetic predisposition and modifiable risk factors, highlighting potential avenues for targeted interventions to mitigate the gastric cancer risk in genetically susceptible individuals. These results underscored the need for further research to validate and expand upon these observations, potentially informing future strategies for early prevention and risk reduction in high-risk populations.

2. Methods

2.1. Study Population

This study utilized data from the Korea Genomic and Epidemiological Study (KoGES), a large urban hospital cohort study conducted by the Korea Disease Control and Prevention Agency (KDCA) between 2004 and 2013. The cohort included Korean adults aged ≥40 years (*n* = 58,701) who volunteered to participate. The KoGES protocol was approved by the institutional review boards of the Korean National Institute of Health (Approval Code: KBP-2015-055, approved on 20 August 2015), and all participants provided written informed consent.

2.2. Case-Control Selection

Participants were categorized into two groups based on their self-reported gastric cancer diagnosis. The gastric cancer group (GC, cases; *n* = 312) comprised individuals who reported a physician-diagnosed gastric cancer, and subjects with any history of cancer other than gastric cancer were excluded (*n* = 10,395) (Figure 1). The non-gastric cancer group (N-GC, controls; *n* = 47,994) included participants without any reported cancer diagnoses. While different cancers might share some biological pathways or risk factors, their effects could act as a bias in determining gastric cancer risk. By excluding other cancers, we reduced the risk of these shared factors obscuring gastric cancer-specific genetic associations. This targeted approach allowed for a more precise investigation into the genetic basis and risk factors unique to gastric cancer, potentially leading to clearer and more robust results. Ultimately, this exclusion criterion helped to isolate the genetic factors specifically associated with gastric cancer, enabling a more accurate analysis of its pathogenesis.

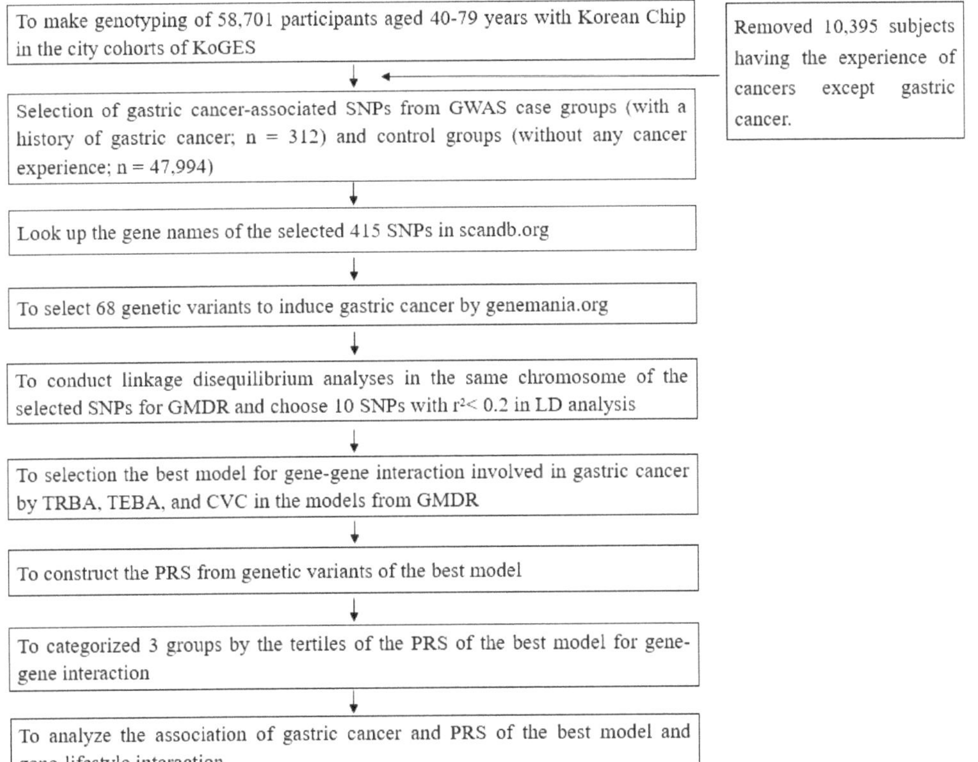

Figure 1. Flow chart for the generation of polygenic risk scores (PRSs) that influence gastric cancer risk and their interaction with metabolic parameters and lifestyles. KoGES, Korean Genome and Epidemiology Study; SNP, single nucleotide polymorphism; GWAS, genome-wide association study; LD, linkage disequilibrium; GMDR, generalized multifactor dimensionality reduction; TRBA, trained balanced accuracy; TEBA, test balance accuracy; CVC, cross-validation consistency; PRS, polygenic risk scores.

2.3. Anthropometric and Biochemical Measurements

Demographic information, including age, income, education, alcohol consumption, smoking history, and physical activity, was collected through health interviews [18]. Education levels were categorized as less than high school, high school, and college or higher. Monthly household income was classified as low (<$2000), medium ($2000–4000), and high (>$4000). Smoking status was categorized as non-smoker, former smoker, or current smoker, while alcohol consumption was classified as light drinker (0–20 g/day) or moderate drinker (>20 g/day) [19].

Anthropometric measurements, including weight, height, and waist circumference, were taken by trained specialists using standardized procedures. Body mass index (BMI) was calculated as weight (kg) divided by height (m) squared. Blood samples were collected after a minimum 12 h fast to ensure accurate biochemical analysis [20]. A Hitachi 7600 automatic analyzer (Hitachi, Tokyo, Japan) was used to measure fasting glucose, serum total cholesterol, triglycerides, and high-density lipoprotein cholesterol (HDL-C) levels. White blood cell (WBC) counts were determined from ethylenediaminetetraacetic acid (EDTA)-treated blood samples. Blood pressure measurements were taken with participants in a seated position, with the right arm properly supported at heart level.

2.4. Dietary Assessment

Dietary intake was evaluated using a semiquantitative food frequency questionnaire (SQFFQ) developed and validated for the KoGES. The SQFFQ covered 106 food items, assessing long-term dietary patterns based on portion size and frequency of consumption. Nutrient intake was estimated using the computer-aided nutritional analysis program CAN-pro version 3.0, developed by the Korean Nutrition Society [21]. The dietary inflammatory index (DII) was calculated to assess the pro-inflammatory potential of participants' diets. The index was computed using 38 food and nutrient components, excluding garlic, ginger, saffron, and turmeric due to a lack of intake data. DII scores were determined by multiplying component-specific inflammatory scores by daily intake, summing these products, and dividing by 100.

2.5. Genotyping and Quality Control

Genomic DNA extraction from whole blood and genotyping were performed by the Center for Genome Science at the National Institute of Health in Korea using a Korean-specific gene chip produced by Affymetrix (Santa Clara, CA, USA) [22]. Genotyping accuracy was determined using the Bayesian robust linear modeling with Mahalanobis distance genotyping algorithm [22]. Strict quality control parameters were applied to ensure data accuracy and representativeness, including genotyping accuracy ($\geq$98%), genotype missing rate (<4%), heterozygosity (<30%), Hardy–Weinberg equilibrium ($p > 0.05$), and minor allele frequency (MAF > 0.01) [23].

2.6. Selection of Interacting Genetic Variants for Gastric Cancer

A polygenic risk score (PRS) for gastric cancer was developed using a multi-step process (Figure 1). Genome-wide association study (GWAS) methods were employed to identify genetic loci significantly associated with gastric cancer risk after adjusting for age, gender, region of residence, survey year, BMI, daily energy intake, education level, and income. The statistical significance was used for a more liberal cut-off of Bonferroni correction ($p < 5 \times 10^{-4}$) since there was a limited number of participants with gastric cancer. This approach allowed us to identify potentially important SNPs that might have been missed with a stricter threshold, while still accounting for multiple comparisons. From 415 initially identified variants, gene names were determined using scandb.org (accessed on 5 December 2021) and further screened using genemania.org. Linkage disequilibrium (LD) analysis was performed using Haploview 4.2 in the PLINK toolset to identify and exclude strongly linked SNPs ($D' < 0.4$) among the selected 68 SNPs. Finally, 10 potential genetic variants from the best model and on the same chromosome were selected.

The best model for SNP–SNP interactions influencing gastric cancer risk was determined using generalized multifactor dimensionality reduction (GMDR) [24], based on trained balanced accuracy (TRBA), test balance accuracy (TEBA), and cross-validation consistency (CVC). The significance threshold was set at $p < 0.001$ to account for multiple tests. The number of risk alleles for each SNP in the best model was added to obtain the PRS for each individual. The calculated PRS was classified into tertiles, that is, the population was divided into three risk levels: low risk, medium risk, and high risk. A higher PRS value indicates that the individual has more risk alleles in the best gene interaction model and thus has a higher risk of gastric cancer.

2.7. Molecular Docking and Molecular Dynamics Simulation (MDS) of Semaphorin-3C (SEMA3C)

The I-TASSER website (https://zhanggroup.org/I-TASSER/ (accessed on 13 January 2023)) was used to predict the Protein Data Bank (PDB) structures of wild-type and mutant *SEMA3C* (Semaphorin 3C) proteins. The PDB format of *SEMA3C* protein and food components (n = 20,000) was converted to PDBQT files using AutoDock Tools 1.5.6 (Molecular Graphics Laboratory, Scripps Research Institute, Jupiter, FL, USA) [25]. The active sites, active functional pockets of SEMA3C, and mutated sites were found using the ProteinsPlus website (https://proteins.plus/ (accessed on 30 January 2023)) and included in molecular docking. Water molecules attached to the protein and food components were removed [25]. After docking was completed, the output binding energy estimates for each docked pose were analyzed, and those food components with binding energies less than −10.0 kcal/mol were selected as potential binding partners [26]. Binding affinity is a measure of the strength of binding between two molecules. It is usually inversely proportional to the binding free energy, that is, the lower the binding free energy, the higher the binding affinity, making the interaction between the two molecules more stable [26].

Molecular dynamics simulation (MDS) was used to study the conformational changes of protein structures after binding to specific food ingredients. After determining the optimal docking pose, molecular dynamics simulation was used to simulate the dynamic behavior of the complex after *SEMA3C* bound to the food ingredient. During the simulation, the Chemistry at Harvard Macromolecular Mechanics (CHARMM) force field was applied. In order to make the simulation closer to the real situation and more accurately reflect the behavior of the protein under physiological conditions, the protein was solvated, that is, placed in an environment of water molecules (or other solvents). The simulation was carried out for 10 nanoseconds (ns), and parameters such as the root mean square deviation (RMSD), root mean square fluctuation (RMSF), and hydrogen bond value were calculated during the simulation. These parameters are used to evaluate the stability of a protein conformation, the volatility of an atomic position, and the formation and breaking of intermolecular hydrogen bonds.

2.8. Statistical Analyses

Statistical analyses were performed using PLINK version 2.0 and SPSS version 24.0 (IBM SPSS Statistics, New York, NY, USA). SNP–SNP interactions were screened using GMDR, and the significance of SNP–SNP interactions was assessed by the signed-rank test of TRBA and TEBA. The best SNP–SNP interaction model with a p-value < 0.05 was selected. Covariates such as age, sex, BMI, region of residence, physical activity education, income level, smoking, drinking, and energy intake were adjusted or not [20]. Ten-fold cross-validation is a commonly used method to evaluate model performance, especially for cases with large sample sizes (n > 1000). This method ensures that every sample in the dataset has the opportunity to be used as a test set, thereby more comprehensively evaluating the generalization ability of the model and more accurately evaluating the reliability of the CVC [27]. Using the best model, as determined by GMDR analysis, the risk allele of each SNP in the selected best model was counted as 1 [28]. For example, when the G allele was associated with an increased risk of gastric cancer, the TT, GT, and GG were assigned scores of 0, 1, and 2. PRSs were calculated by summing the risk allele scores of each SNP. The best model with 8 SNPs was divided into three categories (0–6, 7–8, and $\geq$9) by tertile, that is, into low-, medium-, and high-PRS groups, respectively. Adjusted odds ratios (ORs) and 95% confidence interval (CI) for gastric cancer risk with PRS were calculated after adjusting for covariates. The covariates included were age, gender, BMI, region of residence, physical activity, education, income level, smoking, alcohol consumption, years with gastric cancer, and energy intake.

Descriptive statistical analyses were performed for categorical variables, such as sex and lifestyle, which were calculated based on the frequency distribution of the PRS tertiles (i.e., low-, medium-, and high-PRS groups). For the frequency distribution of categorical variables, the Chi-squared test was used for analysis. For quantitative variables, the

Kolmogorov–Smirnov test was used to check their normality due to the large sample size, and was achieved by the proc univariate procedure. For variables that met the normal distribution, the means and standard errors were calculated according to the PRS tertile categories or the presence or absence of gastric cancer. To determine whether the differences were significant, a one-way analysis of variance (ANOVA) with covariance adjustment was used, and multiple comparisons between groups were performed using the Tukey test. In addition, to account for the interaction between PRSs and dietary intake parameters, participants were divided into high-intake and low-intake groups. After adjusting for covariates, two-way ANOVA with main effects and interaction terms was used to explore the interaction between PRSs and lifestyle parameters. Throughout the statistical analysis, a *p*-value of <0.05 was used as the criterion for statistical significance. This means that only when the observed differences reach or exceed this statistical significance level are they considered to be real and not due to random errors or sampling variations. Such statistical analysis methods help ensure the accuracy and reliability of research results.

3. Results

3.1. Comparison of the General Characteristics of the Participants

Table 1 describes the demographic and clinical characteristics of the participants, including 312 participants with gastric cancer and 47,994 without cancer. The average age of the GC group was 58 years, significantly higher than that of the N-GC group. The risk of gastric cancer in men was 3.37 times higher than in women ($p < 0.001$). The BMI ($p < 0.001$), plasma concentrations of total cholesterol and triglycerides ($p < 0.01$), and WBC counts ($p < 0.05$) were significantly lower in the GC group than in the N-GC group. The amounts of participants with high income (>$4000) and education levels (high school or > College) were also lower in the GC group than in the N-GC group (Table 1).

Table 1. Socio-economic and metabolic characteristics of the participants according to gastric cancer.

	Non-Gastric Cancer (*n* = 47,994)	Gastric Cancer (*n* = 312)	Adjusted OR (95% CI)
Age (years) [1]	53.48 ± 8.04	58.12 ± 7.85 ***	1.455 (0.987~2.145)
Genders (men: N, %)	16,808 (35.0)	168 (53.8) ***	3.369 (2.173~5.225)
Initial menstruation age [2]	15.10 ± 1.76	15.40 ± 1.83 *	1.606 (0.771~3.345)
Menopause age [3]	49.30 ± 4.81	49.50 ± 4.29	0.902 (0.518~1.571)
Pregnancy experience (Yes, %) [4]	30,076 (96.6)	137 (95.8)	0.543 (0.162~1.822)
Hormone replacement therapy (Yes, %)	4963 (26.4)	28 (26.2)	0.620 (0.324~1.188)
Oral contraceptive (Yes, %)	4816 (15.5)	22 (15.4)	1.277 (0.649~2.509)
Breastfeeding (Yes, %)	25,453 (85.8)	122 (89.1)	1.234 (0.575~2.651)
Ovariectomy (Yes, %)	664 (7.5)	5 (10.4)	1.217 (0.312~4.743)
Hysterectomy (Yes, %)	3434 (11.1)	19 (13.3)	1.012 (0.519~1.976)
Body mass index (BMI, kg/m^2) [5]	24.00 ± 2.88	22.40 ± 3.12 ***	0.353 (0.222~0.563)
Waist circumference (cm) [6]	80.90 ± 8.65	78.00 ± 9.04 ***	1.422 (0.636~3.179)
Plasma total cholesterol (mg/dL) [7]	197.6 ± 35.7	186.3 ± 36.3 ***	0.492 (0.277~0.874)
Plasma triglyceride (mg/dL) [8]	119.4 ± 64.9	99.6 ± 52.3 ***	0.606 (0.380~0.968)
Hypertension (N, %) [9]	13,709 (28.6)	70 (22.4) *	0.757 (0.497~1.152)
Type 2 diabetes (N, %) [10]	4256 (9.1)	28 (9.2)	0.755 (0.431~1.324)
White blood cell counts (10^9/L) [11]	5.73 ± 1.55	5.37 ± 1.40 ***	0.426 (0.237~0.765)
Plasma hs-CRP (mg/dL) [12]	0.14 ± 0.36	0.15 ± 0.47	2.080 (0.937~4.615)
Education (Number, %) [13]			
<High school	14,110 (29.7)	122 (39.2) *	
High school	20,658 (43.4)	110 (35.4)	0.602 (0.388~0.935)
College more	12,778 (26.9)	79 (25.4)	0.480 (0.301~0.764)

Table 1. *Cont.*

	Non-Gastric Cancer (*n* = 47,994)	Gastric Cancer (*n* = 312)	Adjusted OR (95% CI)
Income (Number, %) [14]			
<$2000/month	13,851 (30.5)	125 (42.7) ***	
$2000–4000	27,761 (61.1)	156 (53.2)	0.803 (0.547~1.181)
>$4000	3851 (8.5)	12 (4.1)	0.288 (0.097~0.855)

The values represent means ± standard errors or number of the adults aged ≥40 (percentage of each group). The cutoff points of the reference were as follows: [1] <55 years old for age, [2] <14 years old for initial menstruation age, [3] <50 years old for menopause age, [4] non-pregnancy experience, [5] <25 kg/m^2 BMI, [6] <90 cm for men and 85 cm for women waist circumferences, [7] <230 mg/dL plasma total cholesterol concentrations, [8] <150 mg/dL plasma triglyceride concentrations, [9] <140 mmHg systolic blood pressure, and <90 mmHg diastolic blood pressure plus hypertension medication, [10] <126 mL/dL fasting serum glucose plus diabetic drug intake, [11] <4 × 10^9/L white blood cell counts, [12] <0.5 mg/dL serum high sensitive-C-reactive protein (hs-CRP) concentrations, [13] high school graduation, and [14] <$2000/month income. Adjusted odds ratios (ORs) are shown after adjusting for covariates, including age, gender, body mass index (BMI), residence area, physical activity, education, smoking, years with gastric cancer, and intake of alcohol and energy by logistic regression models. * Significant differences by the non-gastric cancer group at $p < 0.05$, *** $p < 0.001$.

3.2. Comparison of Nutrient Intakes of the GC and N-GC Groups

Table 2 describes the nutrient intakes of the participants with and without gastric cancers. There was no difference in the intake of energy, carbohydrates, proteins, fats, sodium, and fiber between the GC and N-GC groups. The prevalence of previous smoking was higher in the GC group ($p < 0.001$), but the alcohol and coffee intake ($p < 0.05$) were lower than in the N-GC group. The incidence of gastric cancer may be possibly related to reduced alcohol and coffee consumption (Table 2). The total phenol intake was not significantly different between the N-GC and GC groups. However, the DII was lower in the GC group than in the N-GC group, indicating that the individuals experiencing GC had a better diet (Table 2).

Table 2. Nutrient intake and dietary patterns of the participants according to gastric cancer presence.

	Non-Gastric Cancer (*n* = 47,994)	Gastric Cancer (*n* = 312)	Adjusted OR (95% CI)
Energy intake [1] (%)	98.70 ± 31.5	91.80 ± 32.6 ***	0.976 (0.683~1.397)
Carbohydrate intake (En%) [2]	71.53 ± 7.01	73.24 ± 7.16 ***	0.928 (0.552~1.561)
Protein intake (En%) [3]	13.45 ± 2.59	13.17 ± 2.63	1.050 (0.749~1.472)
Fat intake (En%) [4]	14.00 ± 5.43	12.63 ± 5.56 ***	0.744 (0.502~1.103)
Na intake (mg/day) [5]	2454 ± 1389	2387 ± 1549	0.940 (0.640~1.380)
Fiber intake(g/day) [6]	5.71 ± 2.83	5.90 ± 3.27	0.895 (0.198~4.051)
Exercise (Number, %)			
No	21,927 (45.8)	121 (38.9) *	
Yes	25,932 (54.2)	190 (61.1)	1.136 (0.801~1.612)
Smoking (Number, %)			
No	34,996 (73.1)	185 (59.7) ***	
Former smoking	7484 (15.6)	101 (32.6)	2.715 (1.558~4.731)
Smoking	5383 (11.3)	24 (7.7)	0.628 (0.282~1.396)
Alcohol intake (Number, %)			
Mild drink (0–20 g)	45,383 (95.2)	307 (98.4) **	
Moderate drink (≥20 g)	2291 (4.8)	5 (1.6)	0.181 (0.039~0.840)
Coffee intake (Number, %) [7]			
Low	15,427 (32.4)	136 (43.7) ***	
High	32,145 (67.6)	175 (56.3)	0.658 (0.467~0.927)

Table 2. *Cont.*

	Non-Gastric Cancer (n = 47,994)	Gastric Cancer (n = 312)	Adjusted OR (95% CI)
Multivitamin			
No	43,157 (89.9)	281 (90.1)	
Yes	4837 (10.1)	31 (9.9)	0.775 (0.450~1.333)
Total phenol (g/day)	2.51 ± 0.005	2.52 ± 0.041	1.204 (0.999~1.451)
Dietary inflammatory index	−19.9 ± 0.067	−21.5 ± 0.56 **	0.857 (0.716~1.026)
Fried food (Number, %) [8]			
Low	45,184 (94.8)	300 (96.5)	
High	2481 (5.2)	11 (3.5)	1.647 (0.645~4.210)

The values represent means ± standard errors or number of the adults aged ≥40 (percentage of each group). Adjusted odds ratios (ORs) are shown after adjusting for covariates, including age, gender, BMI, residence area, physical activity, education, smoking, years with gastric cancer, and intake of alcohol and energy by logistic regression models. The cutoff points of the reference were as follows: [1] <estimated energy intake, [2] <65 energy % carbohydrate intake, [3] <13 energy % protein intake, [4] <20 energy % fat intake, [5] <1600 sodium intake, [6] <14 fiber intake, [7] <3 g/day coffee drinking, and [8] <1 time/week fried food. En%, energy percent. * Significant differences by the non-gastric cancer group at $p < 0.05$, ** at $p < 0.01$, *** $p < 0.001$.

3.3. Genetic Variants for Gastric Cancer Risk and the Best Model for Gene-Gene Interactions

The best gene variant–gene variant interaction model associated with gastric cancer risk was evaluated by the GMDR method. In order to select the best model among the 10 models shown in Table S1, statistical indicators such as the TRBA, TEBA, and CVC were tested. These indicators help to evaluate the predictive performance and stability of a model. When selecting the best model among the 10 models, the covariates listed in Table S1 were adjusted or not. By adjusting these covariates, the association between gene variants and gastric cancer risk can be more accurately evaluated and the influence of confounding factors can be reduced. The 10 selected SNPs were as follows: rs7521784 of disabled-1 (*DAB1*) on chromosome 1, rs12693006 of pyruvate dehydrogenase kinase-1 (*PDK1*) on chromosome 2, rs1045653 of dedicator-of-cytokinesis-10 (*DOCK10*) on chromosome 2, rs9835646 of zinc finger and BTB domain 20 (*ZBTB20*) on chromosome 3, rs630760 of Kalirin RhoGEF Kinase (*KALRN*) on chromosome 3, rs11946315 of a disintegrin and metalloprotease 29 (*ADAM29*) on chromosome4, rs1207808 of membrane-associated guanylate kinase 2 (*MAGI2*) on chromosome 7, rs58499534 of cub sushi multiple domains-1 (*CSMD1*) on chromosome 8, rs10831776 of molecule interacting with CasL-2 (*MICAL2*) on chromosome 11, and rs205881 of casein kinase IIA1 (*CSNK2A1*) on chromosome 20 (Table 3). Each genetic variant was significantly associated with gastric cancer (OR = 0.61–1.59; $p = 1.70 \times 10^{-6}$ to 0.0008587). The genotype frequency distribution met the HWE criteria ($p > 0.05$), and the minor allele frequency (MAF) value was $p > 0.01$ (Table 3).

Table 3. The characteristics of the 10 genetic variants of genes in gastric cancer used for the generalized multifactor dimensionality reduction analysis in adults aged >40.

Chr [1]	SNP [2]	Position	Mi [3]	Ma [4]	OR [5]	[6] p Value Adjusted	[7] MAF	[8] p Value for HWE	Gene	Functional Consequence
1	rs7521784	58175325	A	G	1.38	3.99×10^{-4}	0.4178	0.7795	*DAB1*	Upstream of transcript
2	rs12693006	173467213	C	T	1.59	1.70×10^{-6}	0.2374	0.6649	*PDK1*	3′ UTR
2	rs1045653	225630435	T	C	0.63	1.90×10^{-5}	0.3389	0.2458	*DOCK10*	3′ UTR
3	rs9835646	114148557	A	C	0.61	4.44×10^{-5}	0.196	0.4959	*ZBTB20*	Upstream of transcript
3	rs630760	124149174	G	A	1.48	2.53×10^{-4}	0.1762	0.3554	*KALRN*	Downstream of transcript
4	rs11946315	175870844	C	T	0.69	4.59×10^{-4}	0.2759	0.2781	*ADAM29*	Intron
7	rs1207808	78496427	C	G	0.66	2.28×10^{-4}	0.2762	0.3319	*MAGI2*	Upstream of transcript
7	rs1527482	80427530	T	C	1.93	2.60×10^{-5}	0.055	0.2334	*SEMA3C*	Missense
8	rs58499534	3471561	G	A	1.58	3.85×10^{-6}	0.2156	0.2413	*CSMD1*	Upstream of transcript
11	rs10831776	12297403	G	A	0.68	3.46×10^{-4}	0.2622	0.5937	*MICAL2*	Intron
20	rs205881	486771	T	C	1.47	1.03×10^{-4}	0.2407	0.4981	*CSNK2A1*	Intron

[1] Chromosome; [2] single-nucleotide polymorphism; [3] minor allele; [4] major allele; [5] odds ratio; [6] p-value for OR after adjusting for age, gender, body mass index, residence area, physical activity, education, smoking, and intake of alcohol and energy; [7] minor allele frequency; [8] Hardy–Weinberg equilibrium.

This study listed multiple models in Table S1 and found that the eight-SNP model had the lowest p-value among all models. This suggested that the eight-SNP model was the best in revealing the association between SNP–SNP interaction and gastric cancer risk. In addition, the cross-validation consistency (CVC) value of the model was 10/10, which further confirmed the stability and reliability of the model. The CVC value reflects the prediction consistency of the model on different data subsets, and a high CVC value indicates that the model can maintain stable prediction performance under different conditions. As a result, this model, which included eight SNPs, including *DAB1_rs7521784*, *PDK1_rs12693006*, *DOCK10_rs1045653*, *MAGI2_rs1207808*, *CSMD1_rs58499534*, *MICAL2_rs10831776*, *CSNK2A1_rs205881*, and *ADAM29_rs11946315*, was selected as the best model (Table S1). The TRBA, TEBA, and CVC values of this eight-SNP model were 0.8474, 0.5377, and 10/10, respectively, after adjusting for age, gender, BMI, residence area, physical activity, education, income, smoking, and alcohol and energy intake.

3.4. PRSs Obtained by the Summation of Risk Alleles in the Best Model for Gastric Cancer Risk

A model containing eight SNPs was used to evaluate the association between polygenic risk scores (PRSs) and gastric cancer risk. The high-PRS group and the low-PRS group were compared, and the adjusted odds ratios (ORs) and their 95% confidence intervals (CIs) were calculated. After adjusting for the first set of covariates (covariate set 1), the adjusted OR for gastric cancer in the high-PRS group was 4.04 (95% CI: 2.68–6.11) (Table S2). This indicated that, after adjusting for other potential influencing factors, the risk of gastric cancer in the high-PRS group was more than four times that in the low-PRS group, and this association was statistically significant. In addition, the above analysis was repeated after adjusting for the second set of covariates (covariate set 2). The results showed that the adjusted OR for gastric cancer in the high-PRS group was 4.12 (95% CI: 2.71–6.27) (Table S2), which was similar to the results after adjustment for the first set of covariates. This shows that, regardless of which set of covariates was adjusted, the risk of gastric cancer in the high-PRS group was significantly higher than that in the low-PRS group. The stability of this association was further verified. Such results are of great significance for understanding the role of genetic mutation in the occurrence of gastric cancer. They help develop personalized risk assessment and intervention strategies based on genetic information. At the same time, they also emphasize the importance of considering multiple covariates when conducting genetic epidemiological studies to ensure the accuracy and reliability of the research results. These results indicated that subjects in the high-PRS group, adjusting for covariate sets 1 and 2, were at a 4.04- and 4.12-fold higher risk of gastric cancer, respectively, than subjects in the low-PRS group ($p < 0.001$). However, in covariate sets 1 and 2, no significant correlation was found between the PRS and serum total cholesterol, TG, LDL, CRP, and HDL, as well as waist circumference, hypertension, and type 2 diabetes risk ($p > 0.05$), indicating that

PRSs may only be significantly associated with gastric cancer, while metabolic markers such as cholesterol and triglycerides are affected by multiple factors such as genetics, diet, and lifestyle, and the complexity and diversity of these factors may make it difficult to simply summarize the relationship between metabolic markers and gastric cancer risk. Therefore, abnormalities in metabolic markers such as serum cholesterol and triglyceride concentrations might not be directly associated with gastric cancer risk, or this association might be masked by other stronger influencing factors.

3.5. Interaction between the PRSs and Biochemical Parameters Influencing Gastric Cancer Risk

This study investigated the relationship between WBC count and gastric cancer risk. The risk of gastric cancer under different PRS group and WBC count combinations was analyzed. In the high-PRS group, individuals with higher WBC counts had a lower risk of gastric cancer than those with lower WBC counts (as shown in Table 4, Figure 2A, and Supplementary Figure S1A). This finding suggests that, even in people with a higher genetic risk, elevated WBC counts may have a certain protective effect on gastric cancer risk. In addition, we paid special attention to the risk of gastric cancer under the combination of a high-PRS group and a low WBC count. This study found that individuals in the high-PRS group with lower WBC counts had a 5.13-fold-increased risk of gastric cancer compared with individuals in the low-PRS group with lower WBC counts ($p = 0.014$; as shown in Table 4). This finding emphasizes that reduced WBC counts may be an important risk factor for gastric cancer in individuals with a high genetic risk. These results suggest that the WBC count may be a biomarker associated with gastric cancer risk, especially when combined with genetic information.

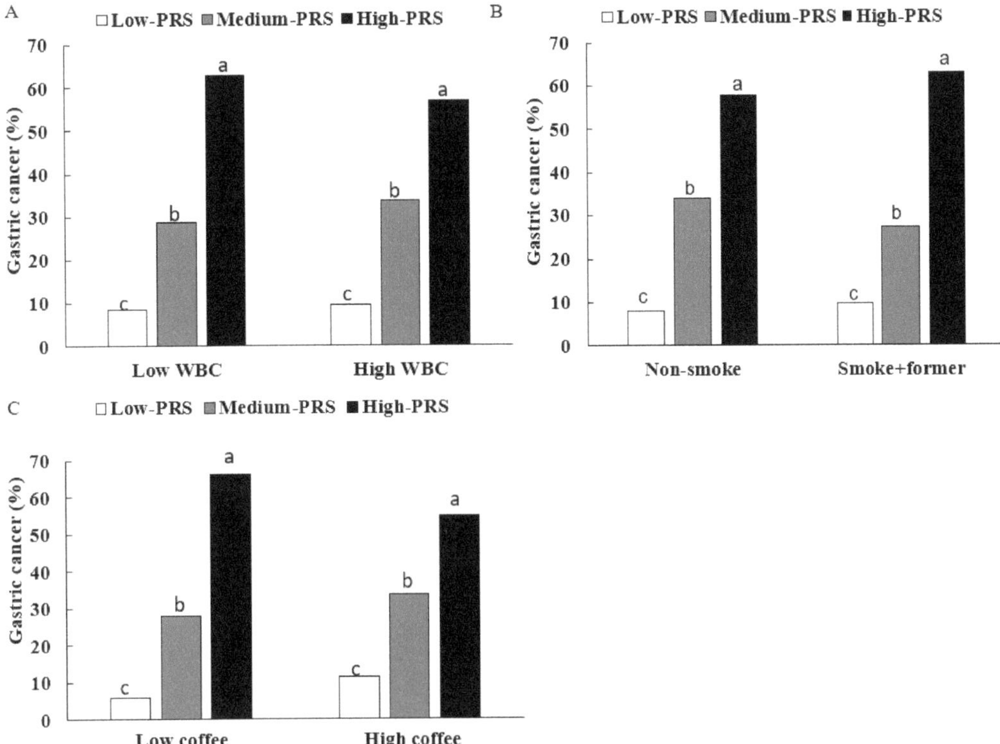

Figure 2. Incidence of gastric cancer according to the parameters to interact with polygenic risk scores (PRS). (**A**) Gastric cancer incidence according to their white blood cell counts (WBC, cutoff value: 4×10^9/L). (**B**) Gastric cancer incidence according to their smoking status. (**C**) Gastric cancer incidence according to their coffee intake (cutoff value: 3 g/day). PRS interacted with white blood cell (WBC) counts, smoking status, and coffee intake. The participants with high-PRS were higher in the low WBC group than in the high WBC group, in the non- and former smokers than in the smokers, and in the low coffee intake (<3 cup times/week) than in the high coffee intake. *p* value indicated the interaction between PRS with designated parameters. a,b,c Different alphabets indicated significant difference among the groups at $p < 0.05$.

Table 4. Adjusted odds ratios (ORs) for the risk of gastric cancer by polygenetic risk scores (PRSs) of the best model after covariate adjustments according to low- and high-lifestyle factors.

	Low-PRS (n = 10,166)	Medium-PRS (n = 20,168)	High-PRS (n = 17,972)	Gene-Nutrient Interaction p Value
Low WBC [1] High WBC	1	2.355(0.604~9.181) 1.780(1.020~3.105)	5.126(1.415~18.567) 3.506(2.063~5.959)	0.014
Low energy [2] High energy	1	1.661(1.000~2.760) 2.709(1.044~7.033)	3.400(2.104~5.493) 7.355(2.956~18.302)	0.244
Low CHO [3] High CHO	1	2.004(0.421~9.529) 1.837(1.154~2.923)	7.552(1.769~32.236) 3.817(2.456~5.931)	0.298
Low protein [4] High protein	1	1.790(0.962~3.329) 1.909(1.007~3.622)	4.200(2.340~7.538) 4.097(2.231~7.525)	0.945
Low fat [5] High fat	1	1.830(1.083~3.091) 1.929(0.830~4.484)	4.294(2.617~7.044) 3.829(1.716~8.543)	0.400
No exercise Exercise	1	1.453(0.748~2.822) 2.212(1.207~4.054)	3.195(1.718~5.939) 4.985(2.799~8.878)	0.795
Non-smoke Smoke + former	1	2.208(1.232~3.957) 1.376(0.685~2.763)	4.295(2.453~7.521) 3.825(2.019~7.249)	$p < 0.0001$
Low Coffee [6] High Coffee	1	2.299(1.065~4.964) 1.669(0.964~2.889)	6.301(3.039~13.07) 3.129(1.858~5.267)	0.04

Values represent odds ratios and 95% confidence intervals of the adults aged $\geq$40. PRSs with eight SNPs were divided into three categories (1–6, 7–8, and $\geq$9) by tertiles as the low, medium, and high groups of the best model of GMDR. The cutoff point was as follows: [1] $<4 \times 10^9$/L white blood cell (WBC) counts, [2] <estimated energy intake, [3] <65% carbohydrate (CHO) intake, [4] <13% protein intake, [5] <20% fat intake, and [6] <3 g/day coffee drinking. Values represent adjusted odds ratios and 95% confidence intervals. Covariates included age, gender, BMI, residence area, physical activity, education, smoking, years with gastric cancer, and intake of alcohol and energy. The reference was the low-PRS group.

3.6. Interaction between PRSs and Lifestyle Factors Influencing Gastric Cancer Risk

The smoking and coffee intake interacted with the PRS to affect gastric cancer risk ($p < 0.0001$ and 0.04, respectively). The incidence of gastric cancer was higher in participants who smoked than in those who did not, regardless of the PRS (Figure 2B and Supplementary Figure S1B). Smokers in the high-PRS group had a higher incidence of gastric cancer than non-smokers (Table 4, Figure 2B and Supplementary Figure S1B). The smokers and non-smokers in the high-PRS group had a 3.83- and 4.29-fold higher risk of gastric cancer than those in the low-PRS group ($p < 0.0001$; Table 4). The gastric cancer incidence was higher in the high-PRS group than in the low-PRS group in participants with both a low and high coffee intake. However, the gastric cancer incidence was much higher in the participants with a high PRS and a low coffee intake (Figure 2C and Supplementary Figure S1C). The PRS was positively associated with 6.30 and 3.13 times higher risk of gastric cancer in the low and high coffee intake groups (Table 4). Those in the high-PRS group with a high coffee intake had a lower risk of gastric cancer than those with a low coffee intake (Table 4, Figure 2C, and Supplementary Figure S1C). Those in the high-PRS group with a low coffee intake had a 6.30-fold higher risk of gastric cancer than those in the low-PRS group with a low coffee intake ($p = 0.04$; Table 4). However, the rate of gastric cancer was higher in the low-coffee intake group than in the high-coffee intake group, regardless of the PRS.

3.7. Binding Free Energy of Food Components to Wild and Mutated Types of SEMA3C_rs1527482

The wild and mutated types of *SEMA3C*_rs1527482 had various levels of binding free energy for 20000 food components. Tables 5 and S3 present the food components which have a low binding free energy with the wild and mutated types of *SEMA3C*_rs1527482. Some food components, including theaflavate, rugosin E, vitisifuran B, plantacyanin, and (cyanidin 3-O-beta-glucoside) (kaempferol 3-O-(2-O-beta-glucosyl-beta-glucoside)-7-O-beta-glucosiduronic acid) malonate (CK-malonate), lowered the binding energy in both wild and mutated types. Some coffee components and metabolites also contribute to a reduction in binding free energy (Table S3). However, pinotin A, delta-viniferin, sanguiin H6, and quercetin 3-O-rhamnosyl-(1->2)-rhamnosyl-(1->6)-glucoside decreased the binding energy with the wild type of *SEMA3C_rs1527482*. Withanolide B, epitheaflagallin 3-O-gallate, pomolic acid, and epigallocatechin had lower binding energies to the mutated types. *SEMA3C_rs1527482* was positively associated with gastric cancer risk. Food components with low binding free energy may modulate and lower *SEMA3C* activity.

Table 5. Binding energy between the wild (WT, Val337) and mutated type (MT, 337Met) *SEMA3C_rs1527482* and food components.

Both of WT and MT		
Natural compounds	Binding energy (kcal/mol)	Foods containing the selected natural compound
Trisjuglone	−11.1	Juglans regia (walnut) roots.
Rugosin E	−11.8	Cloves
Theaflavate B	−11.3	Black tea (*Camellia sinensis*).
Theaflavate A	−11.4	Black tea (*Camellia sinensis*).
Theaflavin 3′-gallate	−11.3	Black tea and commercial oolong tea
Lettowianthine	−11.7	*Annona glabra* (pond apple).
Vitisifuran B	−11.8	wine grape, *Vitis vinifera* 'Kyohou'
Tragopogonsaponin J	−11.3	*Tragopogon porrifolius* (salsify), green vegetables
Mongolicain A	−11.1	Guava
Plantacyanin	−12.5	Cucumber, green vegetables.
WT only		
Natural compounds	Binding energy (kcal/mol)	Foods containing the selected natural compound
Pinotin A	−10.3	Red wine, including Pinotage (CCD)
Quercetin 3-O-rhamnosyl-(1->2)-rhamnosyl-(1->6)-glucoside	−10.2	Common sage, common thyme, Italian oregano, and rosemary
delta-Viniferin	−10.1	Stressed grapevine (*Vitis vinifera*) leaves
Murrayenol	−10.4	Roots of *Murraya koenigii* (curry leaf tree).
Sanguiin H6	−10.1	*Sanguisorba officinalis* (burnet bloodwort), blackberry, and red raspberry.
Isovitexin 6″-rhamnoside	−10.0	Grape and mung bean.
C-K malonate	−10.5	Chives
MT only		
Natural compounds	Binding energy (kcal/mol)	Foods containing the selected natural compound
Withanolide B	−10.6	Leaves of *Lycium chinense* (Chinese boxthorn)
Epitheaflagallin 3-O-gallate	−10.6	Black tea.
Pomolic acid	−10.8	Apple peel, rosemary, lemon balm, pomes, and spearmint.
19-Dehydroursolic acid	−10.6	*Sanguisorba officinalis* (burnet bloodwort).
Ganosporelactone B	−10.9	Spores of *Ganoderma lucidum* (reishi).
alpha-Amyrone	−10.9	*Sambucus nigra* (elderberry)
3,3′-Bisanigorufone	−11.5	Rhizomes of *Musa acuminata* (dwarf banana)
Epigallocatechin-(4 beta->6)-epicatechin 3,3′-digallate	−11.4	Oolong tea, *Camellia sinensis*
Artomunoxanthentrione epoxide	−10.8	Root bark of *Artocarpus communis* (breadfruit).
Khelmarin D	−10.7	*Citrus paradisi* and *Citrus tangerina* (Rutaceae).

Food components to lower binding energy with WT and MT *SEMA3C* rs1527482 and foods containing the selected food component. C-K malonate: (cyanidin 3-O-beta-glucoside) (kaempferol 3-O-(2-O-beta-glucosyl-beta-glucoside)-7-O-beta-glucosiduronic acid) malonate.

Through calculation and simulation, the binding free energy between the wild-type *SEMA3C* protein and the CK-malonate molecule, as well as the changes in this binding in the mutant *SEMA3C* protein, were analyzed. Figure 3A shows the binding free energy between the wild-type *SEMA3C* protein and CK-malonate through hydrogen bonding, where the pink and green parts represent the donor and acceptor of the hydrogen bond, respectively. Figure 3B provides a two-dimensional image that more intuitively shows their binding positions and intermolecular forces. The binding of CK-malonate to the mutant *SEMA3C* protein was further analyzed, as shown in Figure 3C,D. The binding energy of CK-malonate to the wild-type *SEMA3C* protein was −10.5 kcal/mol, while that to the mutant *SEMA3C* protein was −8.5 kcal/mol. This suggested that the mutation might affect the binding stability between the *SEMA3C* protein and CK-malonate. In order to more comprehensively evaluate the stability of this binding, the root mean square deviation (RMSD) and root mean square fluctuation (RMSF) of the *SEMA3C* protein (whether wild or mutant type) when bound to another molecule, CK-malonate, were also calculated. As shown in Figure 4A,B, the RMSD of wild-type *SEMA3C* protein bound to CK-malonate remained close to 3 Å throughout the simulation, indicating that their binding was relatively stable. Similarly, the RMSF of wild-type *SEMA3C* protein bound to CK-malonate also mostly remained below 3 nm, except for one exception at residue 580 in the RMSF map. These results further support the view that CK-malonate can stably bind to wild-type *SEMA3C* protein.

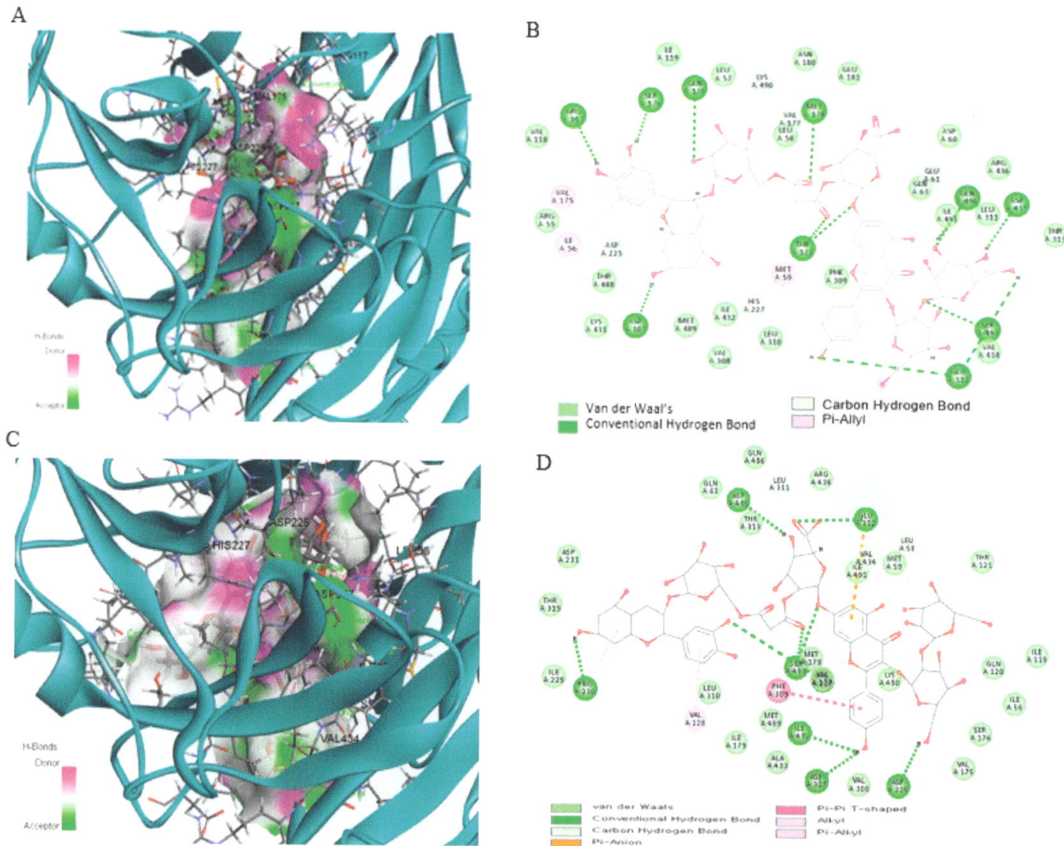

Figure 3. Molecular docking of C-K malonate on *SEMA3C*_rs1527482 wild (WT) and mutated types (MT). (**A**) Molecular docking of (cyanidin 3-O-beta-glucoside)(kaempferol 3-O-(2-O-beta-glucosyl-beta-glucoside)-7-O-beta-glucosiduronic acid) malonate (C-K malonate) on *SEMA3C*_rs1527482 WT.

(**B**) The interaction force between C-K malonate and *SEMA3C*_rs1527482 WT. (**C**) Molecular docking of C-K malonate on *SEMA3C*_rs1527482 MT. (**D**) The interaction force between C-K malonate and *SEMA3C*_rs1527482 MT.

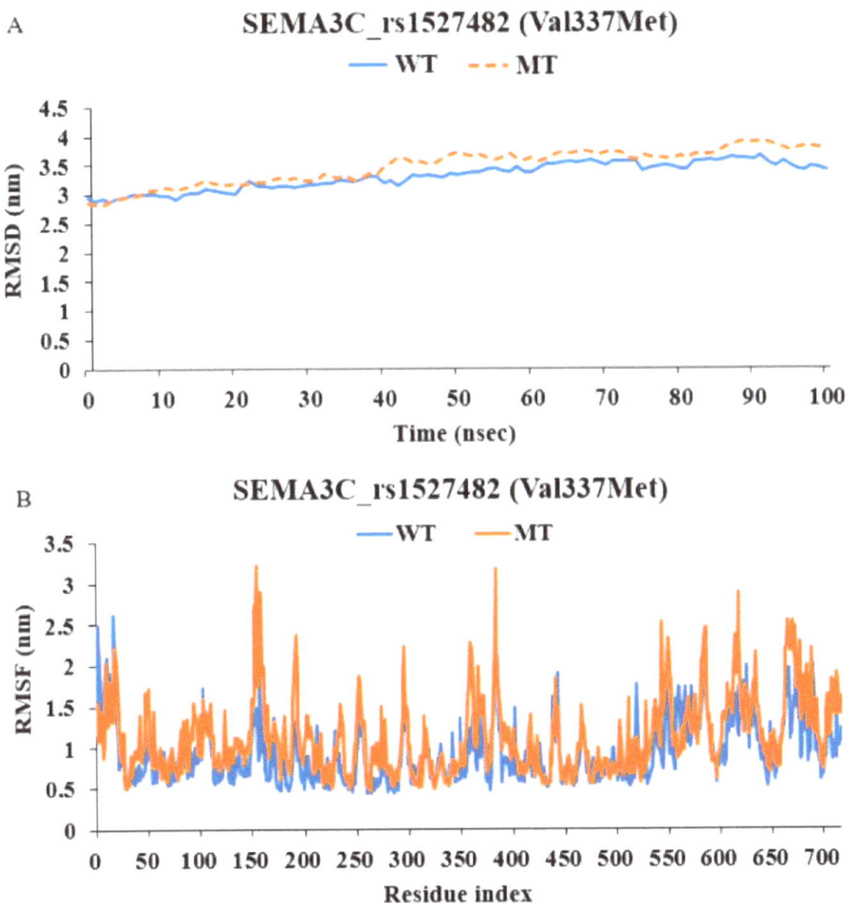

Figure 4. Molecular dynamic simulation (MDS) of C-K malonate on SEMA3C_rs1527482 wild (WT) and mutated types (MT). (**A**) The root-mean-square deviation (RMSD) of (cyanidin 3-O-beta-glucoside)(kaempferol 3-O-(2-O-beta-glucosyl-beta-glucoside)-7-O-beta-glucosiduronic acid) malonate (C-K malonate) on WT and MT of SEMA3C_rs1527482. (**B**) The root-mean-square fluctuation (RMSF) of C-K malonate on WT and MT of SEMA3C_rs1527482.

4. Discussion

In this study, we explored the effects of genetic variants on gastric cancer risk. Through a comprehensive analysis combining GWAS and GMDR, we identified 10 genetic variants significantly correlated with gastric cancer. Further analysis revealed an optimal SNP–SNP interaction model comprising eight SNPs: *DAB1*_rs7521784, *PDK1*_rs12693006, *DOCK10*_rs1045653, *MAGI2*_rs1207808, *CSMD1*_rs58499534, *MICAL2*_rs10831776, *CSNK2A1*_rs205881, and *ADAM29*_rs11946315. The SNPs demonstrated the complex genetic landscape underlying gastric cancer susceptibility. The PRSs derived from these eight SNPs demonstrated interactions between WBC count, smoking status, and coffee consumption. These findings provide novel insights into the complex interplay between genetic and environmental factors in gastric cancer risk. Our in silico analysis focused on *SEMA3C*_rs1527482, a missense mutation. We observed that specific polyphenols altered the binding affinity of this variant, suggesting its potential as a therapeutic target for gastric cancer. This discovery opens new avenues for personalized nutritional interventions in gastric cancer prevention and treatment.

The association of *DAB1_rs7521784* with gastric cancer risk is a novel finding in our study. Previous research has identified *DAB1* mutations in Chinese patients with chronic gastritis and peritoneal metastasis of gastric cancer [29], and reduced *DAB1* mRNA expression has been observed in various cancers [30]. Similarly, our findings regarding *PDK1_rs12693006* align with the known roles of *PDK1* in cancer-related processes and its association with poor gastric cancer prognosis, suggesting this variant's involvement in tumor activity.

The inclusion of *DOCK10_rs1045653* in our model is particularly interesting. DOCK proteins are known to be involved in various pathologies, including cancer, by regulating the actin cytoskeleton, cell adhesion, and migration [31]. *DOCK10*, specifically, has been shown to play roles in immune function and neuroinflammation [32]. Our study is the first to associate this genetic variant with gastric cancer risk, potentially highlighting new pathways in gastric cancer development. While the hypermethylation of *MAGI2* has been linked to gastric cancer tumorigenesis [33], our study is the first to identify an SNP in this gene associated with gastric cancer risk. This finding may provide new avenues for understanding the genetic basis of gastric cancer development.

The involvement of CUB and Sushi Multiple Domains 1(*CSMD1*)_rs58499534 in our model aligns with previous research showing the crucial roles of *CSMD1* in cancer-related processes [34]. Our study extends these findings to include a specific genetic variant associated with gastric cancer risk. Similarly, our identification of microtubule-associated monooxygenase, calponin, and LIM domain containing 2 (*MICAL2*)_rs10831776 as risk factors is consistent with previous research showing elevated *MICAL2* mRNA expression in gastric cancer tissues [35,36]. *CSNK2*, or casein kinase 2 (*CK2*), is involved in various cellular processes and has been implicated in tumor development, with *CSNK2A1* overexpression shown to promote gastric cancer progression [37]. *ADAM29* has been demonstrated to promote gastric cancer cell proliferation, migration, and invasion, with increased expression associated with poor patient survival [38]. These findings contribute to our understanding of the polygenic nature of gastric cancer risk. While each genetic variant may have a minor individual effect, their combination can significantly increase the associated risk [39].

The WBC count is a systemic inflammatory biomarker associated with an increased risk of several chronic diseases. Chronic inflammation is also known to play a role in cancer pathogenesis. A Japanese study reported that a high WBC count was a risk factor for gastric cancer in *H. pylori*-infected subjects. However, no association was observed in the *H. pylori*-negative group [40]. In this study, the incidence of gastric cancer was higher in participants with a low WBC count, and the low count of WBCs interacted with the PRS to increase the risk of gastric cancer. In the low WBC count group, individuals with a high PRS had a 5.13-fold higher risk of gastric cancer than those with a low PRS in subjects whose *H. pylori* infection status was unknown.

The International Agency for Research on Cancer (IARC) classified smoking as a carcinogen in 2004, confirming its role as a significant risk factor for gastric cancer [41]. The carcinogenic process is believed to involve gastric atrophy induced by substances such as nitrosamines and other nitroso compounds present in tobacco smoke [42]. Our study builds upon this knowledge by demonstrating an interaction between smoking status and the PRS in influencing gastric cancer risk. Notably, individuals with a high PRS who were former or current smokers exhibited a 3.83-fold-increased risk of gastric cancer compared to those with a low PRS.

Coffee's relationship with gastric cancer is more complex and controversial. As an intricate mixture of compounds, coffee contains both potential carcinogens and anti-cancer agents. Antioxidants like phenolic compounds, diterpenes, melanoidins, and vitamin precursors may offer protective effects, while trace amounts of aromatic hydrocarbons and heterocyclic amines formed during processing could potentially promote carcinogenesis [43]. Some studies have reported a modest 7% reduction in gastric cancer risk associated with coffee consumption [44], while others have found no significant association [43]. It is important to note that the observed lower coffee consumption in the gastric cancer group may be partially attributed to dietary changes following diagnosis, rather than being solely a contributing factor to cancer development. This potential reverse causality highlights the need for prospective studies to further elucidate the relationship between coffee consumption and gastric cancer risk.

SEMA3C, a secreted glycoprotein of the semaphorin class 3 family, has been implicated in gastric cancer progression [45]. This protein promotes cancer growth and treatment resistance by activating signaling cascades involving the epidermal growth factor receptor (EGFR), erythroblastic oncogene B2 (ErbB2), and mesenchymal-epithelial transition (MET). These pathways are independently transactivated via plexin B1 by cognate ligands [46]. Elevated expression and activity of *SEMA3C* have been associated with increased cancer cell invasion and adhesion [45]. Additionally, *plexin B1* plays a role in modulating immune responses, which may influence cancer development. Our findings align with previous research, demonstrating that the *SEMA3C*_rs1527482 variant is positively associated with gastric cancer risk. Specifically, the minor allele of this SNP appears to confer increased susceptibility to gastric cancer. As a missense mutation, the activity of this genetic variant may be modulated by interactions with dietary components.

Molecular docking studies revealed potential interactions between small molecule food compounds and *SEMA3C*_rs1527482 (wild type and mutant), providing a quantitative metric (binding energy) for evaluating compound–protein interactions. A low binding energy means stronger interactions, which may improve or regulate protein function. Although metabolism may affect the compounds' structure, the docking results still provide a key starting point for understanding how dietary components could have different effects based on individual genetic variations. Despite its limitations, molecular docking lays the foundation for exploring genotype-specific nutritional interventions and their impacts on health. Therefore, molecular docking research is of great significance in accelerating the development of new drugs, guiding the optimization of drug molecular structures, revealing the interaction between drugs and targets, and predicting drug metabolic pathways. It is an indispensable technology in the field of modern drug development.

An in silico analysis revealed that certain food components bind to the SEMA3C protein with binding energies below -10 kcal/mol, suggesting the potential modulation of SEMA3C activity. Interestingly, the binding affinities differed between the wild-type and mutated forms of the protein. For both variants, tea components exhibited strong binding. However, the wild-type protein showed preferential binding to components from grapes and wine, while the mutated form demonstrated stronger interactions with compounds from tea and fruit peels. These findings suggest that specific dietary elements, particularly those found in tea, grapes, and fruit peels, might differentially suppress SEMA3C activity in individuals carrying the wild-type or mutated rs1527482 allele. This potential gene–diet interaction could implicate personalized nutrition strategies in gastric cancer prevention. However, it is crucial to note that these computational predictions require validation through rigorous experimental studies.

The strengths of this study are as follows: (1) This study utilized a large sample size, ensuring strong statistical power and improving the generalizability of our findings to Korean adults. (2) We employed multiple aspects of genetic analysis, polygenic risk scores, and lifestyle to provide a more nuanced understanding of gastric cancer risk factors, thereby improving the validity and relevance of our results. (3) The interactions observed between specific food components and genetic variants provided potential practical applications. These results might help develop personalized gastric cancer prevention and management strategies. The limitations of this study are as follows: (1) The cross-sectional nature of this study limited our ability to establish temporal relationships between variables. Therefore, we could not directly infer causal relationships or track changes in disease status over time. (2) Our study population was recruited from urban hospitals, and because the study samples were mainly from urban hospitals, the results might not be applicable to a wider population in rural or remote areas. In addition, participants might have been more or less inclined to participate in the study due to factors such as health status, knowledge level, or socioeconomic status, which could also introduce selection bias. (3) The reliability of the self-reported data was often affected by environmental factors such as the memory, understanding, and honesty of the participants. In this study, gastric cancer diagnosis was self-reported and not independently verified, which might lead to inaccurate or biased information. In addition, we did not distinguish between gastric cancer subtypes because they might have different risk factors depending on the location of the tumor. (4) Patients' lifestyles and nutrient intake were self-reported based on individual estimates of their usual intake [21]. The food intake measured by SQFFQ might not fully capture long-term dietary habits, and the collection process might be subject to bias, similar to other self-report methods. (5) *H. pylori* infection has been widely recognized as an important risk factor for gastric cancer. Failure to adjust for the confounding factor of *H. pylori* infection might exaggerate or underestimate the associations between other risk factors and gastric cancer. Such bias might affect the reliability and accuracy of the study results. Despite these limitations, our study provided valuable insights into the complex interactions between genetic and environmental factors in gastric cancer risk, laying the foundation for future research and potential prevention strategies.

5. Conclusions

Our study identifies a novel eight-SNP PRS model that significantly elevates the gastric cancer risk by 4.12-fold and highlights the potential role of SEMA3C_rs1527482 in gastric cancer susceptibility. We found evidence suggesting that specific components in tea, grapes, and fruit peels might differentially affect wild-type and mutated SEMA3C protein activity. Important interactions between white blood cell counts, PRSs, coffee consumption, and smoking status were revealed, amplifying the genetic susceptibility to gastric cancer and underscoring the complex interplay between genetic and environmental factors in cancer development. While these findings contribute significantly to our understanding of gastric cancer risk, it is important to acknowledge the limitations of our study, including its cross-sectional nature, reliance on self-reported food intakes and lifestyles, and lack of differentiation between gastric cancer subtypes.

Based on our results, we propose that customized nutritional plans to potentially reduce gastric cancer risk could include increasing coffee and polyphenol-rich food consumption, especially for individuals with a high PRS. Theaflavate, rugosin E, vitisifuran B, and plantacyanin could be recommended regardless of *SEMA3C_rs1527482* variant status. Additionally, immune-boosting foods and smoking cessation strategies could be emphasized for participants with high PRSs. However, these dietary recommendations are preliminary and require further clinical validation. Future research should focus on validating these findings in larger, more diverse populations, conducting long-term clinical trials to assess the efficacy of targeted dietary interventions, and integrating other relevant genetic markers and environmental factors to develop more comprehensive and personalized prevention strategies for gastric cancer.

Supplementary Materials: The following supporting information can be downloaded at: https://www.mdpi.com/article/10.3390/nu16193263/s1, Table S1: Generalized multifactor dimensionality reduction (GMDR) results of multi-locus interaction with genes in gastric cancer risk. Table S2: Odds ratios for gastric cancer risk-adjusted for alleles of GMDR after adjustment for covariates. Table S3: Biding energy between the wild (WT, Val337) and mutated type (MT, 337Met) of SEMA3C based on rs1527482 and coffee components and its metabolites. Figure S1: Interaction between PRSs and lifestyle factors influencing gastric cancer incidence.

Author Contributions: M.L.: Conceptualization, Investigation, Formal Analysis, Visualization. S.-S.S.: Investigation, Formal analysis, Writing—original draft. S.P.: Conceptualization, Methodology, Resources, Supervision, Validation, Funding acquisition, Writing—review & editing. All authors have read and agreed to the published version of the manuscript.

Funding: This study was supported by a grant from the National Research Foundation of Korea (NRF) funded by the Ministry of Science and ICT (RS-2023-00208567).

Institutional Review Boards Statement: The KoGES protocol received approval from the institutional review boards of the Korean National Institute of Health (Approval Code: KBP-2015-055; approved on 20 August 2015).

Informed Consent Statement: We had written informed consent from each participant.

Data Availability Statement: The data were deposited in the Korean biobank (Osong, Republic of Korea).

Conflicts of Interest: The authors declare no conflicts of interest.

Abbreviations

GC, gastric cancer group; N-GC, non-gastric cancer group; SNP, single nucleotide polymorphism; GWAS, genome-wide association study; GMDR, generalized multifactor dimensionality reduction; PRS, polygenic risk score; OR, odds ratio; *CDH1*, cadherin 1; *TP53*, tumor protein 53; *H. pylori*, *Helicobacter pylori*, *IL1B*, interleukin-1 beta; *IL1RN*, interleukin-1 receptor antagonist protein; *TNF-α*, tumor necrosis factor-α; *CYP2E1*, cytochrome P450 family 2 subfamily E member 1; *GSTM1*, glutathione S-transferase mu 1; *MTHFR*, methylenetetrahydrofolate reductase; *XRCC1*, X-ray repair cross-complementing group 1; *PRKAA1*, protein kinase AMP-activated catalytic subunit alpha 1; *MUC1*, mucin 1; *PLCE1*, phospholipase C epsilon 1; *PSCA*, prostate stem cell antigen; KoGES, Korean genomic epidemiology survey; KDCA, Korea Disease Control and Prevention Agency; BMI, body mass index; HDL-C, high-density lipoprotein cholesterol; WBC, white blood cell; SQFFQ, semiquantitative food frequency questionnaire; HWE, Hardy–Weinberg Equilibrium; LD, linkage disequilibrium; TRBA, trained balanced accuracy; TEBA, test balance accuracy; CVC, cross-validation consistency; *SEMA3C*, semaphorin-3C; PDB, protein data bank; PDBQT, protein data bank, partial charge (Q), and atom type (T); CHARMM, Chemistry at Harvard Macromolecular Mechanics; RMSD, root mean square deviation; RMSF, root mean square fluctuations; ANOVA, one-way analysis of variance; *DAB1*, disabled-1, *PDK1*, pyruvate dehydrogenase kinase-1, *DOCK10*, dedicator-of-cytokinesis-10; *ZBTB20*, zinc finger and BTB domain 20; *KALRN*, Kalirin RhoGEF Kinase; *ADAM29*, disintegrin and metalloprotease 29; *MAGI2*, membrane-associated guanylate kinase 2; *CSMD1*, cub sushi multiple domains-1; *MICAL2*, molecule interacting with CasL-2; *CSNK2A1*, casein kinase IIA1; MAF, minor allele frequency.

References

1. Sung, H.; Ferlay, J.; Siegel, R.L.; Laversanne, M.; Soerjomataram, I.; Jemal, A.; Bray, F. Global Cancer Statistics 2020: GLOBOCAN Estimates of Incidence and Mortality Worldwide for 36 Cancers in 185 Countries. *CA Cancer J. Clin.* **2021**, *71*, 209–249. [CrossRef] [PubMed]
2. Colquhoun, A.; Arnold, M.; Ferlay, J.; Goodman, K.J.; Forman, D.; Soerjomataram, I. Global patterns of cardia and non-cardia gastric cancer incidence in 2012. *Gut* **2015**, *64*, 1881–1888. [CrossRef] [PubMed]
3. Tsugane, S.; Sasazuki, S. Diet and the risk of gastric cancer: Review of epidemiological evidence. *Gastric Cancer* **2007**, *10*, 75–83. [CrossRef]
4. Choi, S.; Jang, J.; Heo, Y.J.; Kang, S.Y.; Kim, S.T.; Lee, J.; Kang, W.K.; Kim, J.W.; Kim, K.M. CDH1 mutations in gastric cancers are not associated with family history. *Pathol. Res. Pract.* **2020**, *216*, 152941. [CrossRef] [PubMed]
5. Moon, S.; Balch, C.; Park, S.; Lee, J.; Sung, J.; Nam, S. Systematic Inspection of the Clinical Relevance of TP53 Missense Mutations in Gastric Cancer. *IEEE/ACM Trans. Comput. Biol. Bioinform.* **2019**, *16*, 1693–1701. [CrossRef]
6. Chang, Y.W.; Jang, J.Y.; Kim, N.H.; Lee, J.W.; Lee, H.J.; Jung, W.W.; Dong, S.H.; Kim, H.J.; Kim, B.H.; Lee, J.I.; et al. Interleukin-1B (IL-1B) polymorphisms and gastric mucosal levels of IL-1beta cytokine in Korean patients with gastric cancer. *Int. J. Cancer* **2005**, *114*, 465–471. [CrossRef] [PubMed]

7. Garza-González, E.; Bosques-Padilla, F.J.; El-Omar, E.; Hold, G.; Tijerina-Menchaca, R.; Maldonado-Garza, H.J.; Pérez-Pérez, G.I. Role of the polymorphic IL-1B, IL-1RN and TNF-A genes in distal gastric cancer in Mexico. *Int. J. Cancer* **2005**, *114*, 237–241. [CrossRef]
8. Park, G.T.; Lee, O.Y.; Kwon, S.J.; Lee, C.G.; Yoon, B.C.; Hahm, J.S.; Lee, M.H.; Hoo Lee, D.; Kee, C.S.; Sun, H.S. Analysis of CYP2E1 polymorphism for the determination of genetic susceptibility to gastric cancer in Koreans. *J. Gastroenterol. Hepatol.* **2003**, *18*, 1257–1263. [CrossRef]
9. Chen, Z.H.; Xian, J.F.; Luo, L.P. Association between GSTM1, GSTT1, and GSTP1 polymorphisms and gastric cancer risk, and their interactions with environmental factors. *Genet. Mol. Res.* **2017**, *16*, gmr16018877. [CrossRef]
10. Han, Z.; Sheng, H.; Gao, Q.; Fan, Y.; Xie, X. Associations of the MTHFR rs1801133 polymorphism with gastric cancer risk in the Chinese Han population. *Biomed. Rep.* **2021**, *14*, 14. [CrossRef]
11. Putthanachote, N.; Promthet, S.; Hurst, C.; Suwanrungruang, K.; Chopjitt, P.; Wiangnon, S.; Chen, S.L.; Yen, A.M.; Chen, T.H. The XRCC 1 DNA repair gene modifies the environmental risk of stomach cancer: A hospital-based matched case-control study. *BMC Cancer* **2017**, *17*, 680. [CrossRef] [PubMed]
12. Kim, Y.D.; Yim, D.H.; Eom, S.Y.; Moon, S.I.; Yun, H.Y.; Song, Y.J.; Youn, S.J.; Hyun, T.; Park, J.S.; Kim, B.S.; et al. Risk of gastric cancer is associated with PRKAA1 gene polymorphisms in Koreans. *World J. Gastroenterol.* **2014**, *20*, 8592–8598. [CrossRef] [PubMed]
13. Song, H.R.; Kim, H.N.; Kweon, S.S.; Choi, J.S.; Shim, H.J.; Cho, S.H.; Chung, I.J.; Park, Y.K.; Kim, S.H.; Choi, Y.D.; et al. Common genetic variants at 1q22 and 10q23 and gastric cancer susceptibility in a Korean population. *Tumour Biol.* **2014**, *35*, 3133–3137. [CrossRef] [PubMed]
14. Park, B.; Yang, S.; Lee, J.; Woo, H.D.; Choi, I.J.; Kim, Y.W.; Ryu, K.W.; Kim, Y.I.; Kim, J. Genome-Wide Association of Genetic Variation in the PSCA Gene with Gastric Cancer Susceptibility in a Korean Population. *Cancer Res. Treat.* **2019**, *51*, 748–757. [CrossRef] [PubMed]
15. Liu, Y.Q.; Wang, T.P.; Yan, C.W.; Zhu, M.; Yang, M.; Wang, M.Y.; Hu, Z.B.; Shen, H.B.; Jin, G.F. Association between polygenic risk score and age at onset of gastric cancer. *Zhonghua Liu Xing Bing. Xue Za Zhi* **2021**, *42*, 1092–1096.
16. Park, B.; Yang, S.; Lee, J.; Choi, I.J.; Kim, Y.I.; Kim, J. Gastric Cancer Risk Prediction Using an Epidemiological Risk Assessment Model and Polygenic Risk Score. *Cancers* **2021**, *13*, 876. [CrossRef]
17. Jin, G.; Lv, J.; Yang, M.; Wang, M.; Zhu, M.; Wang, T.; Yan, C.; Yu, C.; Ding, Y.; Li, G.; et al. Genetic risk, incident gastric cancer, and healthy lifestyle: A meta-analysis of genome-wide association studies and prospective cohort study. *Lancet Oncol.* **2020**, *21*, 1378–1386. [CrossRef]
18. Liu, M.; Jin, H.S.; Park, S. Protein and fat intake interacts with the haplotype of PTPN11_rs11066325, RPH3A_rs886477, and OAS3_rs2072134 to modulate serum HDL concentrations in middle-aged people. *Clin. Nutr.* **2020**, *39*, 942–949. [CrossRef]
19. Park, S.; Ahn, J.; Lee, B.K. Self-rated Subjective Health Status Is Strongly Associated with Sociodemographic Factors, Lifestyle, Nutrient Intakes, and Biochemical Indices, but Not Smoking Status: KNHANES 2007-2012. *J. Korean Med. Sci.* **2015**, *30*, 1279–1287. [CrossRef]
20. Kim, Y.; Han, B.G. Cohort Profile: The Korean Genome and Epidemiology Study (KoGES) Consortium. *Int. J. Epidemiol.* **2017**, *46*, e20. [CrossRef]
21. Ahn, Y.; Kwon, E.; Shim, J.E.; Park, M.K.; Joo, Y.; Kimm, K.; Park, C.; Kim, D.H. Validation and reproducibility of food frequency questionnaire for Korean genome epidemiologic study. *Eur. J. Clin. Nutr.* **2007**, *61*, 1435–1441. [CrossRef] [PubMed]
22. Lee, S.K.; Kim, M.K. Relationship of sodium intake with obesity among Korean children and adolescents: Korea National Health and Nutrition Examination Survey. *Br. J. Nutr.* **2016**, *115*, 834–841. [CrossRef] [PubMed]
23. Rabbee, N.; Speed, T.P. A genotype calling algorithm for affymetrix SNP arrays. *Bioinformatics* **2006**, *22*, 7–12. [CrossRef] [PubMed]
24. Chen, G.-B.; Xu, Y.; Xu, H.-M.; Li, M.D.; Zhu, J.; Lou, X.-Y. Practical and theoretical considerations in study design for detecting gene-gene interactions using MDR and GMDR approaches. *PLoS ONE* **2011**, *6*, e16981. [CrossRef]
25. Yuan, H.; Liu, L.; Zhou, J.; Zhang, T.; Daily, J.W.; Park, S. Bioactive Components of Houttuynia cordata Thunb and Their Potential Mechanisms Against COVID-19 Using Network Pharmacology and Molecular Docking Approaches. *J. Med. Food* **2022**, *25*, 355–366. [CrossRef]
26. Yang, Y.; Shi, C.-Y.; Xie, J.; Dai, J.-H.; He, S.-L.; Tian, Y. Identification of potential dipeptidyl peptidase (DPP)-IV inhibitors among Moringa oleifera phytochemicals by virtual screening, molecular docking analysis, ADME/T-based prediction, and in vitro analyses. *Molecules* **2020**, *25*, 189. [CrossRef]
27. Uma Jyothi, K.; Reddy, B.M. Gene-gene and gene-environment interactions in the etiology of type 2 diabetes mellitus in the population of Hyderabad, India. *Meta Gene* **2015**, *5*, 9–20. [CrossRef]
28. Hong, K.W.; Kim, S.H.; Zhang, X.; Park, S. Interactions among the variants of insulin-related genes and nutrients increase the risk of type 2 diabetes. *Nutr. Res.* **2018**, *51*, 82–92. [CrossRef]
29. Zhang, J.; Huang, J.Y.; Chen, Y.N.; Yuan, F.; Zhang, H.; Yan, F.H.; Wang, M.J.; Wang, G.; Su, M.; Lu, G.; et al. Whole genome and transcriptome sequencing of matched primary and peritoneal metastatic gastric carcinoma. *Sci. Rep.* **2015**, *5*, 13750. [CrossRef]
30. McAvoy, S.; Zhu, Y.; Perez, D.S.; James, C.D.; Smith, D.I. Disabled-1 is a large common fragile site gene, inactivated in multiple cancers. *Genes Chromosomes Cancer* **2008**, *47*, 165–174. [CrossRef]
31. Gadea, G.; Blangy, A. Dock-family exchange factors in cell migration and disease. *Eur. J. Cell Biol.* **2014**, *93*, 466–477. [CrossRef] [PubMed]

32. Namekata, K.; Guo, X.; Kimura, A.; Azuchi, Y.; Kitamura, Y.; Harada, C.; Harada, T. Roles of the DOCK-D family proteins in a mouse model of neuroinflammation. *J. Biol. Chem.* **2020**, *295*, 6710–6720. [CrossRef] [PubMed]
33. Qu, Y.; Gao, N.; Wu, T. Expression and clinical significance of SYNE1 and MAGI2 gene promoter methylation in gastric cancer. *Medicine* **2021**, *100*, e23788. [CrossRef]
34. Chen, X.L.; Hong, L.L.; Wang, K.L.; Liu, X.; Wang, J.L.; Lei, L.; Xu, Z.Y.; Cheng, X.D.; Ling, Z.Q. Deregulation of CSMD1 targeted by microRNA-10b drives gastric cancer progression through the NF-κB pathway. *Int. J. Biol. Sci.* **2019**, *15*, 2075–2086. [CrossRef]
35. Knobel, M.; Medeiros-Neto, G. Relevance of iodine intake as a reputed predisposing factor for thyroid cancer. *Arq. Bras. Endocrinol. Metabol.* **2007**, *51*, 701–712. [CrossRef] [PubMed]
36. Qi, C.; Min, P.; Wang, Q.; Wang, Y.; Song, Y.; Zhang, Y.; Bibi, M.; Du, J. MICAL2 Contributes to Gastric Cancer Cell Proliferation by Promoting YAP Dephosphorylation and Nuclear Translocation. *Oxidative Med. Cell. Longev.* **2021**, *2021*, 9955717. [CrossRef]
37. Jiang, C.; Ma, Z.; Zhang, G.; Yang, X.; Du, Q.; Wang, W. CSNK2A1 Promotes Gastric Cancer Invasion Through the PI3K-Akt-mTOR Signaling Pathway. *Cancer Manag. Res.* **2019**, *11*, 10135–10143. [CrossRef]
38. Chen, H.; Wang, S. Clinical significance of ADAM29 promoting the invasion and growth of gastric cancer cells in vitro. *Oncol. Lett.* **2018**, *16*, 1483–1490. [CrossRef]
39. Park, S.; Kang, S. A Western-style diet interacts with genetic variants of the LDL receptor to hyper-LDL cholesterolemia in Korean adults. *Public Health Nutr.* **2021**, *24*, 2964–2974. [CrossRef]
40. Iida, M.; Ikeda, F.; Ninomiya, T.; Yonemoto, K.; Doi, Y.; Hata, J.; Matsumoto, T.; Iida, M.; Kiyohara, Y. White blood cell count and risk of gastric cancer incidence in a general Japanese population: The Hisayama study. *Am. J. Epidemiol.* **2012**, *175*, 504–510. [CrossRef]
41. Tobacco smoke and involuntary smoking. *IARC Monogr. Eval. Carcinog. Risks Hum.* **2004**, *83*, 1–1438.
42. Hishida, A.; Matsuo, K.; Goto, Y.; Naito, M.; Wakai, K.; Tajima, K.; Hamajima, N. Smoking behavior and risk of Helicobacter pylori infection, gastric atrophy and gastric cancer in Japanese. *Asian Pac. J. Cancer Prev.* **2010**, *11*, 669–673. [PubMed]
43. Martimianaki, G.; Bertuccio, P.; Alicandro, G.; Pelucchi, C.; Bravi, F.; Carioli, G.; Bonzi, R.; Rabkin, C.S.; Liao, L.M.; Sinha, R.; et al. Coffee consumption and gastric cancer: A pooled analysis from the Stomach cancer Pooling Project consortium. *Eur. J. Cancer Prev.* **2022**, *31*, 117–127. [CrossRef] [PubMed]
44. Xie, Y.; Huang, S.; He, T.; Su, Y. Coffee consumption and risk of gastric cancer: An updated meta-analysis. *Asia Pac. J. Clin. Nutr.* **2016**, *25*, 578–588.
45. Miyato, H.; Tsuno, N.H.; Kitayama, J. Semaphorin 3C is involved in the progression of gastric cancer. *Cancer Sci.* **2012**, *103*, 1961–1966. [CrossRef]
46. Peacock, J.W.; Takeuchi, A.; Hayashi, N.; Liu, L.; Tam, K.J.; Al Nakouzi, N.; Khazamipour, N.; Tombe, T.; Dejima, T.; Lee, K.C.; et al. SEMA3C drives cancer growth by transactivating multiple receptor tyrosine kinases via Plexin B1. *EMBO Mol. Med.* **2018**, *10*, 219–238. [CrossRef]

Article

Mushroom against Cancer: Aqueous Extract of *Fomitopsis betulina* in Fight against Tumors

Paulina Nowotarska [1,*], Maciej Janeczek [1] and Benita Wiatrak [2,*]

1 Department of Biostructure and Animal Physiology, Wroclaw University of Environmental and Life Sciences, Norwida 25/27, 50-375 Wroclaw, Poland; maciej.janeczek@upwr.edu.pl
2 Department of Pharmacology, Wroclaw Medical University, Mikulicza-Radeckiego 2, 50-345 Wroclaw, Poland
* Correspondence: paulina.nowotarska@upwr.edu.pl (P.N.); benita.wiatrak@umw.edu.pl (B.W.)

Abstract: Background/Objectives: This study investigated the anticancer potential of an aqueous extract of the fungus *Fomitopsis betulina*. **Methods:** The study assessed the effect of the extract on nine cancer cell lines, including melanoma (LM-MEL-75), lung cancer (A549), and colorectal cancer (HT29, LoVo), and four normal cell lines. The cytotoxicity of the extract was evaluated using MTT, sulforhodamine-B (SRB), and clonogenic viability assays. Additionally, the study examined the effect of the extract on plant model organisms, garden cress (*Lepidium sativum*) and common onion (Allium cepa), to further investigate its biological activity. **Results:** The assays demonstrated selective cytotoxicity of the extract toward cancer cells, while sparing normal cells. The extract induced significant cytotoxic effects at lower concentrations in lung cancer, melanoma, and colon cancer cells, showing promise as a potential anticancer agent. The results also revealed that the extract inhibited seed germination and root growth, suggesting its potential to disrupt cell cycles and induce apoptosis. **Conclusions:** This study highlights the therapeutic potential of *F. betulina* and highlights the need for further research to identify the active ingredients and mechanisms underlying its anticancer effects.

Keywords: *Fomitopsis betulina*; in vitro; anticancer activity; plant model organisms; cytotoxicity

Citation: Nowotarska, P.; Janeczek, M.; Wiatrak, B. Mushroom against Cancer: Aqueous Extract of *Fomitopsis betulina* in Fight against Tumors. *Nutrients* **2024**, *16*, 3316. https://doi.org/10.3390/nu16193316

Academic Editors: Yi-Wen Liu and Ching-Hsein Chen

Received: 21 August 2024
Revised: 24 September 2024
Accepted: 24 September 2024
Published: 30 September 2024

1. Introduction

Cancer poses a significant challenge for modern medicine, necessitating continuous research to discover new and effective therapies. In vitro studies on compounds of both plant and fungal origin play a crucial role in identifying potential anticancer drugs. These studies allow for precise assessments of how these compounds impact the growth, proliferation, and survival of cancer cells.

Plant compounds, abundant in nature, have long intrigued researchers due to their potential anticancer properties. Flavonoids, alkaloids, polyphenols, and other phytochemicals found in many plants exhibit anticancer activity [1]. In vitro investigations enable us to understand the cellular mechanisms of action of these compounds, aiding in the selection of promising candidates for further clinical research [2]. Fungi, with their genetic and metabolic diversity, represent another promising source of potential anticancer compounds. In recent years, scientists have focused on identifying biologically active fungal compounds for cancer therapy [3]. *Fomitopsis betulina* (formerly known as *Piptoporus betulinus*) has garnered scientific interest due to its potential health benefits. However, clinical research on this mushroom remains limited, and preclinical studies have shed light on its promising properties. *Fomitopsis betulina* is commonly found on birch trees. An intriguing historical connection lies in the mummified remains of Ötzi (the Iceman), who lived approximately 5300 years ago. Ötzi carried fragments of *Fomitopsis betulina*, which is commonly found on birch trees, suggesting its use for medicinal purposes even in ancient times [4].

It occurs in the northern hemisphere [5] and is considered a common species in Poland. It grows on both living tree trunks and dead wood fragments as well as on branches. The birch polypore exclusively grows on trees from the birch family, such as silver birch, downy birch, dark birch, and paper birch [4,6]. Its presence spans from August to October, and the dead fruiting bodies remain on the tree throughout the year [7]. However, attention must be paid to the location of the trees from which the fungus is collected as environmental pollutants in industrial zones may lead to contamination with heavy metals and toxins. Such factors have been known to impact the chemical composition and medicinal properties of fungi.

This fungus is an annual organism. After penetrating a wound site in healthy birch bark, it can remain in a state of hibernation for many years. Interestingly, studies on other medicinal fungi, such as *Ganoderma lucidum* and *Trametes versicolor*, have noted similar resilience and adaptive strategies in nature, which may contribute to their therapeutic effects [8,9]. Only when the tree weakens (due to excessive shade, drought, or fire) does the birch polypore activate, breaking down wood and causing white rot—a feeding strategy that classifies it as a parasite [10]. The cap of the birch polypore is semicircular and gray-brown, droops 3–15 cm, has a width of 4.5–12.5 cm, and has a thickness of 1.7–3.5 cm at the base. The mushroom tubes reach a length of up to 8 mm, and in mature specimens, they easily detach from the cap flesh [5]. In mature specimens, the hymenium appears white to ocher, and the pores are round or angular and occur at a rate of 3–5 per millimeter. Initially, they are thick and filled with secretions, but over time, they rupture, tear, and eventually develop a hydnoid structure. The tube layer can be easily separated from the surrounding context and may reach a thickness of up to 10 mm [5]. The flesh is white, bitter, and astringent in taste. Some sources suggest that young specimens are edible [5,11].

The infusion of *F. betulina* basidiom gained popularity, particularly in Russia, the Baltic countries, Hungary, and Romania, for its calming effects, nutritional benefits, and anti-fatigue properties [12–14]. This is consistent with the use of other medicinal mushrooms such as Chaga in Siberian and Baltic traditional medicine, which also offers antioxidant and immune-modulating properties [15]. Additionally, mushroom tea served as an antiparasitic remedy and was used to address gastrointestinal issues [12–14]. In the Czech Republic, *Fomitopsis betulina* was utilized for treating rectal cancer and stomach ailments [13]. Interestingly, the polypore fungus has been recognized for its antiseptic and analgesic properties in both Europe and the USA [13,16]. Traditionally, *F. betulina* has been employed as an antiparasitic and antimicrobial agent to address wounds and control bleeding [17]. Fresh fruiting bodies of *F. betulina* were used to create antiseptic and anti-bleeding dressings, while the powder derived from dried material served as a painkiller [12,18,19]. Additionally, it found use as tinder and an anesthetic [20]. *Fomitopsis betulina* has exhibited antibacterial activity, particularly against Gram-positive bacteria such as *Streptococcus* and *Staphylococcus*. It is also used in folk medicine, especially in Baltic countries, due to its antifungal and antiparasitic properties [11,20]. The specific active compounds responsible for this effect are still under investigation. Additionally, preparations derived from *Fomitopsis betulina* have shown immunomodulatory effects and potential neuroprotective properties.

In vitro studies have revealed that extracts from *Fomitopsis betulina* contain triterpenes and polysaccharides, which exhibit anticancer activity by inducing apoptosis, inhibiting angiogenesis, and blocking cell signaling pathways. These discoveries suggest that *F. betulina* could be a valuable component or source of anticancer drugs. In particular, the presence of triterpenes, similar to those found in *Ganoderma* and Chaga mushrooms, may explain the apoptotic and cytostatic effects observed in cancer cell lines [15,21]. While clinical trials involving this fungus are scarce, there is ongoing research that aims to explore its potential applications in health and biotechnology.

Previous studies have investigated the anticancer properties of ethanolic extracts of *F. betulina*, mainly focusing on ethanolic extracts [8,9]. This study aims to address this gap by investigating the effect of an aqueous extract from *F. betulina* on a panel of cancer cell lines.

In this context, the current study seeks to evaluate the selective cytotoxicity of aqueous extracts from *Fomitopsis betulina* on several cancer cell lines while also assessing its effect on normal cell lines. Additionally, we explore the biological activity of the extract using plant model organisms, such as *Lepidium sativum* (garden cress) and *Allium cepa* (onion), to assess its broader biological impact. This research contributes to the growing body of evidence on natural extracts in cancer therapy, offering insights into the therapeutic potential of aqueous extracts that may be safer than alcohol-based alternatives.

2. Results

2.1. Viability Assays

2.1.1. MTT Assay

To investigate the impact of the aqueous extract of *F. betulina* on the cytotoxic potential of cancer cells, cell viability studies were conducted by incubating the extract for 24 h using the MTT assay on nine cancer cell lines. The MTT test allows the assessment of mitochondrial activity in cells. Figure 1 illustrates the effect of the extract on cells pre-incubated with either the tested extract or cisplatin (a chemotherapeutic agent commonly used in cancer treatment). Traditional cytostatics are highly toxic not only to cancer cells but also to normal cells. Therefore, the influence of the tested extract was also evaluated on four normal cell lines. The study revealed that the extract, up to a concentration of 10,000 µg/mL, did not exhibit cytotoxic potential in normal mice (L929), hamsters (V79), and human (NHDF) fibroblasts. Moreover, in the case of the mouse fibroblast line, in the concentration range of 1–10,000 ug/mL, a significant increase in mitochondrial activity was observed compared to the control. In contrast, cisplatin demonstrated cytotoxicity in human and mouse fibroblast lines at a concentration as low as 10 µg/mL. Compared to skin cancer cells (melanoma—LM-MEL-75 and epidermoid carcinoma—A431), the aqueous extract of *F. betulina* exhibited cytotoxicity at lower concentrations, which were non-toxic to normal human dermal fibroblasts. Notably, the melanoma cell line demonstrated statistically significant cytotoxic potential for the tested extract even at a concentration of 100 µg/mL.

Importantly, the tested extract did not exhibit cytotoxic activity toward normal kidney cells (Vero). This lack of cytotoxicity is particularly significant for drug elimination from the body. In contrast, cisplatin demonstrated cytotoxic potential even at a concentration of 0.1 µg/mL.

Interestingly, up to a concentration of 10,000 µg/mL, we observed increased mitochondrial activity in normal intestinal epithelial cells (CCD 841 CoN). Simultaneously, there was a concentration-dependent inhibition of activity in both HT29 and LoVo colorectal cancer cells. Notably, the LoVo cell line exhibited cytotoxic potential at a concentration of 10 µg/mL.

In the case of lymphoblastic leukemia tumors (CCRF-CEM)—these cells come from lymphoblastic leukemia and express certain receptors such as CXCR4 [18]). The impact of the tested extract was observed only at the highest tested concentration. However, in the case of fibrosarcoma, cytotoxic potential was observed already at a concentration of 100 µg/mL. The tested extract demonstrated a significant impact in the context of lung cancer (A549) at a concentration of 10 µg/mL. No antitumor activity was observed in breast (MCF7) and ovarian (HeLa) cancer cell lines, and there was even an increase in mitochondrial activity.

IC_{50}, which refers to the concentration of a substance required to inhibit the growth of 50% of cells, allows for the evaluation of the effectiveness of these substances in inhibiting cell viability. These values were determined for the tested extract and cisplatin in various cell lines, both cancerous and normal (Table 1).

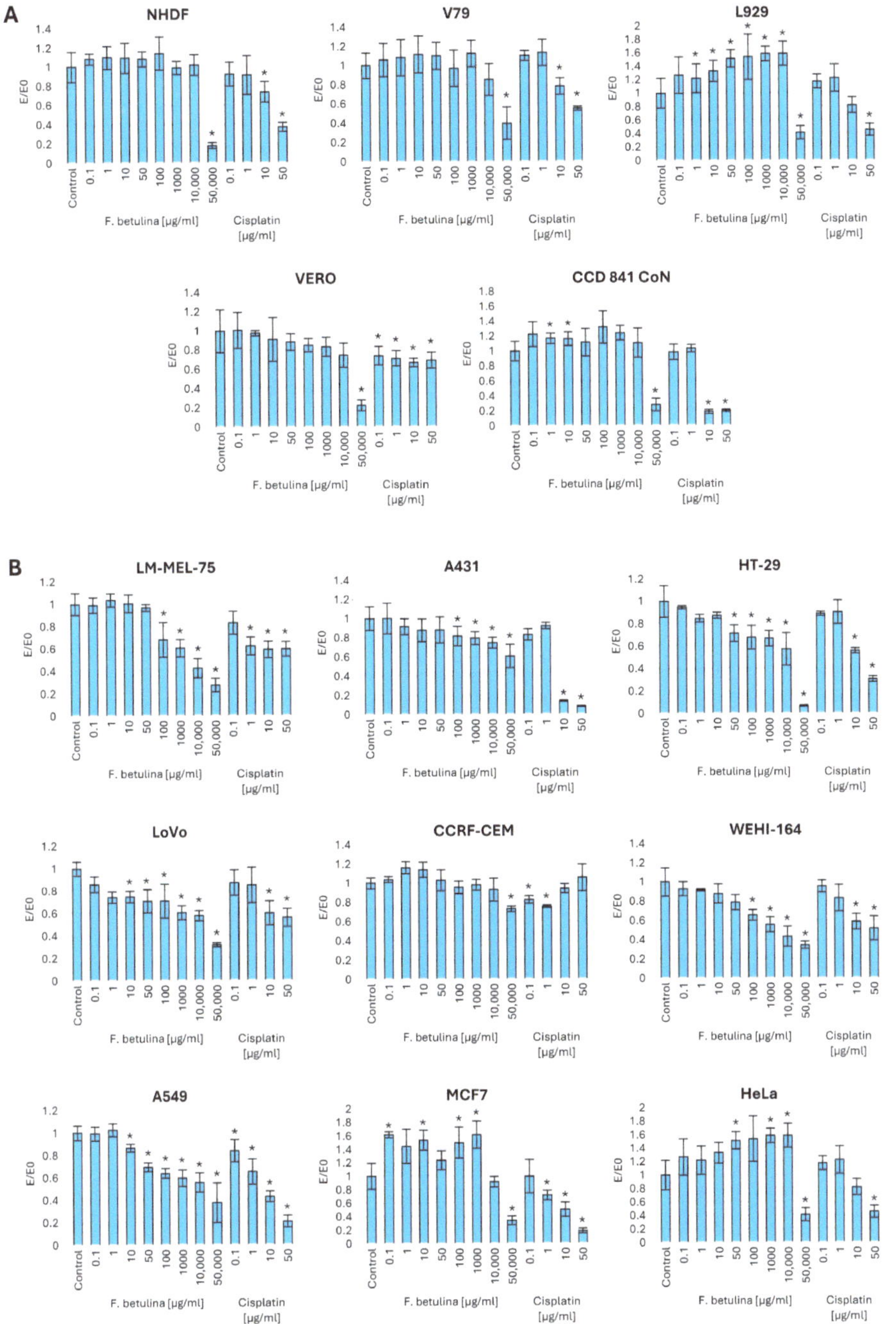

Figure 1. Mitochondrial activity after incubation with aqueous extract of *F. betulina*: (**A**) Normal cell lines; (**B**) Cancer cell lines. Data are presented as E/E_0 (where E—the average number of colonies in the tested extract concentration and E_0—the average number in the control culture and standard deviation (SD); * statistically significant differences in mitochondrial activity ($p < 0.05$) when comparing treated versus control groups, with error bars representing standard deviation.

Table 1. The IC$_{50}$ values (μg/mL) for *F. betulina* extract and cisplatin across various cell lines, both cancerous and normal. IC$_{50}$ represents the concentration of a substance required to inhibit the growth of 50% of cells, which allows for the evaluation of cytotoxicity and the potential selectivity of the substance toward cancer cells in comparison to normal cells.

Cell Line	IC$_{50}$ [μg/mL]	
	F. betulina	Ciaplatin
NHDF	27,409.79	31.337
V79	22,002.85	>50
L929	47,995.60	20.61
Mel75	2367.68	>50
A431	>50,000	6.04
Vero	23,833.79	>50
CCD-841	>50,000	3.54
HT-29	2996.55	13.58
LoVo	7204.86	>50
CCRF	>50,000	>50
WEHI-164	2175.99	>50
A549	3256.86	5.02
MCF-7	>50,000	20.61
HeLa	16,693.67	8.31

For the *F. betulina* extract, it can be observed that it has relatively low toxicity toward normal dermal fibroblast cells (NHDF) and fibroblasts (L929), with IC$_{50}$ values of 27,409.79 μg/mL and 47,995.60 μg/mL, respectively. Similarly, in the case of the Vero cell line (monkey kidney cells), the IC$_{50}$ is 23,833.79 μg/mL, indicating low toxicity toward normal cells. For the CCD-841 cell line, which represents normal intestinal epithelium, the extract was non-toxic, which is an interesting observation since, for cancerous intestinal lines such as HT-29 and LoVo, the IC$_{50}$ values were 2996.55 μg/mL and 7204.86 μg/mL, respectively. This indicates that *F. betulina* exhibits some selectivity toward cancerous intestinal cells compared to normal intestinal epithelium.

Regarding the Mel75 cell line (melanoma), the *F. betulina* extract showed relatively high cytotoxicity (IC$_{50}$ = 2367.68 μg/mL), which, when compared to the values for normal dermal fibroblast cells (NHDF), indicates significant selectivity toward cancerous cells. This suggests that the extract may be promising in the context of melanoma therapy, as its toxicity to normal dermal fibroblast cells is considerably lower.

On the other hand, cisplatin, a widely used anticancer drug, demonstrated significantly higher cytotoxicity in cancer cell lines. For A431 cells (squamous cell carcinoma of the skin), the IC$_{50}$ was 6.04 μg/mL, and for the A549 cell line (lung cancer), it was 5.02 μg/mL, which reflects its effectiveness in targeting cancer cells. In the case of the colorectal cancer cell line (HT-29), cisplatin showed moderate toxicity (IC$_{50}$ = 13.58 μg/mL), while it was even more toxic to normal intestinal epithelial cells (CCD-841), with an IC$_{50}$ of 3.54 μg/mL.

On the other hand, in cisplatin's action on melanoma cells (Mel75), the substance was found to be non-toxic, suggesting that cisplatin may not be an effective drug in treating this type of cancer, in contrast to the *F. betulina* extract. Cisplatin also showed no toxicity toward several other cell lines, such as fibroblasts (L929) and osteosarcoma cells (WEHI-164).

In conclusion, the *F. betulina* extract exhibits interesting selectivity toward cancerous cells, particularly in the case of skin and intestinal cancers, while being less toxic to normal dermal fibroblast cells, fibroblasts, and intestinal epithelium. Cisplatin, on the other hand, shows high efficacy against most tested cancer cell lines, especially skin and lung cancers, but lacks activity against melanoma.

2.1.2. SRB Assay

To assess the total protein content in cells treated with birch stump aqueous extract, we conducted a sulforhodamine B (SRB) assay. SRB binds stoichiometrically to proteins under slightly acidic conditions, allowing us to measure the overall protein content within the cell. This measurement reflects both the structural and functional activity of the cell. To evaluate the impact of administering the *Fomitopsis betulina* extract, we analyzed cells that had been preincubated with the aqueous extract for 48 h. Additionally, we examined control cells treated solely with medium and a cytostatic agent commonly used in cancer therapy—cisplatin. Figure 2 presents the results of the SRB assay.

Regarding a 48 h incubation of normal cells, both fibroblasts (NHDF, V79, L929) and colon epithelium (CCD 841 CoN) resulted in an increase in the amount of total protein in the culture, which may indicate increased proliferation of normal cells. At the same time, no statistically significant cytotoxic effects were observed in normal kidney cells.

Significantly strong cytotoxicity was observed toward melanoma cells, with statistical significance evident from a concentration of 10 µg/mL. Similarly, in the case of the colorectal cancer cell line HT-29, a potent cytotoxic effect was observed across the entire concentration range, comparable to concentrations of 0.1–1 µg/mL cisplatin. However, cytotoxicity toward A431 and LoVo cells was only evident at the highest tested concentrations. It is worth noting that in the case of hormone-dependent cancer—MCF7 and cervical cancer (HeLa), an increase in the amount of cellular protein (increase in proliferation) was observed. There was no significant effect of the tested extract on lymphoblastic leukemia cells (CCRF-CEM), and there was a strong cytotoxic effect on fibrosarcoma (WEHI-164). In the case of lung cancer (A549), statistically significant cytotoxic effects are observed at levels as low as 1 µg/mL.

2.1.3. Clonogenic Assay

The clonogenic test allows for assessing differences in cell ability to produce offspring after treatment with various bioactive compounds [22]. Figure 3 illustrates the colony formation properties of cells following incubation with an aqueous extract of *F. betulina* and cisplatin as a control.

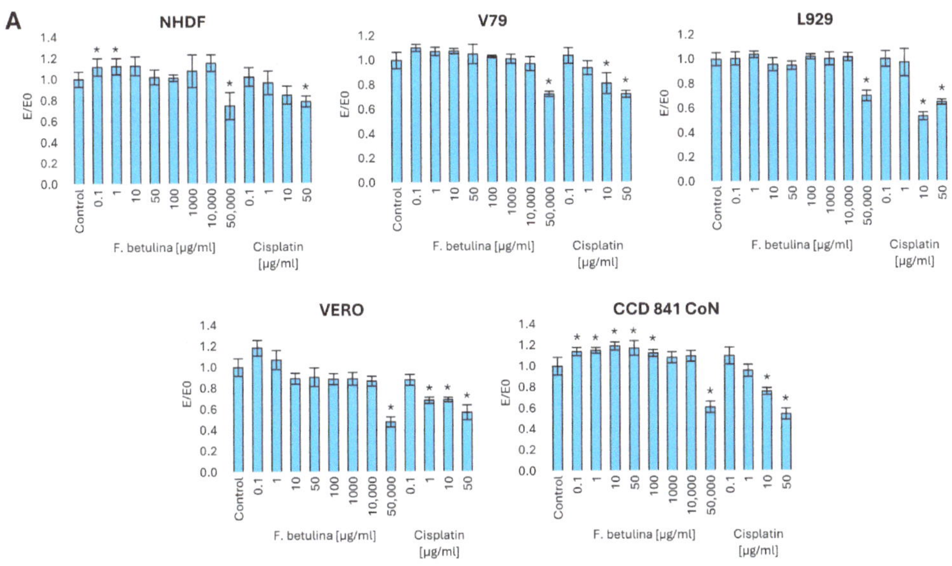

Figure 2. *Cont.*

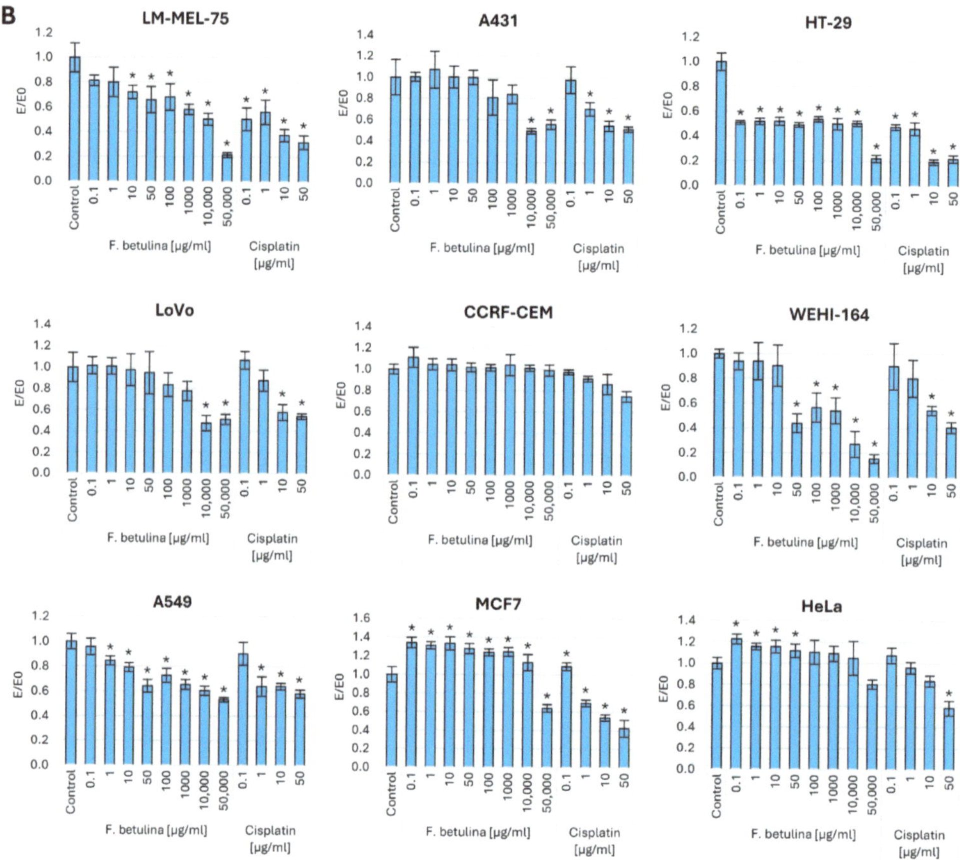

Figure 2. Cell viability after incubation with an aqueous extract of *F. betulina*: (**A**) Normal cell lines; (**B**) Cancer cell lines. Data are presented as E/E_0 (where E—the average number of colonies in the tested extract concentration and E_0—the average number in the control culture and standard deviation (SD); * statistically significant differences in cell viability ($p < 0.05$) when comparing treated versus control groups, with error bars representing standard deviation.

2.2. Plant Model Organisms

Cisplatin, within the tested concentration range of 0.1–10 µg/mL, completely inhibited the formation of colonies in all tested cell lines—both normal and cancers. However, the aqueous extract from *Fomitopsis betulina*, tested at concentrations ranging from 1 to 50 µg/mL, did not exhibit any cytotoxic effects on colony formation. Specifically, in the recommended cell line (V79), according to ISO 10993-5 standard [23], there was a statistically significant reduction in colonies at concentrations of 10 and 50 µg/mL compared to the control. Similarly, colorectal cancer cells (LoVo) and fibrosarcoma (WEHI-164) also showed statistically significant reductions in colonies at the same concentrations. Notably, across the entire tested concentration range, melanoma (LM-MEL-75), colorectal cancer (HT29), ovarian cancer (HeLa), and lung cancer (A549) all exhibited statistically significant reductions in colony formation (Figure 3). In the case of breast cancer (MCF7) and epidermoid carcinoma (A431), an increase in the number of daughter colonies was observed after 7 days of incubation, even at the highest concentration tested compared to the control.

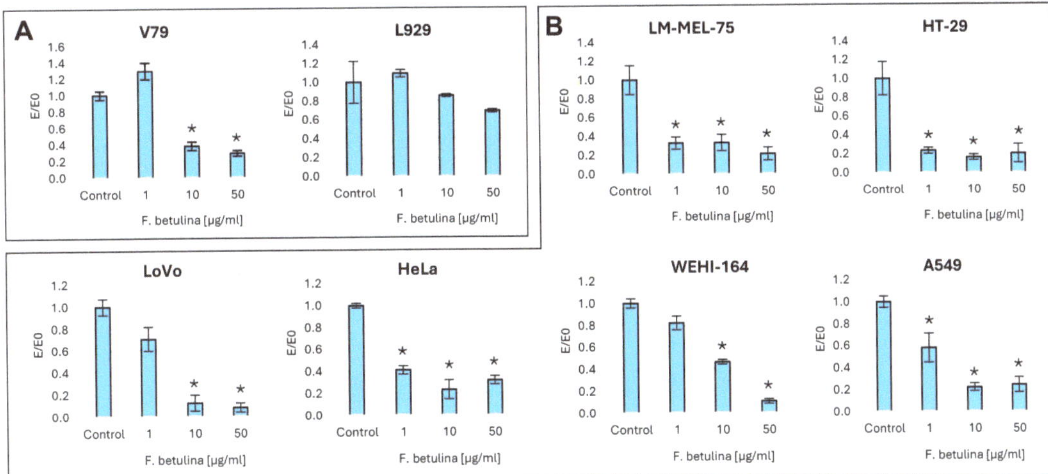

Figure 3. Cell colony formation properties after incubation with an aqueous extract of *F. betulina*: (**A**) Normal cell lines; (**B**) Cancer cell lines. Data are presented as E/E_0 (where E—the average number of colonies in the tested extract concentration and E_0—the average number in the control culture and standard deviation (SD); * statistically significant differences in cell colony formation ($p < 0.05$) when comparing treated versus control groups, with error bars representing standard deviation.

2.2.1. Garden Cress (*Lepidium sativum*)

While cress is not directly employed in human cancer research, it serves as a valuable model for investigating biological processes that could have implications for anticancer therapies. For instance, studies on cell cycle regulation, apoptosis, and stress responses may yield insights into potential anticancer compounds [24,25].

In this study, the aqueous extract of *Fomitopsis betulina* demonstrated statistically significant inhibition of cress germination within the concentration range of 10–50,000 µg/mL. Notably, transferring the seeds to fresh paper soaked only in water did not lead to an increase in the number of germinating seeds (Figures 4 and 5). These findings suggest that the tested aqueous extract containing *F. betulina* may potentially impact apoptosis and cell cycle inhibition.

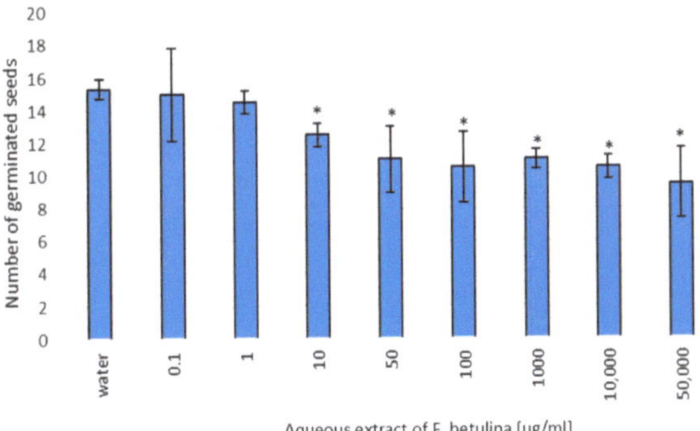

Figure 4. Assessment of the genotoxic effect of an aqueous extract of *F. betulina* on cress seeds; * statistically significant differences in number of germinated seeds ($p < 0.05$) when comparing treated versus control groups, with error bars representing standard deviation.

Figure 5. Representative photos from a test of cress during growth and exposure to an aqueous extract of *F. betulina* on cress seeds.

2.2.2. Modified *Allium cepa* L. Test

Due to the potential impact of the extract on the cell cycle, a commonly used test was performed to assess the disruption of cell division in a plant model organism. In studies on compounds with anticancer activity, assessing the inhibition of the mitotic index in common onion (*Allium cepa* L.) is crucial. The mitotic index serves as an indicator of mitotic activity, representing cell division. In the case of onions, inhibiting this index may suggest potential anticancer effects [26].

Direct exposure of onion roots to aqueous extracts of *F. betulina* resulted in a statistically significant reduction in root growth at concentrations of 10,000–50,000 µg/mL. Furthermore, statistically significant prolonged cell cycle arrest was observed at concentrations as low as 50 µg/mL of the aqueous extract (Figures 6 and 7). This indicates that the aqueous extract of *F. betulina* affects the cell cycle, which is relevant both in drug development for cancer therapy and chemoprevention.

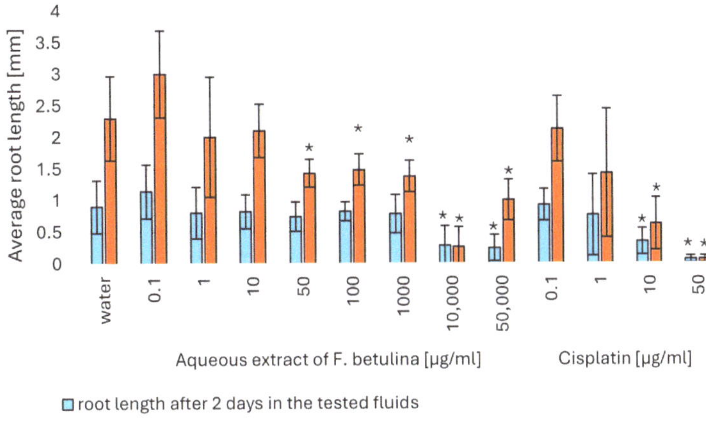

Figure 6. Average root length in Modified *Allium cepa* L. test; * statistically significant differences in average root length ($p < 0.05$) when comparing treated versus control groups, with error bars representing standard deviation.

Figure 7. Representative photos from a Modified *Allium cepa* L. test during growth and exposure to an aqueous extract of *F. betulina*.

3. Discussion

Numerous studies have explored the potential anticancer activity of birch tree stump extracts. These investigations primarily focus on analyzing the composition of substances found in water and alcohol extracts. However, their assessment is often limited to specific cancer cell lines, evaluating the cytotoxic effects on individual cell types. Previous research on fungi like *Ganoderma lucidum* and *Trametes versicolor* has highlighted the importance of understanding the broader range of effects that fungal extracts can have on various cancer types, including the selective cytotoxicity toward tumor cells [8,9]. Given the growing interest in understanding the biological activities of natural compounds, especially in the context of anticancer properties, we deemed it crucial to characterize the impact of *Fomitopsis betulina* extract across a broad spectrum of cancer and normal cell lines. Additionally, an essential aspect of our research involves employing model organisms. While cell cultures serve as an initial screening platform for assessing potential bioactive compounds, they do not fully represent complex tissues and organs. Consequently, false positive results may arise. Ethically exploring research with plant model organisms, before resorting to in vivo studies using laboratory animals, holds promise. The study investigated the anticancer activity of an aqueous extract from *Fomitopsis betulina* and explored its potential for further research. Specifically, we examined how this extract affected nine different cancer cell lines. Additionally, we assessed its impact on four normal cell lines to determine the selectivity index. Viability experiments were conducted to analyze the extract's effects on total protein content (SRB) and mitochondrial activity (MTT). Furthermore, we investigated its influence on cell progeny formation using a clonogenic assay. Finally, we compared the potential mutagenic effects of the extract on cell cycle disruption in a model of common onion roots.

The study of the effect on cells and the anticancer activity of the tested aqueous extract from *F. betulina* was conducted on continuous cell lines. An initial assessment of the toxicity of the tested compounds for non-cancer (normal) cells was performed on L929 and V79 mouse fibroblast cultures, following the recommendations of ISO 10993-5 [23], as well as on Vero kidney cells (normal kidney cells). It is widely known that cytostatics affect the proper functioning of the kidneys. Additionally, studies were carried out on human skin fibroblasts (NHDF) to determine the selectivity index compared to melanoma and epidermoid carcinoma lines. Furthermore, to analyze the selectivity index for colorectal cancer, investigations were conducted on normal intestinal epithelial cells (CDD841). The selection of cancer cell lines was guided by the National Cancer Institute's guidelines to assess the impact of the tested compounds on various types of cancer. According to the latest World Health Organization (WHO) cancer report, updated in 2020 by the International Agency for Research on Cancer (IARC), the most commonly diagnosed cancers that continue to cause high mortality include lung cancer (A459 lines selected here), colorectal cancer (HT29 and LoVo lines selected here) in women breast cancer (MCF7 cell line) and ovarian cancer (HeLa). Skin cancer (melanoma—LM-MEL-75 and epidermoid cancer—A431) is an equally serious problem. Lymphoblastic leukemia (CCRF-CEM) and fibrosarcoma (WEHI 164) are also a big problem. The incidence of each type of cancer varies according to the region of the world, risk factors, and accessibility to healthcare and screening [27].

The choice of water extract for this study is crucial, as it reflects a safer alternative to alcohol-based solvents, which have been shown to pose carcinogenic risks, according to the World Health Organization (WHO) [28]. Water is a natural biological solvent, making it ideal for in vitro and in vivo testing. It is also safe and non-toxic compared to alcohol. Ethanol, on the other hand, is more effective than water in dissolving many organic compounds and can, therefore, be used to extract specific components [29]. However, the WHO underlines that alcohol (ethanol) is a carcinogen, and there is no safe amount to consume in terms of cancer risk [28]. Therefore, if one wishes to study extracts from natural products in the context of anticancer effects, one should avoid those that may be damaging to the body.

Previous studies on the anticancer activity of *F. betulina* extract and isolated compounds (betulin, betulinic acid, and triterpenes) were carried out using ethanol extract, which resulted in increased cytotoxicity toward melanoma lines (A375, Hs895) and glioma lines (U251MG, U343MG) [8,9]. In comparison to previous studies using ethanol extracts of *Fomitopsis betulina*, which showed strong cytotoxicity toward glioma cells [9], our use of aqueous extracts presents a safer alternative with selective toxicity toward lung and colon cancer cells. This highlights the potential application of aqueous extracts for cancer treatment with lower side effects. Our research shows the effect of *F. betulina* extract on cancer and normal cells. We demonstrated that the aqueous extract had a weaker effect than cisplatin on the tested cancer cells, but there was, statistically, significantly better selectivity toward normal cells compared to the cisplatin effect. At the same time, in the case of the lung cancer cell line (A549) and colorectal cancer cell line (LoVo and HT29), the cytotoxic potential was observed for 24 h incubation.

The chemical composition of *F. betulina* is complex, comprising various bioactive compounds. The main constituents identified in the mushroom are polysaccharides, terpene compounds, fatty acids, and sterols. These compounds contribute to its numerous therapeutic properties such as antimicrobial, anti-inflammatory, antioxidant, immunomodulatory, and anticancer effects [15]. The presence of triterpenes in *F. betulina*, such as betulin and betulinic acid, is consistent with findings in other medicinal mushrooms like Ganoderma lucidum, where similar compounds have demonstrated anticancer properties through apoptosis induction and cell cycle arrest [10,30]. Our own research demonstrated that the tested extract exhibited its most potent effect on the A549 cell line at a concentration of 50 µg/mL for 24h incubation and 1 µg/mL for 48h incubation. However, other researchers examining isolated compounds, including terpene, did not observe any substantial cytotoxic effects [9]. The cytotoxic effects observed in melanoma and lung cancer cells may be attributed to the presence of triterpenes, which have been documented to induce apoptosis through mitochondrial pathways. The lack of activity in breast and ovarian cancer cells suggests hormone dependency, which could limit the extract's efficacy. Further mechanistic studies could explore the role of specific compounds in modulating the PI3K/AKT and MAPK pathways. These results underline the importance of investigating the synergistic effects of multiple bioactive compounds within the extract rather than focusing solely on individual components.

What is also crucial is the source of our extract: mycelium or fruiting bodies. In the context of A375 and WM795 cells (melanoma), which exhibit varying metastatic potential, we observed a more favorable effect with the mycelium extract. However, the fruiting body extract showed no impact on the WM795 cell line [21]. Our study also revealed the inhibitory effect of the tested mix mycelial and fruiting body aqueous extract on cell growth, irrespective of the incubation time with the extract.

Previous literature reports suggest that the tested compounds from the extract, as well as the fruiting and/or mycelium extract itself, may impact the cell cycle and influence apoptosis in the context of anticancer activity. Consequently, we decided to investigate this potential mechanism using plant model organisms commonly employed for such studies: cress and common onion [26,30]. The use of model organisms like *Lepidium sativum* (cress) and *Allium cepa* (onion) is supported by previous studies on their role in cytotoxicity and genotoxicity assessments, particularly in evaluating natural extracts [30,31]. Garden cress (*Lepidium sativum* L.) serves as a model plant for ecotoxicity assessments in terrestrial ecosystems. This species is particularly suitable for evaluating the cytotoxic effects of tested extracts due to its consistent seed germination and root elongation. These characteristics contribute to reliable and repeatable results, allowing for experimental designs with a limited number of repetitions [24,25]. *Allium cepa*, or common onion, is tested in genotoxicity tests, assessing the inhibition of mitotic division due to several basic features. Onion is characterized by rapid root growth, which allows the genotoxic effect to be observed in a relatively short time. Chromosomal abnormalities can be easily observed under a microscope, making the results easier to view. Moreover, the modified root length growth observation test significantly speeds up the procedure and is also very effective. *Allium cepa* is a popular food plant, which makes it accessible and relatively cheap to study. In short, *Allium cepa* is a universal tool for defining genetic applications [26,31]. After 72 h of incubation, the presence of aqueous extracts had varying effects on the germination of *L. sativum* seeds compared to those germinating in distilled water. No statistically significant germination toxicity was observed in seeds exposed to the lowest two concentrations of the tested extract. However, higher extract concentrations in the medium led to a statistically significant reduction in germination. This observation suggests a potential genotoxic or pro-apoptotic effect of the tested extract. Additionally, inhibition of *Allium cepa* L. root growth occurred upon direct administration of the extract. Interestingly, this effect persisted even after discontinuation of direct administration and subsequent incubation of *Allium cepa* L. in water, indicating an impact on mitotic divisions—the cell cycle.

Limitations: The composition of compounds present in *Fomitopsis betulina* can vary depending on the location and time of collection of this fungus. There are differences in the chemical composition between extracts obtained from fruiting bodies collected from different trees as well as between extracts from fruiting bodies and fungal mycelia.

The location of the mushroom collection can affect the chemical composition due to differences in environmental conditions such as soil type, humidity, sunlight, and the type of host tree [32]. For example, extracts from *Fomitopsis betulina* collected from birch trees growing in different regions may contain different amounts and types of biologically active compounds.

The time of collection also affects the chemical composition of *Fomitopsis betulina*. Differences in weather conditions and developmental stages of the fungus can affect the production and accumulation of different chemical compounds [32]. Although we did not present the chemical composition of the *F. betulina* mushroom in our work, it is worth emphasizing that its composition has been well studied in numerous scientific publications and is relatively constant.

The use of plant model organisms also presents limitations. While it is acknowledged that results from plant model organisms cannot be directly extrapolated to humans, it is important to recognize that these organisms are more complex than single monolayers of cells lacking tissue organization. Furthermore, preliminary research involving plant organisms enables us to investigate specific mechanisms of action without resorting to laboratory animals, which is ethically significant.

The *Fomitopsis betulina* extract has shown promising anticancer properties in in vitro studies, particularly in relation to selected cell lines. However, to assess its potential as an anticancer therapeutic in clinical settings, further in vivo studies and clinical trials are necessary. Such studies would allow for a better understanding of the extract's bioavailability, metabolism, and potential interactions with other drugs used in cancer therapy. Additionally, it will be crucial to determine the molecular mechanisms of the extract's action, including its influence on signaling pathways responsible for apoptosis and cancer cell proliferation. Conducting preclinical studies, including animal models, and then clinical trials will enable the evaluation of the extract's efficacy, safety, and any possible side effects associated with its long-term use.

4. Materials and Methods

4.1. Raw Material

Fomitopsis betulina specimens were collected in October 2021 from the Lower Silesia forests voivodeship near the village of Ławszowa. The mushrooms were carefully detached from the tree trunk to avoid causing damage. Subsequently, the collected mushrooms were cleaned, sliced into 1 cm sections, and weighed. These *F. betulina* slices were then subjected to 72 h of drying in a specialized mushroom dryer. After this drying period, the mushrooms were re-weighed, revealing a weight reduction of approximately 24%. Finally, the dried *Fomitopsis betulina* was ground using a 160 W electric mill.

4.2. Extraction Methods

An amount of 5 g of milled *F. betulina* was weighed and poured over distilled water to reach a final weight of 50g (Figure 8). The mixture was placed on a plate and brought to a boil (95 °C), then it was boiled for 10 min. After this time, the cooled extract was centrifuged at 4500 RPM for 10 min. The supernatant was passed through filters (TPP, Trasadingen, Switzerland; Syringe—filter 0.22 μm, cat. no. 99722).

4.3. Tested Compounds

The extract was prepared immediately before the tests. After appropriate dilution of the stock extract, the following concentrations were tested: 0.1, 1, 10, 50, 100, 1000, 10,000, and 50,000 [μg/mL]. The concentration range (from 0.1 μg/mL to 50,000 μg/mL) was selected based on preliminary cytotoxicity studies and literature data, indicating that concentrations within this range can induce effects in various cancer cell lines while maintaining low toxicity toward normal cells. This allows for the assessment of the dose-response relationship and the determination of the therapeutic window. Cisplatin (Merck Sigma, St. Louis, MO, USA, CAS 15663-27-1) was used as a control in relation to the test substance at concentrations of 0.1, 1, 10, and 50 [μg/mL].

Figure 8. *F. betulina* collected in the Lower Silesian forests and used to prepare the extract came from the same collection.

4.4. Cell Culture and Conditions

The study utilized normal cell lines NHDF, L929, CCD841, V79, and Vero. Also, the compounds were tested on cancer cell lines, including HeLa, WEHI 164, A549, CCRF, MCF7, LoVo, HT29, A375, and A431. All cell lines were maintained in an incubator using special culture bottles containing cell media. For the NHDF line, we used DMEM (cat. no. BE12-917F; Lonza, Basel, Switzerland) without the addition of phenol red. The culture medium for L929, CCD841, MCF7, Vero, HeLa, and A549 lines was EMEM (Eagle's minimum essential medium) (cat. no. 01-025-1A; Biological Industries, Beit-Haemek, Israel). Lovo cells were maintained on DMEM/F12 (HAM) 1:1 medium (cat. no. 01-170-1A; Biological Industries), while HT-29 cells were cultured in Mc Coy's 5A Medium (modified) (cat. no. 01-075-1A; Biological Industries). RPMI 1640 medium (cat. no. 01-100-1A; Biological Industries) was used for WEHI 164 and CCRF lines. Lines A375, A431, and V79 were cultured in DMEM high Glucose medium (cat. no. 01-056-1A; Biological Industries). All culture media were supplemented with 10% FBS (fetal bovine serum) (cat. no. P30-8500; PAN Biotech, Aidenbach, Germany), 2 mM L-glutamine (cat. no. BE17-605E/U1, Lonza), 50 mg/mL Gentamycin Sulfate (cat. no. 03-035-1B; Biological Industries), and amphotericin B (cat. no. 15290-026; Gibco, Thermo Fisher Scientific, Waltham, MA, USA). The incubator maintained 95% humidity, a temperature of 37 °C, and a 5% CO_2, 21% O_2 concentrations. Tank water in the incubator was enriched with Aquaguard-1 solution (cat. no. 01-867-1B; Biological Industries). Cultures were evaluated at least twice a week using an inverted microscope, and cells were used for testing when they were in the logarithmic growth phase with confluence above 70%. Adherent cell lines were trypsinized using TrypLE solution. The doubling times for the cancer and normal cell lines used in this study were 24–72 h, which were taken into consideration for the timing of the assays.

4.5. Viability Assay

The viability assay was conducted following ISO 10993-5:2009 [23], which outlines test methods for assessing the in vitro cytotoxicity. In this case, cultured cells were incubated with the test substance. The following cell lines were used for the test: L929, LM-MEL-

75, LoVo, WEHI 164, A549, V79, CCD841, HT29, NHDF, A431, Vero, HeLa, MCF7, and CCRF-CEM. Cells were Seeded in 96-well plates with 10,000 cells per well. The next day, the test substance was applied to the cells after removing the culture medium. Cells were incubated with the compounds for 24 h. The next day, yellow tetrazolium salt 1 mg/mL (Sigma Aldrich, St. Louis, MO, USA, Cas-No: 298-93-1) dissolved in PBS was added to the cells. Insoluble formazan crystals formed as a result of cellular reactions. Formazan crystals are indicative of viable cells. Dead cells or those with impaired metabolism would not produce significant formazan. The plates were placed in a 37 °C, 5% CO_2, 95% humidity incubator for 2h After incubation, the supernatant was removed, and alcohol—isopropanol (cat. no. 603-117-00-0; STANLAB Sp. z o.o., Lublin, Poland) was added to allow crystals dissolving. Plates were placed on a shaker for 30 min, and absorbance was measured at 570 nm using a microplate reader (Thermo Scientific Multiskan GO, Waltham, MA, USA, type 1510; REF: 51119300).

4.6. The Sulforhodamine-B (SRB) Assay

Assessment of total protein using the SRB test is recommended by the NCI (National Cancer Institute). The following cell lines were used for the test: MCF7, CCD841, HeLa, Vero, A431, HT29, NHDF, V79, A549, and CCRF-CEM. Cells were seeded into 96-well plates at 20,000 cells per well. Incubation with the compounds lasted 48 h. The cells were fixed with cold 50% trichloroacetic acid (TCA); for suspension cells, 30% TCA was used in a volume of 20 μL. The plates were incubated at 4–8 °C for 30 h. The plates were rinsed five times under running water to remove any residual acid and allowed to dry at room temperature. An amount of 100 μL of SRB was added for 30 min, and the plates were left at room temperature. Dye residues were removed by rinsing the plates five times with 1% acetic acid using a sprayer and allowed to air dry. After the plates were dry, 100 μL of Tris base solution was added to all wells to dissolve the dye, and the plates were placed in a shaker for 5 min. Absorbance was measured using a microplate reader (Thermo Scientific Multiskan GO, type 1510) at 485, 489 nm, and 565 nm.

4.7. Clonogenic Assay

To evaluate colony formation properties following therapy, cells were seeded in dilutions (100 cells) in 6-well plates according to ISO 10993-5 annex B [23]. The following cell lines were used for the test: LM-MEL-75, HT29, HeLa, A549, V79, and L929. The plates were placed in an incubator and left undisturbed for 7 days until colonies were observed in control samples. After incubation, DMEM was removed, and cells were washed with PBS. Cells were fixed with frozen 80% methanol. The wells were then washed with PSB, and the plates were allowed to dry. Staining of the clones was carried out using 0.5% crystal violet for 20 min. Excess stain was removed by washing with water, and the plates were allowed to dry at room temperature. Only colonies visible to the naked eye (>~0.02 cm) were manually counted. The colony counting process was unbiased, as the counter was unaware of the sample identifiers. The resulting data were then plotted.

4.8. Garden Cress (Lepidium sativum)

Garden cress is an annual plant belonging to the cabbage family. It serves as a model species due to its rapid growth and high sensitivity to phytophysical extract. A germination test was conducted to evaluate the biological substances. Seeds of garden cress from Toraf company, Barcelona, Sapain (lot number: PL211604336/045TB) were purchased from an online store (www.ogrodniczowiniarski.pl (accessed on 26 June 2023)). These seeds underwent selection to remove any that were damaged, discolored, or differed in size from the others. Each dish was lined with an 80 g/m^2 filter paper, saturated with the appropriate concentration of test extract and distilled tap water. Twenty garden cress seeds were placed on each dish, which were then covered. The control and test groups were kept in a sunny location without direct light exposure at room temperature. Observations were conducted for 3 days, after which the seeds placed on the test extract were transferred to dishes with blotting papers soaked in tap water. Observations continued for another 3 days. The test extract was prepared on the first day of the experiment and stored in the refrigerator. Throughout the experiment, 3 mL of the appropriate liquid at room temperature was added to each blotting paper daily.

4.9. Modified Allium cepa L. Test

The common onion (*Allium cepa*) belongs to the Amaryllidaceae family and is cultivated and consumed worldwide. It is valued for its flavor and health-promoting properties. Depending on the variety, it can be a biennial or annual plant. Onion roots are commonly used in biological tests, such as cytotoxicity and mutagenicity tests. The tests use onion roots, which are often the first to be exposed to chemicals in the environment. The root tip of onions provides results similar to those of animal cytotoxicity tests. The International Program on Plant Bioassays (IPPB) approves onion tests for chromosomal aberrations related to environmental contaminants. Bulbs of spring onions of the Senshyu Yellow winter variety (producer "PIASECKI", Essington, PA, USA, lot number: PL 14/14/SENS/00006) were purchased. Individuals of similar diameters were selected. Before the test, the scales and the brown pallor were removed from each onion. The root bud (Latin primordium) was left behind. Shriveled, moldy, or spoiled onions, as well as those that had begun to shoot green leaves, were discarded. A modified version of the onion test was used to check the inhibitory effect of the mitotic activity of the tested extracts. The root apices and the base of the onions were placed directly in the test extract at the appropriate concentration and in tap water, bypassing the root germination stage. Five tubers were used for each extract and concentration. Observations lasted 48 h, where two complete mitotic cycles were 30 h (Figure 9). Root evaluation was carried out based on photos taken and length measurements. After this time, the test extract was replaced with tap water for another 48 h. The test extract and tap water were at room temperature. At the end of the experiment, the roots of all bulbs were taken, measured, and evaluated.

Figure 9. Conducting Modified *Allium cepa* L. test.

4.10. Statistical Analysis

Each experiment was performed in triplicate with three biological replicates per assay. The data were reported as the mean ± standard deviation. We assessed the normal distribution and equality of variances using the Shapiro–Wilk and Levene tests, respectively. Additionally, we performed Tukey's post hoc analysis for ANOVA. The significance level was set at 0.05.

5. Conclusions

In summary, our experiments demonstrated that the aqueous extract of *Fomitopsis betulina* exhibits anticancer activity in several cancer cell lines. While its effects were weaker than cisplatin, the extract showed significantly less cytotoxicity toward normal cells. Further research is needed to identify molecular targets and optimize bioactive component extraction. In vivo studies are crucial to assess the extract's therapeutic potential and safety, especially its effects on metastasis and immune modulation. The extract showed potential activity against lung cancer (A549), colorectal cancer (LoVo, HT29), and melanoma (LM-MEL-75) but had a pro-proliferative effect on hormone- and HPV-dependent cells like breast (MCF7) and ovarian cancer (HeLa). Additionally, it demonstrated selective cytotoxicity, sparing fibroblasts, intestinal epithelium, and kidney cells. Plant model organism studies suggest that the extract may inhibit mitotic divisions, leading to genotoxic and pro-apoptotic effects (Figure 10).

In vitro, the extract acts as a cytostatic rather than a cytotoxic agent on malignant cells. While promising, natural treatments should always be used with caution, complementing rather than replacing conventional therapies.

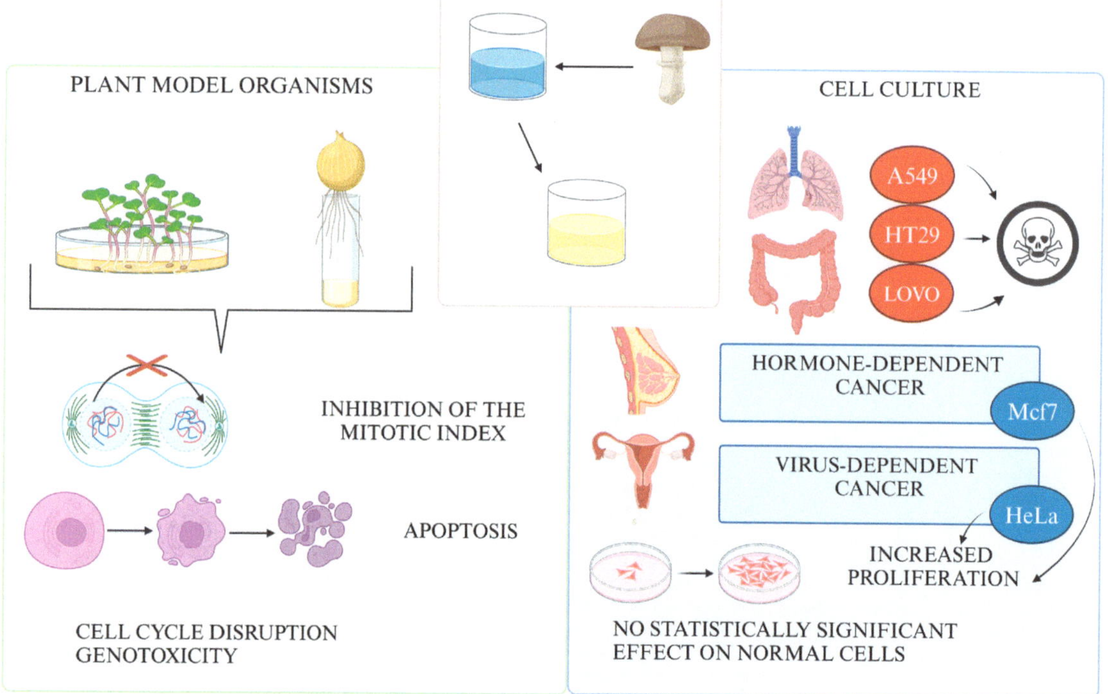

Figure 10. Properties of *F. betulina* aqueous extract in in vitro assays and with model plant organisms.

Author Contributions: Conceptualization, P.N. and B.W.; methodology, P.N. and B.W.; validation, P.N., M.J. and B.W.; formal analysis, P.N.; investigation, P.N.; resources, M.J.; data curation, P.N.; writing—original draft preparation, P.N.; writing—review and editing, B.W.; visualization, P.N.; supervision, M.J. and B.W.; project administration, P.N.; funding acquisition, M.J. All authors have read and agreed to the published version of the manuscript.

Funding: The APC is financed by Wrocław University of Environmental and Life Sciences.

Institutional Review Board Statement: Not applicable.

Informed Consent Statement: Not applicable.

Data Availability Statement: Raw data are available after contacting the authors by e-mail.

Conflicts of Interest: The authors declare no conflicts of interest.

References

1. Russo, E.B. Taming THC: Potential Cannabis Synergy and Phytocannabinoid-Terpenoid Entourage Effects. *Br. J. Pharmacol.* **2011**, *163*, 1344. [CrossRef]
2. Newman, D.J.; Cragg, G.M. Natural Products as Sources of New Drugs over the 30 Years from 1981 to 2010. *J. Nat. Prod.* **2012**, *75*, 311–335. [CrossRef] [PubMed]
3. Arshadi, N.; Nouri, H.; Moghimi, H. Increasing the Production of the Bioactive Compounds in Medicinal Mushrooms: An Omics Perspective. *Microb. Cell Factories* **2023**, *22*, 11. [CrossRef] [PubMed]
4. Pleszczyńska, M.; Lemieszek, M.K.; Siwulski, M.; Wiater, A.; Rzeski, W.; Szczodrak, J. *Fomitopsis betulina* (Formerly *Piptoporus betulinus*): The Iceman's Polypore Fungus with Modern Biotechnological Potential. *World J. Microbiol. Biotechnol.* **2017**, *33*, 83. [CrossRef] [PubMed]
5. Gafforov, Y.; Rašeta, M.; Rapior, S.; Yarasheva, M.; Wang, X.; Zhou, L.; Wan-Mohtar, W.A.A.Q.I.; Zafar, M.; Lim, Y.W.; Wang, M.; et al. Macrofungi as Medicinal Resources in Uzbekistan: Biodiversity, Ethnomycology, and Ethnomedicinal Practices. *J. Fungi* **2023**, *9*, 922. [CrossRef]
6. Sułkowska-Ziaja, K.; Fijałkowska, A.; Muszyńska, B. Selected Species of Medicinal/Arboreal Mushrooms as a Source of Substances with Antioxidant Properties. In *Plant Antioxidants and Health*; Springer: Cham, Switzerland, 2021; pp. 1–27. [CrossRef]
7. Gumińska, B.; Wojewoda, W. *Grzyby i Ich Oznaczanie*; Państwowe Wydawnicto Rolnicze i Leśne: Warsaw, Poland, 1983; p. 503.
8. Groszek, A.; Kaczmarczyk-Sedlak, I. Rola Brzozy w Wierzeniach Ludowych Oraz w Medycynie Ludowej i Współczesnej Fitoterapii. *Herbalism* **2023**, *7*, 106–128. [CrossRef]
9. Sofrenić, I.; Anđelković, B.; Todorović, N.; Stanojković, T.; Vujisić, L.; Novaković, M.; Milosavljević, S.; Tešević, V. Cytotoxic Triterpenoids and Triterpene Sugar Esters from the Medicinal Mushroom *Fomitopsis betulina*. *Phytochemistry* **2021**, *181*, 112580. [CrossRef] [PubMed]
10. Meuninck, J. *Basic Illustrated Edible and Medicinal Mushrooms*; Rowman & Littlefield: Lanham, MD, USA, 2015.
11. Wasser, S.P. Medicinal Mushroom Science: History, Current Status, Future Trends and Unsolved Problems. *Int. J. Med. Mushrooms* **2010**, *12*, 1–16. [CrossRef]
12. Papp, N.; Rudolf, K.; Bencsik, T.; Czégényi, D. Ethnomycological Use of *Fomes fomentarius* (L.) Fr. and *Piptoporus betulinus* (Bull.) P. Karst. in Transylvania, Romania. *Genet. Resour. Crop Evol.* **2017**, *64*, 101–111. [CrossRef]
13. Grienke, U.; Zöll, M.; Peintner, U.; Rollinger, J.M. European Medicinal Polypores—A Modern View on Traditional Uses. *J. Ethnopharmacol.* **2014**, *154*, 564–583. [CrossRef]
14. Peintner, U.; Kuhnert-Finkernagel, R.; Wille, V.; Biasioli, F.; Shiryaev, A.; Perini, C. How to Resolve Cryptic Species of Polypores: An Example in Fomes. *IMA Fungus* **2019**, *10*, 17. [CrossRef] [PubMed]
15. Sułkowska-Ziaja, K.; Motyl, P.; Muszyńska, B.; Firlej, A. *Piptoporus betulinus* (Bull.) P. Karst.—Bogate Źródło Związków Aktywnych Biologicznie. *Post Fitoter* **2015**, *2*, 89–95.
16. Hobbs, C. *Medicinal Mushrooms: An Exploration of Tradition, Healing, and Culture*; Book Publishing Company: Summertown, TN, USA, 2002.
17. Stamets, P. *Growing Gourmet and Medicinal Mushrooms*; Ten Speed Press: Berkeley, CA, USA, 2011.
18. Vunduk, J.; Klaus, A.; Kozarski, M.; Petrović, P.; Žižak, Ž.; Nikšić, M.; Van Griensven, L.J.L.D. Did the Iceman Know Better? Screening of the Medicinal Properties of the Birch Polypore Medicinal Mushroom, *Piptoporus betulinus* (Higher Basidiomycetes). *Int. J. Med. Mushrooms* **2015**, *17*, 1113–1125. [CrossRef] [PubMed]
19. Rutalek, R. Ethnomykologie–Eine Übersicht. *Österr. Z. Pilzkd.* **2002**, *11*, 79–94.
20. Lemieszek, M.K.; Langner, E.; Kaczor, J.; Kandefer-Szerszen, M.; Sanecka, B.; Mazurkiewicz, W.; Rzeski, W. Anticancer Effect of Fraction Isolated from Medicinal Birch Polypore Mushroom, *Piptoporus betulinus* (Bull.: Fr.) P. Karst. (Aphyllophoromycetideae): In Vitro Studies. *Int. J. Med. Mushrooms* **2009**, *11*, 351–364. [CrossRef]
21. Sułkowska-Ziaja, K.; Szewczyk, A.; Galanty, A.; Gdula-Argasińska, J.; Muszyńska, B. Chemical Composition and Biological Activity of Extracts from Fruiting Bodies and Mycelial Cultures of *Fomitopsis betulina*. *Mol. Biol. Rep.* **2018**, *45*, 2535–2544. [CrossRef] [PubMed]

22. Rafehi, H.; Orlowski, C.; Georgiadis, G.T.; Ververis, K.; El-Osta, A.; Karagiannis, T.C. Clonogenic Assay: Adherent Cells. *J. Vis. Exp.* **2011**, *49*, e2573. [CrossRef]
23. *ISO 10993-5:2009*; Biological Evaluation of Medical Devices—Part 5: Tests for In Vitro Cytotoxicity. PN-EN ISO: Brussels, Belgium, 2009.
24. Noorhosseini, S.A.; Jokar, N.K.; Damalas, C.A. Improving Seed Germination and Early Growth of Garden Cress (*Lepidium sativum*) and Basil (*Ocimum basilicum*) with Hydro-Priming. *J. Plant Growth Regul.* **2018**, *37*, 323–334. [CrossRef]
25. Iannilli, V.; D'Onofrio, G.; Marzi, D.; Passatore, L.; Pietrini, F.; Massimi, L.; Zacchini, M. Lithium Toxicity in *Lepidium sativum* L. Seedlings: Exploring Li Accumulation's Impact on Germination, Root Growth, and DNA Integrity. *Environments* **2024**, *11*, 93. [CrossRef]
26. Bonciu, E.; Firbas, P.; Fontanetti, C.S.; Wusheng, J.; Karaismailoğlu, C.; Liu, D.; Menicucci, F.; Pesnya, D.S.; Popescu, A.; Romanovsky, A.V.; et al. An Evaluation for the Standardization of the Allium Cepa Test as Cytotoxicity and Genotoxicity Assay. *Caryologia Int. J. Cytol. Cytosyst. Cytogenet.* **2018**, *71*, 191–209. [CrossRef]
27. World Cancer Report: Cancer Research for Cancer Prevention—IARC. Available online: https://www.iarc.who.int/featured-news/new-world-cancer-report/ (accessed on 8 July 2024).
28. International Agency for Research on Cancer. *Personal Habits and Indoor Combustions*; International Agency for Research on Cancer, Weltgesundheits Organisation, Eds.; IARC: Lyon, France, 2012; ISBN 978-92-832-1322-2.
29. Marcus, Y. Extraction by Subcritical and Supercritical Water, Methanol, Ethanol and Their Mixtures. *Separations* **2018**, *5*, 4. [CrossRef]
30. OECD. *Test No. 208: Terrestrial Plant Test: Seedling Emergence and Seedling Growth Test*; OECD Publishing: Paris, France, 2006. [CrossRef]
31. Cabuga, C.C.; Joy, J.; Abelada, Z.; Rose, R.; Apostado, Q.; Joy, B.; Hernando, H.; Erick, J.; Lador, C.; Lloyd, O.; et al. Allium Cepa Test: An Evaluation of Genotoxicity. *Proc. Int. Acad. Ecol. Environ. Sci.* **2017**, *7*, 12–19.
32. Adams, R.P. *Identification of Essential Oil Components by Gas Chromatography/Mass Spectrometry*, 4th ed.; Allured Publishing: Carol Stream, IL, USA, 2007.

MDPI AG
Grosspeteranlage 5
4052 Basel
Switzerland
Tel.: +41 61 683 77 34

Nutrients Editorial Office
E-mail: nutrients@mdpi.com
www.mdpi.com/journal/nutrients

www.ingramcontent.com/pod-product-compliance
Lightning Source LLC
LaVergne TN
LVHW071930160726
843515LV00011B/2564